高等学校计算机科学与技术**项目驱动案例实践**规划教材

Android
应用开发案例教程

毋建军 徐振东 林瀚 编著

清华大学出版社

北京

内 容 简 介

本书应用“项目驱动(Project-Driven)”最新教学模式,通过完整的项目案例系统地介绍了使用Android技术设计与开发应用系统的理论和方法。全书论述了Android开发概述,Android应用程序组成,Android UI(用户界面)基础,Android UI系统控件基础,Android UI系统控件进阶,Android UI菜单、对话框,Android组件广播消息与服务,Android数据存储与访问,手机通信服务,Google API服务等内容。

本书注重理论与实践相结合,内容详尽,提供了大量实例,突出应用能力的培养,将一个实际项目的知识点分解在各章作为案例讲解,是一本实用性突出的教材。本书可作为普通高等学校计算机专业本、专科生Android应用开发课程的教材,也可供设计开发人员参考使用。

图书在版编目(CIP)数据

Android应用开发案例教程/毋建军,徐振东,林瀚编著. --北京:清华大学出版社,2013.3(2018.2重印)
高等学校计算机科学与技术项目驱动案例实践规划教材
ISBN 978-7-302-31100-3

Ⅰ.①A… Ⅱ.①毋… ②徐… ③林… Ⅲ.①移动终端-应用程序-程序设计-高等学校-教材 Ⅳ.①TN929.53

中国版本图书馆CIP数据核字(2012)第309143号

责任编辑:张瑞庆 顾 冰
封面设计:常雪影
责任校对:时翠兰
责任印制:杨 艳

出版发行:清华大学出版社
网　　址:http://www.tup.com.cn,http://www.wqbook.com
地　　址:北京清华大学学研大厦A座　　**邮　　编**:100084
社 总 机:010-62770175　　**邮　　购**:010-62786544
投稿与读者服务:010-62776969,c-service@tup.tsinghua.edu.cn
质量反馈:010-62772015,zhiliang@tup.tsinghua.edu.cn
课件下载:http://www.tup.com.cn,010-62795954
印 装 者:虎彩印艺股份有限公司
经　　销:全国新华书店
开　　本:185mm×260mm　　**印　　张**:28　　**字　　数**:682千字
版　　次:2013年3月第1版　　**印　　次**:2018年2月第5次印刷
印　　数:4801~5300
定　　价:59.00元

产品编号:049059-02

高等学校计算机科学与技术项目驱动案例实践规划教材

编写指导委员会

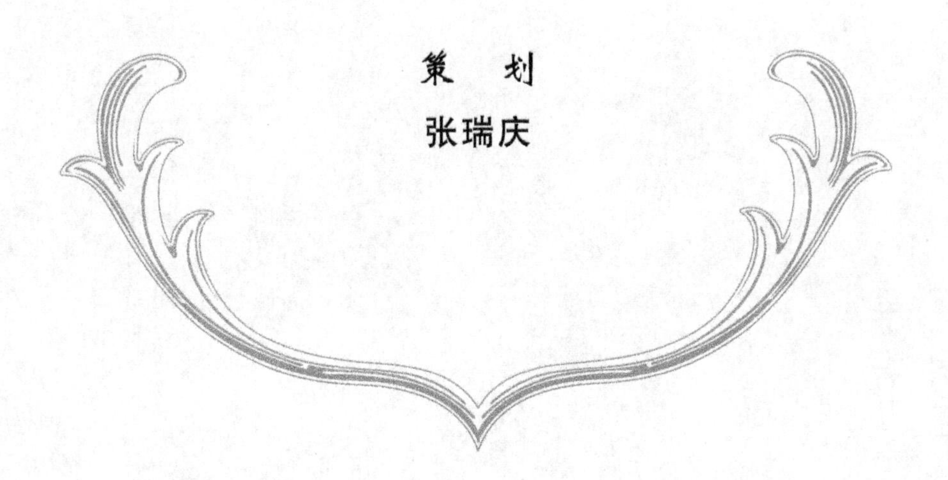

FOREWORD

序 言

作为我们高等学校计算机科学与技术教学指导委员会的工作内容之一，自从 2003 年参与清华大学出版社的“21 世纪大学本科计算机专业系列教材”的组织工作以来，陆续参加或见证了多个出版社的多套教材的出版，但是现在读者看到的这一套“高等学校计算机科学与技术项目驱动案例实践规划教材”有着特殊的意义。

这个特殊性在于其内容。这是第一套我所涉及的以项目驱动教学为特色，实践性极强的规划教材。如何培养符合国家信息产业发展要求的计算机专业人才，一直是这些年人们十分关心的问题。加强学生的实践能力的培养，是人们达成的重要共识之一。为此，高等学校计算机科学与技术教学指导委员会专门编写了《高等学校计算机科学与技术专业实践教学体系与规范》(清华大学出版社出版)。但是，如何加强学生的实践能力培养，在现实中依然遇到种种困难。困难之一，就是合适教材的缺乏。以往的系列教材，大都比较“传统”，没有跳出固有的框框。而这一套教材，在设计上采用软件行业中卓有成效的项目驱动教学思想，突出“做中学”的理念，突出案例(而不是“练习作业”)的作用，为高校计算机专业教材的繁荣带来了一股新风。

这个特殊性在于其作者。本套教材目前规划了十余本，其主要编写人不是我们常见的知名大学教授，而是知名软件人才培训机构或者企业的骨干人员，以及在该机构或者企业得到过培训的并且在高校教学一线有多年教学经验的大学教师。我以为这样一种作者组合很有意义，他们既对发展中的软件行业有具体的认识，对实践中的软件技术有深刻的理解，对大型软件系统的开发有丰富的经验，也有在大学教书的经历和体会，他们能在一起合作编写教材本身就是一件了不起的事情，没有这样的作者组合是难以想象这种教材的规划编写的。我一直感到中国的大学计算机教材尽管繁荣，但也比较“单一”，作者群的同质化是这种风格单一的主要原因。对比国外英文教材，除了 Addison Wesley 和 Morgan Kaufmann 等出版的经典教材长盛不衰外，我们也看到 O'Reilly“动物教材”等的异军突起——这些教材的作者，大都是实战经验丰富的资深专业人士。

这个特殊性还在于其产生的背景。也许是由于我在计算机技术方面的动手能力相对比较弱，其实也不太懂如何教学生提高动手能力，因此一直希望有一个机会实际地了解所谓“实训”到底是怎么回事，也希望能有一种安排让现在教学岗位的一些青年教师得到相关的培训和体会。于是作为2006—2010年教育部高等学校计算机科学与技术教学指导委员会的一项工作，我们和教育部软件工程专业大学生实习实训基地（亚思晟）合作，举办了6期“高等学校青年教师软件工程设计开发高级研修班”，时间虽然只是短短的1～2周，但是对于大多数参加研修的青年教师来说都是很有收获的一段时光，在对他们的结业问卷中充分反映了这一点。从这种研修班得到的认识之一，就是目前市场上缺乏相应的教材。于是，这套“高等学校计算机科学与技术项目驱动案例实践规划教材”应运而生。

当然，这样一套教材，由于“新”，难免有风险。从内容程度的把握、知识点的提炼与铺陈，到与其他教学内容的结合，都需要在实践中逐步磨合。同时，这样一套教材对我们的高校教师也是一种挑战，只能按传统方式讲软件课程的人可能会觉得有些障碍。相信清华大学出版社今后将和作者以及高等学校计算机科学与技术教学指导委员会一起，举办一些相应的培训活动。总之，我认为编写这样的教材本身就是一种很有意义的实践，祝愿成功。也希望看到更多业界资深技术人员加入到大学教材编写的行列中来，和高校一线教师密切合作，将学科、行业的新知识、新技术、新成果写入教材，开发适用性和实践性强的优秀教材，共同为提高高等教育教学质量和人才培养质量做出贡献。

李晓明

教育部高等学校计算机科学与技术教学指导委员会副主任

2011年8月 于北京大学

PREFACE

前 言

21世纪,什么技术将影响人类的生活?什么产业将决定国家的发展?信息技术与信息产业是首选的答案。大专院校学生是企业和政府的后备军,国家教育部门计划在大专院校中普及政府和企业信息技术与软件工程教育。经过多所院校的实践,信息技术与软件工程教育受到同学们的普遍欢迎,取得了很好的教学效果。然而也存在一些不容忽视的共性问题,其中突出的是教材问题。

从近两年信息技术与软件工程教育研究来看,许多任课教师提出目前教材不合适。具体体现在:第一,来自信息技术与软件工程专业的术语很多,对于没有这些知识背景的同学学习起来具有一定难度;第二,书中案例比较匮乏,与企业的实际情况相差太远,致使案例可参考性差;第三,缺乏具体的课程实践指导和真实项目。因此,针对大专院校信息技术与软件工程课程教学特点与需求,编写适用的规范化教材已是刻不容缓。

本书就是针对以上问题编写的,作者希望推广一种最有效的学习与培训的捷径,这就是 Project-Driven Training,也就是用项目实践来带动理论的学习(或者叫作"做中学")。基于此,作者围绕一个艾斯医药移动商务系统项目案例来贯穿 Android 应用开发各个模块的理论讲解,包括 Android 开发概述,Android 应用程序组成,Android UI(用户界面)基础,Android UI 系统控件基础,Android UI 系统控件进阶,Android UI 菜单、对话框,Android 组件广播消息与服务,Android 数据存储与访问,手机通信服务,Google API 服务等。通过项目实践,可以对技术应用有明确的目的性(为什么学),对技术原理更好地融会贯通(学什么),也可以更好地检验学习效果(学得怎样)。

本书特色:

1. 重项目实践

作者多年项目开发经验的体会是"IT 是做出来的,不是想出来的",理论虽然重要,但一定要为实践服务。以项目为主线,带动理论的学习是最好、最快、最有效的方法。本书的特色是提供了一个完整的医药商务系统项目。通过此书,作者希望读者对 Android 开发技术和流程有一个整体了解,减少对项目的盲目感和神秘感,能够根据本书的体系循序渐进地动手做出自己的真实项目来。

PREFACE

2. 重理论要点

本书是以项目实践为主线的，着重介绍 Android 开发理论中最重要、最精华的部分，以及它们之间的融会贯通；而不是面面俱到，没有重点和特色。读者首先通过项目把握整体概貌，再深入局部细节，系统学习理论；然后不断优化和扩展细节，完善整体框架和改进项目。既有整体框架，又有重点理论和技术。一书在手，思路清晰，项目无忧。

本书由梁立新审稿、统稿并定稿。

为了便于教学，本书配有教学课件，读者可从清华大学出版社的网站下载。

鉴于编者的水平有限，书中难免有不足之处，敬请广大读者批评指正。

编　者

2012 年 10 月

CONTENTS

目 录

第 1 章　Android 开发概述 …… 1

1.1　智能手机发展 …… 1

1.2　Android 简介 …… 7

1.3　搭建 Android 开发环境 …… 11

1.3.1　Android 开发环境系统要求 …… 12

1.3.2　Windows 系统平台下搭建开发环境 …… 12

1.3.3　Linux 系统平台下搭建开发环境 …… 20

1.4　Android SDK 概述 …… 22

1.4.1　Android SDK 目录结构 …… 22

1.4.2　Android 常用开发工具 …… 22

1.4.3　Android SDK 实例 …… 24

1.5　创建 Android 程序 …… 24

1.5.1　创建和使用虚拟设备 …… 24

1.5.2　在 Eclipse 下创建 Android 程序 …… 27

1.5.3　命令行创建 Android 程序 …… 29

1.5.4　调试 Android 程序 …… 36

习题 1 …… 42

第 2 章　Android 在线医药应用——艾斯医药系统开发 …… 43

2.1　系统需求分析设计 …… 44

2.1.1　系统开发背景 …… 44

2.1.2　系统功能需求 …… 44

2.1.3　系统开发及部署平台 …… 45

2.2　系统详细设计分析 …… 46

2.2.1　Web 服务器端系统总体架构设计 …… 46

2.2.2　Web 服务器端系统功能概述 …… 48

2.2.3　Android 手机客户端总体架构设计 …… 55

2.2.4　AscentSys（艾斯医药）移动客户端系统功能概述 …… 55

2.3　数据库详细设计分析 …… 57

2.3.1　数据库平台环境及要求 …… 57

2.3.2　数据库及表设计 …… 58

2.4　Web 服务器端功能模块开发 …… 61

CONTENTS

2.4.1 服务器端开发准备 …… 61
2.4.2 注册登录模块 …… 62
2.4.3 购物模块 …… 62
2.4.4 订单模块 …… 63
2.5 AscentSys 医药商务系统移动客户端功能模块开发 …… 64
2.6 AscentSys 移动客户端打包、签名、发布 …… 65
2.7 AscentSys 医药系统部署 …… 68
习题 2 …… 70

第 3 章 Android 应用程序 …… 71

3.1 Android 项目构成 …… 71
3.1.1 目录结构 …… 71
3.1.2 AndroidManifest.xml 文件简介 …… 72
3.1.3 gen 目录 …… 74
3.1.4 res 目录 …… 75
3.1.5 default.properties 文件 …… 76
3.2 Android 应用程序组成 …… 76
3.2.1 Android 应用程序概述 …… 76
3.2.2 Activity 组件 …… 77
3.2.3 Service 组件 …… 77
3.2.4 Intent 和 IntentFilter 组件 …… 77
3.2.5 BroadcastReceiver 组件 …… 81
3.2.6 ContentProvider 组件 …… 82
3.3 Android 生命周期 …… 82
3.3.1 程序生命周期 …… 82
3.3.2 组件生命周期 …… 84
3.4 项目案例 …… 94
习题 3 …… 98

第 4 章 Android UI(用户界面)基础 …… 99

4.1 Android UI 简介 …… 99
4.2 Android UI 框架 …… 100
4.2.1 Android 与 MVC 设计 …… 100
4.2.2 视图树模型(View 和 Viewgroup) …… 101
4.3 Android UI 控件类简介 …… 101
4.3.1 View 类 …… 101
4.3.2 ViewGroup 类 …… 102

CONTENTS

4.3.3 界面控件 …… 103

4.4 Android UI 布局 …… 103

4.4.1 线性布局 …… 103

4.4.2 线性布局应用案例 …… 105

4.4.3 相对布局 …… 107

4.4.4 相对布局应用案例 …… 108

4.4.5 表格布局 …… 110

4.4.6 表格布局应用案例 …… 112

4.4.7 帧布局 …… 115

4.4.8 帧布局应用案例 …… 115

4.4.9 绝对布局 …… 118

4.4.10 绝对布局应用案例 …… 119

4.5 项目案例 …… 121

习题 4 …… 125

第 5 章 Android UI 系统控件基础 …… 126

5.1 文本控件简介 …… 127

5.1.1 文本框 …… 127

5.1.2 TextView 应用案例 …… 128

5.1.3 编辑框 …… 130

5.1.4 EditText 应用案例 …… 131

5.2 按钮控件简介 …… 132

5.2.1 按钮 …… 132

5.2.2 Button 应用案例 …… 133

5.2.3 图片按钮 …… 135

5.2.4 ImageButton 应用案例 …… 136

5.3 单选与复选按钮简介 …… 138

5.3.1 单选按钮 …… 138

5.3.2 复选按钮 …… 140

5.3.3 RadioButton 和 CheckBox 综合应用案例 …… 142

5.4 时间与日期控件简介 …… 144

5.4.1 时间选择器 …… 144

5.4.2 日期选择器 …… 144

5.4.3 时间与日期控件综合应用案例 …… 145

5.5 图片控件简介 …… 149

5.5.1 图片控件 …… 149

5.5.2 ImageView 应用案例 …… 150

CONTENTS

5.5.3 切换图片控件 ImageSwitcher、Gallery …… 153
5.5.4 ImageSwitcher、Gallery 综合应用案例 …… 154
5.6 时钟控件简介 …… 159
5.6.1 模拟时钟与数字时钟 …… 159
5.6.2 AnalogClock 和 DigitalClock 应用案例 …… 160
5.7 项目案例 …… 163
习题 5 …… 167

第 6 章 Android UI 系统控件进阶 …… 168

6.1 列表控件简介 …… 168
6.1.1 列表控件 …… 168
6.1.2 ListView 应用案例 …… 170
6.1.3 下拉列表控件 …… 172
6.1.4 Spinner 应用案例 …… 174
6.2 进度条与滑块控件简介 …… 176
6.2.1 进度条 …… 176
6.2.2 ProgressBar 应用案例 …… 177
6.2.3 滑块 …… 179
6.2.4 SeekBar 应用案例 …… 181
6.3 评分控件简介 …… 183
6.3.1 评分控件 …… 183
6.3.2 RatingBar 应用案例 …… 183
6.4 自动完成文本控件简介 …… 185
6.4.1 自动完成文本控件 …… 185
6.4.2 AutoCompleteTextView 应用案例 …… 186
6.5 Tabhost 控件简介 …… 188
6.5.1 Tabhost 控件 …… 188
6.5.2 Tabhost 应用案例 …… 189
6.6 视图控件简介 …… 191
6.6.1 滚动视图控件 …… 191
6.6.2 ScrollView 应用案例 …… 191
6.6.3 网格视图控件 …… 193
6.6.4 GridView 应用案例 …… 194
6.7 Android 事件处理 …… 196
6.7.1 Android 事件和监听器 …… 196
6.7.2 Android 事件处理机制 …… 197
6.7.3 Android 事件处理机制应用案例 …… 201

CONTENTS

6.7.4 按键事件应用案例 …… 206
6.7.5 触摸事件应用案例 …… 208
6.8 项目案例 …… 210
习题 6 …… 213

第 7 章 Android UI 菜单、对话框 …… 214

7.1 菜单控件 Menu …… 214
7.1.1 Menu 简介 …… 214
7.1.2 选项菜单 …… 215
7.1.3 选项菜单应用案例 …… 216
7.1.4 子菜单 …… 219
7.1.5 子菜单应用案例 …… 219
7.1.6 快捷菜单 …… 222
7.1.7 快捷菜单应用案例 …… 223
7.2 对话框控件 Dialog …… 226
7.2.1 Dialog 简介 …… 226
7.2.2 警告(提示)对话框 AlertDialog …… 227
7.2.3 AlertDialog 应用案例 …… 227
7.2.4 日期选择对话框 DatePickerDialog …… 229
7.2.5 DatePickerDialog 应用案例 …… 230
7.2.6 时间选择对话框 TimePickerDialog …… 233
7.2.7 TimePickerDialog 应用案例 …… 233
7.2.8 进度对话框 ProgressDialog …… 236
7.2.9 ProgressDialog 应用案例 …… 236
7.3 信息提示控件 …… 239
7.3.1 Toast 控件简介 …… 239
7.3.2 Toast 应用案例 …… 239
7.3.3 Notification 控件简介 …… 242
7.3.4 Notification 应用案例 …… 243
7.4 项目案例 …… 246
习题 7 …… 250

第 8 章 Android 组件广播消息与服务 …… 251

8.1 Intent 消息通信 …… 251
8.1.1 Intent 简介 …… 251
8.1.2 使用 Intent 进行组件通信 …… 254
8.1.3 使用 Intent 启动 Activity …… 254

CONTENTS

8.1.4 获取 Activity 返回值 ………… 260
8.1.5 Intent Filter 原理与匹配机制 ………… 263
8.2 Intent 广播消息 ………… 267
8.2.1 广播消息 ………… 267
8.2.2 BroadcastReceiver 监听广播消息 ………… 267
8.2.3 Broadcast Receiver 应用案例 ………… 270
8.3 Service 组件服务 ………… 273
8.4 项目案例 ………… 274
习题 8 ………… 281
第 9 章 Android 数据存储与访问 ………… 282
9.1 SharedPreferences ………… 283
9.1.1 SharedPreferences 简介 ………… 283
9.1.2 读取应用程序数据案例 ………… 286
9.1.3 读取其他应用程序数据案例 ………… 289
9.2 文件存储 ………… 290
9.2.1 文件存储简介 ………… 291
9.2.2 文件存储应用案例 ………… 293
9.2.3 SDCard 存储简介 ………… 301
9.2.4 SD 卡存储应用案例 ………… 302
9.3 SQLite 数据库存储 ………… 307
9.3.1 SQLite 数据库简介 ………… 307
9.3.2 创建 SQLite 数据库方式 ………… 310
9.3.3 SQLite 数据库操作 ………… 313
9.3.4 SQLite 数据库管理 ………… 317
9.3.5 SQLite 数据库应用案例 ………… 319
9.4 数据共享 ………… 328
9.4.1 ContentProvider 简介 ………… 328
9.4.2 Uri、UriMatcher 和 ContentUris 简介 ………… 329
9.4.3 创建 ContentProvider ………… 332
9.4.4 ContentResolver 操作数据 ………… 333
9.4.5 ContentProvider 应用案例 ………… 334
9.5 网络存储 ………… 337
9.5.1 网络存储简介 ………… 337
9.5.2 网络存储应用案例 ………… 337
9.6 数据存储项目案例 ………… 339
习题 9 ………… 361

CONTENTS

第 10 章　手机通信服务 …… 362

10.1　短信服务 …… 362

10.1.1　短信服务简介 …… 362

10.1.2　短信发送与提示案例 …… 363

10.1.3　短信发送状态查询案例 …… 367

10.2　电话服务 …… 370

10.2.1　电话服务简介 …… 370

10.2.2　接打电话案例 …… 371

10.3　E-mail 服务 …… 374

10.3.1　SMTP 简介 …… 374

10.3.2　发送邮件案例 …… 377

10.4　网络资源访问与处理 …… 382

10.4.1　使用 URL 读取网络资源 …… 383

10.4.2　使用 URL 访问网络应用案例 …… 384

10.4.3　使用 HTTP 访问网络资源(HttpURLConnection) …… 386

10.4.4　使用 HTTP 访问网络应用案例 …… 388

10.5　项目案例 …… 396

习题 10 …… 413

第 11 章　Google API 服务 …… 414

11.1　地理位置定位服务 …… 414

11.1.1　Android Location API 简介 …… 415

11.1.2　获取位置定位案例 …… 418

11.2　Google Map 服务 …… 422

11.2.1　Google Map API 简介 …… 422

11.2.2　申请 Map API KEY …… 422

11.2.3　使用 Map API 创建 AVD 应用 …… 424

11.3　项目案例 …… 426

习题 11 …… 431

致谢 …… 432

第1章 Android开发概述

学习目标

本章主要介绍智能手机的发展,Android 系统版本的发展历史、过程。同时讲述 Android 开发环境的搭建过程、Android SDK,以及创建 Android 程序的方法和工具。通过本章的学习,使读者达到以下知识要点的学习:

(1) 智能手机的含义、基本构成、特点。

(2) 智能手机操作系统及类别、智能手机未来发展趋势。

(3) Android 系统发展历史、体系结构、特征及未来发展方向。

(4) Android 系统在不同平台下开发环境的搭建。

(5) Android SDK 结构、构成及工具。

(6) 使用不同方式和方法创建、调试 Android 应用程序。

1.1 智能手机发展

在学习和了解 Android 系统平台之前,必须先了解和掌握一些关于智能手机发展及其智能手机有关的基本概念。本节将介绍有关智能手机的基础知识,主要涉及智能手机基本概念、特点、常用的智能手机操作系统等方面。

智能手机是由掌上电脑(Pocket PC)演变而来的。最早的掌上电脑是不具备手机的通话功能的,但是随着用户对于掌上电脑的个人信息处理方面功能依赖的提升,又由于人们不习惯于随时都携带手机和 PPC 两个设备,因此厂商将掌上电脑的系统移植到了手机中,于是才出现了智能手机这个概念。

1. 智能手机的含义

智能手机是指像个人计算机一样，具有独立的操作系统，可以由用户自行安装软件、游戏等第三方服务商提供的程序，通过此类程序来不断对手机的功能进行扩充，并可以通过移动通信网络来实现无线网络接入的这样一类手机的总称。从广义上说，智能手机除了具备手机的通话功能外，还具备了 PDA 的大部分功能，特别是个人信息管理以及基于无线数据通信的浏览器和电子邮件功能。第三方可根据操作系统提供的应用编程接口为手机开发各种扩展应用和提供各种扩展硬件。也有人把智能手机简单定位为产品、操作系统和网络的集成，如图 1-1 所示。

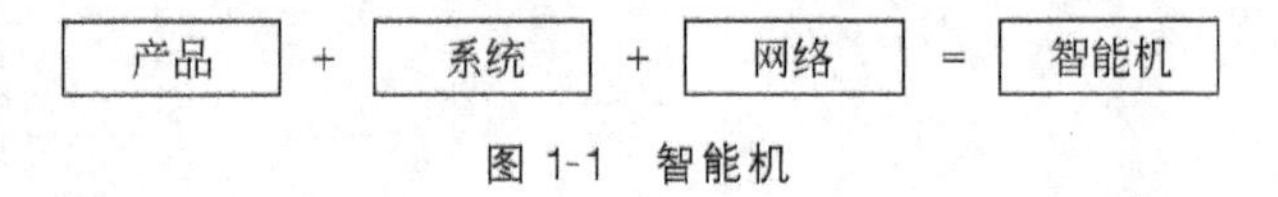

图 1-1 智能机

在智能手机已经广泛应用的今天，用户的很多增值业务，如股票、新闻、天气、交通、商品、应用程序下载、音乐图片下载、收发邮件、办公等都已经成为智能必备的功能。在 3G 通信网络的支持下，智能手机势必将成为一个功能强大，集通话、短信、网络接入、影视娱乐为一体的综合性个人手持终端设备。

新一代的智能手机未来发展的目标是成为因特网智能终端，能为用户带来畅快的网络体验，界面更为简洁，可以自由加载增值应用的智能终端，如图 1-2 所示。

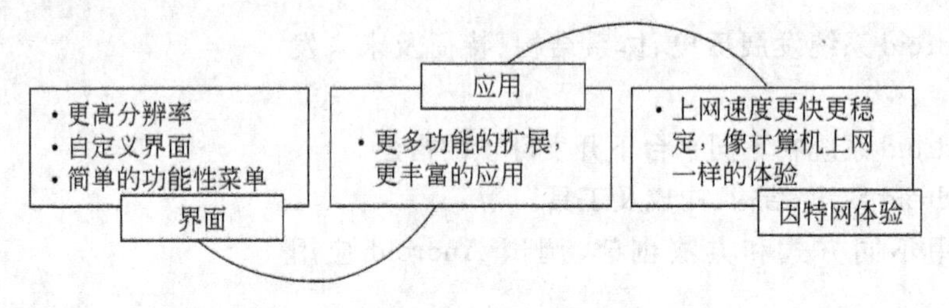

图 1-2 智能手机涉及的功能

2. 智能手机基本构成

智能手机通常由硬件、操作系统和网络支持三大部分组成，如图 1-3 所示。

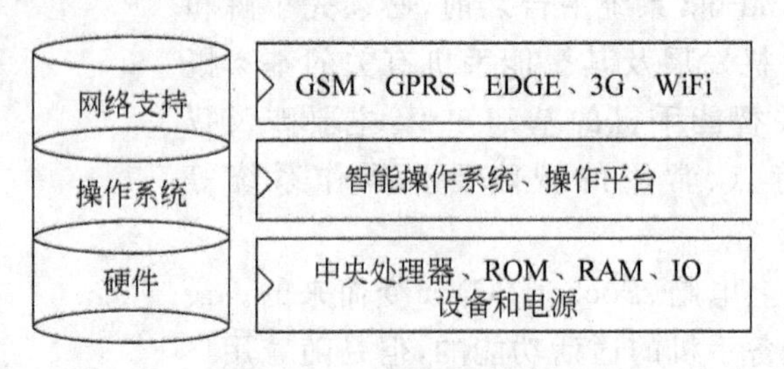

图 1-3 智能手机构成

3. 智能手机的特点

- 通用性。具备普通手机的全部功能，能够进行正常的通话、收发短信等手机功能的应用。

- 扩展能力。具备一个具有开放性的操作系统，在这个平台上可以安装更多的应用程序，从而使智能手机的功能得到无限的扩充。
- 多任务处理。具备 PDA 的功能，同时运行多任务的能力。
- 多媒体处理。具备强大的多媒体处理能力，支持图像拍摄及录制，以及支持各种格式的声音，视频播放，剪辑处理。
- 网络支持。具备支持各种类型的网络接入因特网的能力，以及各种因特网的应用。

4. 智能机与非智能机的区别

(1) 智能机(Smart Phone)相当于一台有通信功能的计算机，可以像计算机一样上网，用户可以安装丰富的应用软件。智能机有 CPU/RAM/ROM，相当于计算机的 CPU/内存/硬盘。

(2) 非智能机(Feature Phone)除基本的电话和通信功能外，扩展能力非常有限。

5. 智能手机操作系统

目前，市场上的智能手机操作系统有很多，曾经以及现在比较有影响力的智能手机操作系统主要有：

1) Symbian 系统

Symbian(塞班)是一个实时性、多任务的操作系统，起源于 1998 年。为了对抗微软即将推出的智能手机系统，诺基亚、摩托罗拉、爱立信和宝意昂公司在英国伦敦共同投资成立了 Symbian 公司。1999 年，支持 Symbian 的爱立信 R380 上市，当时由于 R380 系统正处于实践阶段，并未得到很好的推广。世界上第一款采用 Symbian 操作系统的手机如图 1-4 所示。

2001 年，诺基亚推出了第一款 Symbian PDA 手机，如图 1-5 所示，型号为 9110，采用了 AMD 公司的内嵌式 CPU，内置 8MB 存储空间。9110 已经集成了网络、PIM、网页浏览和电子邮件等功能，并且已经开始支持 Java，这使得它已经能够运行小型的第三方软件。

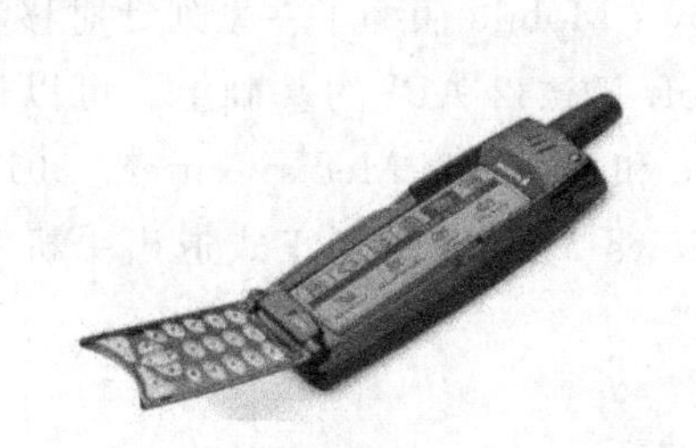

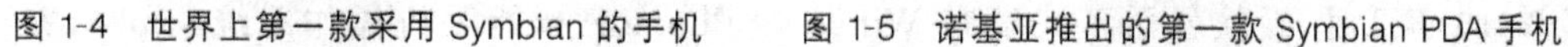

图 1-4　世界上第一款采用 Symbian 的手机　　图 1-5　诺基亚推出的第一款 Symbian PDA 手机

此后，Nokia 主导的 S60v3、S60v5、Symbian^3 相继发布和推广。在 2008 年之前，塞班系统在市场上占主导地位，接着 Symbian 公司被诺基亚全资收购，成为诺基亚旗下公司。参与开发的多家厂商仍然有系统的使用权，但到后来都纷纷宣布退出塞班的阵营，取而代之的是安卓，目前塞班的支持厂商只有诺基亚。

2011 年，诺基亚正式宣布与微软达成全球战略合作伙伴关系，双方在智能手机领域进行深度合作。微软的 Windows Phone 7 系统成为诺基亚的主要手机操作系统。2012 年，诺基亚 Lumia 智能手机——诺基亚 Lumia 820 和 Lumia 920，配置了 Windows Phone 8 操作系统及双核处理器，并在市场推广应用。

2）Android 系统

Android 的原意指“机器人”。Google 于 2007 年推出基于 Linux 平台的开源手机操作系统 Android，Android 系统平台由操作系统、中间件、用户界面和应用软件组成，是首个为移动终端打造的真正开放和完整的移动软件系统平台。

目前，市场上采用 Android 系统的主要手机厂商包括宏达电子（HTC）、三星、摩托罗拉和 LG、Sony Ericsson 等，国内厂商有华为、中兴、联想和酷派等。第一款使用谷歌 Android 操作系统的手机 G1 如图 1-6 所示。Android 系统不但应用于智能手机，也在平板电脑市场急速扩张。目前 Android 成为全球最受欢迎的智能手机平台。

图 1-6　第一款采用 Android 操作系统的手机（G1）

随着 Android 4.1 的诞生及发展，Android 系统的发展及应用越来越广泛。Android 平台资源下载安装与 Symbian 类似，不同的是 Android 系统平台是一个开源系统，所安装的程序不需要进行证书检查。Android 常用的资源网站及管理软件有：

安卓市场：http://www.hiapk.com/bbs

机锋市场：http://www.gfan.com

PC 端 Android 平台管理软件“豌豆荚”：http://wandoujia.com

3）Windows Mobile

Windows Mobile（WM）是微软针对移动设备而开发的操作系统。该操作系统的设计初衷是尽量接近于桌面版本的 Windows。微软按照计算机操作系统的模式来设计 WM，以便能使得 WM 与计算机操作系统一模一样。WM 的应用软件以 Microsoft Win32 API 为基础。在 Windows Mobile 6.5 发布的同时，微软宣布以后的 Windows Mobile 产品将改名为 Windows Phone，以改变其落后的形象。Windows Mobile 捆绑了一系列针对移动设备而开发的应用软件，这些应用软件创建在 Microsoft Win32 API 的基础上。可以运行 Windows Mobile 的设备包括 Pocket PC、Smartphone 和 Portable Media Center。2010 年 10 月，微软宣布终止对 WM 的所有技术支持，Windows Mobile 系列正式退出手机系统市场。

4）Windows Phone

2010 年 2 月，微软公司正式发布 Windows Phone 7 智能手机操作系统（wp7），并于 2010 年年底发布了基于此平台的移动设备。全新的 Windows Phone 把网络、个人计算机和手机的优势集于一身，让人们可以随时随地享受到想要的体验，它具有桌面定制、图标拖曳、滑动控制等一系列前卫的操作体验。其主屏幕通过提供类似仪表盘的体验来显示新的电子邮件、短信、未接来电和日历约会等，让人们对重要信息保持时刻更新。Windows Phone 力图打破人们与信息和应用之间的隔阂，提供适用于人们包括工作和娱乐在内完整生活的方方面面，最优秀的端到端体验。

2012 年 9 月，诺基亚与微软在纽约召开联合发布会，发布 Windows Phone 8 系统手机

Lumia 920 及 Lumia 820。Lumia 920 实现了手机摄影、高清屏显示、无线充电三大技术突破，这也是诺基亚旗下首款 Windows Phone 8 系统手机，如图 1-7 所示。

图 1-7　诺基亚 Lumia 920 和 Lumia 820

5）iOS

iPhone OS 是由苹果公司为 iPhone 开发的操作系统。后来套用到 iPod touch、iPad 以及 Apple TV 产品上使用。iPhone 将移动电话、可触摸宽屏以及具有桌面级电子邮件、网页浏览、搜索和地图功能的因特网通信设备这三种产品完美地融为一体，重新定义了移动电话的功能。iOS 拥有简单易用的界面，有良好的操作体验。就像其基于 Mac OS X 操作系统一样，它也是以 Darwin 为基础的。在 2010 年 6 月 WWDC 大会上，苹果宣布 iPhone、iPod touch 和 iPad 使用的 iPhone OS 操作系统更名为 iOS，统一了苹果的移动设备名称。

iOS 的系统架构分为 4 个层次：核心操作系统层（the Core OS layer）、核心服务层（the Core Services layer）、媒体层（the Media layer）和可轻触层（the Cocoa Touch），如图 1-8 所示。

2012 年 9 月 21 日，苹果 iPhone 5 在全球 9 个国家和地区同步发售，iPhone 5 采用的是全新 Nano-SIM 卡，同时苹果正式放出了 iOS 6 的正式版。

6）黑莓（BlackBerry）

黑莓是加拿大一家手提无线通信设备品牌，于 1999 年创立。其系统特色是支持 PushMail 电子邮件、移动电话、文字短信、因特网传真、网页浏览及其他无线资讯服务。

2007 年 7 月，在中国大陆地区引进第一款设备 Blackberry 8700g，如图 1-9 所示，由 TCL 代工生产，中国移动同时也向企业用户推广 Blackberry 业务。

图 1-8　iOS 6 运行于 iPhone 4S

图 1-9　Blackberry 8700g

黑莓由于与中国移动推出手机邮箱有冲突、gprs 的网络速度限制及中国用户当时没有使用电子邮件的习惯，因此当时中移动引进黑莓并没有大范围的得到推广。

7）其他手机操作系统

- OMS：中国移动在 Android 系统上定制开发的系统；

- OS：加拿大 RIM 公司开发；
- MeeGo：Nokia 与 Intel 联合开发；
- Bada：Samsung 研发；
- BrewMP：高通公司开发。

6. 手机操作系统与智能机

上述内容讲解了智能手机的操作系统，不同的操作系统有不同手机厂商采用并支持，具体如图 1-10 所示。

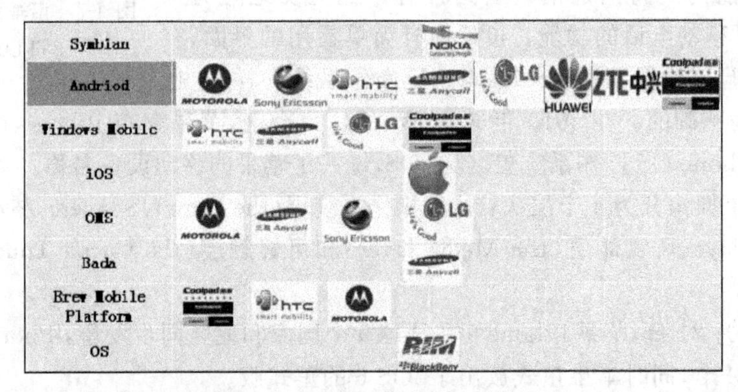

图 1-10　使用不同操作系统的智能品牌手机对应图

7. 智能手机的未来发展趋势

(1) GPS。目前，GPS 功能越来越普遍。宏达电、RIM 和其他智能手机厂商均推出了支持 GPS 功能的手机产品。它不仅可以帮助用户从 A 点走到 B 点，更重要的是，GPS 服务商也推出了各种各样的服务。

(2) 开源。开源是智能手机发展的一个新趋势，目前智能手机厂商和运营商都宣布了自己的开源战略或产品。

(3) 电池寿命。电池续航时间是衡量智能手机的一个重要指标，Wi-Fi、蓝牙、彩屏和免提等均消耗不小的电量。因此，为延长续航时间，尽量关闭不常用功能。

(4) Wi-Fi。新的 Wi-Fi 芯片，如 Atheros AR6002 系列可以有效降低能耗，延长电池续航时间。

(5) 验证接入。T-Mobile 曾经推出 Hotspot@Home 服务，允许用户通过手机拨打 VOIP 电话，但只支持 Wi-Fi 手机。RIM 推出支持该服务的智能手机。

(6) 安全。智能手机面临着各种安全威胁，如设备锁定、功能锁定、加密、验证、远程删除数据、防火墙和 VPN 等。

8. 智能机的缺点

(1) 病毒。

智能手机与台式机、笔记本一样，都是基于软件应用的系统平台，所以针对手机操作系统的病毒也日渐盛行，为了避免病毒造成资料丢失或损坏，需要经常备份重要资料。

(2) 死机。

使用开放式操作系统的智能机，经常会受到非法程序干扰而死机。建议最好不要任意安装未经认证的各种应用软件。此外，给智能机安装过多的软件及文档存储过大，智能机的数据读写速度会变慢，从而影响用户的使用。

(3) 耗电。

智能手机的 CPU、屏幕等硬件的耗电量相对较大，容易将电量耗尽，尤其现在的智能手机追求时尚轻薄，屏幕大，所以小的电池容量就显得更加不可续航、更加费电。

1.2 Android 简介

1. Android 的发展历史

Android 一词的本义指“机器人”，是 Google 于 2007 年推出以 Linux 为基础的开放源代码操作系统，主要使用于便携设备。目前尚未有统一的中文名称，中国大陆地区较多人使用“安卓”或“安致”。Android 操作系统最初由 Andy Rubin 开发，最初主要支持手机。2011 年第一季度，Android 在全球的市场份额首次超过塞班系统，跃居全球第一。2012 年 7 月数据显示，Android 占据全球智能手机操作系统市场 59%的份额，中国市场占有率为 76.7%。

Android 平台由操作系统、中间件、用户界面和应用软件组成。目前，最新版本为 Android 3.0 Honeycomb、Android 4.1 Jelly Bean。

2003 年 10 月，Andy Rubin 等人创建 Android 公司，并组建 Android 团队。

2005 年 8 月，Google 低调收购了成立仅 22 个月的高科技企业 Android 及其团队。安迪鲁宾成为 Google 公司工程部副总裁，继续负责 Android 项目。

2007 年 11 月 5 日，谷歌公司推出 Android 操作系统，并且在这一天谷歌宣布建立一个全球性的联盟组织，该组织由 34 家手机制造商、软件开发商、电信运营商以及芯片制造商共同组成，并与 84 家硬件制造商、软件开发商及电信营运商组成开放手持设备联盟(Open Handset Alliance)来共同研发改良 Android 系统，这一联盟将支持谷歌发布的手机操作系统以及应用软件，Google 以 Apache 免费开源许可证的授权方式发布了 Android 的源代码。

2008 年，在 Google I/O 大会上谷歌提出了 Android HAL 架构图，在同年 8 月 18 号，Android 获得了美国联邦通信委员会(FCC)的批准，在 2008 年 9 月，谷歌正式发布了 Android 1.0 系统，这也是 Android 系统最早的版本。

2009 年 4 月，谷歌正式推出了 Android 1.5 这款手机，从 Android 1.5 版本开始，谷歌开始将 Android 的版本以甜品的名字命名，Android 1.5 命名为 Cupcake(纸杯蛋糕)。该系统与 Android 1.0 相比有了很大的改进。

2009 年 9 月，谷歌发布了 Android 1.6 的正式版 Donut(甜甜圈)，并且推出了搭载 Android 1.6 正式版的手机 HTC Hero(G3)，凭借着出色的外观设计以及全新的 Android 1.6 操作系统，HTC Hero(G3)成为当时全球最受欢迎的手机。

2010 年 2 月，Linux 内核开发者 Greg Kroah-Hartman 将 Android 的驱动程序从 Linux 内核“状态树(Staging Tree)”上除去，从此，Android 与 Linux 开发主流分开发展。在同年 5 月，谷歌正式发布了 Android 2.2 操作系统。

2010 年 12 月，谷歌正式发布了 Android 2.3 操作系统 Gingerbread(姜饼)。

2011 年 1 月，谷歌称每日的 Android 设备新用户数量达到了 30 万部，到 2011 年 7 月，这个数字增长到 55 万部，而 Android 系统设备的用户总数达到了 1.35 亿，Android 系统已经成为智能手机领域占有量最高的系统。

2011 年 8 月 2 日，Android 手机已占据全球智能机市场 48%的份额，并在亚太地区市场占据统治地位，终结了 Symbian(塞班系统)的霸主地位，跃居全球第一。

2011 年 9 月，Android 系统的应用数目已经达到了 48 万，而在智能手机市场，Android 系统的占有率已经达到了 43%，继续排在移动操作系统首位。接着谷歌发布全新的 Android 4.0 操作系统 Ice Cream Sandwich(冰激凌三明治)。

2012 年 1 月 6 日，谷歌 Android Market 目前已有 10 万开发者推出超过 40 万活跃的应用，大多数的应用程序免费。Android Market 应用程序商店目录在新年首周周末突破 40 万基准，距离突破 30 万应用仅 4 个月。

2. Android 系统版本及功能发展

Android 在正式发行之前，最开始拥有两个内部测试版本，并且以著名的机器人名称来对其进行命名，分别是阿童木(Android Beta)和发条机器人(Android 1.0)。后来谷歌将其命名规则变更为用甜点作为它们系统版本的代号的命名方法。甜点命名法开始于 Android 1.5 发布。作为每个版本代表的甜点的尺寸越变越大，然后按照 26 个字母数序：纸杯蛋糕(Android 1.5)，甜甜圈(Android 1.6)，松饼(Android 2.0/2.1)，冻酸奶(Android 2.2)，姜饼(Android 2.3)，蜂巢(Android 3.0)，冰激凌三明治(Android 4.0)，而最新一代 Android 版本名为果冻豆(Jelly Bean，Android 4.1)。Android 4.0 之前各版本发布的 Logo 如图 1-11 所示。

图 1-11　Android 4.0 之前各版本发布的 Logo

1) Android 1.5 Cupcake(纸杯蛋糕)

2009 年 4 月 30 日发布，其主要的功能更新有：

- 拍摄/播放影片，并支持上传到 Youtube；支持立体声蓝牙耳机，同时改善自动配对性能；最新的采用 WebKit 技术的浏览器，支持复制/贴上和页面中搜索。
- GPS 性能大大提高。
- 提供屏幕虚拟键盘。
- 主屏幕增加音乐播放器和相框 widgets。
- 应用程序自动随着手机旋转。
- 短信、Gmail、日历，浏览器的用户接口大幅改进，如 Gmail 可以批量删除邮件。
- 相机启动速度加快，拍摄图片可以直接上传到 Picasa；来电照片显示。

2) Android 2.3.xGingerbread(姜饼)

2010 年 12 月 7 日发布，其主要的功能更新有：

- 增加了新的垃圾回收和优化处理事件。
- 原生代码可直接存取输入和感应器事件、EGL/OpenGLES、OpenSL ES。

- 新的管理窗口和生命周期的框架。
- 支持 VP8 和 WebM 视频格式，提供 AAC 和 AMR 宽频编码，提供了新的音频效果器。
- 支持前置摄像头、SIP/VOIP 和 NFC(近场通信)。
- 简化界面、速度提升。
- 更快、更直观的文字输入。
- 一键文字选择和复制/粘贴。
- 改进的电源管理系统。
- 新的应用管理方式。

3) Android 4.1 Jelly Bean(果冻豆)

其功能新特性主要有：

- 更快、更流畅、更灵敏。
- 特效动画的帧速提高至 60fps，增加了三倍缓冲。
- 增强通知栏。
- 全新搜索。
- 搜索将会带来全新的 UI、智能语音搜索和 Google Now 三项新功能。
- 桌面插件自动调整大小。
- 加强无障碍操作。
- 语言和输入法扩展。
- 新的输入类型和功能。
- 新的连接类型。

3. Android 系统的优势及缺点

1) Android 系统与其他系统相比优势

与 Symbian 相比，Android 系统是开源系统，系统发展更具前景；快速增长的海量第三方免费软件；无"证书"限制，安装软件更自由。

与 Iphone 相比，Android 更开放；风格更自由，简捷；开源系统，更多第三方免费软件；软件安装卸载更方便，无需第三方平台软件。

与 Windows mobile 相比，Android 更方便，简捷。

2) Android 系统的缺点

Android 系统手机电池续航普遍不足，由于厂商丰富，产品类型多样，不同厂商之间手机系统应用软件更多依赖第三方厂商，没有统一标准，兼容性不好，其系统开发设计及应用依赖于手机厂商。

4. Android 体系结构

Android 作为一个移动设备系统平台，采用软件堆层的架构，共分为 4 层，自下而上分别是 Linux 内核(操作系统，OS)、中间件层、应用程序框架、应用程序，如图 1-12 所示。

(1) 底层以 Linux 核心为基础，由 C 语言开发，只提供基本功能，是硬件和其他软件堆层之间的一个抽象隔离层，提供安全机制、内存管理、进程管理、网络协议堆栈和驱动程序等，如图 1-13 所示。

(2) 中间层包括函数库 Library 和虚拟机 Virtual Machine，由 C++ 开发，由函数库和 Android 运行时构成，如图 1-14 所示。

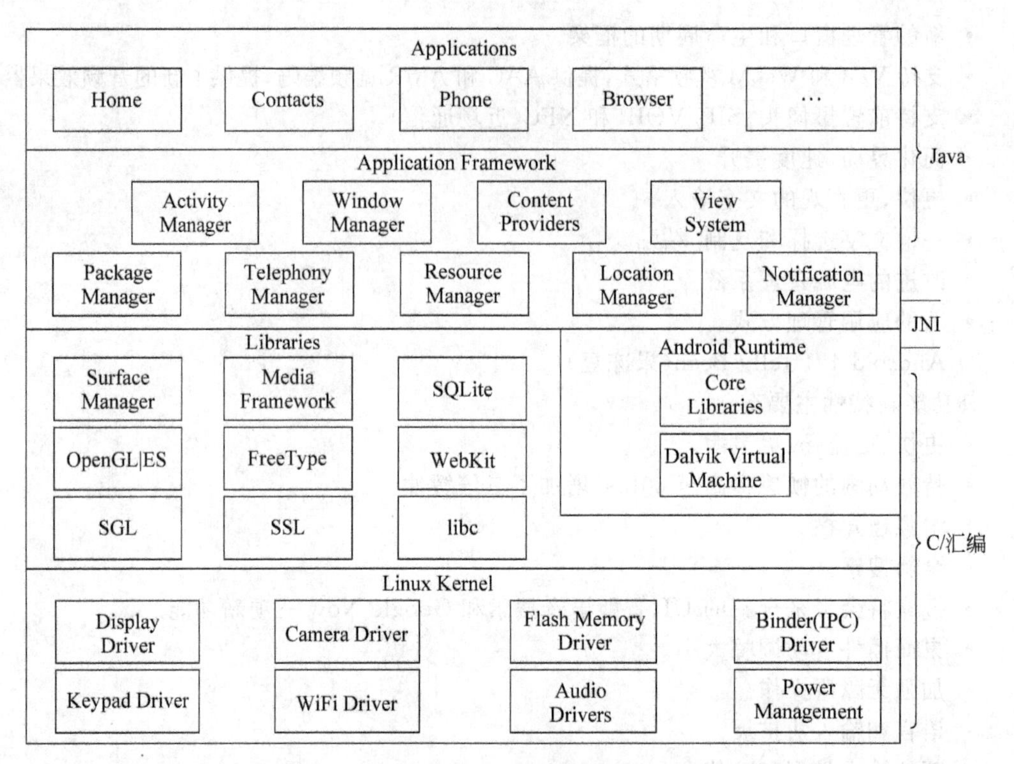

图 1-12 Android 系统结构

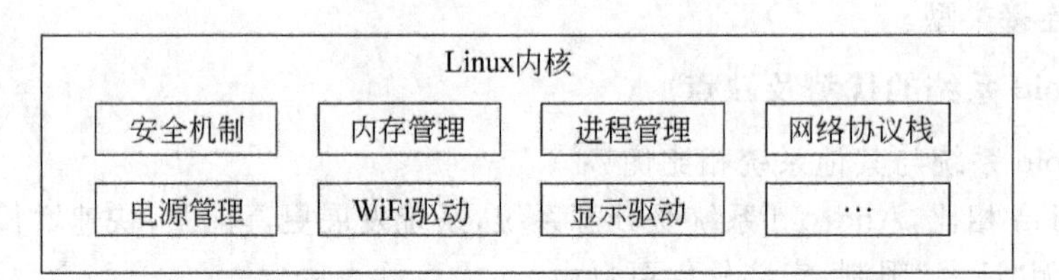

图 1-13 Linux 内核

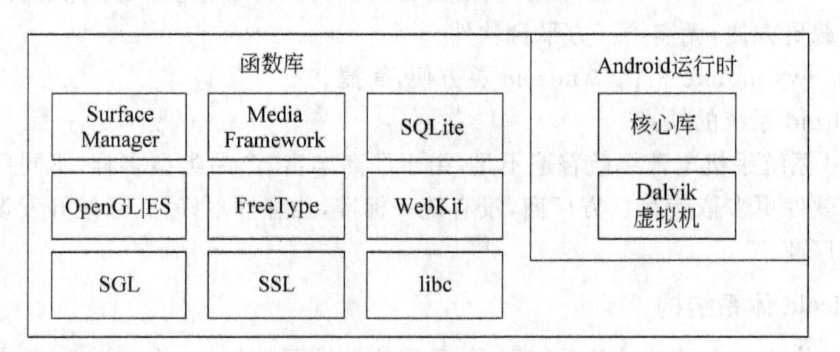

图 1-14 Android 系统中间层

① 函数库。主要提供一组基于 C/C++ 的函数库。主要包括：

- Surface Manager：支持显示子系统的访问，提供应用程序与 2D、3D 图像层的平滑连接。
- Media Framework：实现音视频的播放和录制功能。
- SQLite：轻量级的关系数据库引擎。

- OpenGL ES：基于 3D 图像加速。
- FreeType：位图与矢量字体渲染。
- WebKit：Web 浏览器引擎。
- SGL：2D 图像引擎。
- SSL：数据加密与安全传输的函数库。
- Libc：标准 C 运行库，Linux 系统中底层应用程序开发接口。

② Android 运行时。

- 核心库：提供 Android 系统的特有函数功能和 Java 语言函数功能。
- Dalvik 虚拟机：实现基于 Linux 内核的线程管理和底层内存管理。

(3) 应用框架层包含操作系统的各种管理程序，提供 Android 平台基本的管理功能和组件重用机制，包含：

- Activity Manager：管理应用程序的生命周期。
- Windows Manager：启动应用程序的窗体。
- Content Provider：共享私有数据，实现跨进程的数据访问。
- Package Manager：管理安装在 Android 系统内的应用程序。
- Teleghony Manager：管理与拨打和接听电话的相关功能。
- Resource Manager：允许应用程序使用非代码资源。
- Location Manager：管理与地图相关的服务功能。
- Notification Manager：允许应用程序在状态栏中显示提示信息。

如图 1-15 所示。

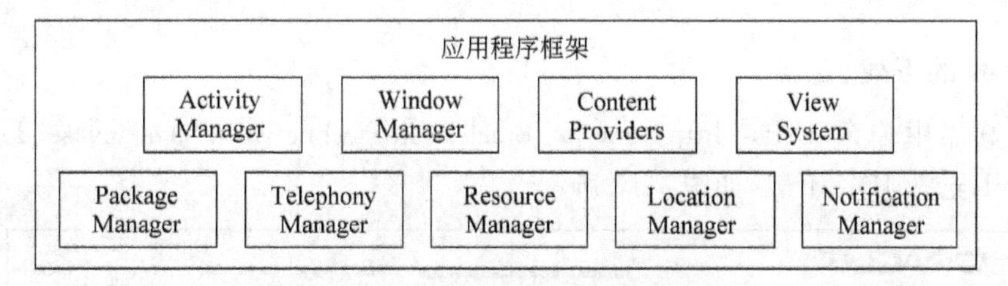

图 1-15 Android 系统应用程序框架

(4) 应用层是最上层，包括各种应用软件，如通话程序、短信程序、邮件客户端、浏览器、通讯录和日历等。应用软件则由各公司自行开发，以 Java 编写，如图 1-16 所示。

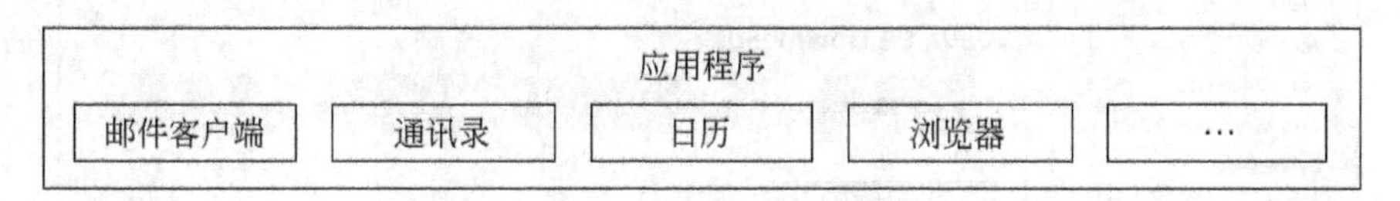

图 1-16 Android 系统应用层

1.3 搭建 Android 开发环境

由于操作系统的不同，关于 Android 系统开发环境的搭建过程也有很大的不同，本部分针对 Windows 系统环境和 Linux 环境下不同的操作系统简要介绍其开发环境的搭建过程。

1.3.1 Android 开发环境系统要求

Android 开发环境的设置，针对不同的 Windows 操作系统、Eclipse 版本、Android 开发工具包以及虚拟机 JDK，有许多可以采用的组合方式，具体选择如表 1-1 所示。

表 1-1 Android 开发环境系统要求

操作系统/IDE	版 本	备 注
Windows 系列	XP、Vista、Windows 7、Windows 8、Server 版本	
Eclipse IDE (for Java Developer)	Eclipse 3. x,3. 3. x(Europa)、3. 4. x(Ganymede)、3. 5. x(Galileo)、3. 6. x(Helios)	
ADT (Android Development Tools)	12. 0. 0～20. 0. 2	20. 0. 2 必须使用 Eclipse 3. 6. x 以上版本
Java SE		JDK 1. 5、JDK 1. 6、JDK 1. 7

1.3.2 Windows 系统平台下搭建开发环境

在 Windows 系列操作系统中搭建配置 Android 开发环境，需要的支持软件有 JDK、Eclipse、Android SDK 和 ADT(Android Development Tools)。具体开发环境搭建配置过程如下。

1. JDK 的下载、安装

在浏览器中输入 URL：http://www. oracle. com/technetwork/java/javase/downloads/index. html，选择 JDK 下载，如图 1-17 所示。

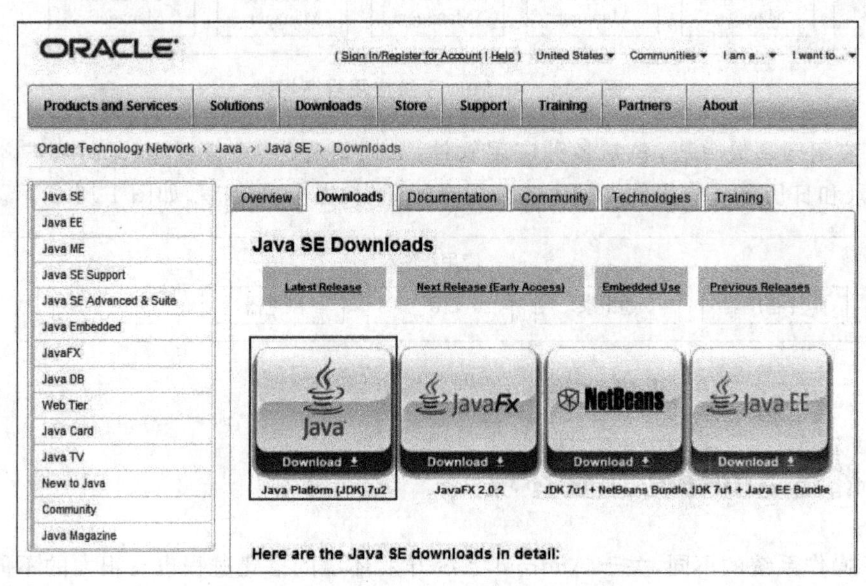

图 1-17 JDK 下载

选择接受协议 Accept License Agreement，然后根据不同的操作系统及处理器选择需要下载的 JDK 版本“jdk-7u2-windows-i586.exe”，如图 1-18 所示。

Java SE Development Kit 7u2

You must accept the Oracle Binary Code License Agreement for Java SE to download this software.

Thank you for accepting the Oracle Binary Code License Agreement for Java SE; you may now download this software.

Product / File Description	File Size	Download
Linux x86	63.62 MB	jdk-7u2-linux-i586.rpm
Linux x86	78.62 MB	jdk-7u2-linux-i586.tar.gz
Linux x64	64.51 MB	jdk-7u2-linux-x64.rpm
Linux x64	77.46 MB	jdk-7u2-linux-x64.tar.gz
Solaris x86	135.87 MB	jdk-7u2-solaris-i586.tar.Z
Solaris x86	81.37 MB	jdk-7u2-solaris-i586.tar.gz
Solaris SPARC	138.94 MB	jdk-7u2-solaris-sparc.tar.Z
Solaris SPARC	86.05 MB	jdk-7u2-solaris-sparc.tar.gz
Solaris SPARC 64-bit	16.13 MB	jdk-7u2-solaris-sparcv9.tar.Z
Solaris SPARC 64-bit	12.31 MB	jdk-7u2-solaris-sparcv9.tar.gz
Solaris x64	14.45 MB	jdk-7u2-solaris-x64.tar.Z
Solaris x64	9.25 MB	jdk-7u2-solaris-x64.tar.gz
Windows x86	84.04 MB	jdk-7u2-windows-i586.exe
Windows x64	87.35 MB	jdk-7u2-windows-x64.exe

图 1-18　JDK 7u2 Windows x86 版本

然后进行默认安装或者自选安装路径，安装完成后在控制台输入命令 java，测试 JDK 是否安装成功，如图 1-19 所示。

```
C:\windows\system32\cmd.exe
(C) 版权所有 1985-2001 Microsoft Corp.

C:\Documents and Settings\zyq>java
Usage: java [-options] class [args...]
           (to execute a class)
   or  java [-options] -jar jarfile [args...]
           (to execute a jar file)

where options include:
    -client       to select the "client" VM
    -server       to select the "server" VM
    -hotspot      is a synonym for the "client" VM  [deprecated]
                  The default VM is client.

    -cp <class search path of directories and zip/jar files>
    -classpath <class search path of directories and zip/jar files>
                  A ; separated list of directories, JAR archives,
                  and ZIP archives to search for class files.
    -D<name>=<value>
                  set a system property
    -verbose[:class|gc|jni]
                  enable verbose output
    -version      print product version and exit
    -version:<value>
                  require the specified version to run
```

图 1-19　测试 Java 安装

2. Eclipse 软件的下载

在 IE 地址栏输入 URL 地址 http://www.eclipse.org/downloads，如图 1-20 所示。选择 Eclipse IDE for Java Developers 进行下载，然后进行解压即可。注意：Eclipse 软件不需要安装，在安装完 JDK 后只需解压 Eclipse 即可运用。

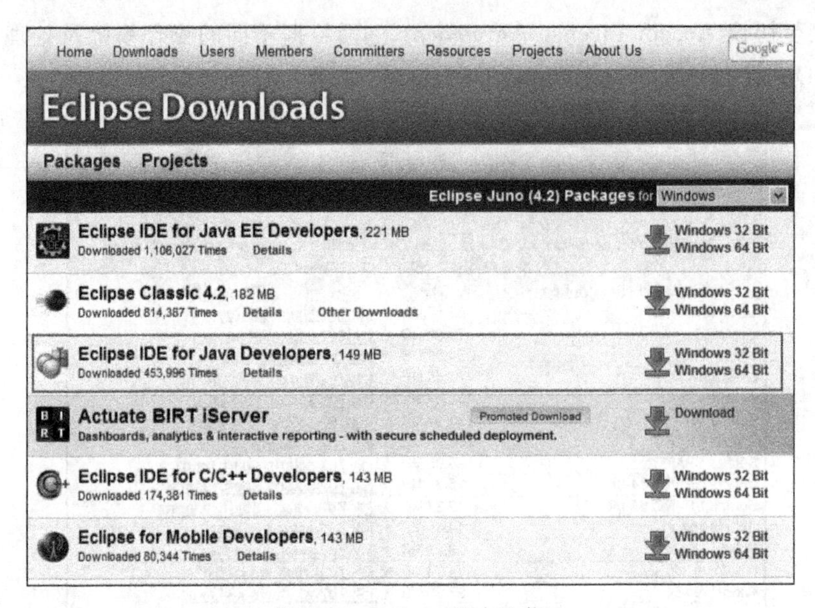

图 1-20 Eclipse 版本下载

3. Android SDK 的下载、配置

(1) 在地址栏输入 http://developer.android.com/sdk/index.html，如图 1-21 所示。选择 Android SDK 成熟版本 Android 3.0 Platform 进行下载，并在本地目录 h:\android\ 下进行解压，以便后续开发使用。

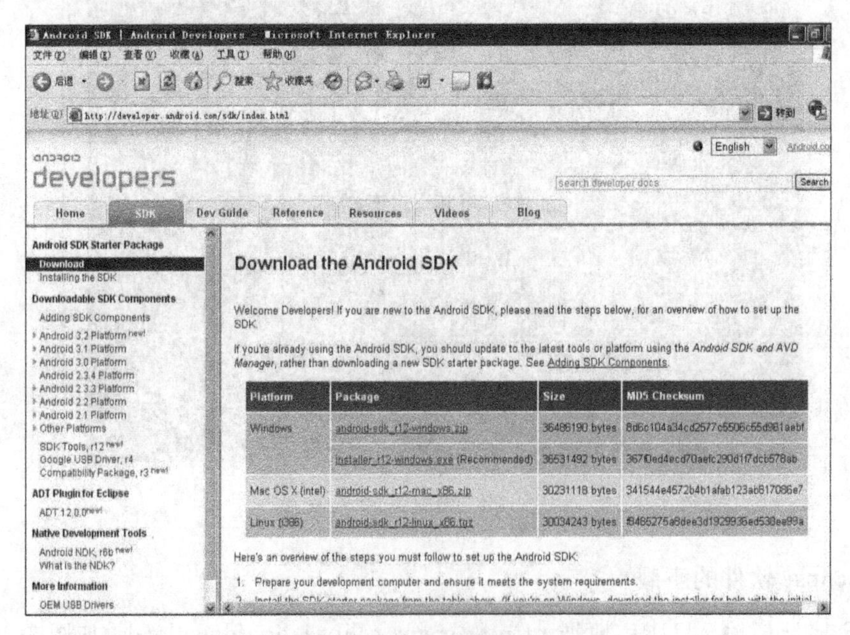

图 1-21 Android SDK 的下载

(2) 双击 h:\android\android-sdk-windows 文件中的 SDK Manager.exe，SDK 程序自动从因特网上检测当前 SDK 可选安装包的版本，如图 1-22 所示。

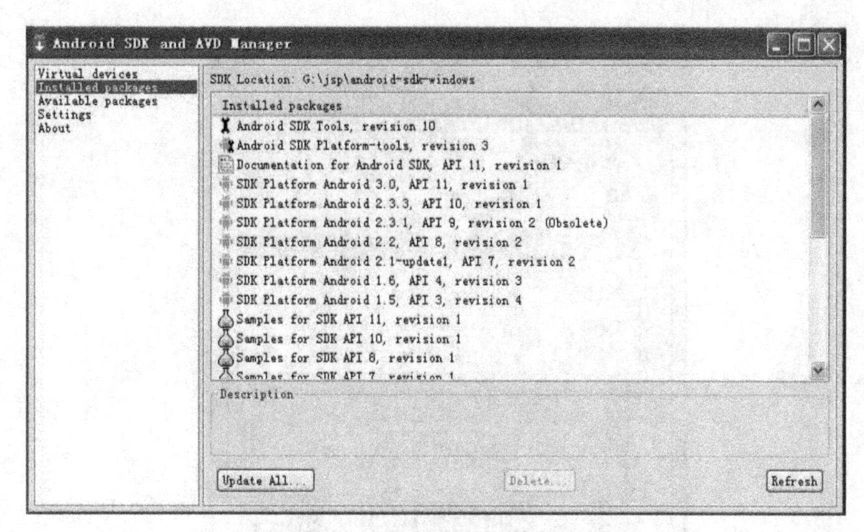

图 1-22　网络检测 SDK 安装包

然后即可以选择部分安装，也可以选择全部安装不同版本的 SDK 软件包。接着单击 Install 按钮进行下载安装选择的 SDK 版本，下载安装到 h:\android\android-sdk-windows 目录文件中，如图 1-23 所示。

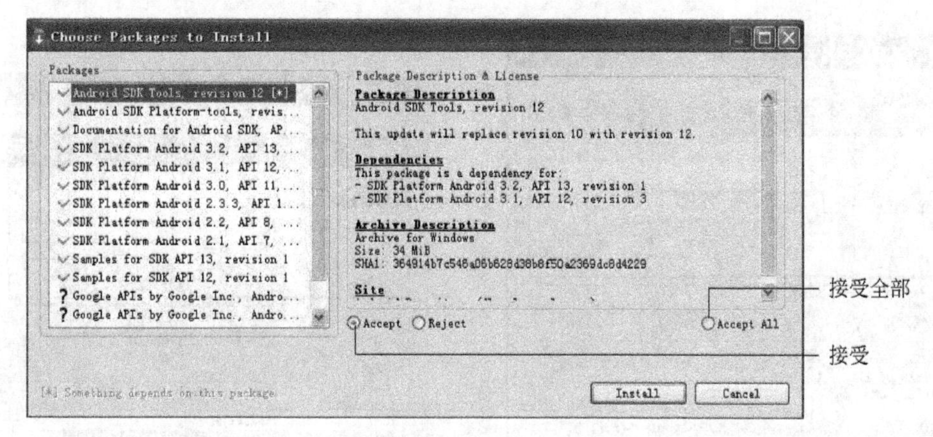

图 1-23　选择 SDK 安装包

4. 设置 Android SDK HOME

如果需要在控制台界面(cmd 运行)运行 SDK 命令，则需要进行环境变量配置。右击"我的电脑"图标，从弹出的快捷菜单中选择"高级"→"环境变量"→"系统变量"菜单项，在新建系统变量 Android_SDK_HOME 中添加变量值为 h:\android\android-sdk-windows，在系统变量 Path 中添加"%Android_SDK_HOME%\tools"和"%Android_SDK_HOME%\platform-tools"，如图 1-24～图 1-26 所示。

5. ADT(Android Development Tools)的安装、配置

ADT 是开发 Android 的工具插件，在 Eclipse 环境下开发 Android 程序必须使用 ADT 插件。

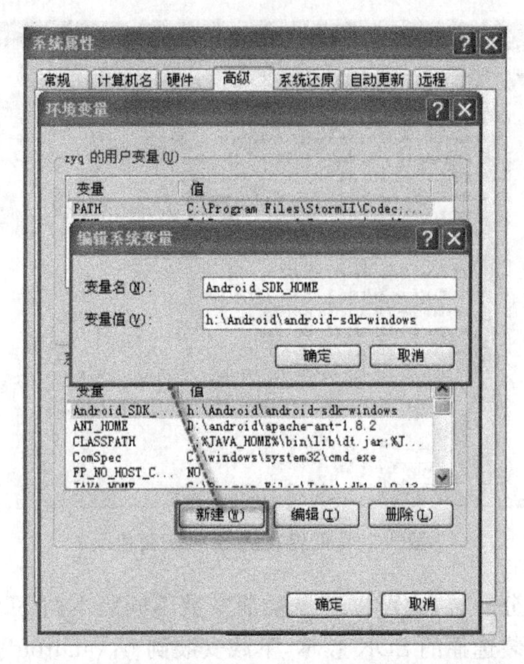

图 1-24 设置 Android SDK HOME 环境变量

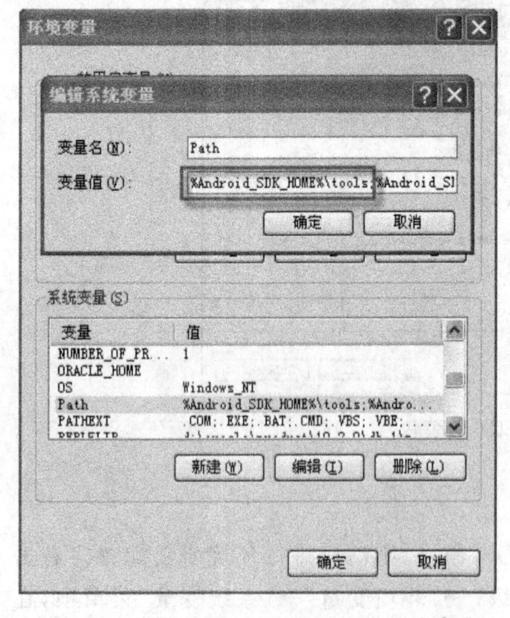

图 1-25 设置 Android SDK HOME 环境变量

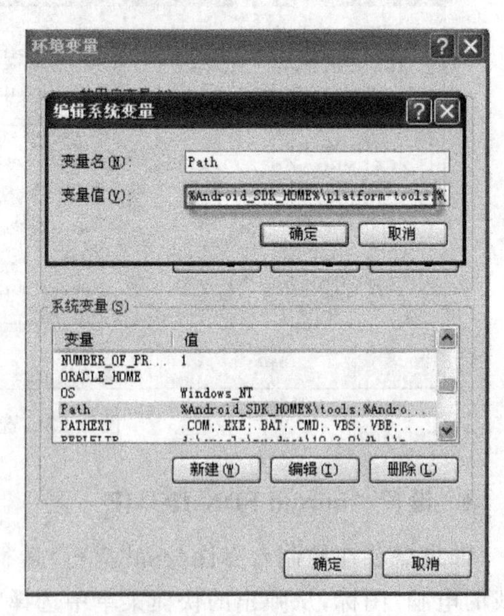

图 1-26 设置 Android SDK HOME 环境变量

ADT 的安装步骤如下：

(1) 打开 Eclipse 软件，选择 Help→Install New Software 菜单项，如图 1-27 所示。

(2) 单击 Add 按钮，弹出添加新站点的界面，在 Name 文本框中填写自己确定的名称，在 Location 文本框中输入"http://dl-ssl.google.com/android/eclipse"，然后单击 OK 按钮，如图 1-28 所示。

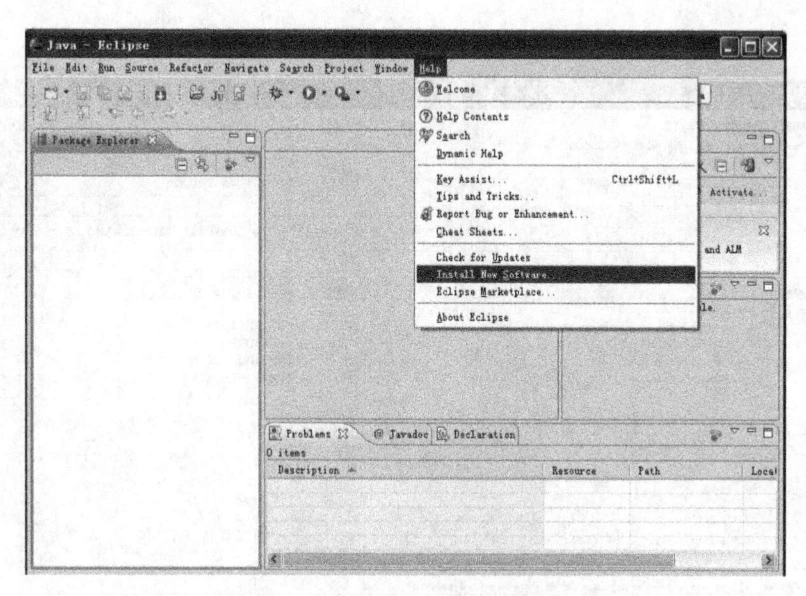

图 1-27　在 Eclipse 中选择安装新软件

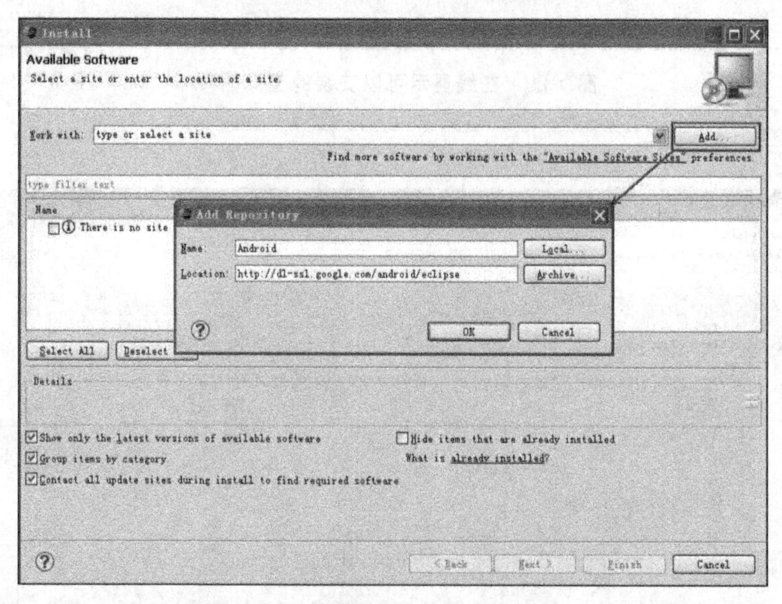

图 1-28　添加下载软件的新站点

(3) Eclipse 远程连接刚才输入的 URL 站点，并在线显示可以安装的工具插件(DDMS、ADT 等)，如图 1-29 所示。注意，一定要联网才能下载安装插件。

(4) 选中 DDMS、ADT 工具插件，单击 Next 按钮，出现插件安装界面，如图 1-30 所示。

(5) 单击 Next 按钮，弹出 Review Licenses 窗口，选择 I accept the terms of the license agreements 单选按钮，单击 Finish 按钮进行安装，如图 1-31 所示。

(6) 在安装过程中会出现安装软件包中包含未签名的内容警告窗口，如图 1-32 所示。单击 OK 按钮继续安装，安装完成后，在弹出的是否重启 Eclipse 对话框中单击 Yes 按钮，以便安装软件生效，如图 1-33 所示。

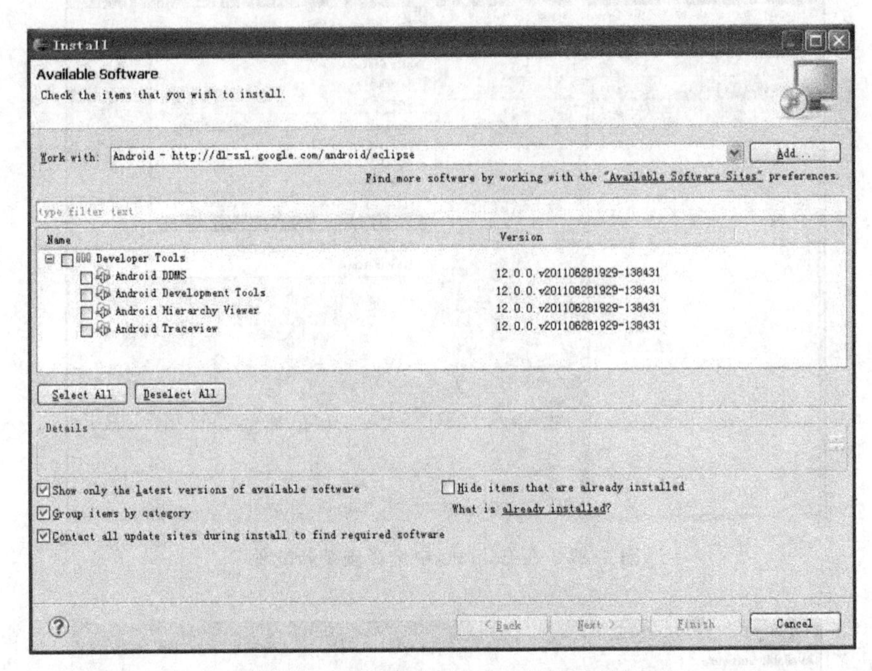

图 1-29　在线显示可以安装的工具插件

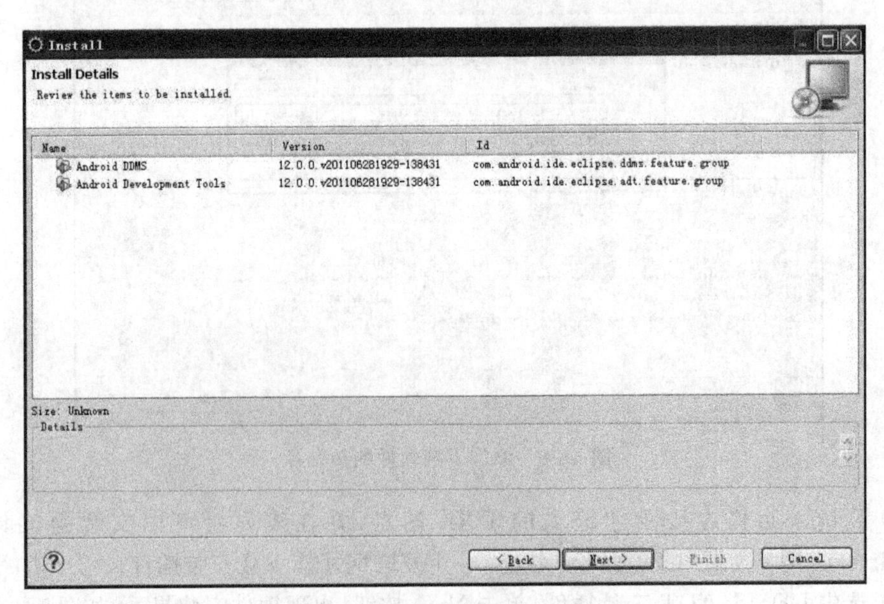

图 1-30　选择后的插件安装界面

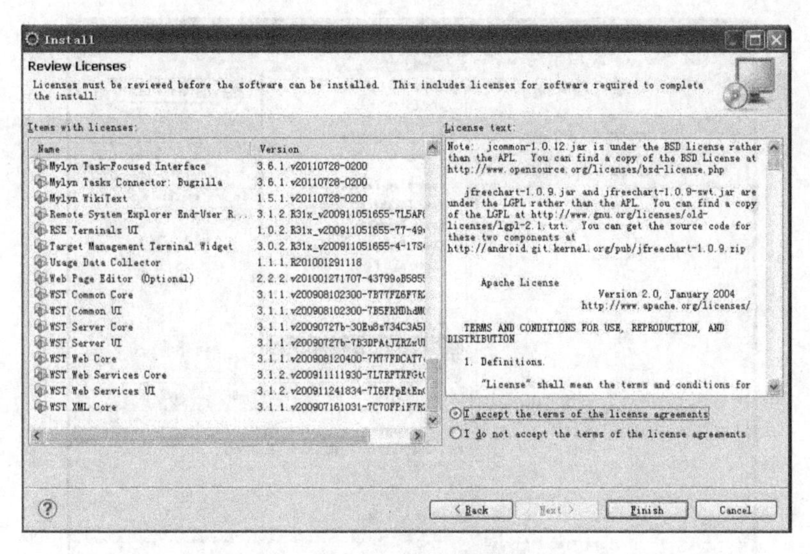

图 1-31　接受安装

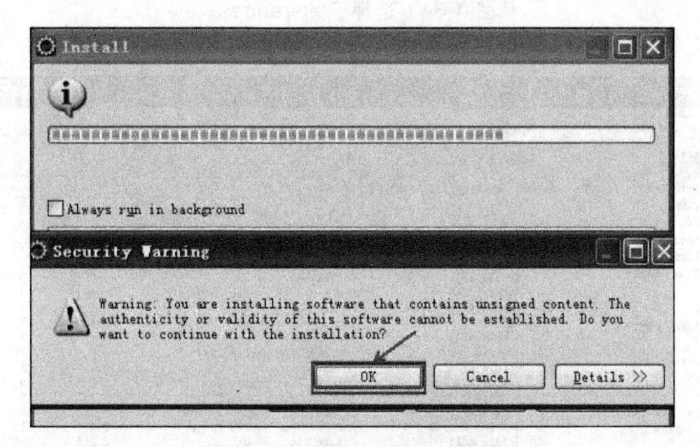

图 1-32　选择插件继续安装

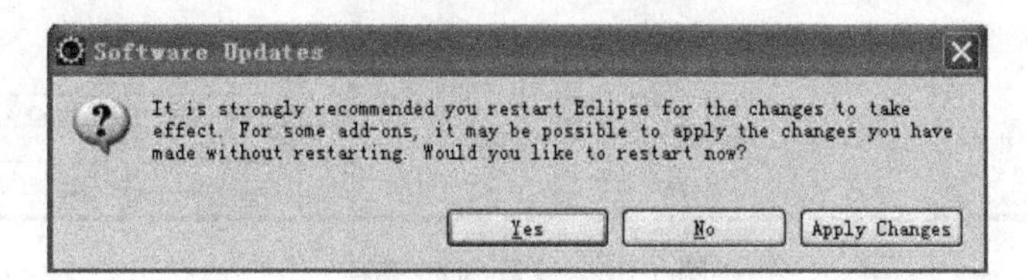

图 1-33　重启生效

6. Eclipse 中设置 Android SDK HOME

(1) 在 Eclipse 中选择 Window→Preferences 菜单项，如图 1-34 所示。

(2) 在弹出的窗口中选择左侧 Android 节点，然后在右侧的 SDK Location 文本框中选择下载的 SDK 解压目录，在窗口的右侧下方会出现 SDK Target 的列表，单击 OK 按钮，完成 SDK 的配置，如图 1-35 所示。

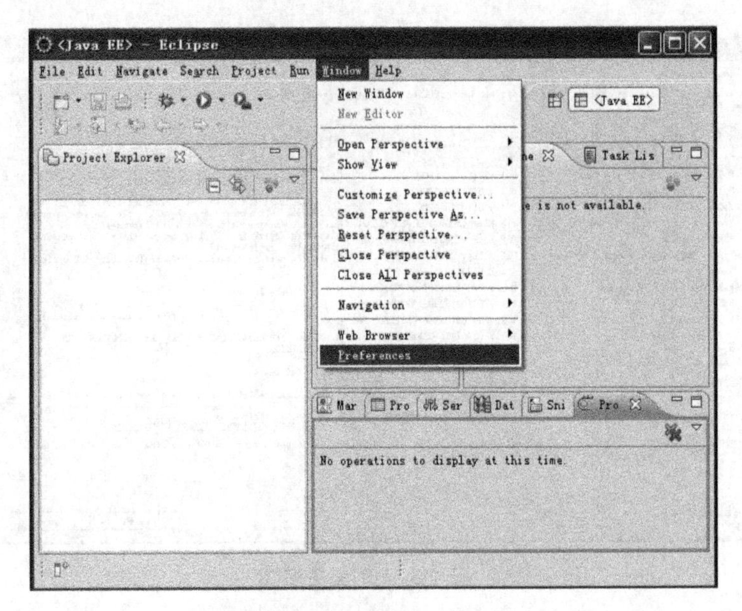

图 1-34　选择 Preferences

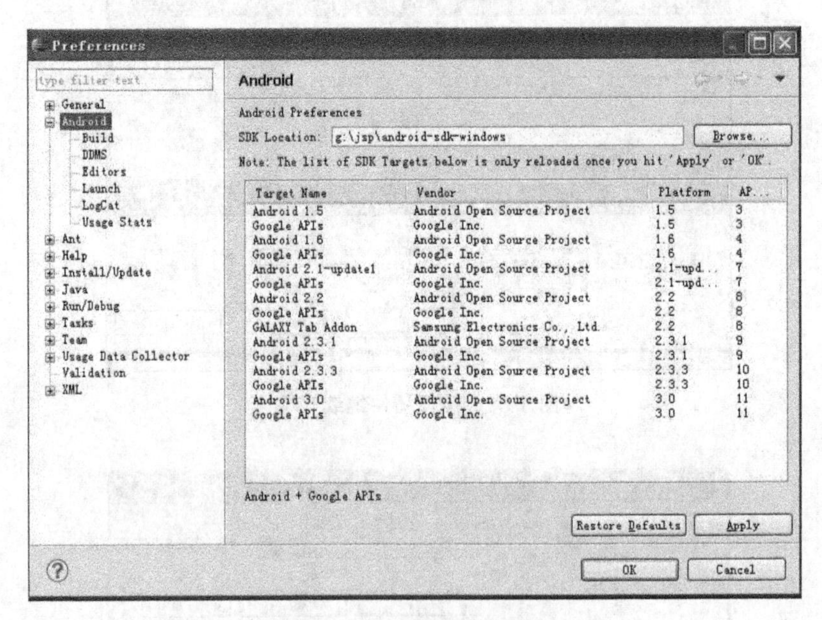

图 1-35　设置 SDK 路径

1.3.3　Linux 系统平台下搭建开发环境

1. Ubuntu Java 安装配置的详细步骤

1) 下载 jdk(Linux 版本)

通过浏览器地址 http://www.oracle.com/technetwork/java/javase/downloads/index.html,选择 jdk-6u33-linux-i586.bin 下载,并将 jdk-6u33-linux-i586.bin 放置于目录 /home/wjj/下。

2）解压文件

打开终端，进入放置 jdk 的目录 cd /home/wjj，使用命令 chmod u+x jdk-6u33-linux-i586. bin，更改文件权限为可执行，然后使用命令 sudo. /jdk-6u33-linux-i586. bin 解压文件，则在 wjj 目录下面可以看到解压的文件夹 jdk1. 6. 0_33。

3）配置环境变量

以 root 身份使用命令 sudo gedit /etc/profile，打开并编辑 profile 文件。在 profile 文件最后添加：

```
#set java environment
JAVA_HOME=/home/wjj/jdk1.6.0_33
export JRE_HOME=/home/wjj/jdk1.6.0_33/jre
export CLASSPATH=$JAVA_HOME/lib:$JRE_HOME/lib:$CLASSPATH
export PATH=$JAVA_HOME/bin:$JRE_HOME/bin:$PATH:$SDK
```

然后保存并关闭文件。

4）重启系统

为使上述配置生效，可使用命令 reboot 重启系统。

5）查看 Java 版本。

在终端输入 java -version 将会显示 Java 版本的相关信息，表明 Ubuntu java 安装成功。

2. 集成开发工具 Eclipse 安装

1）下载 Eclipse

通过下载地址 http://www. eclipse. org/downloads/，下载 eclipse-jee-helios-linux-gtk. tar. gz，放在/home/wjj 目录下。

2）解压

进入放置目录 cd/home/wjj，使用命令解压 tar xvfz eclipse-jee-helios-linux-gtk. tar. gz，在此目录下会解压出一个 eclipse 文件夹，进入双击 eclipse 即可运行。

3. Android 安装配置

1）下载

通过下载地址 http://developer. android. com/sdk/index. html，下载 android-sdk_r20-linux_86. tgz。

2）解压文件放在目录/home/wjj 下

使用命令进入目录 cd/home/wjj，使用命令 tar zxvf android-sdk_r20-linux_86. tgz 解压文件。

3）配置环境变量

以 root 身份使用命令 sudo gedit /etc/profile，打开并编辑 profile 文件。在 profile 文件最后添加：

```
export SDK=${PATH}:<your_sdk_dir>/tools
```

例：

```
export SDK=$/home/wjj/android-sdk-linux_86/tools
```

4）下载和配置 ADT

安装和配置过程与 Windows 环境下配置过程相同。

1.4 Android SDK 概述

SDK(Software Development Kit，软件开发工具包)是指为特定的软件包、软件框架、硬件平台和操作系统等建立应用软件的开发工具的集合。Android SDK 是指专门用于 Android 手机操作系统创建应用软件的软件开发工具包。

Android SDK 采用 Java 语言，所以需要先安装 JDK 5.0 及以上版本。Android SDK 不用安装，下载后直接解压到指定的位置即可。具体安装步骤前面已述，不再赘述。

1.4.1 Android SDK 目录结构

解压完 Android SDK 后，打开目录，目录结构如图 1-36 所示。

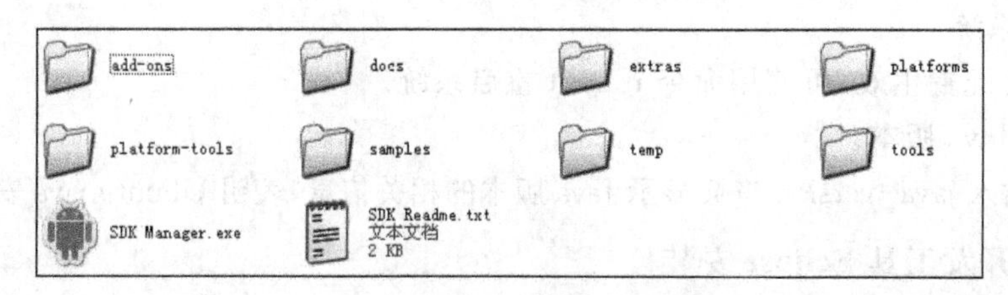

图 1-36 Android SDK 目录结构

- add-ons：目录下是用来开发应用的 Google API，支持基于 Google Map 的地图开发及系统模拟器图片。
- docs：目录下是 Android SDK 的帮助文档和说明文档，通过根目录下的 index.html 文件启动。
- platforms：目录中存在子目录 android-3、android-4 等，分别用来保存 1.5、1.6、2.0、2.1、2.2、2.3、3.0、4.0、4.1 版的 Android SDK 的库函数、外观样式、程序示例和辅助工具等。
- tools：目录下是通用的 Android 开发和调试工具。
- extras：包含 android 和 google 两个目录，分别用来存放 support library 和 usb drivers。usb_driver 目录下存放 amd64 和 x86 平台的 USB 驱动程序。
- platform-tools：目录下包含开发平台需要的开发工具和测试工具。
- temp：目录下包含了一些常用的文件模板。
- samples：目录下包含不同版本的 SDK 演示实例。
- SDK Manager：SDK 管理，用于安装和更新 SDK 组件。

1.4.2 Android 常用开发工具

Android SDK 中包含了许多开发工具，这些工具帮助程序开发者在 Android 开发平台上开发移动应用。这些开发工具被划分为两类：SDK 工具和平台工具，SDK 工具是独立于平台，适用于不同的平台；平台工具通常支持最新的 Android 开发平台。

1. 最重要的 SDK 工具

最重要的 SDK 工具有 SDK Manager（android sdk）、AVD Manager（android avd）、emulator 和 ddms(Dalvik Debug Monitor Server)，下面就简要介绍一些常用的 SDK 工具。

1）Android

用于管理 AVD、工程和安装 SDK 的组件，可以创建、删除 AVD。

在 Eclipse 中选择 Run Configuration→Target 菜单项，单击 Manager 按钮就会调用该脚本。

ddms(调试监视服务)用于调试 Android 应用程序，管理运行在设备或模拟器上的进程，监视 Android 系统中进程、堆栈信息，查看 logcat 日志，实现端口转发服务和屏幕截图功能，模拟器电话呼叫和 SMS 短信，以及浏览 Android 模拟器文件系统等。

2）Android 模拟器(Android Emulator)

Android 模拟器是运行在计算机上的虚拟移动设备，Android SDK 最重要的工具，支持加载 SD 卡映像文件，更改模拟网络状态，延迟和速度，模拟电话呼叫和接收短信等。不支持接听真实电话，USB 连接，摄像头捕获，设备耳机，电池电量和 AC 电源检测，SD 卡插拔检查和使用蓝牙设备。

3）dmtracedump

dmtracedump 工具原设计是将整个调用过程和时间分析相结合，以函数调用图的形式呈现，但目前 dmtracedump 只有-0 选项可使用，其在应用中，需要结合 Graphviz Dot 组件来生成图形，所以要运行 dmtracedump 就必须先安装 Graphviz。

4）Draw 9-patch

NinePatch 是 Android 提供的可伸缩的图形文件格式，基于 PNG 文件。draw 9-patch 工具可以使用 WYSIWYG 编辑器建立 NinePatch 文件。它也可以预览经过拉伸的图像，高亮显示内容区域。

5）Hierarchy Viewer

层级观察器工具允许调试和优化用户界面。它用可视的方法把视图(View)的布局层次展现出来，此外还给当前界面提供了一个具有像素栅格(Grid)的放大镜观察器，以便正确地布局。

6）traceview

traceview 跟踪显示工具是 Android 平台配备的一个很好的性能分析的工具。以图形化的方式显示应用程序，让用户了解要跟踪程序的性能，并且能具体到 method。用来调试应用程序，分析执行效率。

7）mksdcard

创建 SD 卡工具，该工具是创建一个 FAT32 格式的磁盘镜像，主要是用于模拟手机 SD 卡的，在创建 AVD 中可以选择该文件作为 SD 卡。mksdcard [-l label] <size>[K|M] <file>。

8）zipalign. exe

该工具优化了应用程序的打包方式。这样做使 Android 与用户的应用程序交互更加的有效和简便，有可能提高应用程序和整个系统的运行速度。

9）hprof-conv

主要用于转换文件格式，转换 hprof 文件为一个标准的文件格式，以便浏览。

10）layoutopt

用于快速分析开发的应用程序布局，优化、提高效率。

11）Monkey

Android 的压力测试工具，是在模拟器上或设备上运行的一个小程序，它能够产生为随机的用户事件流，例如单击（Click），触摸（Touch），挥手（Gestures），还有一系列的系统级事件。可以使用 Monkey 给正在开发的程序做随机的，但可重复的压力测试。

Monkeyrunner：Android 的自动测试工具。

12）SQLite3

该工具能够让用户方便地访问 SQLite 数据文件。用来创建和管理 SQLite 数据库。

13）ProGuard

这个工具是一个 Java 代码混淆的工具。在 2.3 版本的 sdk 中可以看到在 android-sdk-windows/tools/下面多了一个 proguard 文件夹，Google 已经把 proguard 技术放在了 android sdk 里面，通过正常的编译方式也能实现代码混淆了。

2. Android 平台工具

通常，开发者直接使用的平台工具是 ADB（Android Debug Bridge，Android 调试桥），ADB 工具可以让开发者在模拟器或设备上安装应用程序的.apk 文件，并从命令行访问模拟器或设备。也可以用它把 Android 模拟器或设备上的应用程序代码和一个标准的调试器连接在一起。

其他的平台工具，如 AIDL（Android interface Description Language，Android 接口描述语言）、AAPT（Android Asset Packaging Tool，Android 资源打包工具）、dexdump、dx，很少由开发者直接使用，一般都是通过 android 编译工具或者 ADT 调用来完成任务。

1.4.3 Android SDK 实例

在 Android SDK 目录中的 samples 文件夹下，针对不同的版本有不同的文件夹，在此下面存放了许多示例程序，以帮助入门者理解一些基础的 Android APIs 和代码实践，创建、修改 samples 中的程序，并在模拟器或虚拟设备上运行。此处不再详述，具体见 Android SDK 目录下的 samples 文件夹。

1.5 创建 Android 程序

1.5.1 创建和使用虚拟设备

AVD（模拟器或模拟设备）是 Android 程序测试运行的虚拟平台，每一个 AVD 都模拟了一个独立的虚拟设备来运行 Android 的程序。自 Android SDK1.5 版本之后，Android 程序测试运行必须创建 AVD 虚拟运行平台才能进行运行及测试。本书分别介绍在 Eclipse 环境中和命令行下 AVD 的创建和使用。

1. Eclipse 环境下使用和创建 AVD 模拟器

（1）在 Eclipse 的菜单栏中选择 Window→Android SDK and AVD Manager 菜单项，

在弹出的对话框中默认是选择 Virtual devices，如图 1-37 所示。

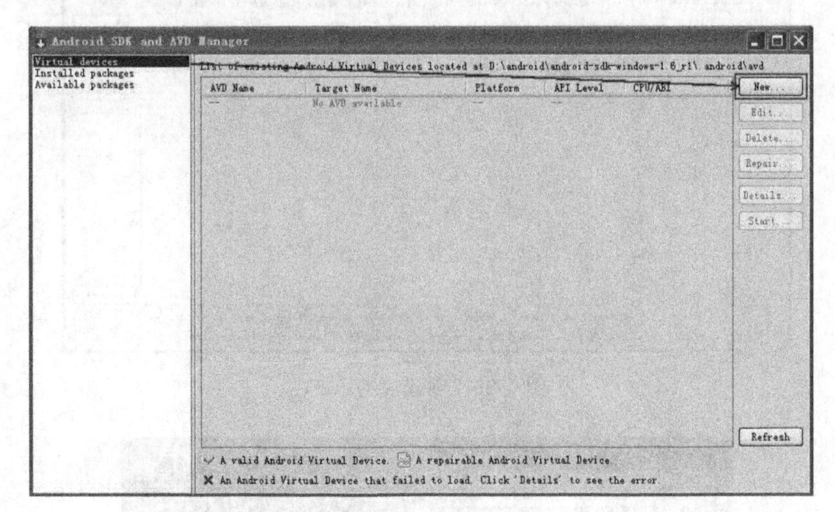

图 1-37 新建 AVD

（2）单击 New 按钮，创建新的 AVD，在弹出的对话框中设置 AVD 的名称、目标平台 API 版本、SD 卡的大小、模拟设备默认的皮肤等参数，如图 1-38 所示。

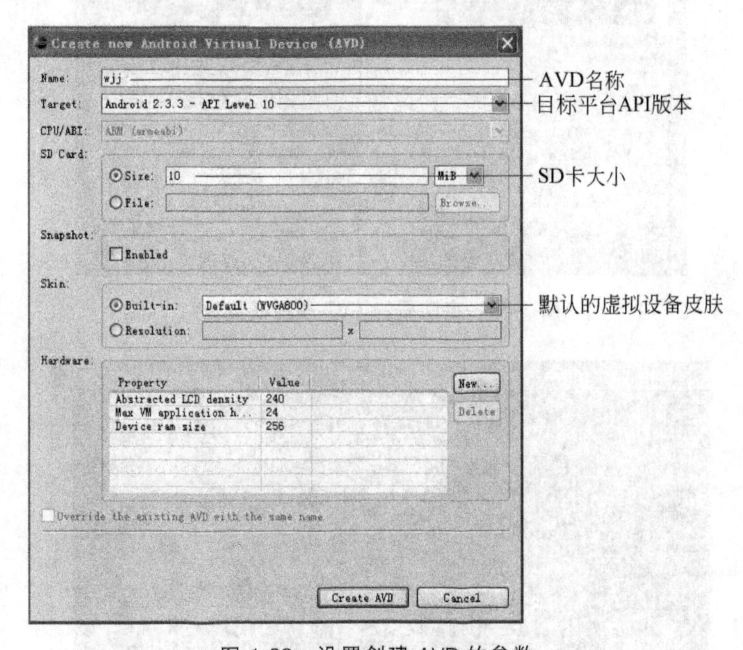

图 1-38 设置创建 AVD 的参数

（3）单击 Create AVD 按钮，创建 AVD，成功创建的 AVD 在对话框右下方显示，如图 1-39 所示。

（4）单击 Start 按钮，启动运行 AVD 模拟设备，运行界面如图 1-40 所示。

2. 在命令行下创建和使用 AVD 模拟器

（1）在控制台窗口中（运行中输入 cmd）输入命令 android list targets，查看可用的目标设备，如图 1-41 所示。

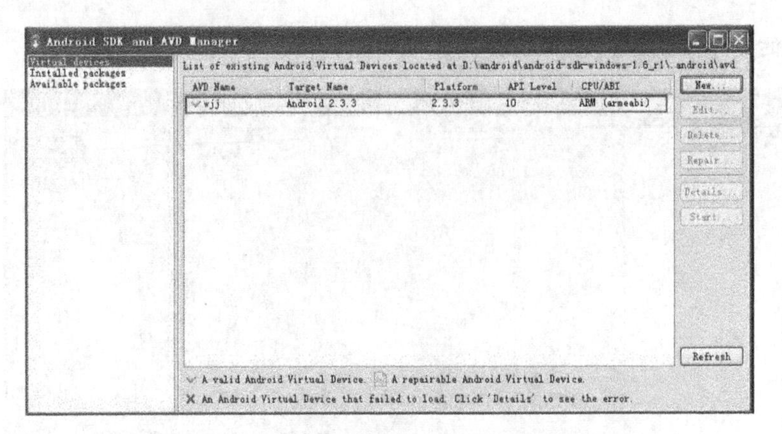

图 1-39　创建后的 AVD

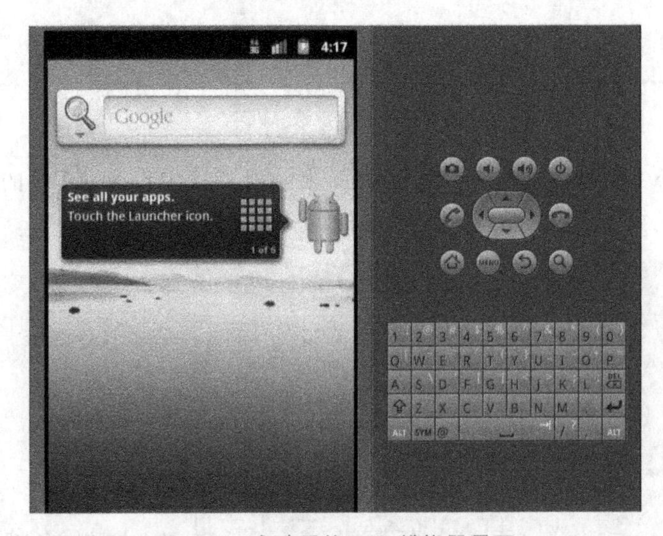

图 1-40　启动后的 AVD 模拟器界面

```
C:\windows\system32\cmd.exe
     Vendor: Samsung Electronics Co., Ltd.
     Revision: 1
     Based on Android 2.2 (API level 8)
     Skins: WVGA854, WQVGA400, GALAXY Tab (default), HVGA, WQVGA432, WVGA800,
GA
id: 10 or "android-9"
     Name: Android 2.3.1
     Type: Platform
     API level: 9
     Revision: 2
     Skins: HVGA, QVGA, WQVGA400, WQVGA432, WVGA800 (default), WVGA854
id: 11 or "Google Inc.:Google APIs:9"
     Name: Google APIs
     Type: Add-On
     Vendor: Google Inc.
     Revision: 2
     Description: Android + Google APIs
     Based on Android 2.3.1 (API level 9)
     Libraries:
      * com.google.android.maps (maps.jar)
          API for Google Maps
     Skins: WVGA854, WQVGA400, HVGA, WQVGA432, WVGA800 (default), QVGA
id: 12 or "android-10"
     Name: Android 2.3.3
     Type: Platform
     API level: 10
     Revision: 1
     Skins: HVGA, QVGA, WQVGA400, WQVGA432, WVGA800 (default), WVGA854
id: 13 or "Google Inc.:Google APIs:10"
```

图 1-41　查看可用的目标平台

(2) 选择其中一个目标平台 API 版本，输入命令 android create avd --name wjj --target 10，其中 wjj 是虚拟设备的名称，10 是选择的目标平台 API 版本 id 值，创建虚拟设备 AVD，如图 1-42 所示。

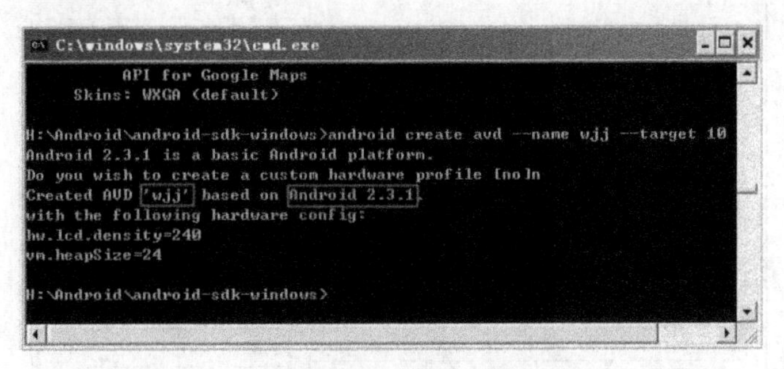

图 1-42　创建后的 AVD

(3) 启动虚拟设备，输入命令 emulator -avd wjj，启动上面创建的虚拟设备 wjj，如图 1-43 所示。

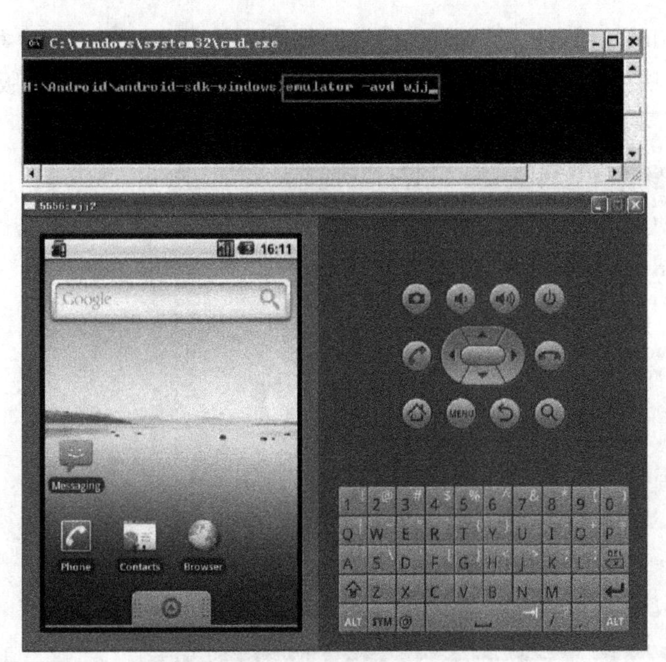

图 1-43　输入命令和启动后的 AVD 界面

1.5.2　在 Eclipse 下创建 Android 程序

前面的章节已经讲解了在 Eclipse 中搭建 Android 开发环境和虚拟设备 AVD 的创建，下面将就在 Eclipse 中如何创建 Android 应用程序的步骤进行介绍。

(1) 打开 Eclipse，选择 File→New→Android Project 菜单项，如图 1-44 所示。或者在右边的面板中单击右键，在弹出的快捷菜单中选择 New→Android Project 命令。

特别提示：如果没有找到 Android Project 项，可以通过选择 File→New→Other 菜单项，在弹出对话框的 Android 下找到 Android Project 项。

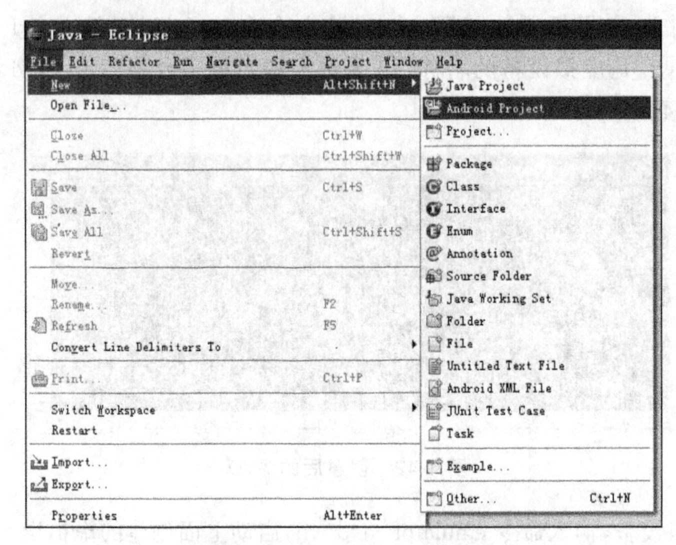

图 1-44　创建 Android 工程

(2) 弹出 Android 项目创建界面,填写项目名称、项目默认的存储路径、目标版本、应用程序名(默认与项目名一致)、包名、创建的 Activity 的名字、最小 SDK 版本(默认与目标版本 API 一致,不要修改),如图 1-45 所示。

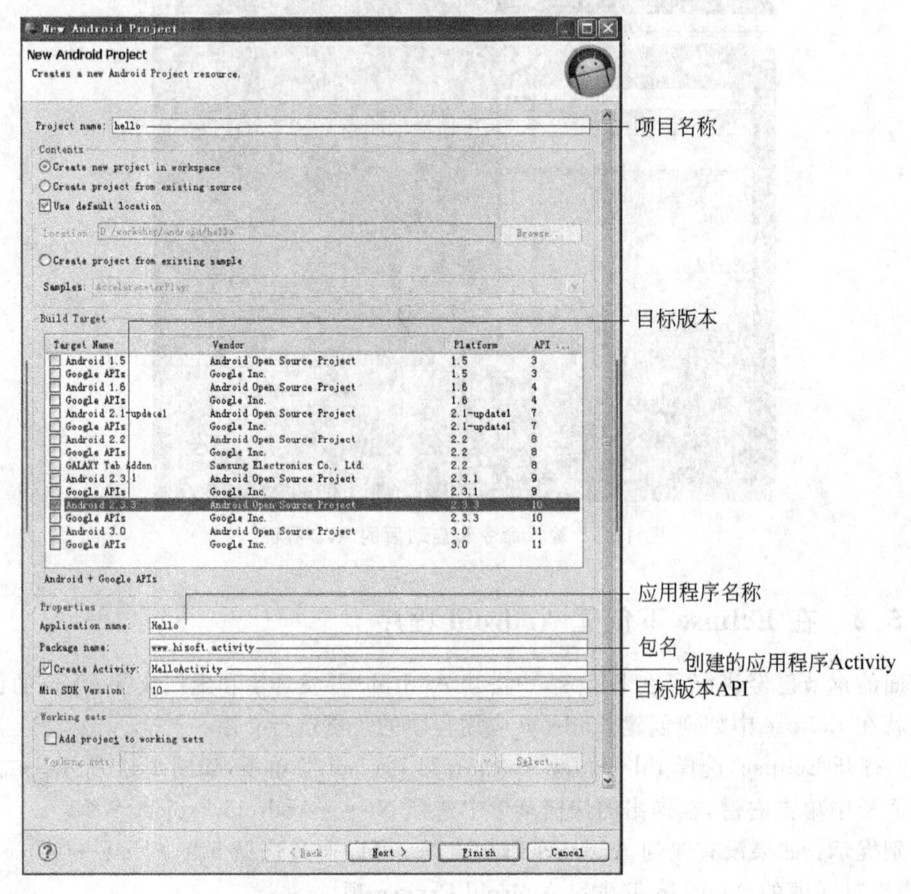

图 1-45　设置工程名、目标平台等参数

特别提示：Min SDK Version(最小 SDK 版本号)中的版本号是根据选择的目标版本 API 自动生成的，不要进行修改。

(3) 单击 Finish 按钮，创建完成项目。在创建完成项目的过程中，ADT(Android Development Tools)会自动生成一些目录和文件，具体后续生成的项目目录结构如图 1-46 所示。

(4) 选中需要运行的 hello 项目右击，在弹出的快捷菜单中选择 Run As→Android Application 菜单项，如图 1-47 所示。如果没有提前启动虚拟设备，系统会默认启动一个已经创建的虚拟设备运行项目，如图 1-48 所示。

特别提示：在运行项目之前，必须先创建虚拟设备 AVD(按前面章节所述方法创建)。创建虚拟设备过程中选择的目标版本和创建项目中选择的目标版本必须一致或者向下兼容，否则会提示没有兼容的目标版本，需要创建新的 AVD。

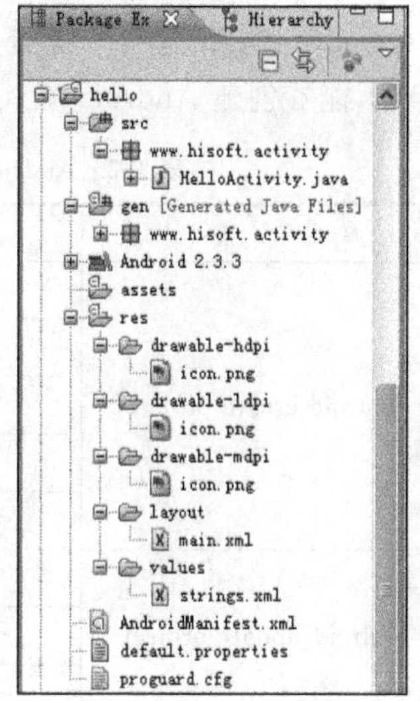

图 1-46　Android 工程目录结构

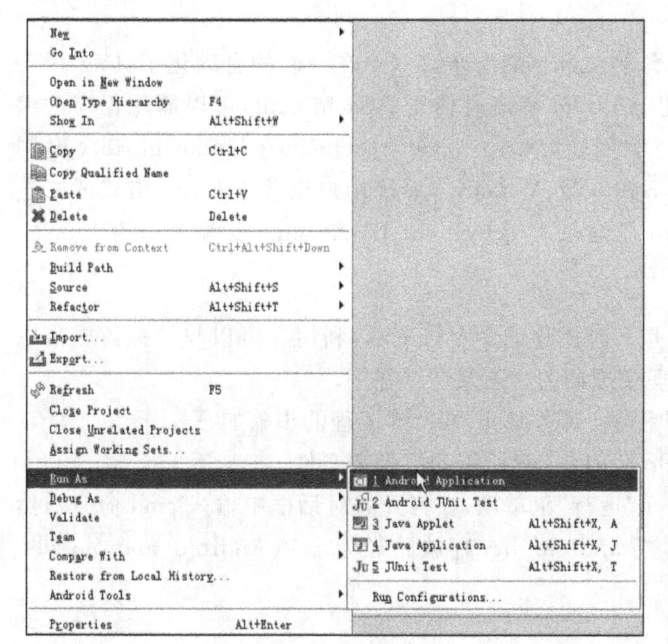

图 1-47　部署工程

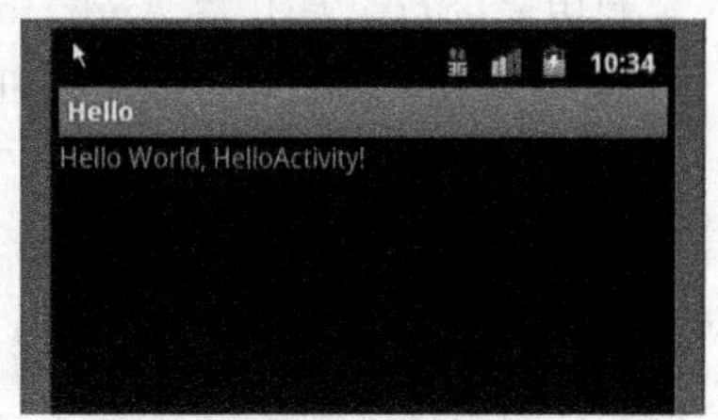

图 1-48　模拟器运行显示

1.5.3　命令行创建 Android 程序

命令行工具保存在<sdk>/platform-tools/和<sdk>/tools/目录下，使用命令行工具创建 Android 程序，需要使用 tools 目录下的 android.bat 和 platform-tools 目录下的 adb.exe 工具及 Apache Ant 软件。

1. Android批处理工具

android.bat是一个批处理文件，可以用来建立和更新Android工程，同时也管理AVD，能够创建Android工程所需要的目录结构和文件。具体命令和参数如表1-2所示。

表1-2　Android.bat建立和更新Android工程的命令和参数说明

命　令	参　数	说　明	备　注
android create project	-k <package>	包名称	必备参数
	-n <name>	工程名称	
	-a <activity>	Activity名称	
	-t <target>	新工程的编译目标	必备参数
	-p <path>	新工程的保存路径	必备参数
android update project	-t <targe>	设定工程的编译目标	必备参数
	-p <path>	工程的保存路径	必备参数
	-n <name>	工程名称	

2. Apache Ant工具

Apache Ant是一个将软件编译、测试和部署等步骤联系在一起的自动化工具，多用于Java环境中的软件开发。若在构建Android程序时使用Apache Ant，可以简化程序的编译和apk打包过程。Apache Ant下载网址为http://ant.apache.org/bindownload.cgi，网站提供zip、tar.gz和tar.bz2三种格式下载，Windows系统用户推荐下载zip格式的二进制包。目前最新下载的Apache Ant压缩包为apache-ant-1.8.2-bin.zip，版本号为1.8.2。

3. ADB工具

ADB(Android Debug Bridge)是多种用途的命令行工具，利用它可以与模拟器或带电的Android设备进行连接通信、管理模拟器的状态及调试程序。

利用命令行工具开发Android程序，创建HelloWorld工程的步骤如下：

1) 使用android.bat建立HelloWorld工程所需的目录和文件

打开cmd控制台(选择“开始”→“运行”菜单项，在打开的对话框中输入cmd命令)，然后进入<sdk>/tools目录下(如本书android.bat存放路径为H:\Android\android-sdk-windows\tools)，输入命令：

```
android create project -n HelloWorld -k www.hisoft.HelloWorld -a HelloWorld -t 9 -p
d:\Android\workplace\HelloWorld
```

或

```
android create project -name HelloWorld --package www.hisoft.HelloWorld --
activity HelloWorldActivity --target 9 --path d:\Android\workplace\HelloWorld
```

上述命令中创建的新工程的名称为HelloWorld，包名称为www.hisoft.HelloWorld，Activity名称是HelloWorldActivity，编译目标的ID为9，新工程的保存路径是d:\Android\workplace\HelloWorld。如图1-49所示，在新建立的工程目录中，发现其中一些是使用 Eclipse 开发环境创建同样的工程不会出现的文件，例如 build. xml、local. properties。这些新文件的出现主要是为了在构建 Android 程序时使用了 Apache Ant。

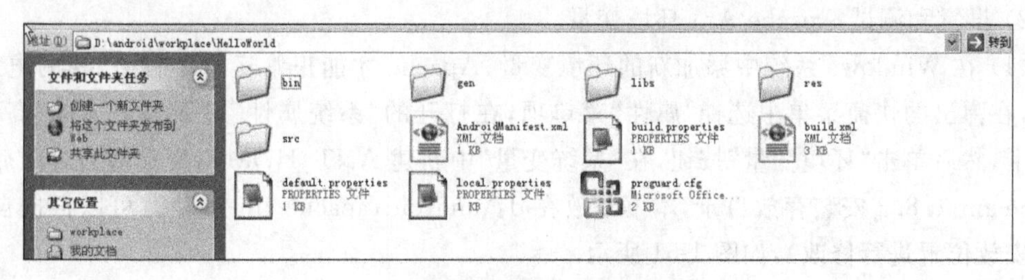

图 1-49　创建的工程目录

新创建的 HelloWorld 工程文件和目录列表说明如表 1-3 所示。

表 1-3　新创建的 HelloWorld 工程文件和目录列表说明

文　件	说　明
AndroidManifest. xml	应用程序声明文件
build. xml	Ant 的构建文件
default. properties	保存编译目标，由 Android 工具自动建立，不可手工修改
build. properties	保存自定义的编译属性
local. properties	保存 Android SDK 的路径，仅供 Ant 使用
src\www\hisoft\HelloWorld\HelloWorld. java	Activity 文件
bin\	编译脚本输出目录
gen\	保存 Ant 自动生成文件的目录，例如 R. java
libs\	私有函数库目录，在工程创建初期是空目录
res\	资源目录
src\	源代码目录

运行结果如图 1-50 所示。

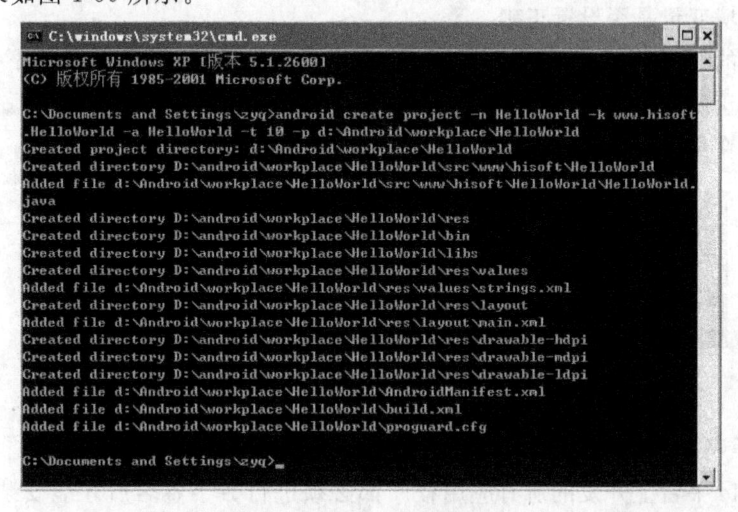

图 1-50　工程运行后的结果

特别注意：如果已经在环境变量中设置了 SDK 路径，则可以直接在 cmd 中输入命令，不用进入<SDK>/tools 后再输入命令。

2）设置和测试 Apache Ant 环境变量

（1）在 Windows 系统中添加新的环境变量，Apache 才能正常运行。右击"我的电脑"图标，在弹出的快捷菜单中选择"属性"菜单项，在打开的"系统属性"对话框中选择"高级"选项卡，然后单击"环境变量"按钮，在"系统变量"中新建 ANT_HOME，变量值为解压后的 apache-ant-1.8.2 安装存放目录，本书是放在 d:\android\apache-ant-1.8.2（可以根据自己实际安放位置进行修改），如图 1-51 所示。

然后在"变量值"文本框中添加"%ANT_HOME%\bin"，如图 1-52 所示。

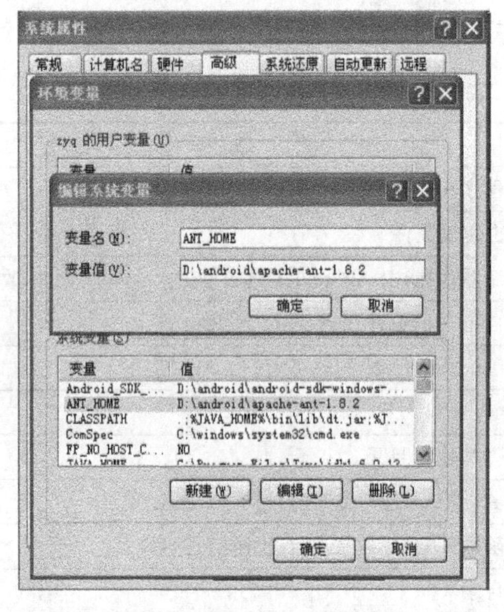

图 1-51　设置 ANT_HOME 变量

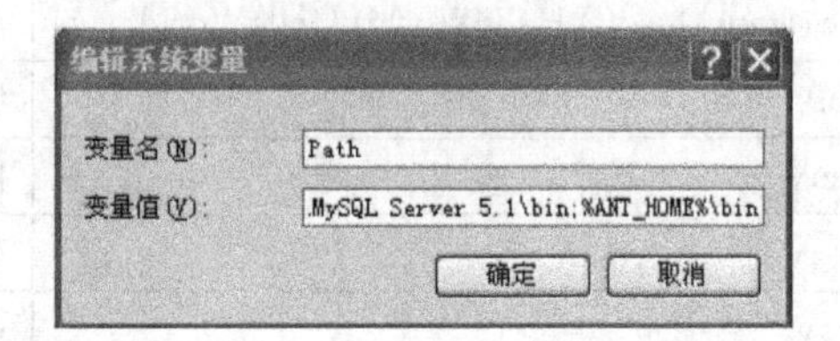

图 1-52　设置 Path 变量

（2）测试判断设置的环境变量正确性。在 CMD 中运行输入 ant 命令，通过命令的输出信息判断环境变量是否设置正确。

如果输出的提示包含"Unable to locate tools.jar. Expected to find it in…"，则表明设置环境变量不正确。

如果环境变量设置正确，ant 命令的输出结果如图 1-53 所示。

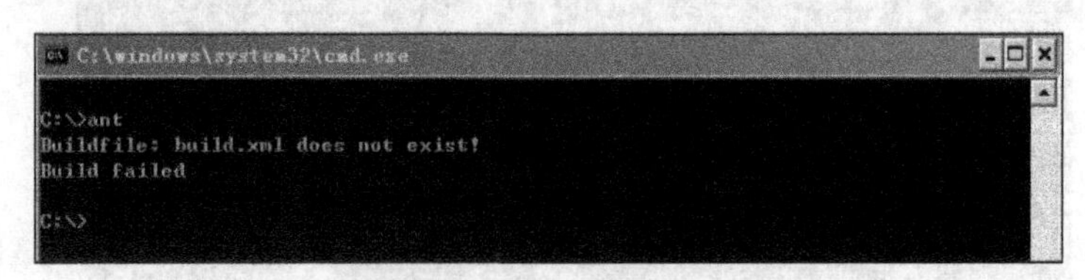

图 1-53　测试 ant 环境变量设置

3）应用程序数字签名

在 Android 平台上开发的所有应用程序都必须进行数字签名后才能安装到模拟器或

手机上，否则将返回错误提示：

```
Failure [INSTALL_PARSE_FAILED_NO_CERTIFICATERS]
```

特别注意：在 Eclipse 开发环境中，ADT 在将 Android 程序安装到模拟器前，已经利用内置的 debug key 为 apk 文件自动做了数字签名，这使用户无需自己生产数字签名的私钥，而能够利于 debug key 快速完成程序调试。但是，如果用户希望正式发布自己的应用程序，则不能使用 debug key，必须使用私有密钥对 Android 程序进行数字签名。

Apache Ant 构建 Android 应用程序支持 Debug 模式和 Release 模式两种构建模式。

Debug 模式是供调试使用的构建模式，用于快速测试开发的应用程序。Debug 模式自动使用 debug key 完成数字签名。

Release 模式是正式发布应用程序时使用的构建模式，生成没有数字签名的 apk 文件。

Debug 模式对 HelloWorld 工程进行编译，生成具有 debug key 的 apk 打包文件。

步骤：使用 CMD，在工程的根目录 D:\android\workplace\HelloWorld 下输入 ant debug 命令，结果如图 1-54 所示。

命令运行后，Apache Ant 在工程 bin 目录中生成打包文件 HelloWorld-debug. apk。

如果需要使用 Release 模式，则需在 CMD 中输入 ant release，运行后会在 bin 目录中生成打包文件 HelloWorld-unsigned. apk。

apk 文件是 Android 系统的安装程序，上传到 Android 模拟器或 Android 手机后可以进行安装。

apk 文件本身是一个 zip 压缩文件，能够使用 WinRAR、UnZip 等软件直接打开。

打开的 HelloWorld-debug. apk 文件如图 1-55 所示。

res\目录用来存放资源文件。

AndroidManifest. xml 是 Android 声明文件。

classes. dex 是 Dalvik 虚拟机的可执行程序。

resources. arsc 是编译后的二进制资源文件。

4) 使用 adb. exe 将 HelloWorld 工程上传到 Android 模拟器中

(1) 启动 AVD

使用命令行启动模拟器时，需要先指定所使用的 AVD。可以使用 android list avds 命令查询当前系统所有已经创建的 AVD，如图 1-56 所示。

在 CMD 中输入命令 emulator -avd wjj，启动 AVD 虚拟设备。

(2) 上传文件

Android 模拟器正常启动后，使用 adb. exe 工具把 HelloWorld-debug. apk 文件上传到模拟器中。

adb. exe 工具除了能够在 Android 模拟器中上传和下载文件外，还能够管理模拟器状态，是调试程序时不可缺少的工具。

在 CMD 中进入工程 HelloWorld/bin 目录，输入命令 adb install HelloWorld-debug. apk。完成 apk 程序上传到模拟器的过程。

如果上传成功，结果如图 1-57 所示。

```
C:\windows\system32\cmd.exe

D:\android\workplace\HelloWorld>ant debug
Buildfile: D:\android\workplace\HelloWorld\build.xml
    [setup] Android SDK Tools Revision 10
    [setup] Project Target: Android 2.3.1
    [setup] API level: 9
    [setup]
    [setup] ------------------
    [setup] Resolving library dependencies:
    [setup] No library dependencies.
    [setup]
    [setup] ------------------
    [setup]
    [setup] WARNING: No minSdkVersion value set. Application will install on all
 Android versions.
    [setup]
    [setup] Importing rules file: tools\ant\main_rules.xml

-debug-obfuscation-check:

-set-debug-mode:

-compile-tested-if-test:

-pre-build:

-dirs:
     [echo] Creating output directories if needed...
    [mkdir] Created dir: D:\android\workplace\HelloWorld\gen
    [mkdir] Created dir: D:\android\workplace\HelloWorld\bin\classes

-aidl:
     [echo] Compiling aidl files into Java classes...

-renderscript:
     [echo] Compiling RenderScript files into Java classes and RenderScript byte
code...

-resource-src:
     [echo] Generating R.java / Manifest.java from the resources...

-pre-compile:

compile:
    [javac] H:\Android\android-sdk-windows\tools\ant\main_rules.xml:384: warning
: 'includeantruntime' was not set, defaulting to build.sysclasspath=last; set to
 false for repeatable builds
    [javac] Compiling 2 source files to D:\android\workplace\HelloWorld\bin\clas
ses

-post-compile:

-obfuscate:

-dex:
     [echo] Converting compiled files and external libraries into D:\android\wor
kplace\HelloWorld\bin\classes.dex...

-package-resources:
     [echo] Packaging resources
     [aapt] Creating full resource package...

-package-debug-sign:
[apkbuilder] Creating HelloWorld-debug-unaligned.apk and signing it with a debug
 key...

debug:
     [echo] Running zip align on final apk...
     [echo] Debug Package: D:\android\workplace\HelloWorld\bin\HelloWorld-debug.
apk

BUILD SUCCESSFUL
Total time: 1 minute 25 seconds

D:\android\workplace\HelloWorld>
```

图 1-54　ant debug 工程

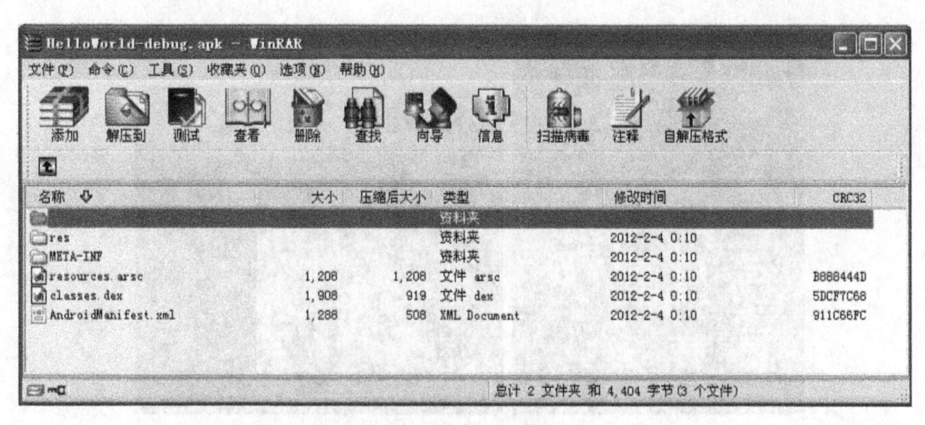

图 1-55 HelloWorld-debug.apk 文件结构

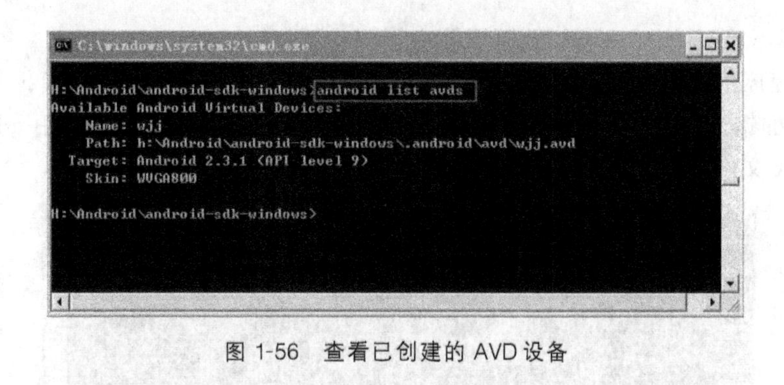

图 1-56 查看已创建的 AVD 设备

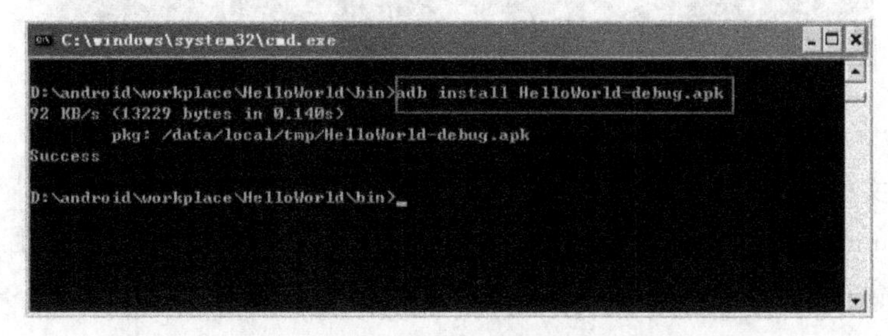

图 1-57 安装新创建的 apk 文件

(3) 启动应用程序

apk 文件上传后，需手工启动 HelloWorld 应用程序。

单击模拟器界面左下角刚安装的 HelloWorld 应用程序图标即可手工启动，如图 1-58 所示。

如果在模拟器界面看不见新安装的程序，可以单击模拟器右侧的 menu 按钮，找到新安装的应用程序图标。如果在模拟器中找不到新安装的程序，可以尝试重新启动 Android 模拟器。因为 Android 的包管理器经常仅在模拟器启动时检查应用程序的 AndroidManifest.xml 文件，这就导致部分上传的 Android 应用程序不能立即启动。

(4) 编译和打包应用程序

修改 HelloWorld 工程代码后，需要使用 Apache Ant 重新编译和打包应用程序，并将

图 1-58　查看程序运行结果

新生成的 apk 文件上传到 Android 模拟器中。

如果新程序的包名称没有改变，则在使用 adb. exe 上传 apk 文件到模拟器时会出现图 1-59 所示的错误提示，此时需要在模拟器中先删除原有 apk 文件，再使用 adb. exe 工具上传新的 apk 文件。

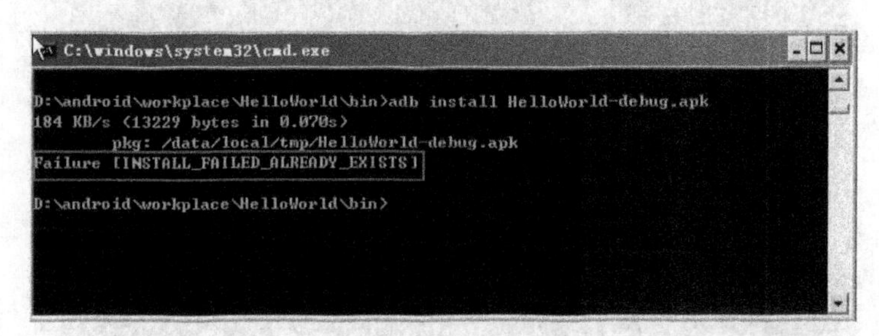

图 1-59　安装失败

删除 apk 文件的方法：

使用 adb uninstall ＜包名称＞的方法，例如删除 HelloCommandline 工程的 apk 文件，可在 CMD 中输入命令 adb uninstall www. hisoft. HelloWorld，提示 Success 则表示成功删除。

使用 adb shell rm /data/app/＜包名称＞-1. apk 的方法，同样以删除 HelloWorld 工程的 apk 文件为例，在 CMD 中输入下面的命令，没有任何提示则表示删除成功。

```
adb shell rm /data/app/edu.hrbeu.HelloWorld-1.apk
```

特别注意：如果仅有一个 Android 模拟器在运行，按照上述操作步骤可以一条命令完成 Android 工程编译、apk 打包和上传过程。如果同时有两个或两个以上的 Android 模拟器存在，这种方法将会失败，因为 adb. exe 不能够确定应该将 apk 文件上传到哪一个 Android 模拟器中。此外，多次使用这种方法时，同样需要先删除模拟器中已有的 apk 文件。

1.5.4　调试 Android 程序

Android SDK 提供了大部分的测试工具，它们分别放在 SDK 的 tools 和 platform-tools 目录下。tools 目录下的测试工具有 DDMS、hierarchyviewer、layoutopt、traceview 和 dmtracedump；platform-tools 目录下的测试工具有 adb，此外还有 Dev Tools Android application 应用测试等。

一个典型的 Android 应用程序测试环境主要由 DDMS(Dalvik Debug Monitor Server)、ADB、设备或者 AVD(Device or Android Virtual Device)、JDWP debugger 这几部分组成。

在一个 Android 应用程序开发后期需要对其进行测试，根据 Android 应用程序开发环境和应用程序本身不同的情况，可以选择不同的测试方法和测试环境工具，主要有 Eclipse 加 ADT 插件开发环境下应用程序测试、其他 Java 集成开发环境(IDEs)的应用程序测试、DDMS 测试、adb 和 logcat 组合测试、用 hierarchyviewer 工具测试优化用户界面、用 layoutopt 工具优化布局、使用 Traceview 和 dmtracdedump 工具以图形的方式展现日志和分析程序性能。

下面对 Android 应用程序的测试所常用的方法和测试工具进行介绍，其他的不再赘述，如感兴趣可以参考 SDK 文档说明。

1. Eclipse 加 ADT 插件开发环境下测试 Android 应用

在 Eclipse 加入 ADT 插件的开发环境下，Android 应用程序的测试可以从两个方面展开：一个是使用 Eclipse 内置的 Java 调试器，进行程序调试、设置断点、查看代码执行中变量变化及使用 logcat 实时查看系统日志；另外一个是使用 DDMS，通过 DDMS 视图中 Devices、Eemulator Control、logcat、Threads、Heap、Allocation Tracker 和 File Explorer 这些面板查看和调试 Android 应用程序。

1) 使用 Eclipse 内置的 Java 调试器

(1) 断点设置

Android 应用程序中断点的设置方法与一般的 Java 程序一样，都是通过在代码区需要设置断点的代码行号前方左侧区域双击或者右击，选择 Toggle Breakpoint 菜单项，设置断点，如图 1-60 所示。

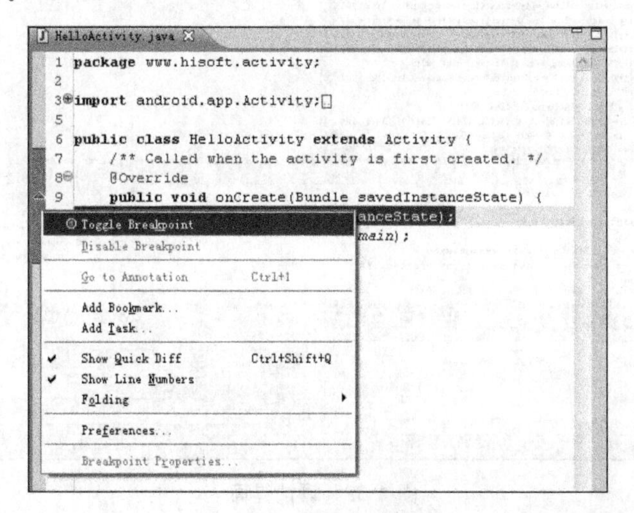

图 1-60 设置调试断点

(2) 运行项目调试

选中上述设置好断点的项目名称右击,从弹出的快捷菜单中选择 Debug As→Android Application 菜单项,如图 1-61 所示。然后执行项目调试,在图 1-62 所示的调试界面中,可以通过 Variables、Breakpoints、Debug 等面板参数查看程序的每一步调试执行情况及变化。

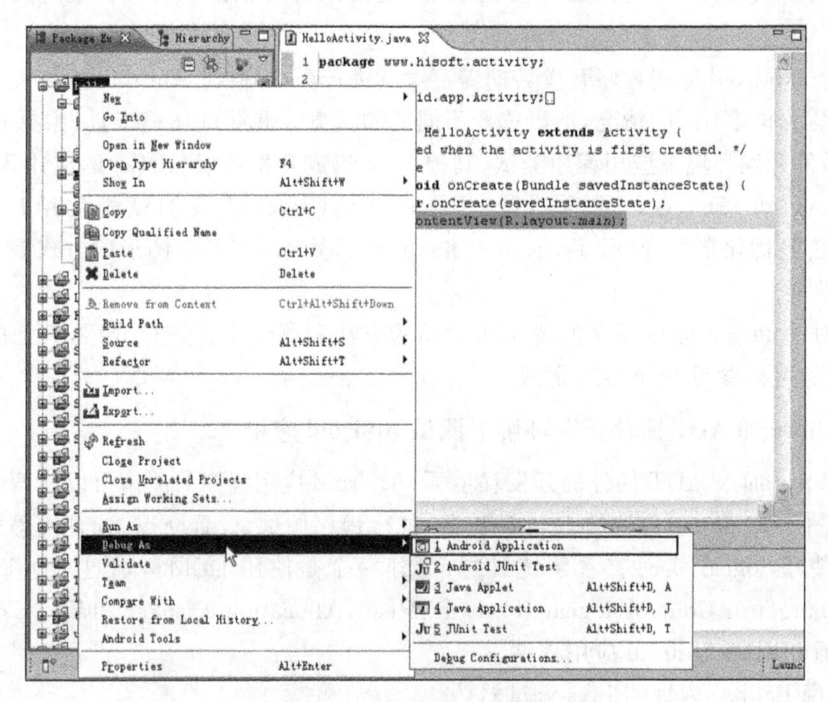

图 1-61 部署调试工程

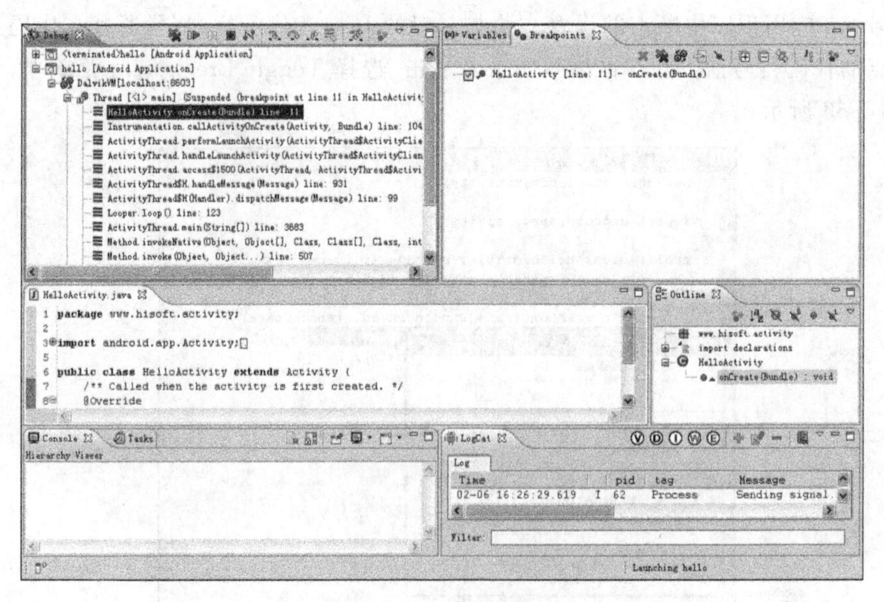

图 1-62 调试界面

2) DDMS 调试

在 Eclipse 集成开发环境中，除了上述的 Java 调试器之外，还可以使用 DDMS 调试视图工具对 Android 应用程序进行调试。DDMS 调试视图工具包含有 Devices 面板、Emulator Control 面板、LogCat、Thread、Heap、Allocation Tracker 和 File Explorer 这些调试面板，如图 1-63 所示。

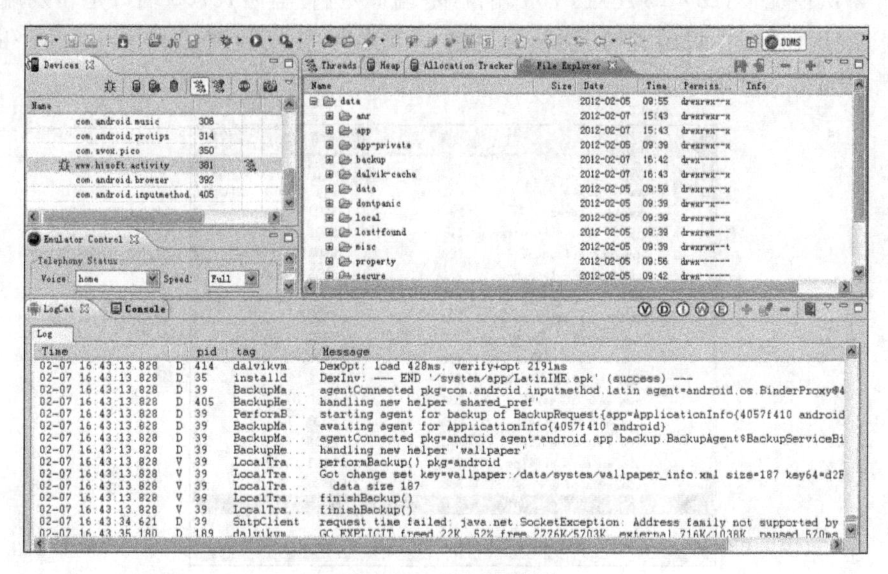

图 1-63 DDMS 界面

一般情况下，可以通过在 Eclipse 中选择 Window→Open Perspective→DDMS 菜单项打开 DDMS 调试视图，或者在 Eclipse 界面的右上角选择 Open Perspective→DDMS 菜单项，如图 1-64 所示。如果没有找到 DDMS，则可以通过 Window→Open Perspective→Other 菜单项，找到 DDMS，如图 1-64 所示。

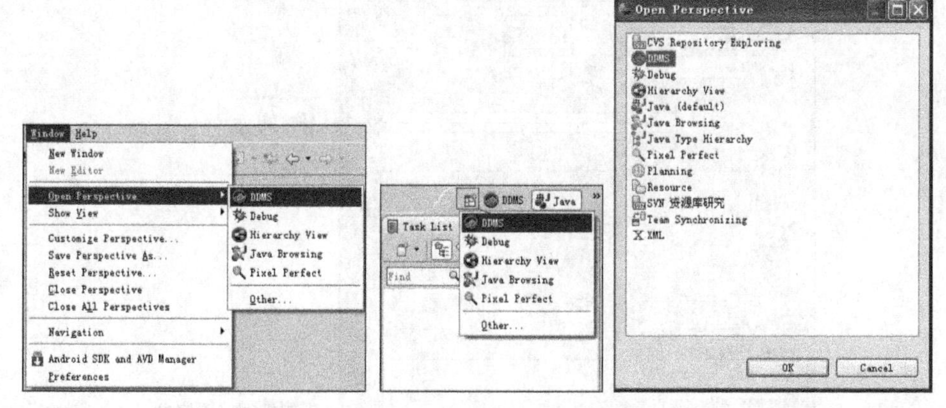

图 1-64 打开 DDMS 调试视图

特别注意：启动 DDMS 的另外一种方式是从命令行启动 DDMS，通过运行保存在 SDK 目录下的 tools 文件夹中的批处理文件 ddms. bat 来启动。

(1) Devices 面板

DDMS 工具中的 Devices 面板显示所有连接到 ADB 的设备、AVD 信息。当 DDMS 启动后，它会与一个设备的 ADB 相连接，并在 DDMS 与 ADB 之间创建一个 VM 监控服务，用于监控当前设备启动或终止时使 DDMS 能够及时得到通知信息。DDMS 通过 ADBD (ADB daemon)获取 VM 处理进程 ID 号，进而连接 VM 调试器。DDMS 给每一个设备的 VM 都分配一个调试端口，通常把 8600 端口分配给第一个调试的 VM，紧接着是 8602，以此类推。默认情况下，DDMS 还对 8700 端口(基端口)进行监听，8700 端口是转发端口，它能接收来自任何调试端口的 VM 流量，并通过 8700 端口转发到调试器。当前程序的调试端口及转发如图 1-65 所示。

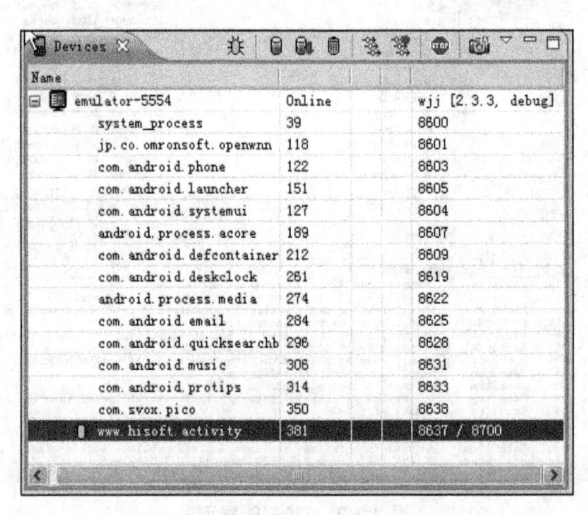

图 1-65　模拟器面板

(2) Emulator Control 面板

在 Emulator Control 面板中可以输入电话号码，向模拟器打电话或发短信息，并可以显示虚拟地理位置信息，如图 1-66 所示。

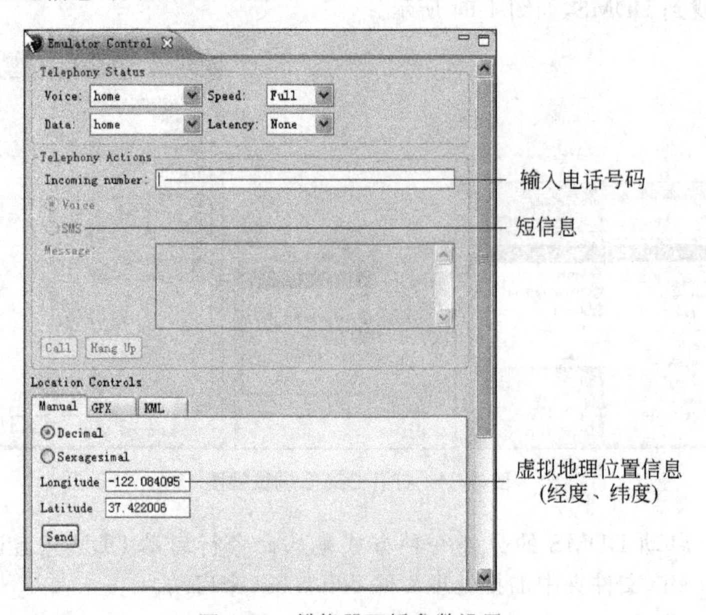

图 1-66　模拟器面板参数设置

(3) Threads

通过选择 Devices 面板上的按钮(Update Threads),在右侧的 Threads 面板上显示应用程序名字及当前执行线程的状态、Tid 号等信息,如图 1-67 所示。

ID	Tid	Status	utime	stime	Name
1	381	suspended	94	40	main
*2	383	vmwait	76	0	HeapWorker
*3	384	vmwait	0	0	GC
*4	385	vmwait	0	0	Signal Catcher
*5	386	running	40	48	JDWP
*6	387	vmwait	0	2	Compiler
7	388	native	0	0	Binder Thread #1
8	389	native	0	0	Binder Thread #2

图 1-67　线程面板

(4) Heap、Allocation Tracker 和 File Explorer

- Heap:在 Devices 面板上选择按钮(Update Heap),在右侧的 Heap 面板上显示堆内存的分配和执行情况。在 Heap 面板上选择 Cause GC 按钮,可以激活堆数据的垃圾回收机制。
- Allocation Tracker:追踪对象的内存分配,并可以看到哪些类、线程分配到对象。在 Allocation Tracker 面板上,通过选择 Start Tracking 按钮和 Get Allocations 按钮,追踪对象的内存分配。
- File Explorer:显示模拟器中的文件,复制、删除。如果启动时加载了 SD 卡,可以查看 SD 卡信息,以及文件的复制操作,如图 1-68 所示。

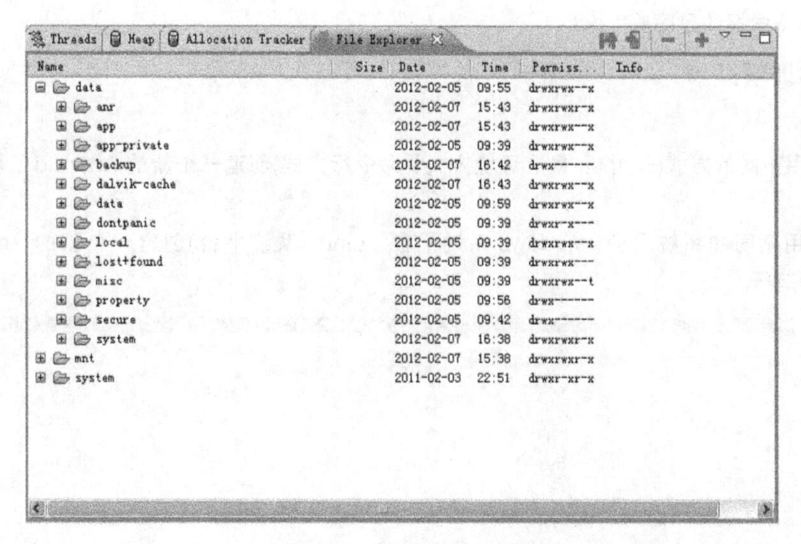

图 1-68　文件浏览面板

注意事项:文件浏览必须在模拟器启动并部署工程后才能查看和浏览。

2. 其他 Java 集成开发环境(IDE)的 Android 应用程序调试

通常这种条件下,Android 应用程序调试主要采用 SDK 提供的调试工具和 Java 调试器,主要有 ADB、DDMS 和 Java 调试器。Java 调试器采用符合 JDWP 规范的调试器,如 JDB。

采用命令行的方式启动调试环境,主要步骤如下。

(1) 启动 AVD。

(2) 进入 SDK 路径下的 tools 目录中,启动 DDMS 和 ADB 工具。

(3) 安装 apk 文件到虚拟设备 AVD。

(4) Java 调试器附加到调试端口 8700,或者应用程序在 DDMS 中的特定端口。

拓展提示:Android 系统版本的发展及其市场应用软件的快速增加,对 Android 学习者而言,既面临着前所未有的机遇,也面临着巨大的挑战。Android 开发的主流涉及哪些方面,哪些是初学者或从事 Android 开发的入门者必备的知识,是值得思考的问题,也是本书之重点。

1. 简答题

(1) 什么是智能机? 智能机的基本构成包含哪些部分?

(2) 智能机与非智能机之间有何差异? 分析智能机与非智能机未来各自的发展方向。

(3) 智能手机操作系统常见的有哪些? 它们之间的异同及决定它们未来市场走向的关键因素可能有哪些?

(4) Android 体系结构包含了哪些方面? 针对不同的部分,其对应的应用方向及市场岗位定位及要求知识有哪些?

(5) Android 在不同系统环境下的搭建有何异同? 它们的优劣是哪些? 对未来从事 Android 应用开发必须掌握的是哪些?

2. 实训项目

要求:

(1) 使用不同的方式(Eclipse 集成环境方式、命令行方式)创建一个新的 Android 工程,并部署运行。

(2) 使用不同的系统平台(Windows 系统平台、Linux 系统平台)创建一个新的 Android 工程,并部署运行。

第2章 Android在线医药应用——艾斯医药系统开发

学习目标

本章主要介绍 Android 在线医药艾斯医药系统的应用，结合艾斯医药系统案例，详细介绍系统需求分析设计、系统详细设计、数据库详细设计、Web 服务器端功能模块、Android 手机客户端功能的设计，以及服务器程序的部署、数据库的部署、客户端程序的打包、签名、发布流程。通过本章的学习，以使读者能够达到以下知识要点的学习：

(1) 系统的需求分析、详细设计。

(2) 数据库表的分析、设计。

(3) Web 服务器端功能的设计、开发、部署。

(4) Android 手机客户端的设计、开发、部署。

(5) Android 手机客户端的打包、签名、发布。

(6) 移动项目开发流程及应用。

本书采用先进的"项目驱动式"教学法，通过一个完整的"艾斯医药系统"项目来贯穿 Android 应用开发的理论学习过程。这个项目的开发过程将会贯穿在之后的各个章节中，结合相关知识点详细讲解和实现。这里先介绍一下"艾斯医药系统"项目的背景知识，为后面的学习做好铺垫。

在实际的 Android 项目开发中，不论是纯 Android 应用还是大型的 Web 项目 Android 客户端应用开发，其开发设计都必须按照软件开发的流程来进行项目的设计与开发。一个完整的软件开发流程通常都必须经过如下几个阶段：软件需求分析、软件概要设计、软件详细设计、数据库设计、软件开发、软件测试。同样，一个 Android 项目的设计与开发也必须符合软件开发的流程和规范。

本书 Android 在线医药应用的设计和开发基本流程分为 6 个阶段，分别是系统需求分析、系统详细设计、数据库详细设计、Web 服务器端功能开发(包含测试)、Android 手机客户端开发、Web 系统部署和 Android 手机客户端打包、发布。

本书介绍的是艾斯医药系统开发项目。由于本书重点是 Android 在线医药应用，因此在系统的设计和开发流程中，对 Web 服务器端和数据库部分只是进行概要的介绍，并把系统概要设计和详细设计两个阶段合并为一个系统详细设计阶段。系统开发和系统测试两个阶段合并为一个系统开发阶段，具体如图 2-1 所示。

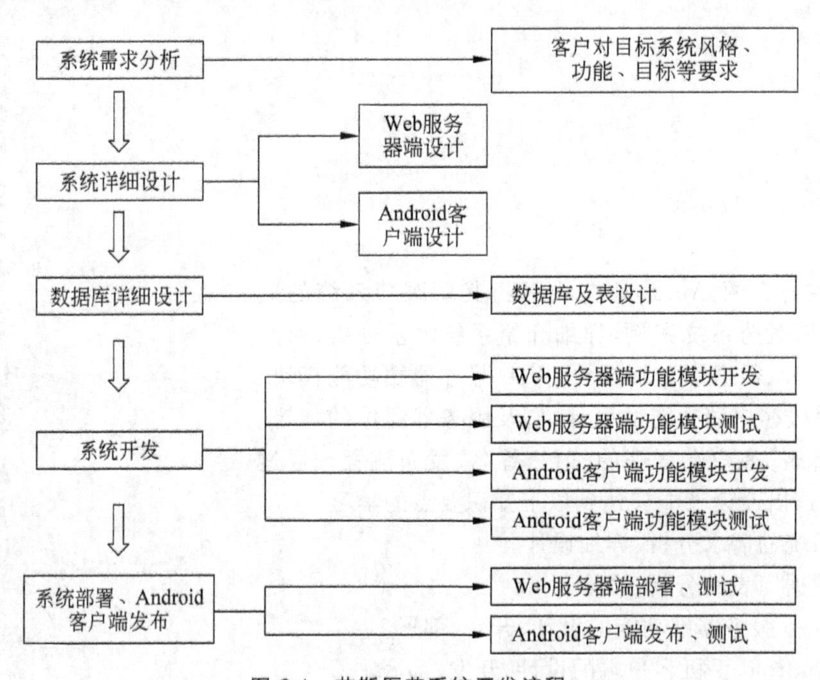

图 2-1　艾斯医药系统开发流程

2.1　系统需求分析设计

在系统需求分析设计阶段通常包括定义潜在的角色（角色指使用系统的人，以及与系统相互作用的软、硬件环境）；识别问题域中的对象和关系，以及基于需求规范说明和角色的需要发现用例（Use Case）和详细描述用例。

2.1.1　系统开发背景

本项目案例艾斯医药系统是基于因特网的应用软件，通过此系统用户可以了解到已公开发布的药品、药品价格查询、药品购买和订单查询。用户可以通过 Web 端或 Android 手机客户端实时方便的查询、购买需要的药品，方便用户购物，实现电子购物方便快捷的功能。

2.1.2　系统功能需求

本书对艾斯医药系统的主要功能需求进行简要介绍，以方便对系统整体了解和掌握使用。

1. 艾斯医药系统功能的规定

艾斯医药系统开发涉及的元素、角色、动作主要包含顾客、管理员、登录、商品浏览、商品查询、购物、订单管理、用户管理、商品管理。其整体功能用例图(Use-Case Diagram)如图 2-2 所示。

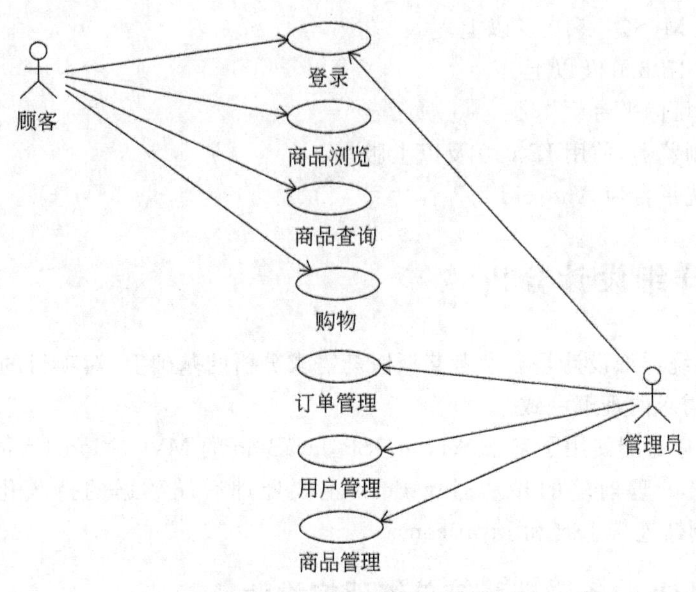

图 2-2 艾斯医药系统用例图

2. 主要功能

(1) 系统管理是给系统管理人员使用的,主要包括以下功能模块:登录、用户管理、商品管理、订单管理等。

(2) (有权限)用户管理主要包括如下功能模块:用户的注册、登录、商品搜索、购物等。

(3) (无权限)游客管理主要包括如下功能模块:商品查询、商品搜索、购物等。

2.1.3 系统开发及部署平台

1. 开发环境

(1) Web 端开发环境

- JDK 1.6 及其以上版本。
- MyEclipse 8.5 及其以上版本。
- MySQL 5.0 及其以上数据库。
- Web 服务器使用 Tomcat 6.0 以上。

(2) Android 客户端开发环境

- JDK 1.6 及其以上版本。
- Eclipse Java EE for Web Developers 3.5(galileo)及其以上版本。
- Android SDK 及其 Eclipse 开发插件 ADT。

2. 部署运行环境

(1) 服务器端为运行本软件所需要的支持软件

- 操作系统：Window Server 2003 以上/Window XP。
- Web 服务器：Tomcat 6.0 及以上。
- 数据库：MySQL 5.0 及以上。
- 客户端：IE 6.0 及以上。

(2) 客户端目标平台

- 客户端浏览器，使用 IE 6.0 及以上版本。
- 手机系统平台为 Android 2.3。

2.2 系统详细设计分析

艾斯医药系统详细设计是在参考艾斯医药需求分析的基础上，对项目的功能设计进行说明，以确保对需求的理解一致。

本项目中服务器端使用了基于 Servlet/JSP/JavaBean 的 MVC(Model View Controller)模式。其中 JSP 处理前端的显示；JavaBean 主要处理后端数据的持久化和业务逻辑；Servlet 作为控制器连接 JSP 和 JavaBean。

2.2.1 Web 服务器端系统总体架构设计

图 2-3 展示了典型的 Web 服务器系统的总体架构设计。

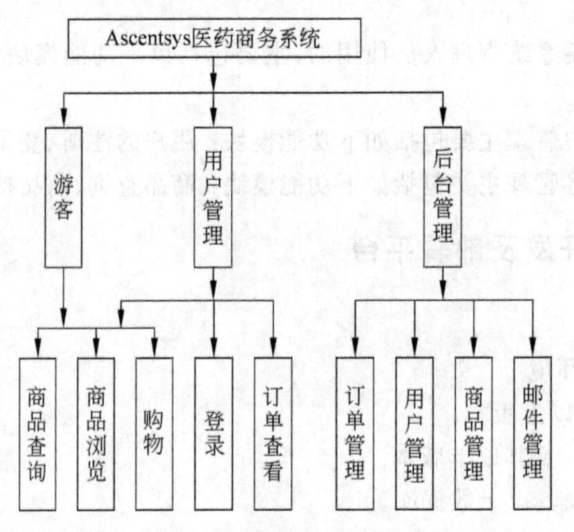

图 2-3 Web 服务器端系统总体架构设计

Web 应用程序的组织结构可以分为如下 5 个部分：

Web 应用根目录下放置用于前端展现的 JSP 文件。

com.ascent.bean 放置处理的 javabean。

com.ascent.servlet 放置处理请求相应的类。

com.ascent.dao 放置处理数据持久化类。

com. ascent. util 放置帮助类和一些其他类。

另外，在 src 下放置了数据库配置文件 datebase. conf. xml。

下面对组织结构中的几个部分分别进行介绍。

(1) JSP 文件。表 2-1 列出了每个 JSP 文件实现的功能。

表 2-1　JSP 文件列表

文件名称	功　能
index. jsp	首页
add_products_admin. jsp	添加商品页面
admin_ordarshow. jsp	管理员订单页面
admin_orderuser. jsp	查看订单用户页面
admin_products_show. jsp	管理员管理商品页面
carthow. jsp	购物车管理页面
changesuperuser. jsp	修改用户角色页面
checkout. jsp	结算页面
checkoutsucc. jsp	结算成功页面
contactUs. jsp	联系我们页面
employee. jsp	管理员添加用户页面
itservice. jsp	修改项目类别页面
mailmamager. jsp	邮件管理页面
orderitem_show. jsp	修订单项查询页面
ordershow. jsp	注册用户订单查看页面
product_search. jsp	商品搜索页面
products_search_show. jsp	商品搜索结果页面
products_showusers. jsp	注册用户管理页面
products. jsp	电子政务介绍页面
register. jsp	注册页面
regist_succ. jsp	注册成功页面
update_products_admin. jsp	修改商品信息页面
updateproductuser. jsp	修改用户信息页面
error. jsp	错误页面

(2) Servlet 中包括的控制器，如表 2-2 所示。

表 2-2　action 列表

文件名称	功　能	文件名称	功　能
LoginServlet. java	用户登录控制器	ProductServlet. java	商品管理控制器
MailServlet. java	邮件管理控制器	ShopCartServlet. java	购物管理控制器
OrderServlet. java	订单管理控制器	UserManagerServlet. java	用户管理控制器

(3) po包括4个逻辑类,如表2-3所示。

表 2-3 JavaBean 列表

文件名称	功能	文件名称	功能
Mailtb. java	邮件类	Product. java	商品类
Orderitem. java	订单项类	Productuser. java	用户类
Orders. java	用订单类	UserProduct. java	用户和商品类

(4) Util类,如表2-4所示。

表 2-4 Util 列表

文件名称	功能
SetCharacterEncodingFilter. java	对提交过来的信息里的特殊字符进行处理
dataAccess. java	数据库连接类
DatabaseConfigParser. java	解析数据库配置文件类
XMLConfigParser. java	解析 XML 类
SendMail. java	发送邮件类
ShopCart. java	购物车类
AuthImg. java	验证码生成类

(5) dao数据层方法类,如表2-5所示。

表 2-5 dao 列表

文件名称	功能
LoginDAO. java	处理登录和退出业务的类
MailDAO. java	处理邮件管理相关功能的类
OrderDAO. java	处理订单管理相关的类(删除、修改和查询等)
ProductDAO. java	处理商品管理相关功能的类
UserManagerDAO. java	处理用户管理相关功能的类

2.2.2 Web服务器端系统功能概述

由于章节篇幅的原因,本书只是对艾斯医药系统的Web服务器主要功能及部分运行效果进行简要介绍。

1. 管理员管理

(1) 启动Web服务器Tomcat,部署运行本系统,打开艾斯医药系统登录页面login. jsp,如图2-4所示,进行管理员登录。

图 2-4　login.jsp 页面

（2）输入正确的用户名和密码后进入系统管理界面，如图 2-5 所示。

用户名	Email	电话	公司名称	角色	修改	角色修改	状态
admin	22	135	aa	管理员			无操作
lixing	lixing@163.com	123456	ascent	普通用户			☑
ascent	lixing@163.com	123456		普通用户			☑
shang	lixing051116@163.com	1213	ascent	管理员			无操作
222222	22@163.com	2	2	普通用户			☑
e11111	11@163.com	82780848	ascent	普通用户			☑
radar	44@hotmai.com	010-62665840	北京	普通用户			☑

图 2-5　管理员登录后的系统管理界面

（3）用户管理。登录进入该管理员管理界面，单击“用户列表”按钮，如图 2-6 所示。进入登录用户管理界面，该模块可以更改用户角色，屏蔽和开启用户以及修改。

2. 注册用户功能

（1）在首页单击“电子商务系统”链接，如图 2-7 所示。

用户名	Email	电话	公司名称	角色	修改	角色修改	状态
admin	22	135	aa	管理员			无操作
lixing	lixing@163.com	123456	ascent	普通用户			☑
ascent	lixing@163.com	123456		普通用户			☑
shang	lixing051116@163.com	1213	ascent	管理员			无操作
222222	22@163.com	2	2	普通用户			☑
e11111	11@163.com	82780848	ascent	普通用户			☑
radar	44@hotmai.com	010-62665840	北京	普通用户			☑

图 2-6　用户列表

图 2-7　电子商务系统登录

(2) 单击“注册”按钮进入注册页面，如图 2-8 所示。

(3) 单击“注册”按钮，验证数据，提示注册成功，如图 2-9 所示。

(4) 注册成功用户登录，在首页输入用户名和密码，进入电子商务介绍界面，当用户进入系统时，应该能看到电子商务信息介绍。

(5) 登录进入电子商务介绍页面，如图 2-10 所示，单击“查询产品浏览产品”链接进入信息商品查询页面，如图 2-11 所示。

图 2-8 注册页面

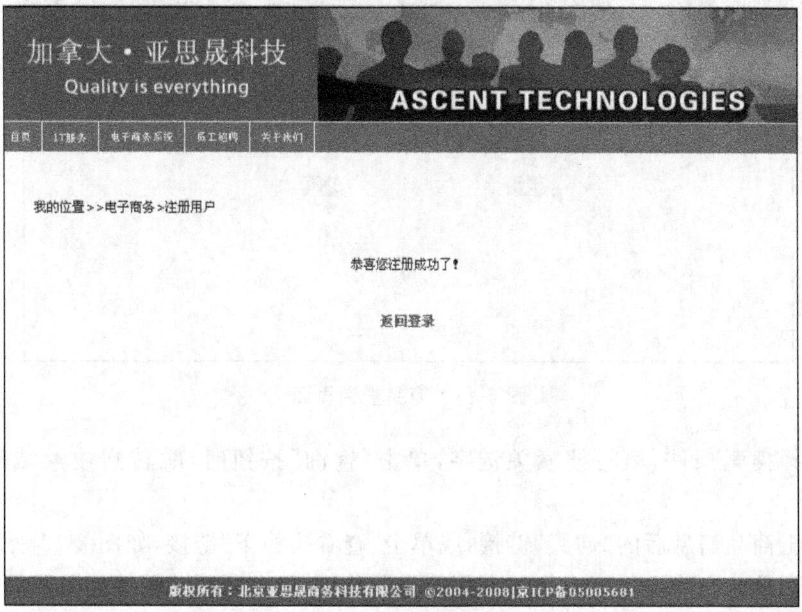

加拿大·亚思晟科技
Quality is everything
ASCENT TECHNOLOGIES
首页
IT服务
电子商务系统
员工招聘
关于我们
我的位置>>电子商务>注册用户
恭喜您注册成功了!
返回登录
版权所有：北京亚思晟商务科技有限公司 ©2004-2008|京ICP备05005681

图 2-9 注册成功页面

图 2-10　电子商务介绍页面

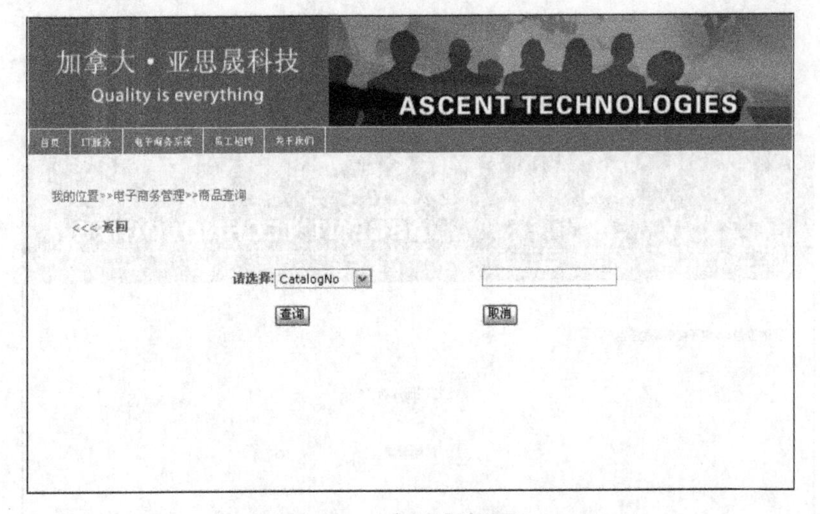

图 2-11　商品查询页面

(6) 选择搜索条件，填写搜索关键字，单击“查询”按钮时，跳转到搜索结果页面，如图 2-12 所示。

(7) 单击商品信息后的“购买”链接，或单击“查看购物车”链接，如图 2-13 所示。

(8) 在商品对应输入框直接修改商品的质量，或者单击“删除”链接删除商品。

单击“结算中心”链接，进入结算页面，如图 2-14 所示。

(9) 填写用户基本信息，单击“提交”按钮，购物成功，生成订单，跳转到提示页面，如图 2-15 所示。

(10) 单击“查看订单”链接，查看该用户的订单详细情况，如图 2-16 所示。

(11) 单击“查看”链接，查看每个订单的详细商品信息，如图 2-17 所示。

图 2-12　搜索结果页面

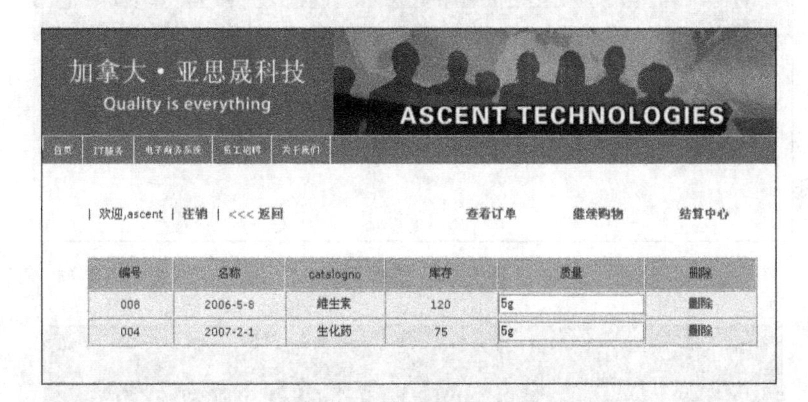

图 2-13　查看购物车

图 2-14　结算页面

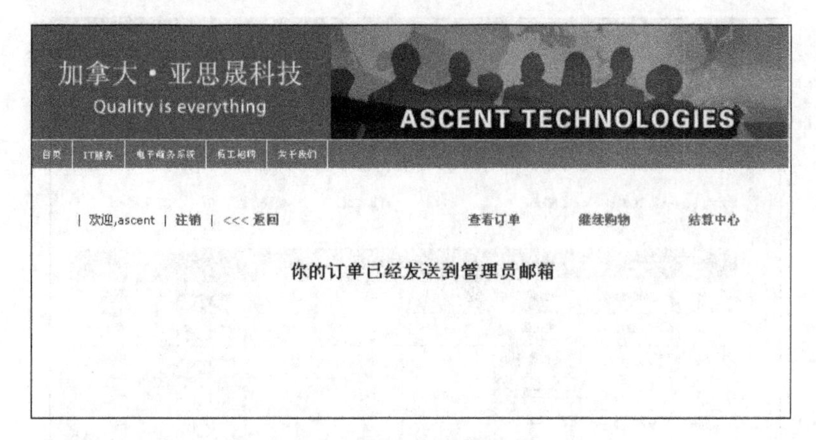

图 2-15　提示页面

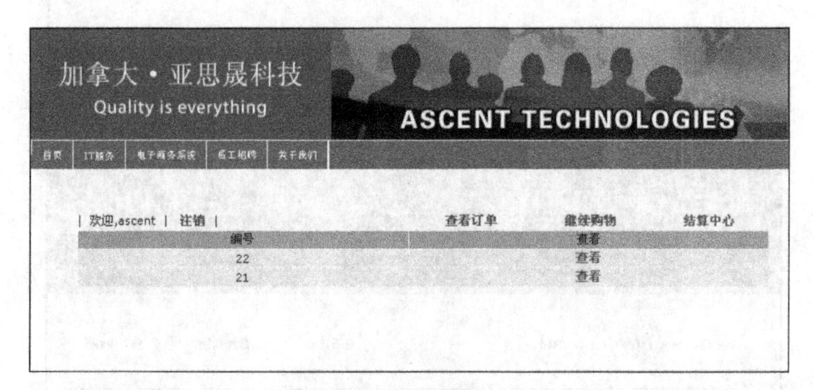

图 2-16　查看订单

编号	名称	catalogno	MDLNumber	价格	库存	图片
004	2007-2-1	生化药	ee	24	100	
008	2006-5-8	维生素	dd	97	520	

图 2-17　详细商品信息

如果是以未注册用户(游客身份)浏览使用本系统,可以进行商品的查询、购买,然后单击“结算中心”链接,进入结算页面,如图 2-18 所示。

图 2-18 未登录用户的结算页面

再填写用户基本信息，单击“提交”按钮，购物成功，生成订单，跳转到提示页面。进行后续的操作。

2.2.3 Android 手机客户端总体架构设计

在了解上述 Web 服务器端系统总体设计之后，下面对艾斯医药系统的 Android 手机客户端应用总体架构设计进行介绍，以方便大家系统全面地掌握艾斯医药系统的 Android 在线医药应用，如图 2-19 所示。

2.2.4 AscentSys(艾斯医药)移动客户端系统功能概述

在运行 AscentSys 移动客户端之前，首先需要在 Tomcat 下部署 AscentSys 系统服务端和导入 aacesys.sql 数据库文件，然后部署运行 AscentSys 移动客户端 ESysClient。

注意：部署运行的 AVD 平台必须是 Google APIs。本书使用的是 Google APIs—API Level 10，以便应用程序中 Map 的应用。

(1) 部署启动 AscentSys 系统，程序运行后的应用图标如图 2-20 所示。

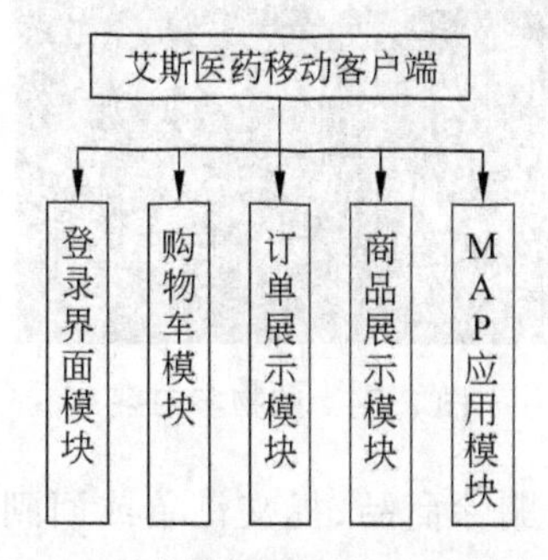

图 2-19 艾斯医药系统移动客户端

图 2-20 Ascent 移动版医药商务系统

(2) 单击上述图标,输入用户名和密码,单击“登录”按钮进行登录,如图 2-21 所示。

(3) 登录成功后,自动从服务器端数据库中提取商品列表及价格供用户选择,如图 2-22 所示。

图 2-21 用户登录界面

图 2-22 商品列表

(4) 用户选择需要购买的商品后单击 menu 按钮,在界面下方出现选择菜单供用户选择,如图 2-23 所示

(5) 单击“添加到购物车”菜单后,再单击 menu 按钮,选择“我的购物车”菜单,登录用户购物车内容显示,如图 2-24 所示。

图 2-23 选择菜单

图 2-24 购物车内容

(6) 单击 menu 按钮,界面下方出现选择菜单,可以删除商品、提交订单或回到商品列表,如图 2-25 所示。

(7) 单击“提交订单”链接，系统自动提取用户相关信息并显示。也可以进行修改，然后单击 menu 按钮，界面下方出现“提交订单”按钮，如图 2-26 所示。

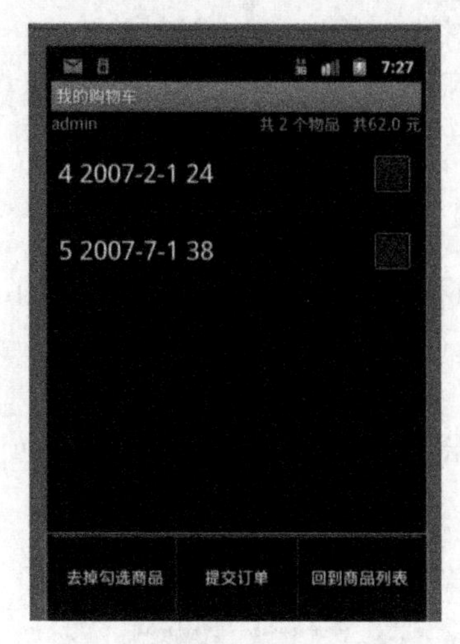

图 2-25 购物车内容选择

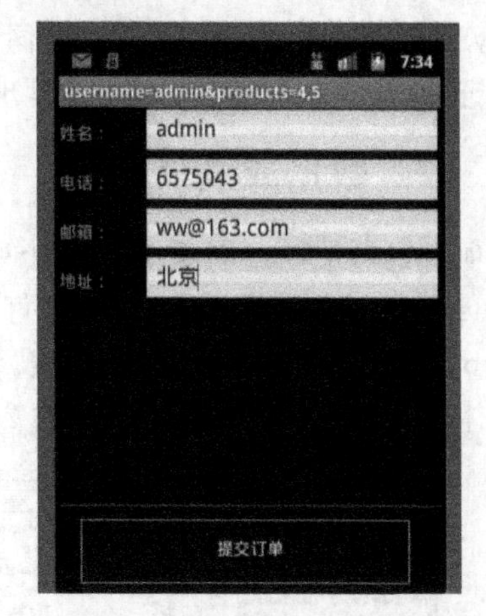

图 2-26 购物用户信息及订单提交

(8) 订单提交成功后，显示提交成功信息提示，并可单击公司地图按钮，查看公司位置，如图 2-27 和图 2-28 所示。

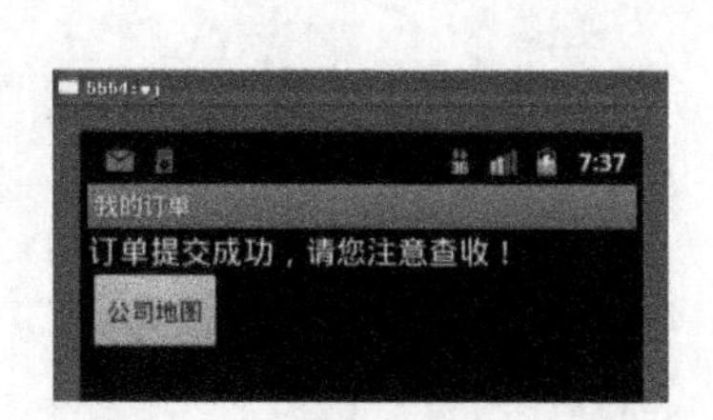

图 2-27 提交成功提示信息

图 2-28 公司位置地图

(9) Ascent 医药移动客户端的其他应用操作，如删除、添加用户等不再一一列举，具体详细操作见程序代码。

2.3 数据库详细设计分析

2.3.1 数据库平台环境及要求

本项目案例系统的运行所需要的数据库为 MySQL。MySQL 是一个多用户、多线程的 SQL 数据库，是一个客户端/服务器结构的应用，它由一个服务器守护程序 mysqld 和很多不同的客户程序和库组成。它是目前市场上运行最快的 SQL（Structured Query Language，结构化查询语言）数据库之一，提供了其他数据库少有的编程工具，而且 MySQL 对于商业和个人用户是免费的。这里使用相对稳定的 5.0.45 版本。

MySQL 的功能特点如下：可以同时处理几乎不限数量的用户；处理多达 50 000 000 以上的记录；命令执行速度快，也许是现今最快的；简单有效的用户特权系统。

2.3.2 数据库及表设计

本项目案例艾斯医药系统所设计和使用的数据库表主要有 6 张，分别是 mailtb（邮件表）、orderitem（订单项）表、Orders（订单）表、product（商品）表、productuser（用户）表和 user_product（用户-产品权限分配）表。

具体表逻辑图和表物理图如图 2-29 和图 2-30 所示。

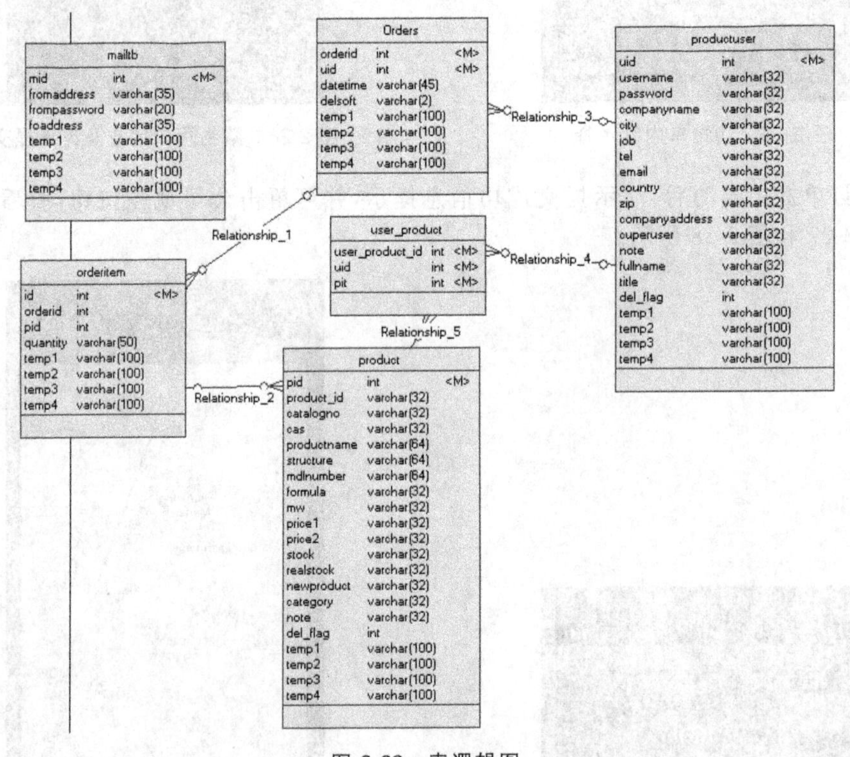

图 2-29 表逻辑图

表结构描述如下。

- mailtb（邮件）表：主要用于记录用户的邮件信息，主要字段有邮件 ID、发邮件地址、收邮件地址和发邮件密码。具体如表 2-6 所示。

表 2-6 mailtb（邮件）表结构

列　　名	类　　型	描　　述
mid	int	表示邮件 ID，是自动递增的主键
fromaddress	varchar(35)	表示发邮件地址
frompassword	varchar(20)	表示发邮件密码

续表

列　　名	类　　型	描　　述
foaddress	varchar(35)	表示收邮件地址
temp1	varchar(100)	表示备用字段 1
temp2	varchar(100)	表示备用字段 2
temp3	varchar(100)	表示备用字段 3
temp4	varchar(100)	表示备用字段 4

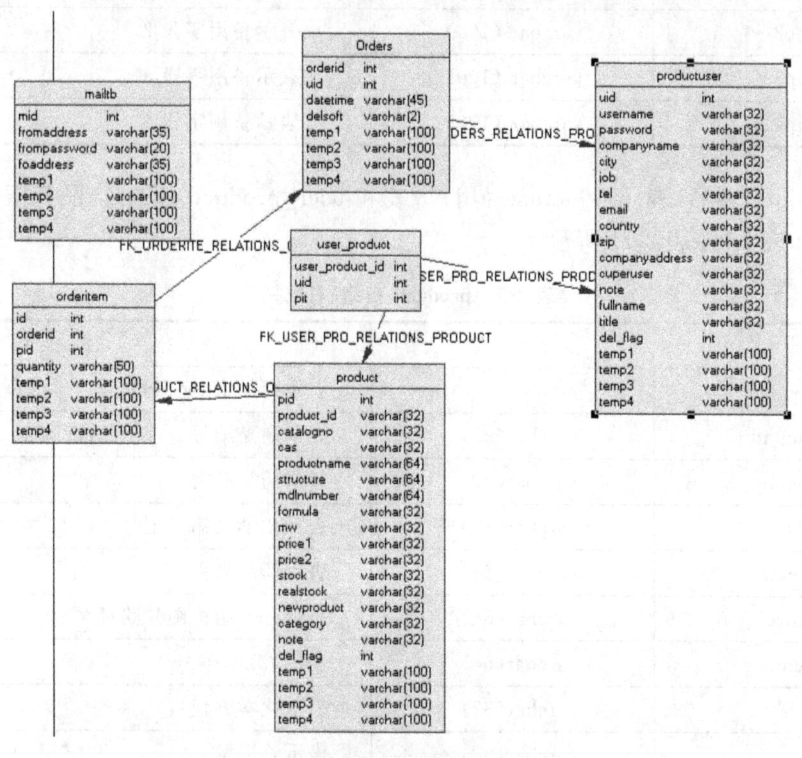

图 2-30　表物理图

- orderitem(订单项)表和 Orders(订单)表：所包含内容如表 2-7 和表 2-8 所示。

表 2-7　orderitem(订单项)表结构

列　　名	类　　型	描　　述
id	int	表示订单项 ID,是自动递增的主键
orderid	int	表示订单 ID
pid	int	表示商品 ID
quantity	varchar (50)	表示商品质量
temp1	varchar (100)	表示备用字段 1
temp2	varchar (100)	表示备用字段 2
temp3	varchar (100)	表示备用字段 3
temp4	varchar (100)	表示备用字段 4

表 2-8　Orders(订单)表结构

列　　名	类　　型	描　　述
orderid	int	表示订单 ID,是自动递增的主键
uid	int	表示客户标识号
datetime	varchar (45)	表示生成订单的时间
delsoft	varchar (2)	软删除(0 为删除,1 为存在)
temp1	varchar (100)	表示备用字段 1
temp2	varchar (100)	表示备用字段 2
temp3	varchar (100)	表示备用字段 3
temp4	varchar (100)	表示备用字段 4

- product(商品)表、productuser(用户)表和 user_product(用户-产品权限分配):具体如表 2-9～表 2-11 所示。

表 2-9　product(商品)表结构

列　　名	类　　型	描　　述
pid	int	表示商品 ID 标识号,是自动递增的主键
product_id	varchar(32)	表示商品编号
catalogno	varchar(32)	表示药品分类
cas	varchar(32)	表示化学文摘登记号
productname	varchar(64)	表示药品名称
structure	varchar(64)	表示分子结构图片路径名称
mdlnumber	varchar(64)	表示 MDL 编号
formula	varchar(32)	表示化学方程式
mw	varchar(32)	表示总重量
price1	varchar(32)	表示普通用户价格
price2	varchar(32)	表示会员优惠价格
stock	varchar(32)	表示库存
realstock	varchar(32)	表示实际库存
newproduct	varchar(32)	表示是否是新产品
category	varchar(32)	表示药品类别
note	varchar(32)	表示备注
del_flag	int	表示删除标志位
temp1	varchar(100)	表示临时字段 1
temp2	varchar(100)	表示临时字段 2
temp3	varchar(100)	表示临时字段 3
temp4	varchar(100)	表示临时字段 4

表 2-10 productuser(用户)表结构

列 名	类 型	描 述
uid	int	表示用户 ID 标识号,是自动递增的主键
username	varchar(32)	表示用户名称
password	varchar(32)	表示用户密码
companyname	varchar(32)	表示用户公司名称
city	varchar(32)	表示用户生活城市
job	varchar(32)	表示用户工作
tel	varchar(32)	表示用户电话
email	varchar(32)	表示用户电子邮件地址
country	varchar(32)	表示用户国家
zip	varchar(32)	表示地区邮政编码
companyaddress	varchar(32)	表示用户公司地址
superuser	varchar(16)	表示用户权限标志(1 为普通注册用户,2 为高权限用户,3 为管理员)
note	varchar(32)	表示备注
fullname	varchar(32)	表示全名
title	varchar(32)	表示称呼
del_flag	int	表示删除标志位
temp1	varchar(100)	表示临时字段 1
temp2	varchar(100)	表示临时字段 2
temp3	varchar(100)	表示临时字段 3
temp4	varchar(100)	表示临时字段 4

表 2-11 user_product(用户-产品权限分配)表结构

列 名	类 型	描 述
user_product_id	int	表示 ID 编号,是自动递增的主键
uid	int	表示客户标识号
pid	int	表示产品标识号

2.4 Web 服务器端功能模块开发

2.4.1 服务器端开发准备

(1) 服务器端开发所需环境:

- JDK 1.6 以上。

- Myeclipse 8.0 版本以上。
- Tomcat 6.0 版本以上。
- MySQL 5.0 版本以上。

(2) 创建工程。在 Myeclipse 中创建 Web 工程 JmAscent，分别创建 com.ascent.bean、com.ascent.dao、com.ascent.servlet、com.ascent.util 等包，它们的含义如下：

- com.ascent.bean：存放数据库表的映射类。
- com.ascent.dao：存放 Dao(Data Access Object)类，用于封装对数据库的操作。
- com.ascent.servlet：存放作为控制器的 Servlet，响应客户端的请求并调用相应的 Dao。
- com.ascent.util：存放各种工具类。

在 src 下创建 database.conf.xml 文件，存放数据库的连接信息。

该工程的结构如图 2-31 所示。

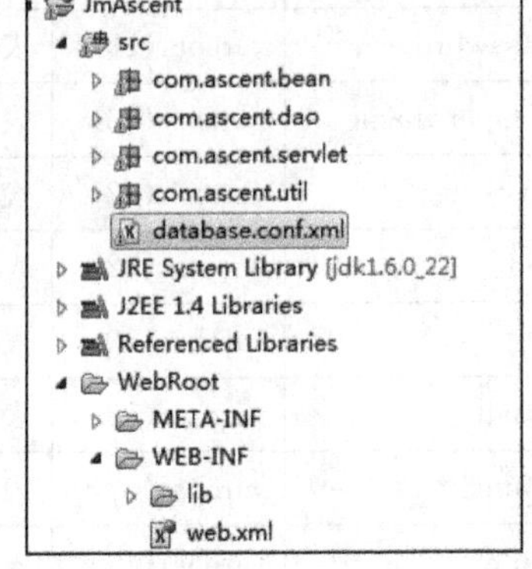

图 2-31　艾斯医药系统 Web 服务器端工程

2.4.2　注册登录模块

在 com.ascent.servlet 包下创建 LoginServlet 类，在 com.ascent.dao 包下创建 LoginDAO 类。LoginServlet 的作用是从 Android 客户端得到表单数据，调用 LoginDAO 对数据库表进行查询，得到结果后将信息以流的方式写回到客户端。流程逻辑如图 2-32 所示。

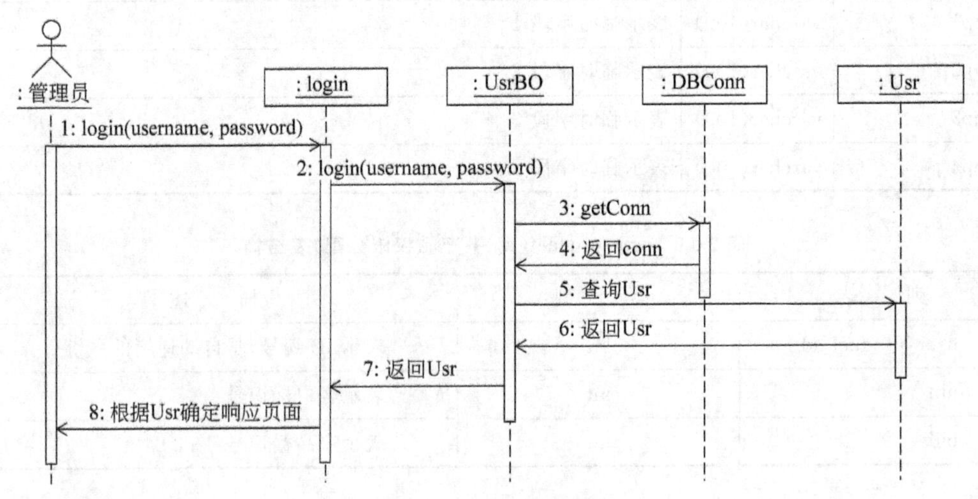

图 2-32　注册登录流程

2.4.3　购物模块

在 com.ascent.servlet 包下创建 ProductServlet 类，在 com.ascent.dao 包下创建 ProductDAO 类。ProductServlet 的作用是根据 Android 客户端的请求调用 ProductDAO 中的相应方法进行药品查询，并将查询结果输出到 Android 客户端。添加商品、删除商品的具体流程如图 2-33 和图 2-34 所示。

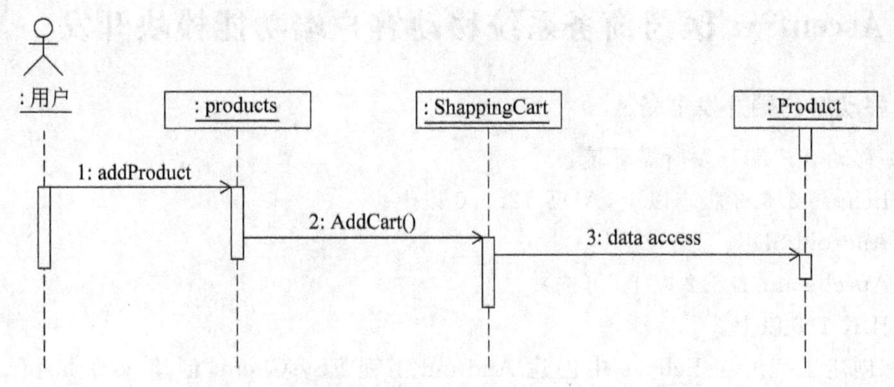

图 2-33　购物模块添加商品到购物车中序列图

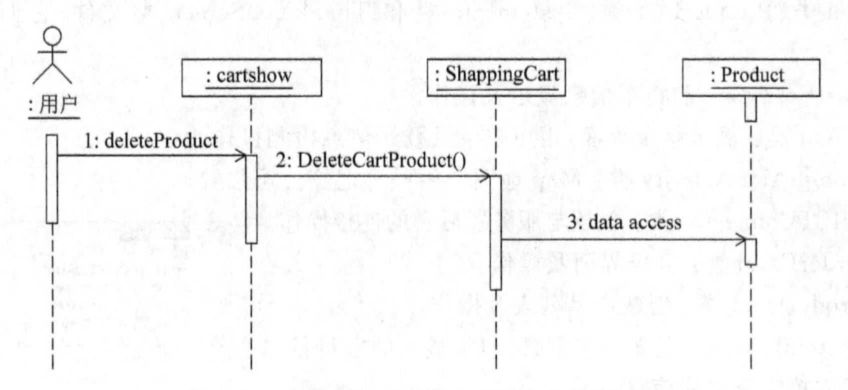

图 2-34　购物模块从购物车中移除商品序列图

2.4.4　订单模块

在 com. ascent. servlet 包下创建 OrderServlet 类，该类的作用是当用户在 Android 客户端单击“提交”订单时生成订单，并对相应的表进行操作，如图 2-35 所示。

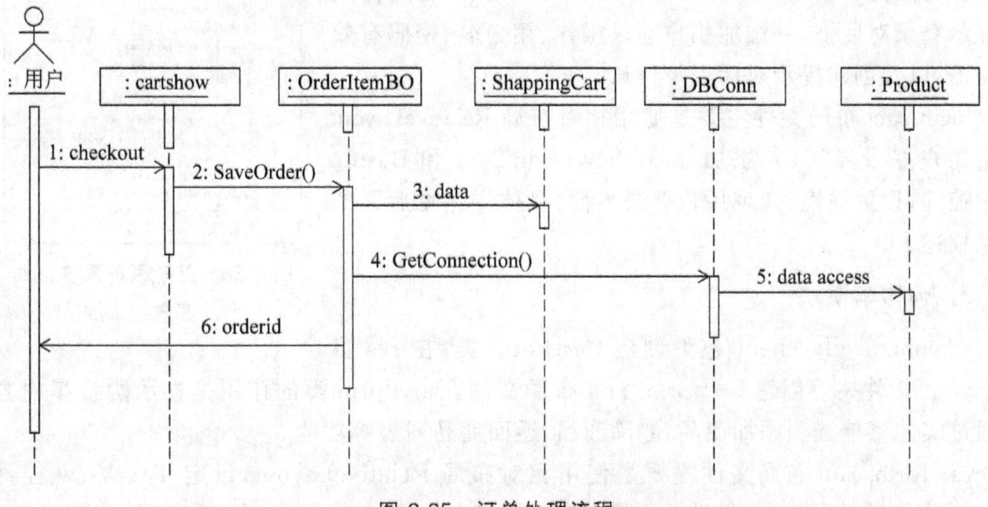

图 2-35　订单处理流程

2.5 AscentSys 医药商务系统移动客户端功能模块开发

1. 移动客户端开发准备

(1) 移动客户端开发所需环境：

- Eclipse 3.5.2 版本以上，ADT 12.0.0 以上。
- Android SDK。
- Apache-ant-1.8.2 以上(可选)。
- JDK 1.6 以上。

(2) 创建工程。在 Eclipse 中创建 Android 工程 ESysClient，创建 com.hisoft.client 包，在包下分别创建 CartForm 类、Client 类、GoogleMapActivity 类、MIDPConnector 类、OrderForm 类、ProductList 类、SystemInfo 类和 ThankYouScreen 类文件，它们的含义如下：

- CartForm 类：购物车信息显示及操作。
- Client 类：显示登录界面，用户登录及登录信息检测操作。
- GoogleMapActivity 类：Map 地图应用——定位公司位置。
- MIDPConnector 类：创建与服务器后台的连接操作。
- OrderForm 类：订单界面及操作。
- ProductList 类：创建产品列表及操作。
- SystemInfo 类：定义一些常量，包括显示的字符信息和连接字符串的信息。
- ThankYouScreen 类：创建界面，以及初始化信息。

该工程的结构如图 2-36 所示。

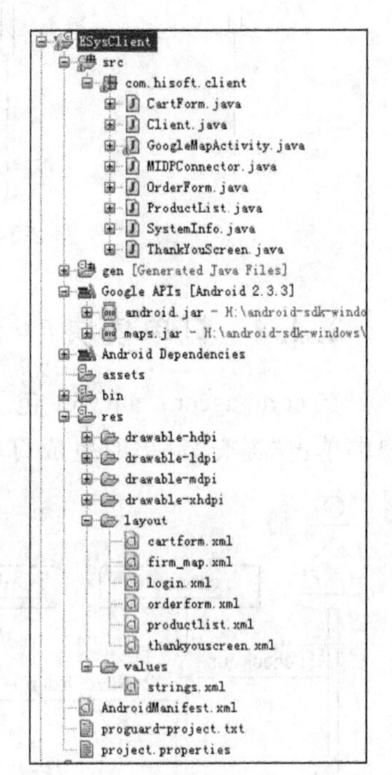

图 2-36 艾斯医药系统移动客户端工程

2. 登录界面模块

在 com.hisoft.client 包下创建 Client 类，在 res 目录 layout 文件夹下创建 login.xml 布局文件，Client 类的作用是显示登录对话框，并添加用户登录操作、用户名、密码有效性检测，以及创建提示对话框等信息。

login.xml 布局文件主要是使用相对布局 RelativeLayout 设定用户登录界面，并添加 TextView、EditText 和 Button 按钮控件，设置属性，实现用户登录界面，具体详述见后面的案例描述。

3. 购物车模块

在 com.hisoft.client 包下创建 CartForm 类，在 res 目录 layout 文件夹下创建 cartform.xml 布局文件，CartForm 类的作用是显示购物车内容，并通过实现菜单选项添加商品、删除商品、返回商品列表等功能。

cartform.xml 布局文件主要是使用相对布局 RelativeLayout，设定 TextView 控件、ListView 控件的属性，实现购物车商品信息的显示。

4. 订单模块

在com. hisoft. client包下创建OrderForm类，在res目录layout文件夹下创建orderform. xml布局文件，OrderForm类的作用是显示订单用户信息、创建订单界面，同时获取并显示购物车信息，然后把订单提交下一个流程处理。

cartform. xml布局文件主要是使用线性布局LinearLayout，并在其中使用TableRow，然后添加TextView控件、EditText控件并设定它们的属性，实现商品订单信息的显示。

5. 商品列表模块

在com. hisoft. client包下创建ProductList类，在res目录layout文件夹下创建productlist. xml布局文件，ProductList类的作用是显示服务器后台存放的商品信息，包含商品名称、商品价格，以及翻页显示，菜单选择、查看购物车、与服务器后台连接等功能。

productlist. xml布局文件是使用线性布局RelativeLayout，并在其中添加TextView控件、ListView控件并设定它们的属性，实现服务器后台商品信息的显示。

6. 地图界面模块

在com. hisoft. client包下创建GoogleMapActivity类，在res目录layout文件夹下创建firm_map. xml布局文件，GoogleMapActivity类的作用是建立MapView对象、设定其显示的选项、预设经纬度等功能。

firm_map. xml布局文件是使用绝对布局AbsoluteLayout，并在其中添加Google MapView控件、Button按钮控件并设定它们的属性，其中的com. google. android. maps. MapView控件中设定申请的Map API Key才能实现Google Map信息的显示及应用。

注意：Map API Key的申请步骤在第11章会讲到，此处不再赘述。

2.6 AscentSys移动客户端打包、签名、发布

移动客户端程序在开发、调试完成后，需要进行打包、签名、发布才能在移动终端设备上运行及应用。关于Android移动终端设备程序的打包及发布方式有两种：一种是Android SDK自动系统工具，使用命令行完成上述流程；另外一种是使用Eclipse集成开发工具完成上述打包、发布流程。本书在第1章的1.5.3节中已经讲述了在命令行下如何创建、开发、打包、发布程序的流程，本节只是采用Eclipse集成开发工具完成Ascent移动客户端程序的打包、签名、发布流程，具体步骤如下：

(1) 选中开发完成的Ascent移动客户端项目ESysClient右击，从弹出的快捷菜单中选择Android Tools→Export Signed Application Package菜单项，如图2-37和图2-38所示。

(2) 创建新的keystore。如果已经存在keystore，可以选择使用现有的或者创建新的。输入keystore的存储路径及密码和确认密码，以及相关信息，如图2-39和图2-40所示。

New
Go Into
Open in New Window
Open Type Hierarchy F4
Show In Alt+Shift+W
Copy Ctrl+C
Copy Qualified Name
Paste Ctrl+V
Delete Delete
Remove from Context Ctrl+Alt+Shift+Down
Build Path
Source Alt+Shift+S
Refactor Alt+Shift+T
Import...
Export...
Refresh F5
Close Project
Close Unrelated Projects
Assign Working Sets...
Run As
Debug As
Profile As
Validate
Team
Compare With
Restore from Local History...
Android Tools
New Test Project...
New Resource File...
Export Signed Application Package...
Export Unsigned Application Package...
Display dex bytecode
Rename Application Package
Add Support Library...
Fix Project Properties
Run Lint: Check for Common Errors
Clear Lint Markers
Add Native Support...

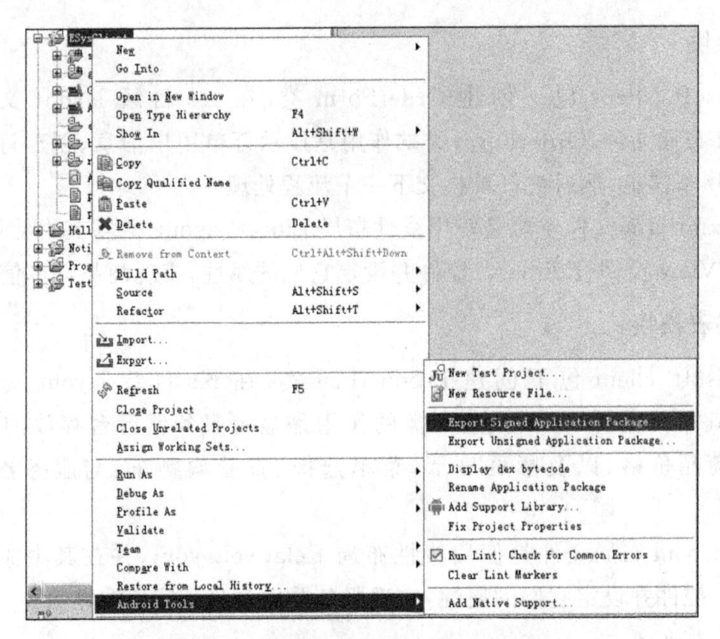

图 2-37 选择导出签名的应用程序包

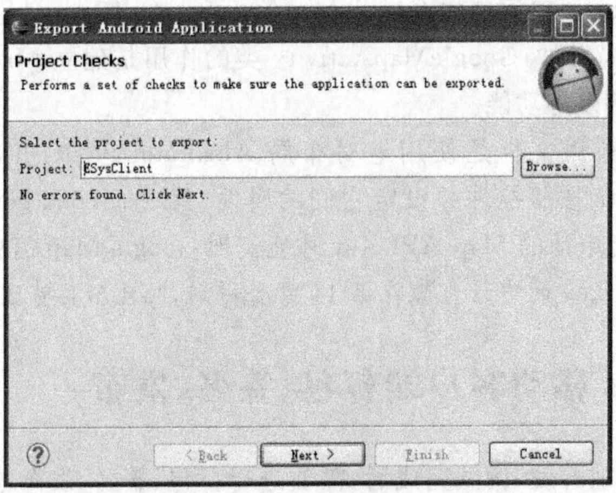

Export Android Application
Project Checks
Performs a set of checks to make sure the application can be exported.
Select the project to export:
Project: ESysClient
Browse...
No errors found. Click Next.
< Back
Next >
Finish
Cancel

图 2-38 导出的项目工程名称

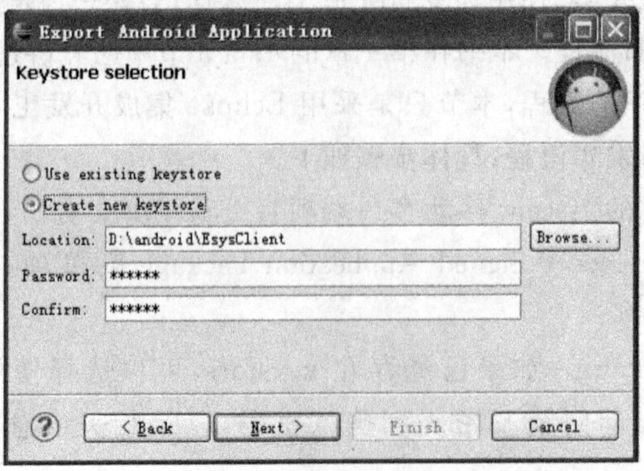

Export Android Application
Keystore selection
Use existing keystore
Create new keystore
Location: D:\android\EsysClient
Browse...
Password: ******
Confirm: ******
< Back
Next >
Finish
Cancel

图 2-39 创建 keystore 界面

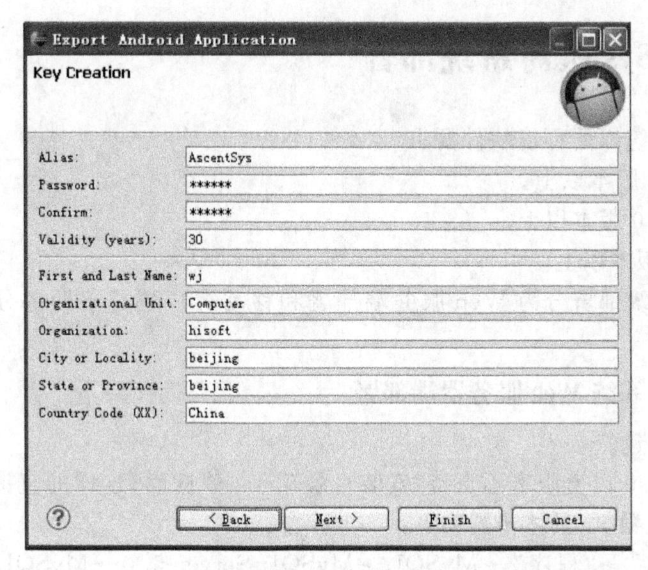

图 2-40　输入 key 的别名、密码、有效期、用户名等信息

（3）输入导出的 apk 文件的存储路径及文件名称和导出的 apk 文件和 key，如图 2-41 和图 2-42 所示。

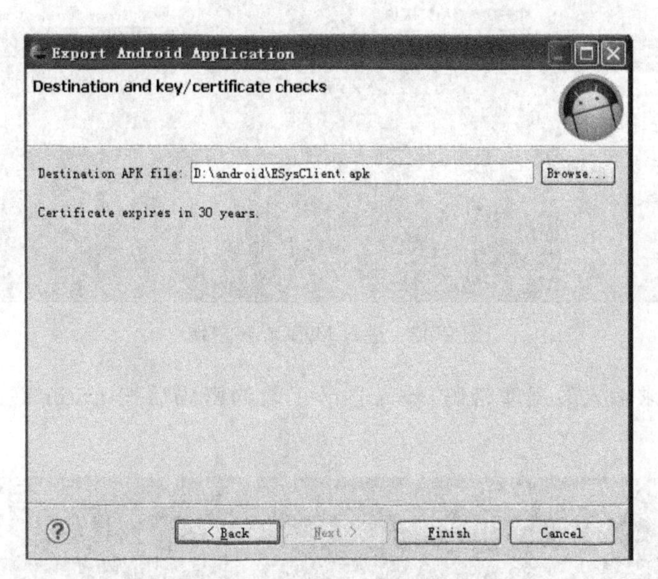

图 2-41　导出的 apk 路径及名称

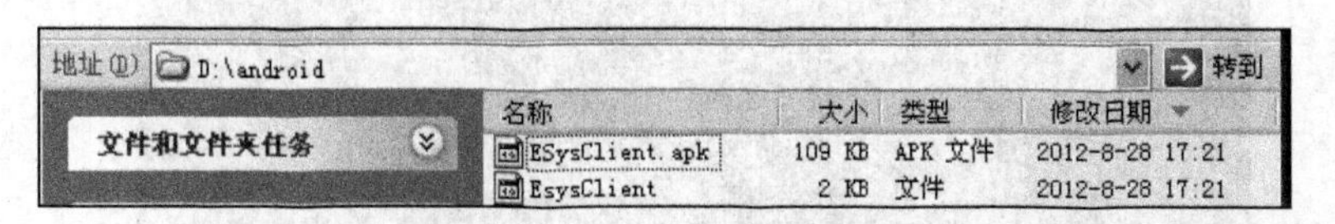

图 2-42　导出的 apk 文件和 key 文件

2.7 AscentSys 医药系统部署

AscentSys 系统部署环境软件要求：

• Mysql 5.0 版本以上。
• Tomcat 6.0 版本以上。
• JDK 1.6 版本以上。

AscentSys 系统部署分为 Web 服务端部署和移动客户端部署两部分，具体部署步骤如下：

1. AscentSys 系统 Web 服务器端部署

（1）数据库创建。

由于 Mysql 5.0 以上版本不支持“安装目录/data/数据库”这样的直接备份，需要自己建立数据库并导入数据，具体步骤如下：

① 选择“开始”→“程序”→MySQL→MySQL Server 5.0→MySQL Command Line Client 菜单项，具体如图 2-43 所示。

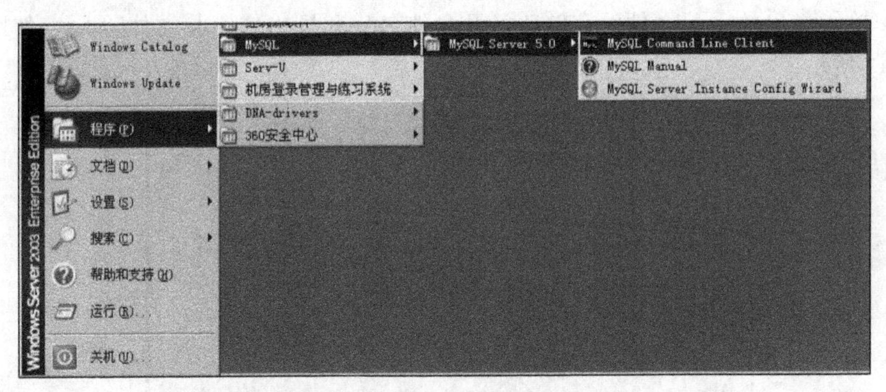

图 2-43　进入 MySQL 客户端

② 进入后要求输入数据库密码，输入自己正确的密码后按 Enter 键进入 MySQL，如图 2-44 所示。

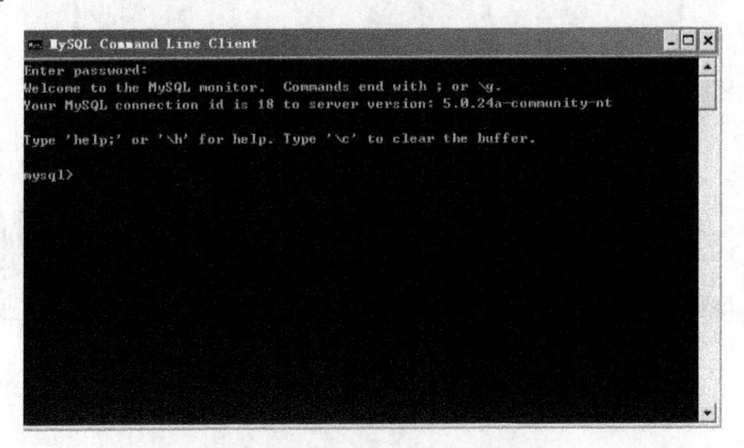

图 2-44　进入 MySQL 数据库

③ 创建 aacesys 数据库，并使用 aacesys 数据库，具体如图 2-45 所示。

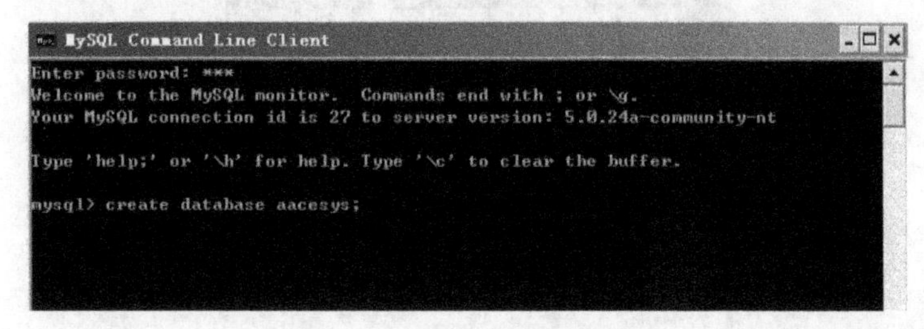

图 2-45　创建 aacesys 数据库

④ 执行导入命令 mysql>source e:/aacesys. sql;，其中 e:/aacesys. sql 是 sql 脚本，可以把它放在任意目录下，本例放在 e 盘下，按 Enter 键执行导入命令，具体如图 2-46 所示。

```
Query OK, 0 rows affected (0.00 sec)
Query OK, 0 rows affected (0.00 sec)
Query OK, 0 rows affected (0.00 sec)
Query OK, 0 rows affected (0.00 sec)
Query OK, 0 rows affected (0.01 sec)
Query OK, 0 rows affected (0.00 sec)
Query OK, 0 rows affected (0.00 sec)
Query OK, 0 rows affected (0.00 sec)
Query OK, 0 rows affected (0.00 sec)
Query OK, 0 rows affected (0.00 sec)
Query OK, 0 rows affected (0.00 sec)
Query OK, 0 rows affected (0.07 sec)
Query OK, 0 rows affected (0.01 sec)
```

图 2-46　导入数据库过程

成功导入后，此时数据库建立成功。

(2) 将 ESysAndroidServer. wa 复制到 tomcat\webapps 下。启动 Tomcat 6.0，放置的文件自动解压生成到 ESysAndroidServer 文件下，然后在路径找到 tomcat\webapps\ESysAndroidServer\WEB-INF\classess\database.conf. xml. xml 文件，打开修改下面代码第 5 行中 user 的值，第 6 行 password 的值，修改为自己数据库的用户名、密码。修改完成即可以启动运行工程。代码如下：

```
1.  <database-conf>
2.  <datasource>
3.    <driver>com.mysql.jdbc.Driver</driver>
4.    <url>jdbc:mysql://localhost:3306/aacesys?useUnicode=true&amp;
      characterEncoding=gb2312</url>
5.    <user>root</user>
6.    <password>wjj</password>
7.  </datasource>
8.  </database-conf>
```

(3) 启动 tomcat，项目 Web 服务器端正确启动运行了。

2. AscentSys 医药系统移动客户端部署

(1) 打开 Eclipse，选择 File→Import 菜单项，导入移动客户端工程 ESysClient，如图 2-47 所示。

(2) 创建 AVD(注意，Target 必须选择 Google APIs，至少为 API Level 10)。

(3) 部署运行 ESysClient 工程，如果打包、发布，具体步骤见本章 2.6 节内容。

拓展提示：Android 系统开发有不同的开发模式，如 B/S 架构，不同的开发模式，Android 系统开发涉及的软件工具、数据内容存储都有很大的不同，尤其是移动客户端对 Web 服务器的页面浏览与通常的浏览器浏览模式在程序开发方面有很大的不同，一般都会有移动服务器版本的系统和 Web 服务器版本两种，不同的客户端访问，服务器会进行判

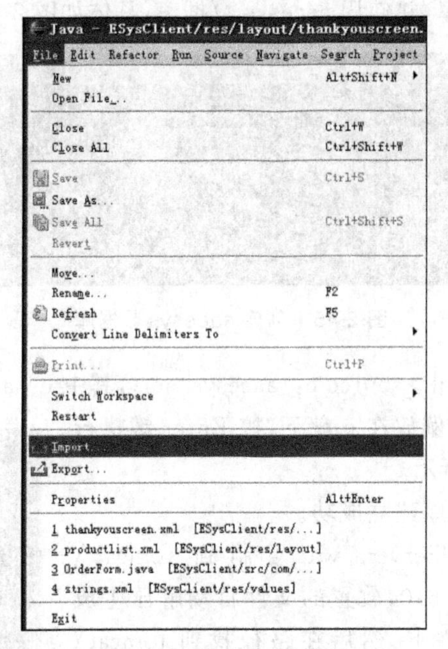

图 2-47 导入 ESysClient 工程

断,重定向发送请求到不同的服务器版本。

1. 简答题

(1) 移动软件开发流程与服务器端软件开发流程的区别有哪里?

(2) Android 移动客户端打包、发布的方式有几种? 它们各自通常包含哪些步骤?

(3) 移动客户端开发的常用软件有哪些? Web 服务器端开发常用的软件有哪些?

2. 完成下面的实训项目

要求: 使用不同的系统平台(Windows 系统平台、Linux 系统平台),部署运行移动客户端 ESysClient、Web 服务器端 ESysAndroidServer 及数据库 MySQL。

第3章 Android应用程序

学习目标

本章主要介绍 Android 项目中的目录结构、AndroidManifest.xml 文件、gen 目录等项目构成文件，同时对 Android 应用程序组件 Activity、Service、Intent 和 IntentFilter、BroadcastReceiver、ContentProvider 进行介绍，并就 Android 程序生命周期和组件生命周期详细讲解。通过本章的学习，以使读者能够达到以下知识要点的学习：

（1）Android 项目文件及应用程序。

（2）Android 系统的进程优先级的变化方式。

（3）Android 系统的基本组件。

（4）Android 程序生命周期及 Activity 的生命周期中各状态的变化关系。

（5）Activity 事件回调函数的作用和调用顺序。

（6）Android 应用程序的调试方法和工具。

在 Android 的应用过程中，对于初学者而言，通常混淆 Android 项目和 Android 应用程序，它们之间既有区别又有联系，要创建 Android 应用，必须先创建 Android 项目，然后才能在项目中创建 Android 应用程序。下面就 Android 项目的构成和 Android 应用程序的组成进行介绍。

3.1 Android 项目构成

3.1.1 目录结构

在建立新项目的过程中，ADT 会自动建立一些目录和文件，这些目录和文件有其固定的作用，有的允许修改，有的不能修改。一个新创建的 Android 项目，项目结构包含 src 目录、gen 目录、assets 目录、res 目录、库文件 android.jar，以及三个项目工程文件 AndroidManifest.xml、default.properties 和 proguard.cfg，如图 3-1 所示，下面逐一进行介绍。

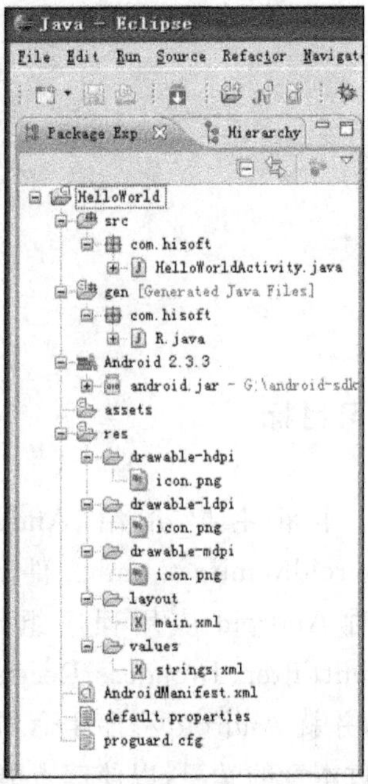

- src 目录：源代码目录，所有允许用户修改的 Java 文件和用户自己添加的 java 文件都保存在这个目录中。如建立 HelloAndroid 工程，ADT 根据用户在工程向导中的 Create Activity 选项自动建立 HelloAndroid.java 文件。
- gen 目录：1.5 版本之后新增的目录，用来保存 ADT 自动生成的 R.java 文件。
- android.jar 文件：Android 程序所能引用的函数库文件，Android 通过平台所支持的 API 都包含在这个文件中。
- assets 目录：用来存放原始格式的文件，例如音频文件、视频文件等二进制格式文件。此目录中的资源不能被 R.java 文件索引，所以只能以字节流的形式读取。一般情况下为空。
- res 目录：资源目录，有 5 个子目录用来保存 Android 程序的所有资源。
- proguard.cfg 文件：Android 混淆器，用来防止程序被反编译。它其实也就是将变量的名称混淆一下，降低程序的可读性。

图 3-1 Android 工程目录结构

特别提醒：

dpi 是 dot per inch 的简称，表示每英寸像素数。

在 Android 中密度分类有 4 种，分别是 ldpi(low)、mdpi(medium)、hdpi(high)和 xhdpi (extra high)。

一般情况下的普通屏幕尺寸：ldpi 是 120，mdpi 是 160，hdpi 是 240，xhdpi 是 320。

需要注意的是：xhdpi 是从 Android 2.2 (API Level 8)版本开始增加的图片分类。xlarge 是从 Android 2.3 (API Level 9)版本开始增加的图片分类。

在 Hello Android 工程中，ADT 在 drawable 目录中自动引入了 icon.png 文件作为 HelloAndroid 程序的图标文件；在 layout 目录生成了 mail.xml 文件，用于描述用户界面。

3.1.2 AndroidManifest.xml 文件简介

AndroidManifest.xml 是 XML 格式的 Android 程序声明文件，是全局描述文件，包含了 Android 系统运行 Android 程序前所必须掌握的重要信息，这些信息包含应用程序名称、图标、包名称、模块组成、授权和 SDK 最低版本等。创建的每个 Android 项目应用程序必须在根目录下包含一个 AndroidManifest.xml 工程文件。

1. AndroidManifest.xml 文件的代码

```
1.   <?xml version="1.0" encoding="utf-8"?>
2.   <manifest xmlns:android="http://schemas.android.com/apk/res/android"
3.       package="com.hisoft"
```

```
4.         android:versionCode="1"
5.         android:versionName="1.0">
6.         <uses-sdk android:minSdkVersion="10" />
7.         <application android:icon="@drawable/icon"android:label="@string/
           app_name">
8.             <activity android:name=".HelloWorldActivity"
9.                 android:label="@string/app_name">
10.                    <intent-filter>
11.                        <action android:name="android.intent.action.MAIN" />
12.                        <category android:name="android.intent.category.LAUNCHER" />
13.                    </intent-filter>
14.            </activity>
15.        </application>
16.    </manifest>
```

AndroidManifest.xml 文件的根元素是 manifest，包含了 xmlns:android、package、android:versionCode 和 android:versionName 共 4 个属性。

- xmlns:android：定义了 Android 的命名空间，值为 http://schemas.android.com/apk/res/android。
- package：定义了应用程序的包名称。
- android:versionCode：定义了应用程序的版本号，是一个整数值，数值越大说明版本越新，但仅在程序内部使用，并不提供给应用程序的使用者。
- android:versionName：定义了应用程序的版本名称，是一个字符串，仅限于为用户提供一个版本标识。

manifest 元素仅能包含一个 application 元素，application 元素中能够声明 Android 程序中最重要的 4 个组成部分，包括 Activity、Service、BroadcastReceiver 和 ContentProvider，所定义的属性将影响所有组成部分。

第 7 行属性 android:icon 定义了 Android 应用程序的图标，其中@drawable/icon 是一种资源引用方式，表示资源类型是图像，资源名称为 icon，对应的资源文件为 res/drawable 目录下的 icon.png。

第 7 行属性 android:label 则定义了 Android 应用程序的标签名称。

activity 元素是对 Activity 子类的声明，必须在 AndroidManifest.xml 文件中声明的 Activity 才能在用户界面中显示。

第 8 行属性 android:name 定义了实现 Activity 类的名称，可以是完整的类名称，也可以是简化后的类名称。

第 9 行属性 android:label 则定义了 Activity 的标签名称，标签名称将在用户界面的 Activity 上部显示。@string/app_name 同样属于资源引用，表示资源类型是字符串，资源名称为 app_name，资源保存在 res/values 目录下的 strings.xml 文件中。

intent-filter 中声明了两个子元素 action 和 category，intent-filter 使 HelloAndroid 程序在启动时将.HelloAndroid 这个 Activity 作为默认启动模块。

2. 可视化编辑器

双击 AndroidManifest.xml 文件，直接进入可视化编辑器，如图 3-2 所示，用户可以直接编辑 Android 工程的应用程序名称、包名称、图标、标签和许可等相关属性。

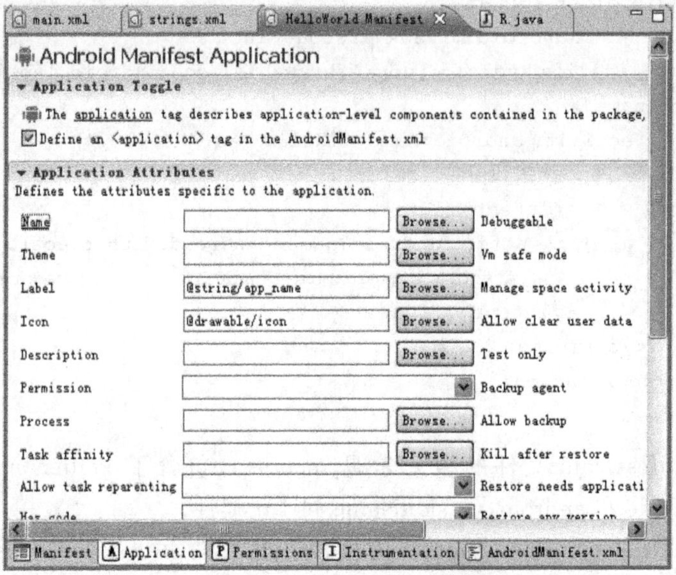

图 3-2　AndroidManifest.xml 文件可视化编辑器

3.1.3　gen 目录

在上述目录结构已讲述，gen 目录下只存放一个由 ADT 自动生成，并不需要人工修改的 R.java 文件。

R.java 文件包含对 drawable、layout 和 values 目录内资源的引用指针，Android 程序能够直接通过 R 类引用目录中的资源。

Android 系统中的资源引用有两种方式：一种是在代码中引用资源；另一种是在资源中引用资源。

代码中引用资源需要使用资源的 ID，可以通过[R.resource_type.resource_name]或[android.R.resource_type.resource_name]获取资源 ID。

resource_type 代表资源类型，也就是 R 类中的内部类名称。

resource_name 代表资源名称，对应资源的文件名或在 XML 文件中定义的资源名称属性。

资源中引用资源，引用格式如下：

```
@[package:]type:name
```

- @表示对资源的引用。
- package 是包名称，如果在相同的包，package 则可以省略。

R.java 文件不能手工修改，如果向资源目录中增加或删除了资源文件，则需要在工程名称上右击，从弹出的快捷菜单中选择 Refresh 菜单项来更新 R.java 文件中的代码。

R 类包含的几个内部类分别与资源类型相对应，资源 ID 便保存在这些内部类中，例如子类 drawable 表示图像资源，内部的静态变量 icon 表示资源名称，其资源 ID 为

0x7f020000。一般情况下,资源名称与资源文件名相同。

HelloAndroid 工程生成的 R. java 文件的代码如下:

```
package com.hisoft;

public final class R {
    public static final class attr {
    }
    public static final class drawable {
        public static final int icon=0x7f020000;
    }
    public static final class layout {
        public static final int main=0x7f030000;
    }
    public static final class string {
        public static final int app_name=0x7f040001;
        public static final int hello=0x7f040000;
    }
}
```

3.1.4 res 目录

res 目录中包含了 5 个子目录,分别是:

- drawable-hdpi 目录:主要放高分辨率的图片,如 WVGA(480x800)、FWVGA(480x854)。默认存放的是 icon. png 图片。
- drawable-mdpi 目录:主要放中等分辨率的图片,如 HVGA(320x480)。默认存放的是 icon. png 图片。
- drawable-ldpi 目录:主要放低分辨率的图片,如 QVGA(240x320)。默认存放的是 icon. png 图片。

系统会根据机器的分辨率来分别到这几个文件夹里面去找对应的图片。

- layout 目录:用来保存与用户界面相关的布局文件,这些布局文件都是 XML 文件。默认存放的是 main. xml 文件。
- valuse 目录:保存文件颜色、风格、主题和字符串等。默认存放的是 strings. xml 文件。

main. xml 文件是界面布局文件,利用 XML 语言描述的用户界面布局的相关内容将在后续章节用户界面设计中进行详细介绍。

(1) main. xml 文件代码:

```
<?xml version="1.0" encoding="utf-8"?>
<LinearLayout xmlns:android="http://schemas.android.com/apk/res/android"
    android:orientation="vertical"
    android:layout_width="fill_parent"
    android:layout_height="fill_parent"
    >
<TextView
    android:layout_width="fill_parent"
```

```
    android:layout_height="wrap_content"
    android:text="@string/hello"
    />
</LinearLayout>
```

第 7 行的代码说明在界面中使用 TextView 控件，TextView 控件主要用来显示字符串文本。

第 10 行代码说明 TextView 控件需要显示的字符串，非常明显，@string/hello 是对资源的引用。

(2) strings. xml 文件代码：

```
<?xml version="1.0" encoding="utf-8"?>
<resources>
    <string name="hello">Hello World, HelloWorldActivity!</string>
    <string name="app_name">HelloWorld</string>
</resources>
```

通过 strings. xml 文件的第 3 行代码分析，在 TextView 控件中显示的字符串应是“Hello World，HelloAndroidActivity!”。

如果读者修改 strings. xml 文件的第 3 行代码的内容，重新编译、运行后，模拟器中显示的结果也应该随之更改。

3.1.5 default. properties 文件

```
#This file is automatically generated by Android Tools.
#Do not modify this file--YOUR CHANGES WILL BE ERASED!
#
#This file must be checked in Version Control Systems.
#
#To customize properties used by the Ant build system use,
#"build.properties", and override values to adapt the script to your
#project structure.

#Project target.
target=android-10
```

default. properties 文件记录 Android 工程的相关设置，该文件不能手动修改，需右键单击工程名称，从弹出的快捷菜单中选择 Properties 菜单项进行修改。

在 default. properties 文件中只有第 12 行是有效代码，说明 Android 程序的编译目标。

3.2 Android 应用程序组成

3.2.1 Android 应用程序概述

Android 应用程序是在 Android 应用框架之上，由一些系统自带和用户创建的应用程序组成。组件是可以调用的基本功能模块，Android 应用程序就是由组件组成的，一个

Android 的应用程序通常包含 4 个核心组件和一个 Intent，4 个核心组件分别是 Activity、Service、BroadcaseReceiver 和 ContentProvider。Intent 是组件之间进行通信的载体，它不仅可以在同一个应用中起传递信息的作用，还可以在不同的应用中传递信息，如图 3-3 所示。

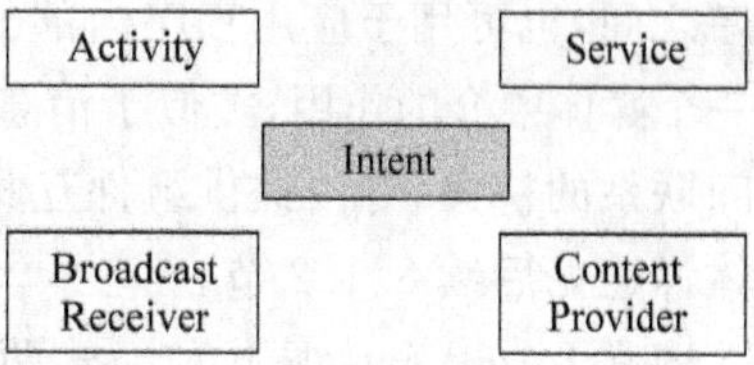

Activity
Service
Intent
Broadcast Receiver
Content Provider

图 3-3 Android 应用程序组件

3.2.2 Activity 组件

Activity 是 Android 程序的呈现层，显示可视化的用户界面，并接收与用户交互所产生的界面事件。一个 Android 应用程序可以包含一个或多个 Activity，其中一个作为 main activity 用于启动显示，一般在程序启动后会呈现一个 Activity，用于提示用户程序已经正常启动。

Activity 通过 View 管理用户界面 UI。View 绘制用户界面 UI 与处理用户界面事件（UI event），View 可通过 xml 描述定义，也可在代码中生成。一般情况下，Android 建议将 UI 设计和逻辑分离，android UI 设计类似 swing，通过布局（layout）组织 UI 组件。

在应用程序中，每一个 Activity 都是一个单独的类，继承实现了 Activity 基础父类，这个类通过它的方法设置并显示由 Views 组成的用户界面 UI，并接受、响应与用户交互产生的界面事件，Activity 通过 startActivity 或 startActivityForResult 启动另外的 Activity。

在应用程序中，一个 Activity 在界面上的表现形式通常有全屏窗体、非全屏悬浮窗体和对话框等。

3.2.3 Service 组件

Service 常用于没有用户界面，但需要长时间在后台运行的应用。与应用程序的其他模块（例如 Activity）一同运行于主线程中。一般通过 startService 或 bindService 方法创建 Service，通过 stopService 或 stopSelf 方法终止 Service。通常情况下，都在 Activity 中启动和终止 Service。

在 Android 应用中，Service 的典型应用是音乐播放器，在一个媒体播放器程序中，大概要有一个或多个活动（Activity）来供用户选择歌曲并播放它。然而，音乐的回放就不能使用活动了，因为用户希望能够切换到其他界面时音乐继续播放。这种情况下，媒体播放器活动要用 Context.startService()启动一个服务来在后台运行保持音乐的播放。系统将保持这个音乐回放服务的运行直到它结束。需要注意，要用 Context.bindService()方法连接服务（如果没有运行，要先启动它）。当连接到服务后，可以通过服务暴露的一个接口和它通信。对于音乐服务，它支持暂停、倒带和重放等功能。

3.2.4 Intent 和 IntentFilter 组件

1. Intent

Android 中提供了 Intent 机制来协助应用间的交互与通信，Intent 负责对应用中一次操作的动作、动作涉及数据、附加数据进行描述，Android 则根据此 Intent 的描述，负责找到对应的组件，将 Intent 传递给调用的组件，并完成组件的调用。Intent 不仅可用于应用程序之间，也可用于应用程序内部的 Activity/Service 之间的交互。因此，Intent 在这里起着一个媒体中介的作用，类似于消息、事件通知，它充当 Activity、Service、broadcastReceiver 之间联系的桥梁，专门提供组件互相调用的相关信息，实现调用者与被调用者之间的解耦。具体详述见后续 8.1.2 节。

通常 Intent 分为显式和隐式两类。显式的 Intent 就是指定了组件的名字，是由程序指定具体的目标组件来处理，即在构造 Intent 对象时就指定接收者，指定了一个明确的组件(setComponent 或 setClass)来使用处理 Intent。

```
Intent intent=new Intent(
    getApplicationContext(),
    Test.class
);
startActivity(intent);
```

特别注意：被启动的 Activity 需要在 AndroidManifest.xml 中进行定义。

隐式的 Intent 就是没有指定 Intent 的组件名字，没有制定明确的组件来处理该 Intent。使用这种方式时，需要让 Intent 与应用中的 IntentFilter 描述表相匹配。需要 Android 根据 Intent 中的 Action、data 和 Category 等来解析匹配。由系统接受调用并决定如何处理，即 Intent 的发送者在构造 Intent 对象时并不知道也不关心接收者是谁，有利于降低发送者和接收者之间的耦合。例如 startActivity(new Intent(Intent.ACTION_DIAL));。

```
Intent intent=new Intent();
intent.setAction("test.intent.IntentTest");
startActivity(intent);
```

目标组件(Activity、Service、Broadcast Receiver)是通过设置它们的 Intent Filter 来界定其处理的 Intent。如果一个组件没有定义 Intent Filter，那么它只能接受处理显式的 Intent，只有定义了 Intent Filter 的组件才能同时处理隐式和显式的 Intent。

一个 Intent 对象包含了很多数据的信息，由 6 个部分组成：

- Action：要执行的动作。
- Data：执行动作要操作的数据。
- Category：被执行动作的附加信息。
- Extras：其他所有附加信息的集合。
- Type：显式指定 Intent 的数据类型(MIME)。
- Component：指定 Intent 的目标组件的类名称，比如要执行的动作、类别、数据和附加信息等。

下面就一个 Intent 中包含的信息进行简要介绍。

(1) Action

一个 Intent 的 Action 在很大程度上说明这个 Intent 要做什么，是查看(View)、删除(Delete)或编辑(Edit)等。Action 是一个字符串命名的动作，Android 中预定义了很多 Action，可以参考 Intent 类查看。表 3-1 是 Android 文档中的几个动作。

表 3-1 Android 动作

Constant	Target component	Action
ACTION_CALL	activity	Initiate a phone call.
ACTION_EDIT	activity	Display data for the user to edit.
ACTION_MAIN	activity	Start up as the initial activity of a task, with no data input and no returned output.
ACTION_SYNC	activity	Synchronize data on a server with data on the mobile device.
ACTION_BATTERY_LOW	broadcast receiver	A warning that the battery is low.
ACTION_HEADSET_PLUG	broadcast receiver	A headset has been plugged into the device, or unplugged from it.
ACTION_SCREEN_ON	broadcast receiver	The screen has been turned on.
ACTION_TIMEZONE_CHANGED	broadcast receiver	The setting for the time zone has changed.

此外,用户也可以自定义 Action,比如 com. flysnow. intent. ACTION_ADD。定义的 Action 最好能表明其所表示的意义,要做什么,这样 Intent 中的数据才好填充。Intent 对象的 getAction()可以获取动作,使用 setAction()可以设置动作。

(2) Data

其实就是一个 URI,用于执行一个 Action 时所用到数据的 URI 和 MIME。不同的 Action 有不同的数据规格,比如 ACTION_EDIT 动作,数据就可能包含一个用于编辑文档的 URI。如果是一个 ACTION_CALL 动作,那么数据就是一个包含了 tel:6546541 的数据字段,所以上面提到的自定义 Action 时要规范命名。数据的 URI 和类型对于 Intent 的匹配是很重要的,Android 往往根据数据的 URI 和 MIME 找到能处理该 Intent 的最佳目标组件。

(3) Component(组件)

指定 Intent 的目标组件的类名称。通常 Android 会根据 Intent 中包含的其他属性的信息,比如 action、data/type 和 category 进行查找,最终找到一个与之匹配的目标组件。

如果设置了 Intent 目标组件的名字,那么这个 Intent 就会被传递给特定的组件,而不再执行上述查找过程。指定了这个属性以后,Intent 的其他所有属性都是可选的。也就是我们说的显式 Intent。如果不设置,则是隐式的 Intent,Android 系统将根据 Intent Filter 中的信息进行匹配。

(4) Category

Category 指定了用于处理 Intent 的组件的类型信息,一个 Intent 可以添加多个 Category,使用 addCategory()方法即可,使用 removeCategory()删除一个已经添加的类别。Android 的 Intent 类里定义了很多常用的类别,可以参考使用。

(5) Extras

有些用于处理 Intent 的目标组件需要一些额外的信息,那么就可以通过 Intent 的 put..()方法把额外的信息塞入到 Intent 对象中,用于目标组件的使用,一个附件信息就是一个 key-value 的键值对。Intent 有一系列的 put 和 get 方法用于处理附加信息的塞入和取出。

2. IntentFilter

应用程序的组件为了告诉 Android 自己能响应、处理哪些隐式 Intent 请求，可以声明一个甚至多个 Intent Filter。每个 Intent Filter 描述该组件所能响应 Intent 请求的能力——组件希望接收什么类型的请求行为，什么类型的请求数据。比如请求网页浏览器这个例子中，网页浏览器程序的 Intent Filter 就应该声明它所希望接收的 Intent Action 是 WEB_SEARCH_ACTION，以及与之相关的请求数据是网页地址 URI 格式。如何为组件声明自己的 Intent Filter？常见的方法是在 AndroidManifest. xml 文件中用属性＜Intent-Filter＞描述组件的 Intent Filter。

Intent 解析机制主要是通过查找已注册在 AndroidManifest. xml 中的所有 IntentFilter 及其中定义的 Intent，最终找到匹配的 Intent。在这个解析过程中，Android 是通过 Intent 的 action、type、category 这三个属性进行判断的，判断方法如下：

(1) 如果 Intent 指明 action，则目标组件的 IntentFilter 的 action 列表中就必须包含有这个 action，否则不能匹配。

(2) 如果 Intent 没有提供 type，系统将从 data 中得到数据类型。和 action 一样，目标组件的数据类型列表中必须包含 Intent 的数据类型，否则不能匹配。

(3) 如果 Intent 中的数据不是 content：类型的 URI，而且 Intent 也没有明确指定它的 type，将根据 Intent 中数据的 scheme（如 http：或者 mailto：）进行匹配。同上，Intent 的 scheme 必须出现在目标组件的 scheme 列表中。

(4) 如果 Intent 指定了一个或多个 category，这些类别必须全部出现在组建的类别列表中。比如 Intent 中包含了两个类别：LAUNCHER_CATEGORY 和 ALTERNATIVE_CATEGORY，解析得到的目标组件必须至少包含这两个类别。

一个 intent 对象只能指定一个 action，而一个 intent filter 可以指定多个 action。action 的列表不能为空，否则它将组织所有的 intent。

一个 intent 对象的 action 必须和 intent filter 中的某一个 action 匹配才能通过测试。如果 intent filter 的 action 列表为空，则不通过。如果 intent 对象不指定 action，并且 intentfilter 的 action 列表不为空，则通过测试。

下面针对 Intent 和 Intent Filter 中包含的子元素 Action（动作）、Data（数据）以及 Category（类别）进行比较检查的具体规则详细介绍。

(1) 动作测试

＜intent-filter＞元素中可以包括子元素＜action＞，例如：

```
<intent-filter>
<action android:name="com.example.project.SHOW_CURRENT" />
<action android:name="com.example.project.SHOW_RECENT" />
<action android:name="com.example.project.SHOW_PENDING" />
</intent-filter>
```

一条＜intent-filter＞元素至少应该包含一个＜action＞，否则任何 Intent 请求都不能和该＜intent-filter＞匹配。如果 Intent 请求的 Action 和＜intent-filter＞中某一条

<action>匹配，那么该 Intent 就通过了这条<intent-filter>的动作测试。如果 Intent 请求或<intent-filter>中没有说明具体的 Action 类型，那么会出现下面两种情况。

① 如果<intent-filter>中没有包含任何 Action 类型，那么无论什么 Intent 请求都无法和这条<intent-filter>匹配。

② 反之，如果 Intent 请求中没有设定 Action 类型，那么只要<intent-filter>中包含有 Action 类型，这个 Intent 请求就将顺利地通过<intent-filter>的行为测试。

(2) 类别测试。

<intent-filter>元素可以包含<category>子元素，例如：

```
<intent-filter…>
<category android:name="android.Intent.Category.DEFAULT" />
<category android:name="android.Intent.Category.BROWSABLE" />
</intent-filter>
```

只有当 Intent 请求中所有的 Category 与组件中某一个 IntentFilter 的<category>完全匹配时，才会让该 Intent 请求通过测试，IntentFilter 中多余的<category>声明并不会导致匹配失败。一个没有指定任何类别测试的 IntentFilter 仅仅只会匹配没有设置类别的 Intent 请求。

(3) 数据测试。

数据在<intent-filter>中的描述如下：

```
<intent-filter…>
<data android:type="video/mpeg" android:scheme="http"… />
<data android:type="audio/mpeg" android:scheme="http"… />
</intent-filter>
```

<data>元素指定了希望接受的 Intent 请求的数据 URI 和数据类型，URI 被分成三部分来进行匹配：scheme、authority 和 path。其中，用 setData()设定的 Inteat 请求的 URI 数据类型和 scheme 必须与 IntentFilter 中所指定的一致。若 IntentFilter 中还指定了 authority 或 path，它们也需要相匹配才会通过测试。

3.2.5 BroadcastReceiver 组件

在 Android 中，Broadcast 是一种广泛运用在应用程序之间传输信息的组件。而 BroadcastReceiver 是接收并响应广播消息的组件，对发送出来的 Broadcast 进行过滤接收并响应，它不包含任何用户界面，可以通过启动 Activity 或者 Notification 通知用户接收到重要信息，在 Notification 中有多种方法提示用户，如闪动背景灯、震动设备、发出声音或在状态栏上放置一个持久的图标。

BroadcastReceiver 过滤接收的过程如下：

在需要发送信息时，把要发送的信息和用于过滤的信息(如 Action、Category)装入一个 Intent 对象，然后通过调用 Context. sendBroadcast()、sendOrderBroadcast() 或 sendStickyBroadcast()方法把 Intent 对象以广播方式发送出去。

当 Intent 发送后，所有已经注册的 BroadcastReceiver 会检查注册时的 IntentFilter 是

否与发送的 Intent 相匹配,若匹配则调用 BroadcastReceiver 的 onReceive()方法。因此在定义一个 BroadcastReceiver 时,通常都需要实现 onReceive()方法。

BroadcastReceiver 注册有两种方式:

一种方式是静态的,在 AndroidManifest. xml 中用<receiver>标签声明注册,并在标签内用<intent- filter>标签设置过滤器。

另一种方式是动态的,在代码中先定义并设置好一个 IntentFilter 对象,然后在需要注册的地方调用 Context. registerReceiver()方法,如果取消时就调用 Context. unregisterReceiver()方法。

不管是用 xml 注册的还是用代码注册的,在程序退出时一般需要注销,否则下次启动程序可能会有多个 BroadcastReceiver。另外,若在使用 sendBroadcast()的方法时指定了接收权限,则只有在 AndroidManifest. xml 中用<uses-permission>标签声明了拥有此权限的 BroadcastReceiver 才会有可能接收到发送来的 Broadcast。

同样,若在注册 BroadcastReceiver 时指定了可接收的 Broadcast 的权限,则只有在包内的 AndroidManifest. xml 中用<uses-permission>标签声明了,拥有此权限的 Context 对象所发送的 Broadcast 才能被这个 BroadcastReceiver 所接收。

3.2.6 ContentProvider 组件

ContentProvider 是 Android 系统提供的一种标准的共享数据的机制。在 Android 中每一个应用程序的资源都为私有,应用程序可以通过 ContentProvider 组件访问其他应用程序的私有数据(私有数据可以是存储在文件系统中的文件,或者是存放在 SQLite 中的数据库),如图 3-4 所示。

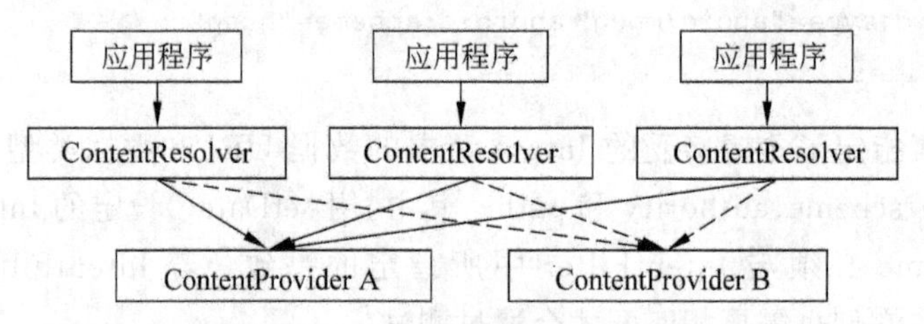

图 3-4 应用程序、ContentResolver 与 ContentProvider

对 ContentProvider 的使用有两种方式:

- ContentResolver 访问。
- Context. getContentResolver()。

Android 系统内部也提供一些内置的 ContentProvider,能够为应用程序提供重要的数据信息。使用 ContentProvider 对外共享数据的好处是统一了数据的访问方式。

3.3 Android 生命周期

3.3.1 程序生命周期

程序的生命周期是指在 Android 系统中进程从启动到终止的所有阶段,也就是 Android 程序启动到停止的全过程。程序的生命周期由 Android 系统进行调度和控制。

Android系统中的进程分为前台进程、可见进程、服务进程、后台进程和空进程。

Android系统中的进程优先级由高到低，如图3-5所示。

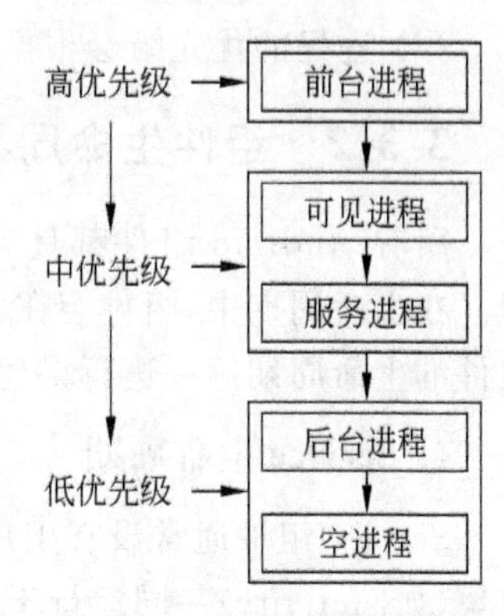

图3-5 Android系统的进程及优先级

1. 前台进程

前台进程是Android系统中最重要的进程，是指与用户正在交互的进程，包含以下4种情况：

(1) 进程中的Activity正在与用户进行交互。

(2) 进程服务被Activity调用，而且这个Activity正在与用户进行交互。

(3) 进程服务正在执行声明周期中的回调方法，如onCreate()、onStart()或onDestroy()。

(4) 进程的BroadcastReceiver正在执行onReceive()方法。

Android系统在多个前台进程同时运行时可能会出现资源不足的情况，此时会清除部分前台进程，保证主要的用户界面能够及时响应。

2. 可见进程

可见进程指部分程序界面能够被用户看见，但不在前台与用户交互，不响应界面事件的进程。如果一个进程包含服务，且这个服务正在被用户可见的Activity调用，此进程同样被视为可见进程。

Android系统一般存在少量的可见进程，只有在特殊的情况下，Android系统才会为保证前台进程的资源而清除可见进程。

3. 服务进程

服务进程是指包含已启动服务的进程，通常特点如下：

- 没有用户界面。
- 在后台长期运行。

Android系统在不能保证前台进程或可视进程所必要的资源，才会强行清除服务进程。

4. 后台进程

后台进程是指不包含任何已经启动的服务，而且没有任何用户可见的Activity的进程。

Android系统中一般存在数量较多的后台进程，在系统资源紧张时，系统将优先清除用户较长时间没有见到的后台进程。

5. 空进程

空进程是指不包含任何活跃组件的进程。空进程在系统资源紧张时会被首先清除。但为了提高Android系统应用程序的启动速度，Android系统会将空进程保存在系统内存中，在用户重新启动该程序时，空进程会被重新使用。

除了以上的优先级外，以下两方面也决定它们的优先级：

(1) 进程的优先级取决于所有组件中优先级最高的部分。

(2) 进程的优先级会根据与其他进程的依赖关系而变化。

3.3.2 组件生命周期

所有 Android 组件都具有自己的生命周期,是指从组件的建立到组件的销毁整个过程。在生命周期中,组件会在可见、不可见、活动、非活动等状态中不断变化。下面就各个组件的生命周期逐一进行讲述。

1. Service 生命周期

Service 组件通常没有用户界面 UI,其启动后一直运行于后台。它与应用程序的其他模块(如 Activity)一同运行于程序的主线程中。

一个 Service 的生命周期通常包含创建、启动、销毁这几个过程。

Service 只继承了 onCreate()、onStart()和 onDestroy()三个方法。当第一次启动 Service 时,先后调用了 onCreate()和 onStart()这两个方法。当停止 Service 时,则执行 onDestroy()方法。需要注意的是,如果 Service 已经启动了,当再次启动 Service 时,不会再执行 onCreate()方法,而是直接执行 onStart()方法。

创建 Service 的方式有两种:一种是通过 startService 创建,另外一种是通过 bindService 创建。两种创建方式的区别在于:startService 是创建并启动 Service;而 bindService 只是创建了一个 Service 实例并取得了一个与该 Service 关联的 binder 对象,但没有启动它,如图 3-6 所示。

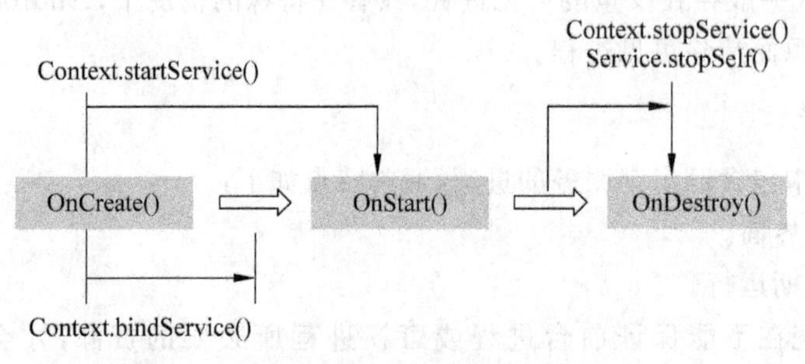

图 3-6 Service 生命周期

如果没有程序停止它或者它自己停止,Service 将一直运行。在这种模式下,Service 开始于调用 Context. startService() ,停止于 Context. stopService()。Service 可以通过调用 StopService()或 Service. stopSelfResult()停止自己。不管调用多少次 startService(),只需要调用一次 StopService()就可以停止 Service。一般在 Activity 中启动和终止 Service。它可以通过接口被外部程序调用。外部程序建立到 Service 的连接,通过连接来操作 Service。建立连接开始于 Context. bindService(),结束于 Context. unbindService()。多个客户端可以绑定到同一个 Service,如果 Service 没有启动,bindService()可以选择启动它。

上述两种方式不是完全分离的。如一个 intent 想要播放音乐,通过 startService()方法启动后台播放音乐的 Service。然后,也许用户想要操作播放器或者获取当前正在播放的乐曲的信息,一个 Activity 就会通过 bindService()建立一个到这个 Service 的连接。这种情况下,stopService()在全部的连接关闭后才会真正停止 Service。

像 Activity 一样，Service 也有可以通过监视状态实现的生命周期。但是比 Activity 要少，通常只有三个方法，而且是 public 的，而不是 protected 的属性，如下：

```
1.   void onCreate()
2.   void onStart(Intent intent)
3.   void onDestroy()
```

通过实现上述三个方法，可以监视 Service 生命周期的两个嵌套循环。

整个生命周期从 onCreate()开始，从 onDestroy()结束。像 Activity 一样，一个 AndroidService 生命周期在 onCreate()中执行初始化操作，在 onDestroy()中释放所有用到的资源。例如，后台播放音乐的 Service 可能在 onCreate()创建一个播放音乐的线程，在 onDestroy()中销毁这个线程。

活动生命周期开始于 onStart()。这个方法处理传入到 startService()方法的 intent。音乐服务会打开 intent 查看要播放哪首歌曲，并开始播放。当服务停止的时候，没有方法检测到(没有 onStop()方法)，onCreate()和 onDestroy()用于所有通过 Context. startService()或 Context. bindService()启动的 Service。onStart()只用于通过 startService()开始的 Service。

如果一个 Android Service 生命周期是可以从外部绑定的，它就可以触发以下的方法：

```
1.   IBinder onBind(Intent intent)
2.   boolean onUnbind(Intent intent)
3.   void onRebind(Intent intent)
```

onBind()回调被传递给调用 bindService 的 intent，onUnbind()被 unbindService()中的 intent 处理。如果服务允许被绑定，那么 onBind()方法返回客户端和 Service 的沟通通道。如果一个新的客户端连接到服务，onUnbind()会触发 onRebind()调用。

后续案例将会讲解说明 Service 的回调方法。将通过 startService 和 bindService()启动的 Service 分开了，但是要注意不管它们是怎么启动的，都有可能被客户端连接，因此都有可能触发到 onBind()和 onUnbind()方法。

2. Service 生命周期应用案例

具体实现步骤如下：

(1) 在 Eclipse 中选择 File→New→Android Project，创建一个新的 Android 工程，项目名为 ServiceTestDemo，目标 API 选择 10(即 Android 2.3.3 版本)，应用程序名为 ServiceTestDemo，包名为 com. hisoft. activity，创建的 Activity 的名字为 ServiceTestDemoActivity，最小 SDK 版本根据选择的目标 API 会自动添加为 10。

(2) 修改 res 目录下 layout 文件夹中的 main. xml 代码，添加 4 个 Button 按钮，代码如下：

```
1.   <?xml version="1.0" encoding="utf-8"?>
2.   <LinearLayout xmlns:android="http://schemas.android.com/apk/res/android"
3.       android:orientation="vertical"
4.       android:layout_width="fill_parent"
5.       android:layout_height="fill_parent"
```

```
6.     >
7.   <TextView
8.     android:id="@+id/text"
9.       android:layout_width="fill_parent"
10.      android:layout_height="wrap_content"
11.      android:text="@string/hello"
12.      />
13.  <Button
14.    android:id="@+id/startservice"
15.    android:layout_width="fill_parent"
16.    android:layout_height="wrap_content"
17.    android:text="启动 Service"
18.  />
19.  <Button
20.    android:id="@+id/stopservice"
21.    android:layout_width="fill_parent"
22.    android:layout_height="wrap_content"
23.    android:text="停止 Service"
24.  />
25.  <Button
26.    android:id="@+id/bindservice"
27.    android:layout_width="fill_parent"
28.    android:layout_height="wrap_content"
29.    android:text="绑定 Service"
30.  />
31.  <Button
32.    android:id="@+id/unbindservice"
33.    android:layout_width="fill_parent"
34.    android:layout_height="wrap_content"
35.    android:text="解除 Service"
36.  />
37. </LinearLayout>
```

(3) 在上述包下新建一个 Service，命名为 MyService.java，代码如下：

```
1.  package com.hisoft.service;
2.  import android.app.Service;
3.  import android.content.Intent;
4.  import android.os.Binder;
5.  import android.os.IBinder;
6.  import android.text.format.Time;
7.  import android.util.Log;
8.  public class MyService extends Service {
9.  //定义一个 Tag 标签
10. private static final String TAG="TestService";
11. //创建一个 Binder 类的子类 MyBinder 对象实例，用在 onBind()方法里，以便 Activity
```

```
    //可以获取到
12.  private MyBinder mBinder=new MyBinder();
13.  @Override
14.  public IBinder onBind(Intent intent) {
15.  Log.e(TAG, "-----start IBinder--------~~");
16.  return mBinder;
17.  }
18.  @Override
19.  public void onCreate() {
20.  Log.e(TAG, "-----start onCreate-----");
21.  super.onCreate();
22.  }
23.
24.  @Override
25.  public void onStart(Intent intent, int startId) {
26.  Log.e(TAG, "-----start onStart-----");
27.  super.onStart(intent, startId);
28.  }
29.
30.  @Override
31.  public void onDestroy() {
32.  Log.e(TAG, "-----start onDestroy-----");
33.  super.onDestroy();
34.  }
35.
36.
37.  @Override
38.  public boolean onUnbind(Intent intent) {
39.  Log.e(TAG, "-----start onUnbind-----");
40.  return super.onUnbind(intent);
41.  }
42.
43.  //定义一个获取当前时间的函数,没有实现格式化
44.  public String getSystemTime(){
45.
46.  Time t=new Time();
47.  t.setToNow();
48.  return t.toString();
49.  }
50.
51.  public class MyBinder extends Binder{
52.  MyService getService()
53.  {
54.  return MyService.this;
55.  }
```

```
56. }
57. }
```

(4) 修改 com. hisoft. activity 包下的 ServiceTestDemoActivity. java 文件，代码如下：

```
1.  package com.hisoft.activity;
2.  import android.app.Activity;
3.  import android.content.ComponentName;
4.  import android.content.Context;
5.  import android.content.Intent;
6.  import android.content.ServiceConnection;
7.  import android.os.Bundle;
8.  import android.os.IBinder;
9.  import android.view.View;
10. import android.view.View.OnClickListener;
11. import android.widget.Button;
12. import android.widget.TextView;
13. public class ServiceTestDemoActivity extends Activity implements
    OnClickListener{
14.   private MyService  mMyService;
15.   private TextView mTextView;
16.   private Button startServiceButton;
17.   private Button stopServiceButton;
18.   private Button bindServiceButton;
19.   private Button unbindServiceButton;
20.   private Context mContext;
21.   //创建 ServiceConnection 类的对象,以供 Context.bindService 和
      //context.unBindService()方法作为参数
22.   private ServiceConnection mServiceConnection=new ServiceConnection() {
23.     //当 bindService 时,使 TextView 显示 MyService 里 getSystemTime()方法的
        //返回值
24.     public void onServiceConnected(ComponentName name, IBinder service) {
25.       //TODO Auto-generated method stub
26.       //mMyService=((MyService.MyBinder)service).getService();
27.     }
28.
29.     public void onServiceDisconnected(ComponentName name) {
30.       //TODO Auto-generated method stub
31.
32.     }
33.   };
34.     public void onCreate(Bundle savedInstanceState) {
35.         super.onCreate(savedInstanceState);
36.         setContentView(R.layout.main);
37.         setupViews();
38.     }
```

```
39.
40.     public void setupViews(){
41.
42.       mContext=ServiceDemo.this;
43.       mTextView=(TextView)findViewById(R.id.text);
44.
45.       startServiceButton=(Button)findViewById(R.id.startservice);
46.       stopServiceButton=(Button)findViewById(R.id.stopservice);
47.       bindServiceButton=(Button)findViewById(R.id.bindservice);
48.       unbindServiceButton=(Button)findViewById(R.id.unbindservice);
49.
50.       startServiceButton.setOnClickListener(this);
51.       stopServiceButton.setOnClickListener(this);
52.       bindServiceButton.setOnClickListener(this);
53.       unbindServiceButton.setOnClickListener(this);
54.     }
55.
56.   public void onClick(View v) {
57.     //TODO Auto-generated method stub
58.     if(v==startServiceButton){
59.       Intent i  =new Intent();
60.       i.setClass(ServiceTestDemoActivity.this, MyService.class);
61.       mContext.startService(i);
62.     }else if(v==stopServiceButton){
63.       Intent i  =new Intent();
64.       i.setClass(ServiceTestDemoActivity.this, MyService.class);
65.       mContext.stopService(i);
66.     }else if(v==bindServiceButton){
67.       Intent i  =new Intent();
68.       i.setClass(ServiceTestDemoActivity.this, MyService.class);
69.       mContext.bindService(i, mServiceConnection, BIND_AUTO_CREATE);
70.     }else{
71.       mContext.unbindService(mServiceConnection);
72.     }
73.   }
74.
75. }
```

(5) 修改 AndroidManifest.xml 代码，在<application>标签的根目录下添加注册新创建的 MyService，如第 14 行代码：

```
1.  <?xml version="1.0" encoding="utf-8"?>
2.  <manifest xmlns:android="http://schemas.android.com/apk/res/android"
3.  package="com.hisoft.activity "
4.  android:versionCode="1"
5.  android:versionName="1.0">
```

```
6.    <application android:icon="@drawable/icon" android:label="@string/
      app_name">
7.    <activity android:name=". ServiceTestDemoActivity"
8.    android:label="@string/app_name">
9.    <intent-filter>
10.   <action android:name="android.intent.action.MAIN" />
11.   <category android:name="android.intent.category.LAUNCHER" />
12.   </intent-filter>
13.   </activity>
14.   <service android:name=".MyService" android:exported="true"></service>
15.   </application>
16.   <uses-sdk android:minSdkVersion="10" />
17.   </manifest>
```

(6) 部署工程,并执行上述工程,运行结果如图 3-7 所示。

① 单击“启动 Service”按钮时,程序先后执行了 Service 中的 onCreate()和 onStart()这两个方法,打开日志界面 Logcat 视窗,如图 3-8 所示。

② 按 Home 键进入 Settings(设置)→Applications(应用)→Running Services(正在运行的服务)查看刚才新启动的服务,如图 3-9 所示。

图 3-7　Service 生命周期运行界面

Log (64)

Time		pid	Message
09-15 16:09:43.306	E	610	------start onCreate------
09-15 16:09:43.306	E	610	------start onStart------

图 3-8　启动 Service 调用顺序

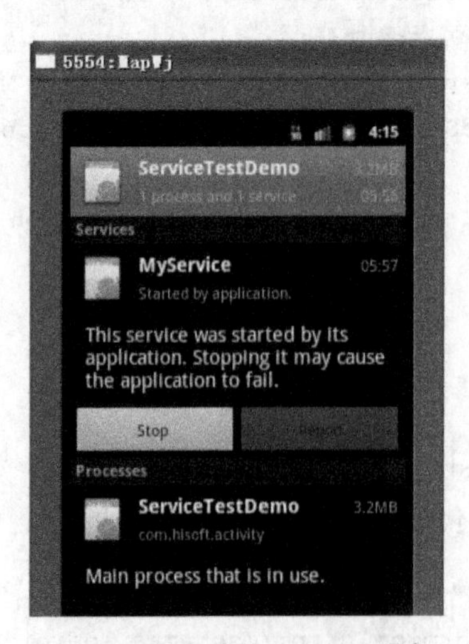

图 3-9　新启动的 Service 服务

③ 单击“停止 Service”按钮时,Service 则执行了 onDestroy()方法,如图 3-10 所示。

Time		pid	Message
09-15 16:16:57.492	E	610	-----start onDestroy-----

图 3-10　停止 Service 服务

④ 再次单击"启动 Service"按钮，然后再单击 bindService 按钮（通常 bindService 都是 bind 已经启动的 Service），查看 Service 的 IBinder()方法执行情况，如图 3-11 所示。

Time		pid	Message
09-15 16:29:56.380	E	713	-----start onCreate-----
09-15 16:29:56.380	E	713	-----start IBinder-----

图 3-11　Service 的 IBinder（ ）方法执行

⑤ 最后单击 unbindService 按钮，则 Service 执行了 onUnbind()方法，如图 3-12 所示。

Time		pid	Message
09-15 16:31:07.054	E	713	-----start onCreate-----
09-15 16:31:07.054	E	713	-----start IBinder-----
09-15 16:31:09.755	E	713	-----start onUnbinder-----
09-15 16:31:09.755	E	713	-----start onDestroy-----

图 3-12　Service 的 onUnbind()方法执行

3. BroadcastReceiver 生命周期

Android 在接收到一个广播 Intent 之后，找到了处理该 Intent 的 BroadcastReceiver，创建一个对象来处理 Intent。然后，调用被创建的 BroadcastReceiver 对象的 onReceive 方法进行处理，然后就撤销这个对象，如图 3-13 所示。只有在执行这个方法时 BroadcastReceiver 才是活动的。当 onReceive()方法执行完，BroadcastReceiver 成为非活动的。

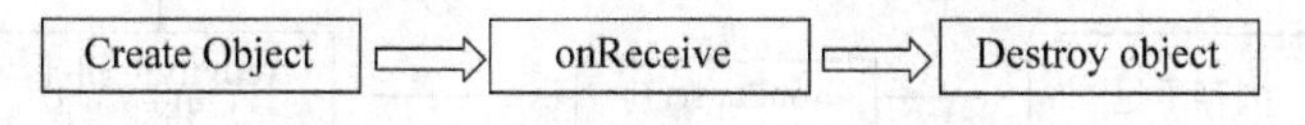

图 3-13　BroadcastReceiver 处理过程

BroadcastReceiver 活动时，它的进程不能被杀掉，而当它的进程中只包含不活动组件时，可能会被系统随时杀掉（其他进程需要消耗它所占用的内存）。解决这个问题的办法是 onReceive()方法启动一个 Android Service 生命周期，让 Service 去做耗时的工作，这样系统就知道此进程中还有活动的工作。

需要注意的是，对象在 onReceive 方法返回之后就被撤销，所以在 onReceive 方法中不宜处理异步的过程。例如弹出对话框与用户交互，可使用消息栏替代。

4. Activity 生命周期

1) Activity 状态

在 Activity 生命周期中，其表现状态有 4 种，分别是活动状态、暂停状态、停止状态和非活动状态。

- Active（活动状态）：是指 Activity 通过 onCreate 被创建。Activity 在用户界面中处于最上层，完全能让用户看到，能够与用户进行交互。

- Pause(暂停状态)：是指当一个 Activity 失去焦点，该 Activity 将进入 Pause 状态。Activity 在界面上被部分遮挡，该 Activity 不再处于用户界面的最上层，且不能够与用户进行交互，系统在内存不足时会将其终止。
- Stop(停止状态)：是指当一个 Activity 被另一个 Activity 覆盖，该 Activity 将进入 Stop 状态。Activity 在界面上完全不能被用户看到，也就是说这个 Activity 被其他 Activity 全部遮挡，系统在需要内存的时候会将其终止。
- 非活动状态不在以上三种状态中的 Activity 则处于非活动状态。

当 Activity 处于 Pause 或者 Stop 状态时，都可能被系统终止并回收。因此，有必要在 onPause 和 onStop 方法中将应用程序运行过程中的一些状态，例如用户输入等保存到持久存储中。如果程序中启动了其他后台线程，也需要注意在这些方法中进行一些处理，例如在线程中打开了一个进度条对话框，如果不在 Pause 或 Stop 中 Cancel 掉线程，则当线程运行完 Cancel 掉对话框时就会抛出异常，如图 3-14 所示。

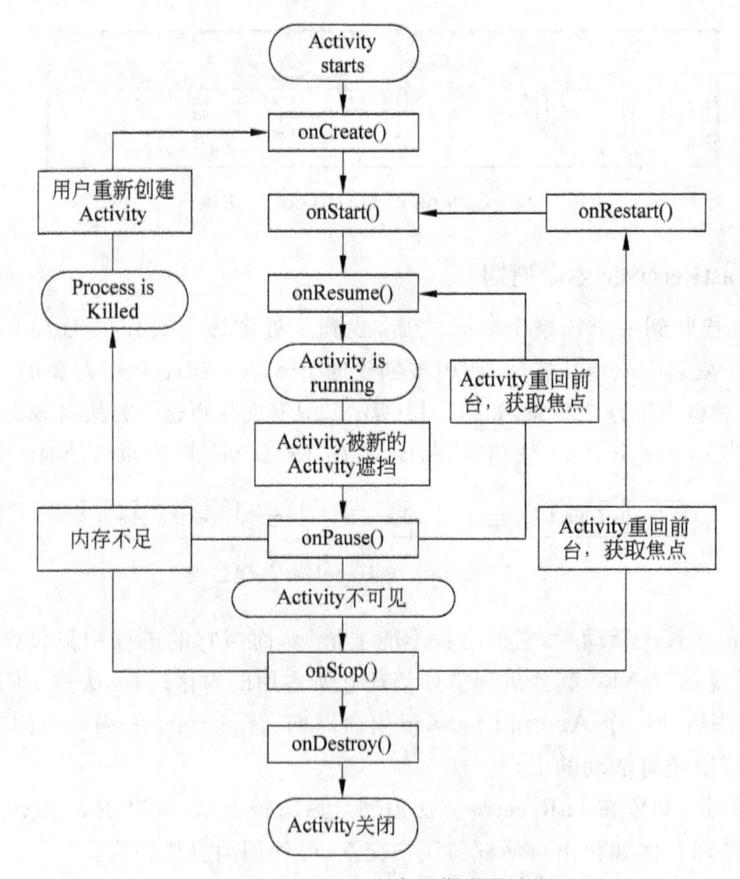

图 3-14　Activity 生命周期调用流程

在 Activity 生命周期中，其事件的回调方法有 7 个，Activity 状态保存/恢复的事件回调方法有两个，如下所示：

```
public class MyActivity extends Activity {
    protected void onCreate(Bundle savedInstanceState);
    public void onRestoreInstanceState(Bundle savedInstanceState);
```

```
    public void onSaveInstanceState(Bundle savedInstanceState);
    protected void onStart();
    protected void onRestart();
    protected void onResume();
    protected void onPause();
    protected void onStop();
    protected void onDestroy();
}
```

具体说明如表 3-2 和表 3-3 所示。

表 3-2　Activity 生命周期的事件回调方法

方　法	是否可终止	说　明
onCreate()	否	Activity启动后第一个被调用的函数，常用来进行 Activity 的初始化，例如创建 View、绑定数据或恢复信息等
onStart()	否	当 Activity 显示在屏幕上时，该函数被调用
onRestart()	否	当 Activity 从停止状态进入活动状态前，调用该函数
onResume()	否	当 Activity 能够与用户交互，接受用户输入时，该函数被调用。此时的 Activity 位于 Activity 栈的栈顶
onPause()	是	当 Activity 进入暂停状态时，该函数被调用。一般用来保存持久的数据或释放占用的资源
onStop()	是	当 Activity 进入停止状态时，该函数被调用
onDestroy()	是	在 Activity 被终止前，即进入非活动状态前，该函数被调用

表 3-3　Activity 状态保存/恢复的事件回调方法

方　法	是否可终止	说　明
onSaveInstanceState()	否	Android 系统因资源不足终止 Activity 前调用该函数，用以保存 Activity 的状态信息，供 onRestoreInstanceState()或 onCreate()恢复之用
onRestoreInstanceState()	否	恢复 onSaveInstanceState()保存的 Activity 状态信息，在 onStart()和 onResume ()之间被调用

2）Activity 生命周期分类

Activity 生命周期指 Activity 从启动到销毁的过程。Activity 的生命周期可分为全生命周期、可视生命周期和活动生命周期。在 Activity 的每个生命周期中包含不同的事件回调方法。

（1）全生命周期。

全生命周期是从 Activity 建立到销毁的全部过程，始于 onCreate()，结束于 onDestroy()。使用者通常在 onCreate()中初始化 Activity 所能使用的全局资源和状态，并在 onDestroy()中释放这些资源。在特殊的情况下，Android 系统会不调用 onDestroy()，而直接终止进程。

（2）可视生命周期。

可视生命周期是 Activity 在界面上从可见到不可见的过程，开始于 onStart()，结束于

onStop()。

在可视生命周期中,onStart()方法一般用来初始化或启动与更新界面相关的资源,onStop()一般用来暂停或停止一切与更新用户界面相关的线程、计时器和服务。onRestart()方法在 onSart()前被调用,用来在 Activity 从不可见变为可见的过程中进行一些特定的处理过程。在可视生命周期中 onStart()和 onStop()一般会被多次调用。此外,onStart()和 onStop()也经常被用来注册和注销 BroadcastReceiver。

(3) 活动生命周期。

活动生命周期是指 Activity 在屏幕的最上层,并能够与用户交互的阶段,开始于 onResume(),结束于 onPause()。在 Activity 的状态变换过程中,onResume()和 onPause()经常被调用,因此应简洁、高效地实现这两个方法。onPause()是第一个被标识为"可终止"的方法,在 onPause()返回后,onStop()和 onDestroy()随时能被 Android 系统终止,onPause()常用来保存持久数据,如界面上用户的输入信息等。

具体 Activity 事件的生命周期划分及回调方法的调用顺序如图 3-15 所示。

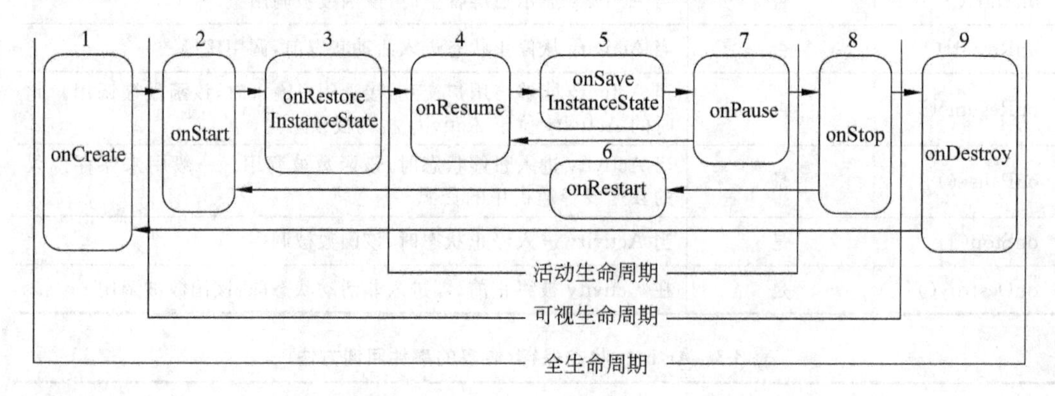

图 3-15 Activity 生命周期划分及事件回调方法的调用顺序

在活动生命周期中,关于 onPause()和 onSaveInstanceState()方法,它们之间的相同之处是这两个方法都可以用来保存界面的用户输入数据,区别在于:

onPause()一般用于保存持久性数据,并将数据保存在存储设备上的文件系统或数据库系统中。

onSaveInstanceState()主要用来保存动态的状态信息,信息一般保存在 Bundle(保存多种格式数据的对象)中,系统在调用 onRestoreInstanceState()和 onCreate()时会同样利用 Bundle 将数据传递给方法。

拓展提示:了解 Android 程序生命周期、组件生命周期、组件状态变化之间的关系,针对 Service 生命周期、Activity 生命周期过程中它自身的组件状态如何进行变化,做一下思考和分析。

3.4 项目案例

学习目标:学习 Activity 生命周期中的事件调用顺序及上述介绍的 Activity 中方法的应用过程,掌握它们的测试及转换过程。

案例描述：使用 Activity 的 onCreate( )、onStart( )、onRestoreInstanceState( )、onResume( )、InstanceState( )、onPause( )、onStop( )、onDestroy( )方法，在不同生命周期中，对相关方法进行调用，并在日志 logcat 中输出其相关调用顺序。

案例要点：采用不同生命周期分类，并就上述方法的调用及调用顺序调试。

案例实施：

(1) 创建一个新的 Android 工程，工程名称为 ActivityLifeCycle，包名称为 com.hisoft.ActivityLifeCycle，Activity 名称为 ActivityLifeCycleActivity，使用 Android 2.3.3 (API Level 10)作为目标平台，创建工程。

(2) 修改 ActivityLifeCycleActivity.java 文件，代码如下：

```
1.  package com.hisoft.ActivityLifeCycle;
2.  import android.app.Activity;
3.  import android.os.Bundle;
4.  import android.util.Log;
5.  public class ActivityLifeCycleActivity extends Activity {
6.  private static String TAG="LIFTCYCLE";
7.  @Override       //完全生命周期开始时被调用,初始化 Activity
8.  public void onCreate(Bundle savedInstanceState) {
9.  super.onCreate(savedInstanceState);
10. setContentView(R.layout.main);
11. Log.i(TAG, "(1) onCreate()");
12. }
13. @Override       //可视生命周期开始时被调用,对用户界面进行必要的更改
14. public void onStart() {
15. super.onStart();
16. Log.i(TAG, "(2) onStart()");
17. }
18. @Override
19. //在 onStart()后被调用,用于恢复 onSaveInstanceState()保存的用户界面信息
20. public void onRestoreInstanceState(Bundle savedInstanceState) {
21. super.onRestoreInstanceState(savedInstanceState);
22. Log.i(TAG, "(3) onRestoreInstanceState()");
23. }
24. @Override
25. //在活动生命周期开始时被调用,恢复被 onPause()停止的用于界面更新的资源
26. public void onResume() {
27. super.onResume();
28. Log.i(TAG, "(4) onResume()");
29. }
30. @Override
31.  //在 onResume()后被调用,保存界面信息
32. public void onSaveInstanceState(Bundle savedInstanceState) {
33. super.onSaveInstanceState(savedInstanceState);
34. Log.i(TAG, "(5) onSaveInstanceState()");
```

```
35. }
36. @Override
37. //在重新进入可视生命周期前被调用,载入界面所需要的更改信息
38. public void onRestart() {
39. super.onRestart();
40. Log.i(TAG, "(6) onRestart()");
41. }
42. @Override
43. //在活动生命周期结束时被调用,用来保存持久的数据或释放占用的资源
44. public void onPause() {
45. super.onPause();
46. Log.i(TAG, " (7) onPause()");
47. }
48. @Override    //在可视生命周期结束时被调用,一般用来保存持久的数据或释放占用的资源
49. public void onStop() {
50. super.onStop();
51. Log.i(TAG, "(8) onStop()");
52. }
53. @Override    //在完全生命周期结束时被调用,释放资源,包括线程、数据连接等
54. public void onDestroy() {
55. super.onDestroy();
56. Log.i(TAG, "(9) onDestroy()");
57. }
58. }
```

上面的程序主要通过在生命周期函数中添加"日志点"的方法进行调试,程序的运行结果将会显示在 LogCat 中。为了观察和分析程序运行结果,在 LogCat 设置过滤器 LifeCycleFilter,过滤方法选择 by Log Tag,过滤关键字为 LIFTCYCLE。

(1) 全生命周期

启动和关闭 ActivityLifeCycleActivity 的 LogCat 输出。

启动 ActivityLifeCycleActivity,按下模拟器的"返回键",关闭 ActivityLifeCycleActivity。LogCat 输出结果如图 3-16 所示。

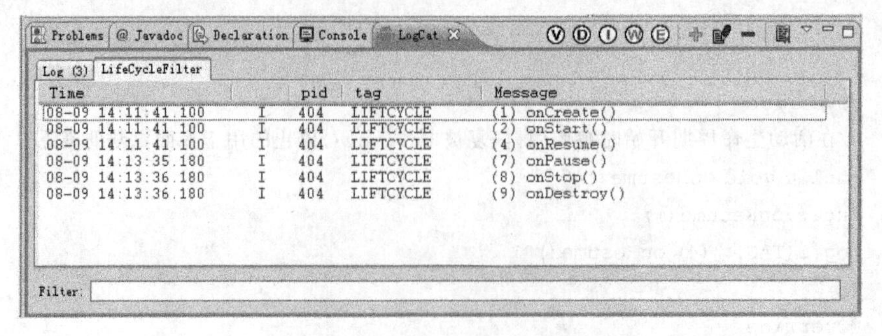

图 3-16 Activity 的全生命周期

从图 3-16 可以看出,方法的调用顺序为 onCreate()→onStart()→onResume()→onPause()→onStop()→onDestroy()。

调用 onCreate()分配资源。

调用 onStart()将 Activity 显示在屏幕上。

调用 onResume()获取屏幕焦点。

调用 onPause()、onStop()和 onDestroy()释放资源并销毁进程。

(2) 可视生命周期

可视生命周期中的状态转换及测试步骤如下：

① 启动 ActivityLifeCycle。

② 按“呼出/接听”键启动内置的拨号程序。

③ 通过“返回”键退出拨号程序。

④ ActivityLifeCycle 重新显示在屏幕中。

可视生命周期的 LogCat 输出结果如图 3-17 所示。

Problems | @ Javadoc | Declaration | Console | LogCat

Log (34) | LifeCycleFilter

Time		pid	tag	Message
08-09 14:23:50.330	I	404	LIFTCYCLE	(1) onCreate()
08-09 14:23:50.441	I	404	LIFTCYCLE	(2) onStart()
08-09 14:23:50.441	I	404	LIFTCYCLE	(4) onResume()
08-09 14:24:15.710	I	404	LIFTCYCLE	(5) onSaveInstanceState()
08-09 14:24:15.720	I	404	LIFTCYCLE	(7) onPause()
08-09 14:24:22.260	I	404	LIFTCYCLE	(8) onStop()
08-09 14:24:37.080	I	404	LIFTCYCLE	(6) onRestart()
08-09 14:24:37.080	I	404	LIFTCYCLE	(2) onStart()
08-09 14:24:37.101	I	404	LIFTCYCLE	(4) onResume()

Filter:

图 3-17 可视生命周期中方法调用过程

其方法的调用顺序为 onSaveInstanceState()→onPause()→onStop()→onRestart()→onStart()→onResume()。每个方法的作用如下：

调用 onSaveInstanceState()函数保存 Activity 状态。

调用 onPause()和 onStop()，停止对不可见 Activity 的更新。

调用 onRestart()恢复载入界面上需要更新的信息。

调用 onStart()和 onResume()重新显示 Activity，并接受用户交互。

(3) 开启 IDA 的可视生命周期

开启步骤：选择 Dev Tools→Development Settings→Immediately destroy activities (IDA)，开启 IDA，如图 3-18 所示。

Problems | @ Javadoc | Declaration | Console | LogCat

Log (17) | LifeCycleFilter

Time		pid	tag	Message
08-09 14:53:05.302	I	404	LIFTCYCLE	(4) onResume()
08-09 14:53:12.150	I	404	LIFTCYCLE	(5) onSaveInstanceState()
08-09 14:53:12.150	I	404	LIFTCYCLE	(7) onPause()
08-09 14:53:14.479	I	404	LIFTCYCLE	(8) onStop()
08-09 14:53:14.479	I	404	LIFTCYCLE	(9) onDestroy()
08-09 14:53:20.009	I	404	LIFTCYCLE	(1) onCreate()
08-09 14:53:20.079	I	404	LIFTCYCLE	(2) onStart()
08-09 14:53:20.230	I	404	LIFTCYCLE	(3) onRestoreInstanceState()
08-09 14:53:20.273	I	404	LIFTCYCLE	(4) onResume()

Filter:

图 3-18 开启 IDA 可视生命周期的调用顺序

开启 IDA 的可视生命周期的方法调用依次顺序为 onSaveInstanceState()→onPause()→onStop()→onDestroy()→onCreate()→onStart()→onRestoreInstanceState()→onResume()。每个方法的作用如下：

调用 onRestoreInstanceState()恢复 Activity 销毁前的状态。

其他的函数调用顺序与程序启动过程的调用顺序相同。

(4) 活动生命周期

活动生命周期的测试步骤及 LogCat 输出如下：

① 启动 ActivityLifeCycle。

② 通过“挂断”键使模拟器进入休眠状态。

③ 通过“挂断”键唤醒模拟器。

LogCat 的输出结果如图 3-19 所示。

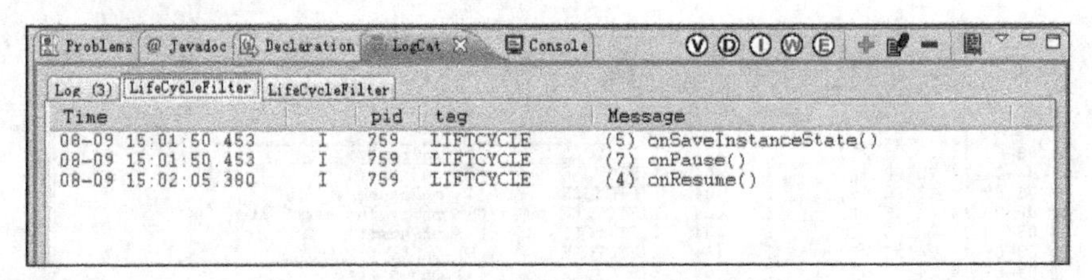

图 3-19 活动生命周期中方法调用过程

方法调用顺序依次如下：onSaveInstanceState()→onPause()→onResume()。每个方法的作用如下：

调用 onSaveInstanceState()保存 Activity 的状态。

调用 onPause()停止与用户交互。

调用 onResume()恢复与用户的交互。

1. 简答题

(1) 简述 AndroidManefiest.xml 及 R.java 文件的作用。

(2) Android 应用程序由哪些部分构成？它们之间的关系是什么？

(3) Android 进程包含哪些？它们之间的优先级别是什么关系？

(4) 简述 Service 生命周期整个过程包含哪些方面。

(5) Activity 生命周期中表现状态分为哪些？其涉及的回调方法有哪些？它们与生命周期之间有什么关系？

2. 完成下面的实训项目

要求：使用两种方式创建 Service 实例，完成 Service 生命周期整个过程的测试，并使用 LogCat 显示 TAG 为自己姓名拼音的生命周期调用。

第4章 Android UI（用户界面）基础

学习目标

本章主要介绍 Android UI 设计原则、UI 框架及视图树模型、UI 控件分类及常用的 Andoird UI 布局，同时就不同的布局 LinearLayout、RelativeLayout、TableLayout、FrameLayout、AbsoluteLayout 应用进行案例讲解。使读者通过本章的学习，能够达到以下知识要点的学习和掌握：

（1）掌握 Android UI 的设计原则、UI 框架及 MVC 设计。

（2）掌握视图树模型、Android UI 控件类的分类。

（3）掌握控件类之间的关系及常用方法。

（4）掌握各种界面布局的特点和使用方法。

4.1 Android UI 简介

用户界面(User Interface，UI)是 Android 系统和用户之间进行信息交换的媒介，它实现了信息的内部表示形式与用户可以接受的形式之间的转换。在计算机出现的早期，批处理界面(1945—1968 年)和命令行界面(1969—1983 年)就已经被广泛地使用。

目前，图形用户界面(Graphical User Interface，GUI)采用图形方式与用户进行交互，是当前用户界面比较流行的一种设计和应用。未来的用户界面的设计发展趋势是将虚拟现实技术应用到界面设计中，使用户能够摆脱键盘与鼠标的交互方式，而通过动作、语言，甚至是脑电波来控制计算机。

在实践中，由于不同移动终端设备厂商生产的屏幕尺寸大小有可能不同，因此针对 Android UI 的设计与 PC 终端的 UI 设计有很大的不同，在设计手机用户界面时，应全面考虑手持移动终端的特点，并坚持界面设计与程序逻辑

松散耦合的理念。因此,在 Android UI 设计中应解决的问题和坚持的原则有:

(1) 需要界面设计与程序逻辑完全分离,这样不仅有利于它们的并行开发,而且在后期修改界面时也不用再次修改程序的逻辑代码。

(2) 根据不同型号手机的屏幕解析度、尺寸和纵横比各不相同,自动调整界面上部分控件的位置和尺寸,避免因为屏幕信息的变化而出现显示错误。

(3) 能够合理利用较小的屏幕显示空间构造出符合人机交互规律的用户界面,避免出现凌乱、拥挤的用户界面。

Android 系统已经解决了前两个问题,使用 XML 文件描述用户界面。资源中的资源文件分为不同的类别独立保存在不同类别的资源文件夹中(如第 3 章讲述的 res 目录下的 drawable-hdpi 等)。对用户界面描述非常灵活,允许不明确定义界面元素的位置和尺寸,仅声明界面元素的相对位置和粗略尺寸。

4.2 Android UI 框架

框架(Framework)的发展已经有 40 多年历史了,最早的是 1980 年的 Smalltalk 语言的 MVC,在框架发展过程中,典型的有 MVC Framework(20 世纪 80 年代初期)、MacApp Framework(20 世纪 80 年代中期)、MFC Framework(20 世纪 90 年代初期)、San Francisco Framework(20 世纪 90 年代中期)、. Net Framework(2000 年)、Android 框架(2007 年)。Android 用户界面框架(Android UI Framework)采用的是比较流行的 MVC(Model View Controller)框架模型,MVC 模型提供了处理用户输入的控制器(Controller),显示用户界面和图像的视图(View),以及保存数据和代码的模型(Model),如图 4-1 所示。

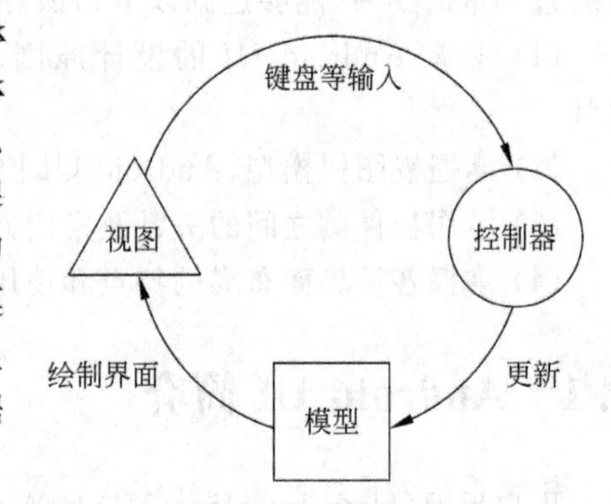

图 4-1 Android MVC 框架模型

4.2.1 Android 与 MVC 设计

MVC 即把一个应用的输入、处理、输出流程按照 Model、View、Controller 的方式进行分离,这样一个应用被分成三个层——模型、视图、控制器。

在 Android 系统中,视图(View)代表用户交互界面,一个应用可能有很多不同的视图。MVC 设计模式对于视图的处理仅限于视图上数据的采集和处理,以及用户的请求,而不包括在视图上的业务流程的处理。业务流程的处理交予模型(Model)处理。比如一个订单的视图只接收来自模型的数据并显示给用户,以及将用户界面的输入数据和请求传递给控制和模型。

模型代表业务逻辑 Bean,就是业务流程/状态的处理以及业务规则的制定。业务流程的处理过程对其他层来说是黑箱操作,模型接收视图请求的数据,并返回最终的处理结果。业务模型的设计可以说是 MVC 最主要的核心。

控制器(Controller)可以理解为从用户接收请求,将模型与视图匹配在一起,共同完成用户的请求。划分控制层的作用明确了它就是一个分发器,选择什么样的模型,选择什么样的视图,可以完成什么样的用户请求。控制层并不做任何的数据处理。例如,用户单击一个连接,控制层接受请求后,并不处理业务信息,它只把用户的信息传递给模型,告诉模型做什么,选择符合要求的视图返回给用户。因此,一个模型可能对应多个视图,一个视图可能对应多个模型。在 Android 系统中,由 Activity 充当控制器的角色。

从开发者的角度,MVC 把应用程序的逻辑层与视图层(界面)完全分开,最大的好处是界面设计人员可以直接参与到界面开发,程序员就可以把精力放在逻辑层处理上,有利于提高效率和明确任务分工。

MVC 模型中的控制器能够接受并响应程序的外部动作,如按键动作或触摸屏动作等。控制器使用队列处理外部动作,每个外部动作作为一个对立的事件被加入队列中,然后 Android 用户界面框架按照"先进先出"的规则从队列中获取事件,并将这个事件分配给所对应的事件处理函数。

4.2.2 视图树模型(View 和 Viewgroup)

Android 用户界面框架采用视图树(View Tree)模型。即在 Android 用户界面框架中界面元素是以一种树型结构组织在一起,并称为视图树。视图树由 View 和 ViewGroup 构成,如图 4-2 所示。

Android 系统会依据视图树的结构从上至下绘制每一个界面元素。每个元素负责对自身的绘制,如果元素包含子元素,该元素会通知其下所有子元素进行绘制。

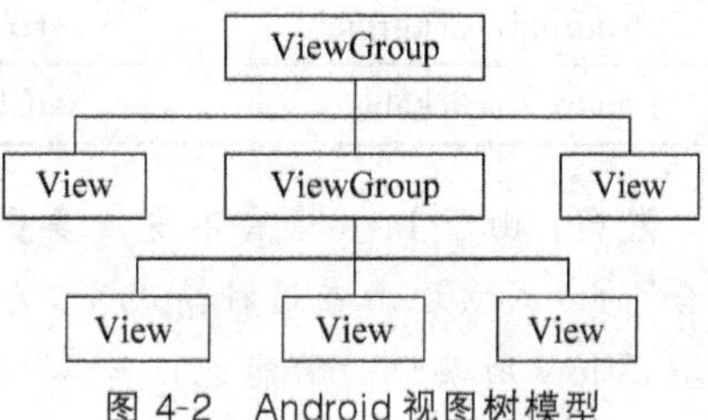

图 4-2 Android 视图树模型

4.3 Android UI 控件类简介

在 Android 中使用各种控件可以实现 UI 的外观,而 View 是各类控件的基类,是创建交互式的图形用户界面的基础。ViewGroup 是布局管理器(Layout)及 View 容器的基类。下面就 View 类和 ViewGroup 类进行简述。

4.3.1 View 类

1. View 类

View 是界面的最基本的可视单元,呈现了最基本的 UI 构造块。一个视图占据屏幕上的一个方形区域,存储了屏幕上特定矩形区域内所显示内容的数据结构,并能够实现所占据区域的界面绘制、焦点变化、用户输入和界面事件处理等功能。

View 是 Android 中最基础的类之一,所有在界面上的可见元素都是 View 的子类,类的视图结构是 android. view. View,如 Button、RadioButton 和 CheckBox 等都是通过继承 View 的方法来实现的。通过继承 View,可以很方便地定制出有个性的控件。

View 有众多的扩展者,它们大部分是在 android. widget 包中,这些继承者实际上就是 Android 系统中的"控件"。View 的直接继承者包括文本视图(TextView)、图像视图(ImageView)和进度条(ProgressBar)等。它们各自又有众多的继承者。每个控件除了继承父类功能之外,一般还具有自己的公有方法、保护方法和 XML 属性等。

一般情况下,在 Android 布局文件中,可以通过设置各种控件的属性实现 UI 的外观,然后在 Java 文件中实现对各种控件的控制动作。控件类的名称也是它们在布局文件 XML 中使用的标签名称。

2. View 类通用行为和属性

View 是 Android 中所有控件类的基类，因此 View 中一些内容是所有控件类都具有的通用行为和属性。

View 作为各种控件的基类，其 XML 属性所有控件通用，几个重要的 XML 属性如表 4-1 所示。

表 4-1　View 中几个重要 XML 属性及其对应的方法

XML 属性名称	Java 中对应方法	描　述
android:visibility	setVisibility(int)	描述 View 的可见性
android:id	setId(int)	设置 View 的标识符，一般通过 findViewById 方法获取
android:background	setBackgroundResource(int)	设置背景
android:clickable	setClickable(boolean)	设置 View 响应单击事件

注意：由于 Java 语言不支持多重继承，因此 Android 控件不可能以基本功能的"排列组合"的方式实现。在这种情况下，为了实现功能的复用，基类的功能比较多，作为控件的父类，View 所实现的功能也较多。

4.3.2　ViewGroup 类

ViewGroup 是一个特殊的 View，它继承于 android.view.View。它的功能就是装载和管理下一层的 View 对象或 ViewGroup 对象，也就是说它是一个容纳其他元素的容器。ViewGroup 是布局管理器(Layout)及 View 容器的基类，如图 4-3 所示。

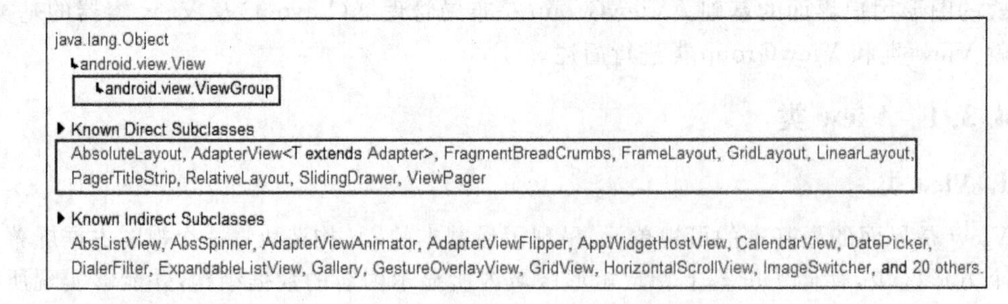

图 4-3　ViewGroup 类继承结构

ViewGroup 中还定义了一个嵌套类 ViewGroup.LayoutParams。这个类定义了一个显示对象的位置、大小等属性，View 通过 LayoutParams 中的这些属性值来告诉父级视图它们将如何放置。

ViewGroup 是一个抽象类，所以真正充当容器的是它的子类们，也就是下面要重点讲述的 LinearLayout、AbsoluteLayout 和 FrameLayout 等布局管理器。

ViewGroup 的功能：一个是承载界面布局，另一个是承载具有原子特性的重构模块。

4.3.3 界面控件

Android 系统的界面控件分为定制控件和系统控件两种。

定制控件是用户独立开发的控件,或通过继承并修改系统控件后所产生的新控件。能够为用户提供特殊的功能或与众不同的显示需求方式。

系统控件是 Android 系统提供给用户已经封装的界面控件。提供在应用程序开发过程中常见功能控件。系统控件更有利于帮助用户进行快速开发,同时能够使 Android 系统中应用程序的界面保持一致性。

常见的系统控件包括文本控件(TextView、EditText)、按钮控件(Button、ImageButton)、单选和复选按钮控件(Checkbox、RadioButton)、Spinner、ListView 和 TabHost 等。界面控件的介绍将在后续的章节中进行详述,此处不再赘述。

4.4 Android UI 布局

UI(用户界面)布局(Layout)是用户界面结构的描述,定义了界面中所有的元素、结构及它们之间的相互关系。

Android 下创建界面布局的方法有三种:

(1) XML 方式。使用 XML 文件描述界面布局。

(2) 程序代码创建。在程序运行时动态添加或修改界面布局。

(3) XML 和程序代码创建相结合。

一般都采用 XML 布局文件创建用户界面,当然,用户既可以独立使用任何一种声明界面布局的方式,也可以同时使用两种方式。与使用代码方式相比较,使用 XML 文件声明界面布局的优势主要有以下三种:

(1) 将程序的表现层和控制层分离。

(2) 在后期修改用户界面时,无需更改程序的源代码。

(3) 用户还能够通过可视化工具直接看到所设计的用户界面,有利于加快界面设计的过程,并且为界面设计与开发带来极大的便利性。

在 Android 系统中,布局管理器是控件的容器,每个控件在窗体中都有具体的位置和大小,在窗体中摆放各种控件时,很难判断其具体位置和大小。不过,使用 Android 布局管理器可以很方便地控制各控件的位置和大小。Android 提供了 5 种布局管理器来管理控件,它们是线性布局管理器(LinearLayout)、表格布局管理器(TableLayout)、帧布局管理器(FrameLayout)、相对布局管理器(RelativeLayout)和绝对布局管理器(AbsoluteLayout)。布局管理器的作用主要有以下三种:

(1) 适应不同的移动设备屏幕分辨率。

(2) 方便横屏和竖屏之间相互切换。

(3) 管理每个控件的大小以及位置。

4.4.1 线性布局

线性布局(LinearLayout)是一种常用的界面布局,也是 RadioGroup、TabWidget、

TableLayout、TableRow 和 ZoomControls 类的父类。在线性布局中，LinearLayout 可以让它的子元素以垂直或水平的方式排成一行（不设置方向的时候默认按照垂直方向排列）。

如果是垂直排列，则每行仅包含一个界面元素，如图 4-4 所示。

如果是水平排列，则每列仅包含一个界面元素，如图 4-5 所示。

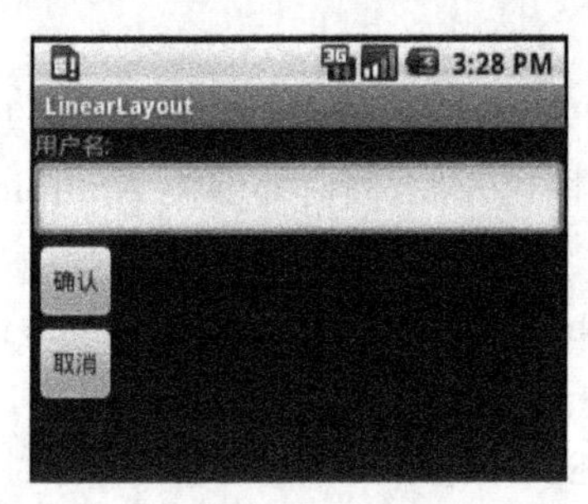

图 4-4　垂直排列

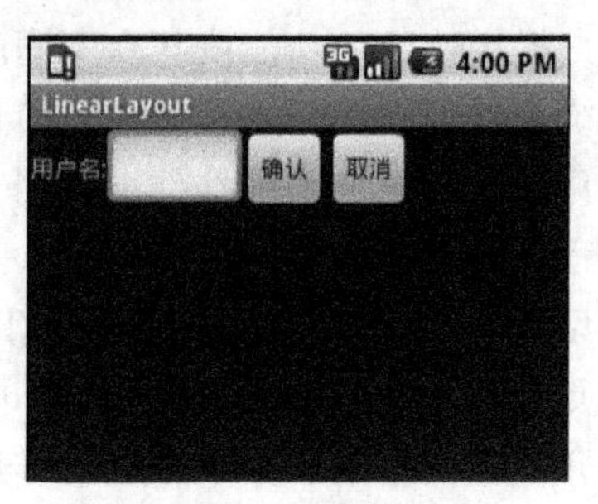

图 4-5　水平排列

LinearLayout 常用属性及对应方法如表 4-2 所示。在 setGravity(int)方法中，可以设置参数，设置线性布局中放置的对象元素的排列对齐方式。如果需要设置多个或者组合设置对齐方式，属性常量由“|”分割。具体属性及描述如表 4-3 所示。

表 4-2　LinearLayout 常用属性及对应方法

XML 属性名	对应的方法	描　述
android:divider	setDividerDrawable(Drawable)	设置用于在按钮间垂直分割的可绘制对象
android:gravity	setGravity(int)	指定在对象内部，横纵方向上如何放置对象的内容
android:orientation	setOrientation(int)	设置线性布局管理器内组件的排列方式，可以设置为 horizontal(水平排列)、vertical(垂直排列、默认值)两个值的其中之一
android:weightSum		定义最大的权值和

表 4-3　setGravity(int)方法可取的属性常量及描述

属性常量	值	描　述
top	0x30	将对象放在其容器的顶部，不改变其大小
bottom	0x50	将对象放在其容器的底部，不改变其大小
left	0x03	将对象放在其容器的左侧，不改变其大小
right	0x05	将对象放在其容器的右侧，不改变其大小
center_vertical	0x10	将对象纵向居中，不改变其大小
fill_vertical	0x70	必要的时候增加对象的纵向大小，以完全充满其容器
center_horizontal	0x01	将对象横向居中，不改变其大小
fill_horizontal	0x07	必要的时候增加对象的横向大小，以完全充满其容器
center	0x11	将对象横纵居中，不改变其大小
fill	0x77	必要的时候增加对象的横纵向大小，以完全充满其容器

续表

属性常量	值	描述
clip_vertical	0x80	附加选项,用于按照容器的边来剪切对象的顶部和/或底部的内容。剪切基于其纵向对齐设置：顶部对齐时,剪切底部;底部对齐时,剪切顶部;除此之外剪切顶部和底部
clip_horizontal	0x08	附加选项,用于按照容器的边来剪切对象的左侧和/或右侧的内容。剪切基于其横向对齐设置：左侧对齐时,剪切右侧;右侧对齐时,剪切左侧;除此之外剪切左侧和右侧

4.4.2 线性布局应用案例

4.4.1 节中对线性布局知识进行了简要介绍,本节将就线性布局在实际程序中如何使用和基本的实现流程进行详述。本案例使用线性布局垂直排列实现了 5 个 button 按钮排列的功能,具体实现步骤如下：

(1) 在 Eclipse 中选择 File→New→Android Project,创建一个新的 Android 工程,项目名为 LinearLayoutDemo,目标 API 选择 10(即 Android 2.3.3 版本),应用程序名为 LinearLayoutDemo,包名为 com.hisoft.activity,创建的 Activity 的名字为 LinearLayoutActivity,最小 SDK 版本根据选择的目标 API 会自动添加为 10,如图 4-6 所示。

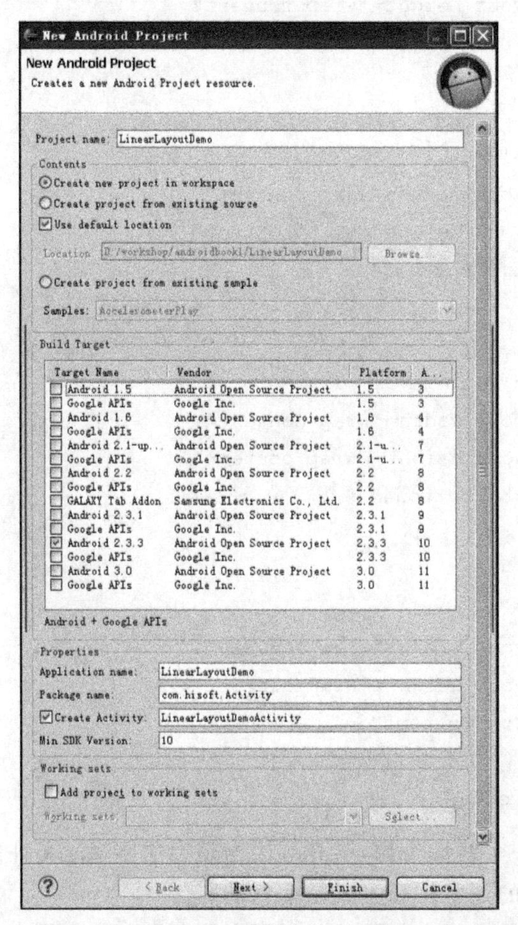

图 4-6 创建 Android 新工程

(2) 打开项目工程中 res→layout 目录下的 main. xml 文件，设置布局，添加 5 个 button 按钮，按钮分别显示 button1、button2 等，代码如下：

```
1.  <?xml version="1.0" encoding="utf-8"?>
2.  <LinearLayout xmlns:android="http://schemas.android.com/apk/res/android"
3.  android:orientation="vertical"
4.  android:layout_width="fill_parent"
5.  android:layout_height="fill_parent"
6.  >
7.  <Button
8.  android:id="@+id/b1"
9.  android:layout_width="wrap_content"
10.   android:layout_height="wrap_content"
11.   android:text="button1"
12.   />
13.   <Button
14.   android:id="@+id/b2"
15.   android:layout_width="wrap_content"
16.   android:layout_height="wrap_content"
17.   android:text="button2"
18.   />
19.   <Button
20.   android:id="@+id/b3"
21.   android:layout_width="wrap_content"
22.   android:layout_height="wrap_content"
23.   android:text="button3"
24.   />
25.   <Button
26.   android:id="@+id/b4"
27.   android:layout_width="wrap_content"
28.   android:layout_height="wrap_content"
29.   android:text="button4"
30.   />
31.   <Button
32.   android:id="@+id/b5"
33.   android:layout_width="wrap_content"
34.   android:layout_height="wrap_content"
35.   android:text="button5"
36.   />
37.   </LinearLayout>
```

第 2～6 行声明了一个线性布局，第 2 行代码是声明 XML 文件的根元素为线性布局，第 3 行设置了线性布局的元素排列方式是垂直排列。

第 4、5 行设置了线性布局在所属的父容器中的布局方式为横向和纵向填充父容器，表示线性布局宽度等于父控件的宽度,就是将线性布局在横向和纵向上占据父控件的所有空间。

第 7～12 行声明了一个 button 按钮控件,第 8 行设置了 ID 为 b1,第 11 行设置了 button 按钮显示为 button1。

第 9 行设置了 button 按钮控件在父容器中的布局方式为只占据自身大小的空间,表示线性布局宽度等于所有子控件的宽度总和,也就是线性布局的宽度刚好将所有子控件包含其中。

第 10 行设置了 button 按钮控件在父容器中的布局方式为只占据自身大小的空间,表示线性布局高度等于所有子控件的高度总和,也就是线性布局的高度刚好将所有子控件包含其中。

(3) src 目录下 com. hisoft. activity 包下的 LinearLayoutActivity. java 文件和 res→values 目录下的 strings. xml 文件都暂不做修改。部署运行项目工程,项目运行效果如图 4-7 所示。

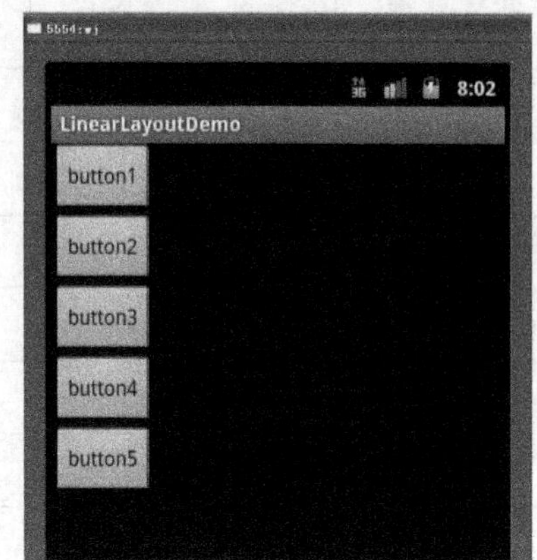

图 4-7 线性布局运行结果

注意:LinearLayout 中的控件按顺序从左到右或从上到下依次排列。

建立横向线性布局与纵向线性布局相似,只需注意将线性布局的 Orientation 属性的值设置为 horizontal 即可。

4.4.3 相对布局

相对布局(RelativeLayout)是一种非常灵活的布局方式,按照控件之间所指定的相对位置参数来自动对控件进行排列,确定界面中所有元素的布局位置。让子元素指定它们相对于其他元素的位置(通过 ID 来指定)或相对于父布局对象,跟 AbsoluteLayout 这个绝对坐标布局是相反的。实际开发中,一般推荐使用这种布局。

相对布局的特点:能够最大程度保证在各种屏幕类型的手机上正确显示界面布局。

在 RelativeLayout 里的控件包含丰富的排列属性,总的可以分为三类:

(1) 以 parent(父控件)为参照物的 XML 属性,属性取值可以为 true 或者 false,如表 4-4 所示。

表 4-4 parent(父控件)为参照物的 XML 属性及描述

XML 属性名称	描 述
android:layout_alignParentTop	如果为 true,将该控件的顶部与其父控件的顶部对齐
android:layout_alignParentBottom	如果为 true,将该控件的底部与其父控件的底部对齐
android:layout_alignParentLeft	如果为 true,将该控件的左部与其父控件的左部对齐
android:layout_alignParentRight	如果为 true,将该控件的右部与其父控件的右部对齐
android:layout_centerHorizontal	如果为 true,将该控件置于父控件的水平居中位置
android:layout_centerVertical	如果为 true,将该控件置于父控件的垂直居中位置
android:layout_centerInParent	如果为 true,将该控件置于父控件的中央

(2) 要指定参照物的 XML 属性,layout_alignBottom、layout_toLeftOf、layout_above、layout_alignBaseline 系列和其他控件 ID,如表 4-5 所示。

表 4-5　参照物的 XML 属性及描述

XML 属性名称	描　　述
android:layout_alignBaseline	将该控件的 baseline 与给定 ID 的 baseline 对齐
android:layout_alignTop	将该控件的顶部边缘与给定 ID 的顶部边缘对齐
android:layout_alignBottom	将该控件的底部边缘与给定 ID 的底部边缘对齐
android:layout_alignLeft	将该控件的左边缘与给定 ID 的左边缘对齐
android:layout_alignRight	将该控件的右边缘与给定 ID 的右边缘对齐
android:layout_above	将该控件的底部置于给定 ID 的控件之上
android:layout_below	将该控件的底部置于给定 ID 的控件之下
android:layout_toLeftOf	将该控件的右边缘与给定 ID 的控件左边缘对齐
android:layout_toRightOf	将该控件的左边缘与给定 ID 的控件右边缘对齐

(3) 指定移动像素的 XML 属性,如表 4-6 所示。

表 4-6　移动像素的 XML 属性及描述

XML 属性名称	描　　述	XML 属性名称	描　　述
android:layout_marginTop	上偏移的值	android:layout_marginLeft	左偏移的值
android:layout_marginBottom	下偏移的值	android:layout_marginRight	右偏移的值

注意事项:

(1) 使用 RelativeLayout 的时候,尽量减少程序运行时做控件布局的更改,因为 RelativeLayout 里面的属性之间很容易冲突。

(2) 在相对布局的大小和它的子控件位置之间要避免出现循环依赖,如设置相对布局高度属性为 WRAP_CONTENT,就不能再设置它的子控件高度属性为 ALIGN_PARENT_BOTTOM。

4.4.4　相对布局应用案例

在 4.4.3 节中对 RelativeLayout 及属性进行了简要介绍后,本节将就 RelativeLayout 在实际程序中的使用和基本的实现进行详述。本案例使用文本控件 TextView 和 EditView 及两个 button 按钮控件实现相对排列的功能,具体实现步骤如下:

(1) 如 4.4.3 节一样,在 Eclipse 中选择 File→New→Android Project,创建一个新的 Android 工程,项目名为 RelativeLayoutDemo,目标 API 选择 10(即 Android 2.3.3 版本),应用程序名为 RelativeLayoutDemo,包名为 com. hisoft. activity, 创建的 Activity 的名字为 RelativeLayoutActivity,最小 SDK 版本根据选择的目标 API 会自动添加为 10,如图 4-8

所示。

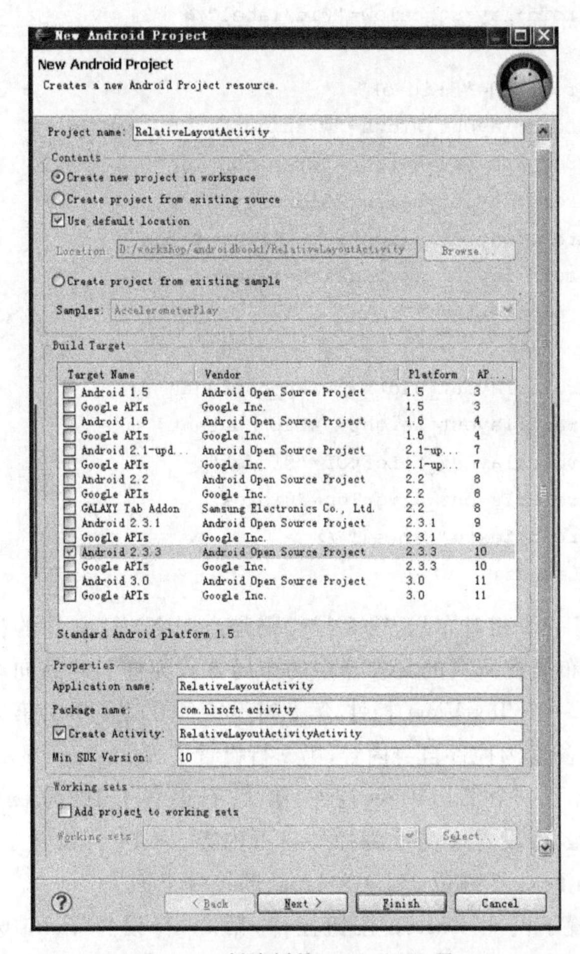

图 4-8 创建新的 Android 工程

(2) 打开项目工程中 res→layout 目录下的 main. xml 文件,设置布局,添加 TextView、EditText 控件和两个 button 按钮控件,代码如下:

```
1.  <?xml version="1.0" encoding="utf-8"?>
2.  <RelativeLayout xmlns:android="http://schemas.android.com/apk/res/android"
3.      android:layout_width="fill_parent"
4.      android:layout_height="fill_parent">
5.      <TextView
6.          android:id="@+id/label"
7.          android:layout_width="fill_parent"
8.          android:layout_height="wrap_content"
9.          android:text="请输入:"/>
10.     <EditText
11.         android:id="@+id/entry"
12.         android:layout_width="fill_parent"
13.         android:layout_height="wrap_content"
```

```
14.            android:background="@android:drawable/editbox_background"
15.            android:layout_below="@id/label"/>
16.        <Button
17.            android:id="@+id/ok"
18.            android:layout_width="wrap_content"
19.            android:layout_height="wrap_content"
20.            android:layout_below="@id/entry"
21.            android:layout_alignParentRight="true"
22.            android:layout_marginLeft="10dip"
23.            android:text="OK" />
24.        <Button
25.            android:layout_width="wrap_content"
26.            android:layout_height="wrap_content"
27.            android:layout_toLeftOf="@id/ok"
28.            android:layout_alignTop="@id/ok"
29.            android:text="Cancel" />
30.    </RelativeLayout>
```

第 2～4 行声明了一个相对布局，第 2 行代码是声明 XML 文件的根元素为相对布局，第 3、4 行设置了相对布局在所属的父容器中的布局方式为横向和纵向填充父容器。

第 5～9 行声明了一个 TextView 控件，第 6 行设置了 ID 为 label，第 7 行设置宽度为填充父容器，第 8 行设置了高度为控件自身内容，第 9 行设置了 TextView 显示的文字内容。

第 10～15 行设置了一个 EditView 控件，第 11 行设置了 ID 为 entry，第 12 行设置宽度为填充父容器，第 13 行设置了高度为控件自身内容，第 14 行设置了背景，第 15 行设置了 ID 为 entry 的 EditText 控件位于 ID 为 label 的控件下面。

第 16～29 行设置了两个 Button 按钮控件，第 20 行设置了 button 按钮位于 ID 为 entry 的控件下方，第 21 行设置了该控件的右部与其父控件的右部对齐，第 22 行设置了按钮从右边框左偏移 10 个 dip，第 23 行设置了按钮显示文字为 OK。

第 27 行设置了 Cancel 按钮控件的右边缘与给定 ID 为 OK 的控件左边缘对齐。

第 28 行设置了 Cancel 按钮控件的顶部边缘与给定 ID 为 OK 的控件的顶部边缘对齐。

第 29 行设置了按钮显示文字为 Cancel。

(3) src 目录下 com. hisoft. activity 包下的 RelativeLayoutActivity. java 文件和 res→values 目录下的 strings. xml 文件都暂不做修改。部署运行 RelativeLayoutDemo 项目工程，项目运行效果如图 4-9 所示。

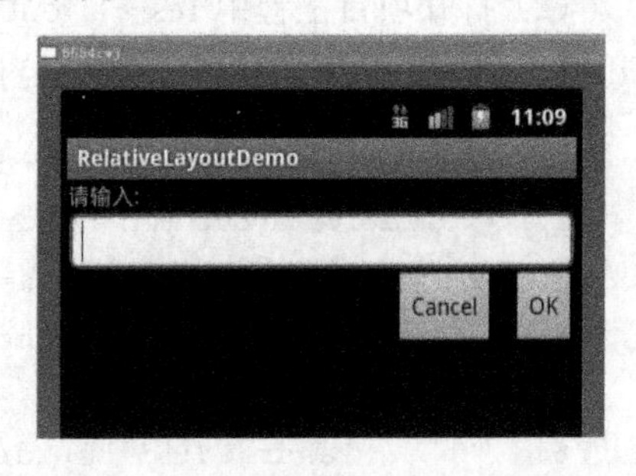

图 4-9 RelativeLayoutDemo 项目运行结果

4.4.5 表格布局

表格布局(TableLayout)也是一种常用的界面布局，采用行，列的形式来管理 UI 组件，它是将屏幕划分成网格单元(网格的边界对用户是不可见的),然后通过指定行和列的方式将界面元素添加到网格中。它并不需要明确地声明包含多少行、列,而是通过添加TableRow、其他组件来控制表格的行数和列数。每次向 TableLayout 中添加一个TableRow,该 TableRow 就是一个表格行,TableRow 也是容器,因此它可以不断地添加其他组件,每添加一个子组件该表格就增加一列。每一行可以有 0 个或多个单元格,每个单元格就是一个 View,一个 Table 中可以有空的单元格,单元格可以像在 HTML 中使用的方式一样,合并多个单元格,跨越多列。这些 TableRow,单元格不能设置 layout_width,宽度属性默认是 fill_parent,只有高度 layout_height 可以自定义,默认值是 wrap_content。

在表格布局中,一个列的宽度由该列中最宽的单元格决定。表格布局支持嵌套,可以将另一个表格布局放置在前一个表格布局的网格中,也可以在表格布局中添加其他界面布局,如线性布局、相对布局等。常用具体属性及方法如表 4-7 所示。

表 4-7 表格布局常用属性及相关方法

属性名称	相关方法	描述
android:collapseColumns	setColumnCollapsed(int,boolean)	设置指定列为 collapse,列索引从 0 开始
android:shrinkColumns	setShrinkAllColumns(boolean)	设置指定列为 shrink,列索引从 0 开始
android:stretchColumns	setStretchAllColumns(boolean)	设置指定列为 stretch,列索引从 0 开始

在表格布局中,如果一个列通过 setColumnShrinkable()方法设置为 shrinkable,则该列的宽度可以进行收缩,使表格能够适应其父容器的大小。

如果一个列通过 setColumnStretchable()方法设置为 stretchable,则该列的宽度可以进行拉伸,扩展它的宽度填充空余的空间。

建立表格布局要注意以下几点:

(1) 向界面中添加一个线性布局,无需修改布局的属性值。其中,Id 属性为 TableLayout01,Layout width 和 Layout height 属性都为 wrap_content。

(2) 向 TableLayout01 中添加两个 TableRow。TableRow 代表一个单独的行,每行被划分为几个小的单元,单元中可以添加一个界面控件。其中,Id 属性分别为 TableRow01 和 TableRow02,Layout width 和 Layout height 属性都为 wrap_content。表格布局示意和布局效果如图 4-10 和图 4-11 所示。

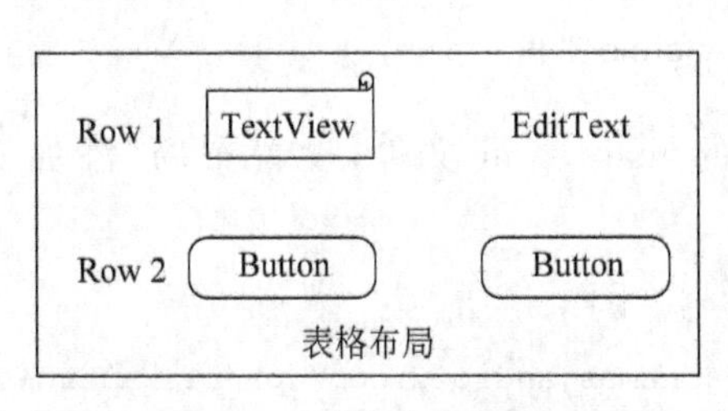

图 4-10 表格布局示意图

图 4-11 表格布局效果图

4.4.6 表格布局应用案例

本节将就 TableLayout 布局的使用和应用进行详述。本案例使用 TableRow 和 View 实现 Table 的功能，具体实现步骤如下：

(1) 在 Eclipse 中选择 File→New→Android Project，创建一个新的 Android 工程，项目名为 TableLayoutDemo，目标 API 选择 10（即 Android 2.3.3 版本），应用程序名为 TableLayoutDemo，包名为 com.hisoft.activity，创建的 Activity 的名字为 TableLayoutActivity，最小 SDK 版本根据选择的目标 API 会自动添加为 10，如图 4-12 所示。

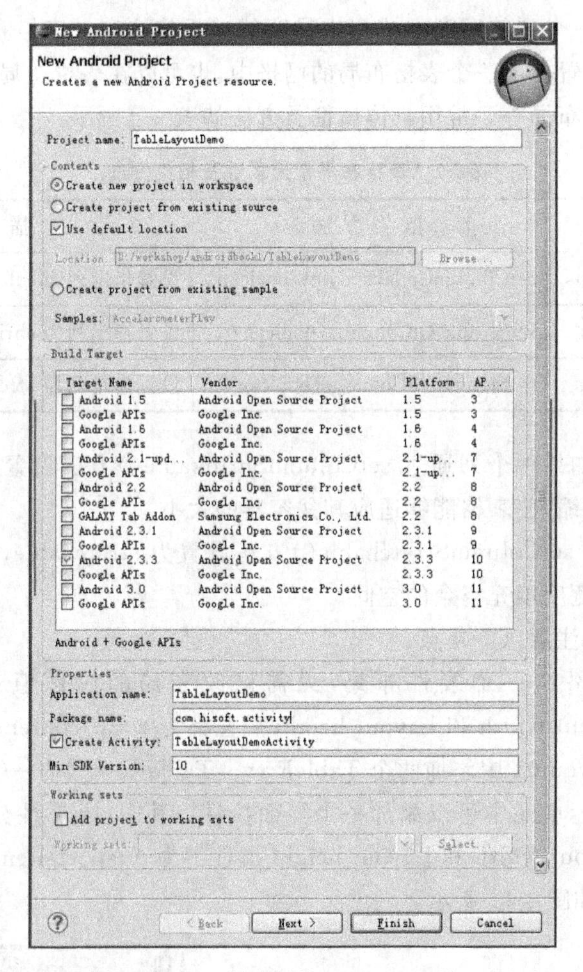

图 4-12 创建新的 Android 工程

(2) 打开项目工程中 res→layout 目录下的 main.xml 文件，设置布局，添加 6 个 TableRow 和两个 View，代码如下：

```
1.  <?xml version="1.0" encoding="utf-8"?>
2.  <TableLayout xmlns:android="http://schemas.android.com/apk/res/android"
3.      android:layout_width="fill_parent"
4.      android:layout_height="fill_parent"
5.      android:stretchColumns="1">
```

```
6.
7.         <TableRow>
8.             <TextView
9.             android:layout_column="1"
10.                android:text="姓名:"
11.                android:padding="3dip" />
12.            <TextView
13.                android:text="张三"
14.                android:gravity="right"
15.                android:padding="3dip" />
16.        </TableRow>
17.
18.        <TableRow>
19.            <TextView
20.                android:layout_column="1"
21.                android:text="年龄:"
22.                android:padding="3dip" />
23.            <TextView
24.                android:text="10"
25.                android:gravity="right"
26.                android:padding="3dip" />
27.        </TableRow>
28.
29.        <TableRow>
30.            <TextView
31.                android:layout_column="1"
32.                android:text="性别:"
33.                android:padding="3dip" />
34.            <TextView
35.                android:text="男"
36.                android:gravity="right"
37.                android:padding="3dip" />
38.        </TableRow>
39.
40.        <View
41.            android:layout_height="2dip"
42.            android:background="#FF909090" />
43.
44.        <TableRow>
45.            <TextView
46.              android:layout_column="1"
47.                android:text="所在城市:"
48.                android:padding="3dip" />
49.            <TextView
50.                android:text="北京"
```

```
51.             android:gravity="right"
52.             android:padding="3dip" />
53.     </TableRow>
54.
55.     <TableRow>
56.         <TextView
57.           android:layout_column="1"
58.             android:text="国籍:"
59.             android:padding="3dip" />
60.         <TextView
61.             android:text="中国"
62.             android:gravity="right"
63.             android:padding="3dip" />
64.     </TableRow>
65.
66.     <View
67.         android:layout_height="2dip"
68.         android:background="#FF909090" />
69.
70.     <TableRow>
71.         <TextView
72.             android:layout_column="1"
73.             android:text="附加信息:"
74.             android:padding="3dip" />
75.     </TableRow>
76. </TableLayout>
```

第 2～5 行声明了一个表格布局，第 2 行代码是声明 XML 文件的根元素为表格布局，第 3、4 行设置了表格布局在所属的父容器中的布局方式为横向和纵向填充父容器。第 5 行是设置 TableLayout 所有行的第 2 列为扩展列，剩余的空间由第 2 列补齐。

第 7～16 行声明了一个 TableRow。

第 8～11 行声明了一个 TextView，第 9 行设置了从第 2 列开始填写(0 是起始列)，第 11 行设置了字符四周到 TextView 的空白边的大小。

第 12～15 行声明了第 2 个 TextView，第 14 行设置了 TextView 内字符的对齐方式，此为右对齐。

第 40～42 行声明了一个 View，加一个分割线，View 是 TextView 的父类，第 41 行设置了线的高度为 2，第 42 行设置了背景颜色。

(3) src 目录下 com. hisoft. activity 包下的 TableLayoutActivity. java 文件和 res→values 目录下的 strings. xml 文件都暂不做修改。部署运行 TableLayoutDemo 项目工程，项目运行效果如图 4-13 所示。

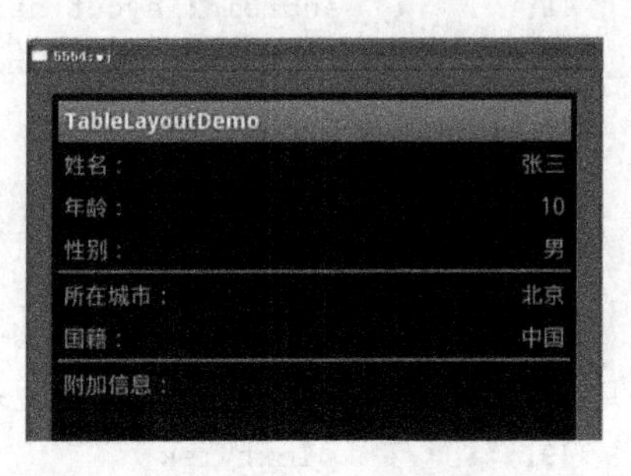

图 4-13 TableLayoutDemo 运行结果

4.4.7 帧布局

帧布局(FrameLayout)是 Android 布局系统中最简单的界面布局,是用来存放一个元素的空白空间,且子元素的位置是不能够指定的,只能够放置在空白空间的左上角。在帧布局中,如果先后存放多个子元素,后放置的子元素将遮挡先放置的子元素。

帧布局由 FrameLayout 所代表,帧布局容器为每个加入其中的组件创建一个空白的区域(成为一帧),每个子组件占据一帧,这些帧都会根据 gravity 属性执行自动对齐。也就是说,把组件一个一个地叠加在一起。

FrameLayout 控件继承自 ViewGroup,它在 ViewGroup 的基础上定义了自己的三个属性,对应的 XML Attributes 分别为 android:foreground、android:foregroundGravity 和 android:measureAllChildren。第一个属性是设置前景色,第二个属性是控制前景色的重心,前两个属性其实是对 android:backgroud 的重写,其目的是可以控制背景的重心。第三个属性如果为 true,则在测量时测量所有的子元素(即使该子元素为 gone)。帧布局的常用属性及方法如表 4-8 所示。

表 4-8 帧布局的常用属性及相关方法

属性名称	相关方法	描述
android:foreground	setForeground(Drawable)	设置绘制在子控件之上的内容,设置前景色
android:foregroundGravity	setForegroundGravity(int)	设置应用于绘制在子控件之上内容的 gravity 属性,控制前景色的重心
android:measureAllChildren	setMeasureAllChildren(boolean)	根据参数值,决定是设置测试所有的元素还是仅仅测量状态是 VISIBLE or INVISIBLE 的元素

4.4.8 帧布局应用案例

前面章节已就 FrameLayout 布局和属性进行了详述。本案例使用 TableRow 和 View 实现 Table 的功能,具体实现步骤如下:

(1) 在 Eclipse 中选择 File→New→Android Project,创建一个新的 Android 工程,项目名为 FrameLayoutDemo,目标 API 选择 10(即 Android 2.3.3 版本),应用程序名为 FrameLayoutDemo,包名为 com.hisoft.activity,创建的 Activity 的名字为 MainActivity,最小 SDK 版本根据选择的目标 API 会自动添加为 10,如图 4-14 所示。

(2) 打开项目工程中 res→layout 目录下的 main.xml 文件,设置线性布局,添加一个 Button 按钮控件,代码如下:

```
1.    <?xml version="1.0" encoding="utf-8"?>
2.    <LinearLayout
3.      xmlns:android="http://schemas.android.com/apk/res/android"
4.      android:orientation="vertical"
5.      android:layout_width="match_parent"
6.      android:layout_height="match_parent">
```

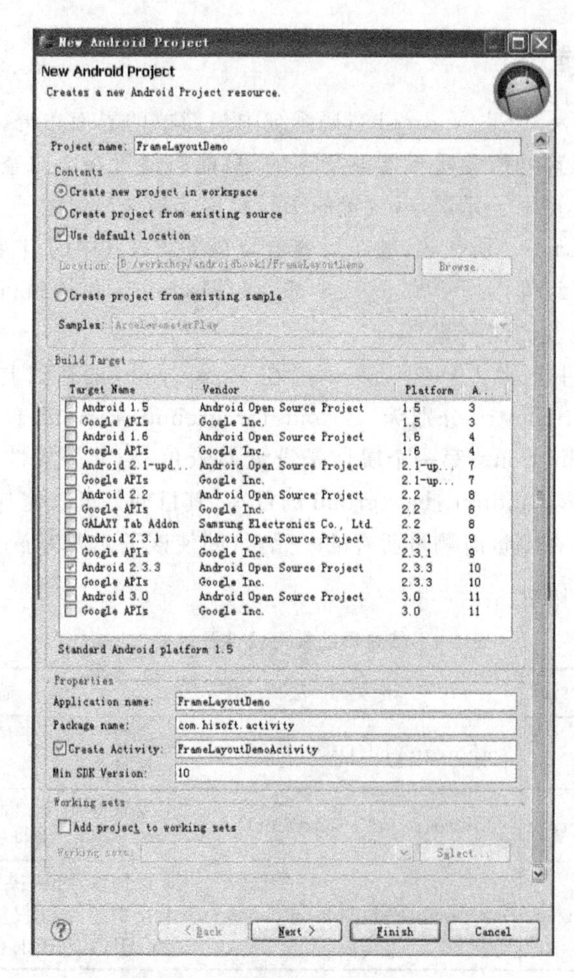

图 4-14　创建新的 Android 工程

```
7.        <Button android:text="Click to Framelayout"
8.            android:id="@+id/button1"
9.            android:layout_width="wrap_content"
10.               android:layout_height="wrap_content">
11.       </Button>
12.   </LinearLayout>
```

(3) 在 src 目录下 com. hisoft. activity 包下的 MainActivity. java 文件中声明创建按钮,为按钮添加监听器,并用 Intent 实现从 MainActivity 到 FrameLayoutActivity 的跳转,代码如下:

```
1.    package com.hisoft.activity;

2.    import android.app.Activity;
3.    import android.content.Intent;
4.    import android.os.Bundle;
5.    import android.view.View;
```

```
6.   import android.view.View.OnClickListener;
7.   import android.widget.Button;

8.   public class MainActivity extends Activity {

9.     private Button button1;

10.    @Override
11.    protected void onCreate(Bundle savedInstanceState) {
12.      super.onCreate(savedInstanceState);
13.      setContentView(R.layout.main);

14.      button1=(Button)findViewById(R.id.button1);
15.      //为按钮绑定一个单击事件的监听器
16.      button1.setOnClickListener(new OnClickListener(){
17.        public void  onClick(View v)
18.        {
19.          //通过 Intent 跳转到 FrameLayoutActivity
20.          Intent intent=new Intent();
21.          intent.setClass(MainActivity.this, FrameLayoutActivity.class);
22.          startActivity(intent);
23.        }
24.      });
25.    }
26.  }
```

(4) 在项目工程 res→layout 目录下新建 framelayout.xml 文件,声明 FrameLayout 布局,并添加两个 ImageView 控件,代码如下:

```
1.   <?xml version="1.0" encoding="utf-8"?>
2.   <FrameLayout xmlns:android="http://schemas.android.com/apk/res/android"
3.     android:id="@+id/frameLayout1"
4.     android:layout_width="fill_parent"
5.     android:layout_height="fill_parent">
6.       <ImageView android:src="@drawable/frame"
7.            android:id="@+id/imageView1"
8.            android:layout_width="wrap_content"
9.            android:layout_height="wrap_content">
10.      </ImageView>
11.  <ImageView android:src="@drawable/icon"
12.           android:id="@+id/imageView2"
13.           android:layout_width="wrap_content"
14.           android:layout_height="wrap_content">
15.      </ImageView>
16.  </FrameLayout>
```

(5) 在 res 目录下,把 frame.jpg 图片添加到 drawable-hdpi、drawable-mdpi、drawable-

ldpi 文件夹中，以供步骤 4 第 6 行控件 ImageView 使用，第 11 行第 2 个 ImageView 控件用系统自带的图片。

(6) 在 src 目录下的 com.hisoft.activity 包下创建 FrameLayoutActivity.java 文件，调用上面创建的帧布局文件 framelayout.xml，代码如下：

```
1.   package com.hisoft.activity;
2.   import android.app.Activity;
3.   import android.os.Bundle;
4.   public class FrameLayoutActivity extends Activity
5.   {
6.       @Override
7.       public void onCreate(Bundle savedInstanceState)
8.       {
9.           super.onCreate(savedInstanceState);
10.            setContentView(R.layout.framelayout);
11.              }
12.      }
```

(7) 在 AndroidManifest.xml 文件中的<application>标签节点下添加新创建的 FrameLayoutActivity 注册声明，代码如下：

```
1.   <activity android:name=".FrameLayoutActivity" android:label="@string/
     app_name" />
```

(8) 部署运行 FrameLayoutDemo 项目工程，项目运行效果如图 4-15 所示。然后单击 Click to Framelayout 按钮，出现图 4-16 所示界面。所有的图片都显示帧布局的左上角，并且第二个覆盖第一个。如果把 framelayout.xml 中的两个 ImageView 控件的描述互换位置，则大图片完全覆盖住小图片。

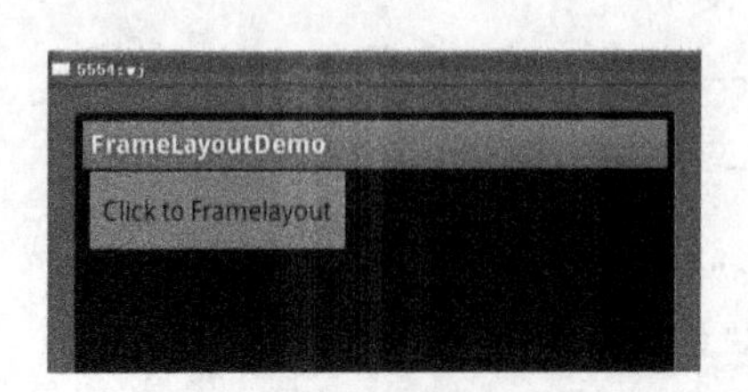

图 4-15　项目运行结果

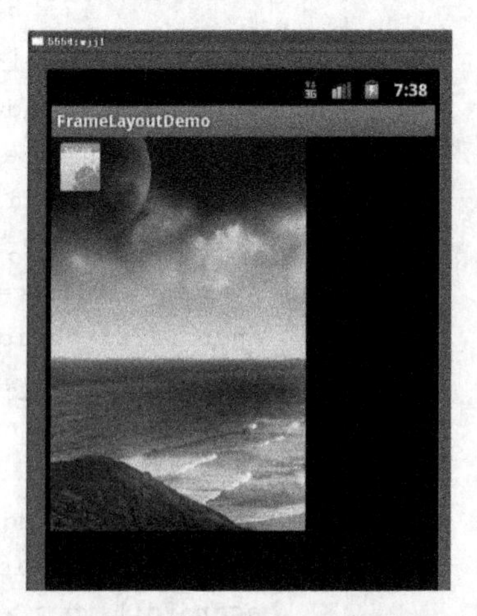

图 4-16　帧布局图片叠加效果

4.4.9　绝对布局

绝对布局(AbsoluteLayout)能通过指定界面元素的坐标位置来确定用户界面的整体

布局。绝对布局是一种不推荐使用的界面布局,因为通过 X 轴和 Y 轴确定界面元素位置后,Android 系统不能够根据不同屏幕对界面元素的位置进行调整,降低了界面布局对不同类型和尺寸屏幕的适应能力。

4.4.10 绝对布局应用案例

上述已就 AbsoluteLayout 布局进行了简介。本节将通过一个案例的使用来介绍 AbsoluteLayout 的使用方法,具体实现步骤如下:

(1) 在 Eclipse 中选择 File→New→Android Project,创建一个新的 Android 工程,项目名为 AbsoluteLayoutDemo,目标 API 选择 10(即 Android 2.3.3 版本),应用程序名为 AbsoluteLayoutDemo,包名为 com.hisoft.activity,创建的 Activity 的名字为 AbsoluteLayoutActivity,最小 SDK 版本根据选择的目标 API 会自动添加为 10,如图 4-17 所示。

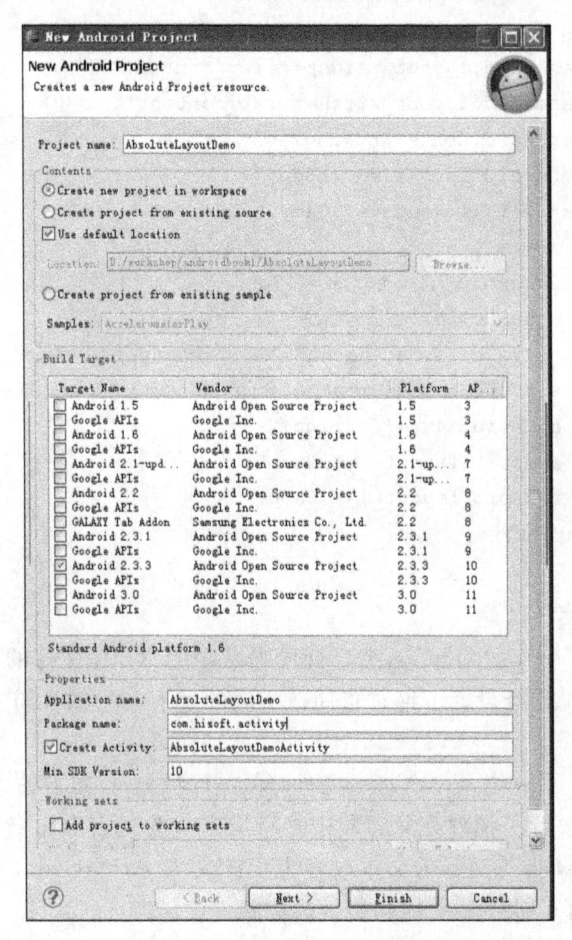

图 4-17 创建新的 Android 工程

(2) 打开项目工程中 res→layout 目录下的 main.xml 文件,设置绝对布局,添加三个 TextView 控件,代码如下:

```
1.    <?xml version="1.0" encoding="utf-8"?>
```

```
2.   <AbsoluteLayout xmlns:android="http://schemas.android.com/apk/res/
     android"
3.           android:id="@+id/absoluteLayout1"
4.           android:layout_width="fill_parent"
5.           android:layout_height="fill_parent">
6.     <TextView android:textSize="18pt"
7.           android:id="@+id/tv1"
8.           android:layout_height="wrap_content"
9.           android:layout_width="wrap_content"
10.          android:text="@string/tv1"
11.          android:layout_x="37dp"
12.          android:layout_y="37dp">
13.    </TextView>
14.       <TextView android:textSize="18pt"
15.             android:id="@+id/tv2"
16.             android:layout_height="wrap_content"
17.             android:layout_width="wrap_content"
18.             android:text="@string/tv2"
19.             android:layout_x="186dp"
20.             android:layout_y="104dp">
21.       </TextView>
22.      <TextView android:textSize="18pt"
23.              android:id="@+id/tv3"
24.              android:layout_height="wrap_content"
25.              android:layout_width="wrap_content"
26.              android:text="@string/tv3"
27.              android:layout_x="106dp"
28.              android:layout_y="188dp">
29.       </TextView>

30.  </AbsoluteLayout>
```

第2～5行声明了一个绝对布局，第2行代码是声明XML文件的根元素为绝对布局，第3行设置了ID，第4、5行设置了绝对布局在所属的父容器中的布局方式为横向和纵向填充父容器

第6～13行声明了一个TextView控件，第6行设置了文本的大小，第7行设置了ID的名称，第8、9行设置了控件在父容器中的布局方式为只占据自身内容大小的空间，第10行设置了TextView控件显示的文本内容为资源文件strings. xml中设置的名称为tv1的值，第11、12行设置了控件显示的起始坐标位置。

第14～21行声明了第二个TextView控件。

第22～29行声明了第三个TextView控件。

(3) 打开res目录下values文件中的strings. xml文件，修改文件代码为：

```
1.    <?xml version="1.0" encoding="utf-8"?>
2.   <resources>
```

```
3.      <string name="app_name">AbsoluteLayoutDemo</string>
4.      <string name="tv1">文本 1</string>
5.      <string name="tv2">文本 2</string>
6.      <string name="tv3">文本 3</string>
7.  </resources>
```

(4) 部署运行 AbsoluteLayoutDemo 项目工程,项目运行效果如图 4-18 所示。

上述使用的是默认屏幕大小为 432 的模拟器或屏幕,如果使用不同屏幕大小的模拟器或设备,如 800 大小,则显示效果将会发生变形。反之,如果使用绝对布局在默认屏幕大小为 800 上显示正常,在屏幕大小为 432 的屏幕上通常会出现遮挡或无法全部显示的效果。如图 4-19 所示。

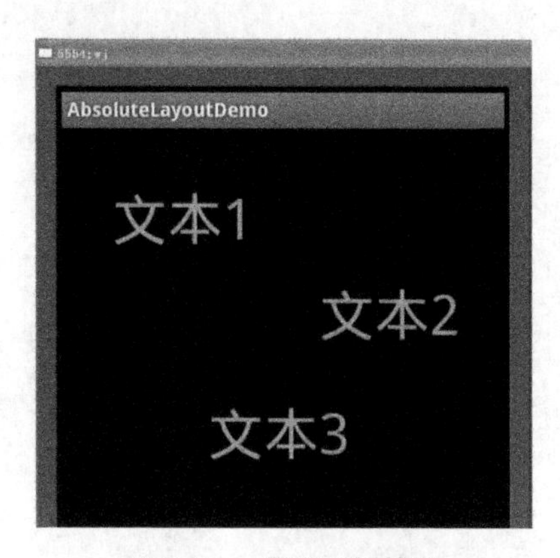

图 4-18 项目运行结果

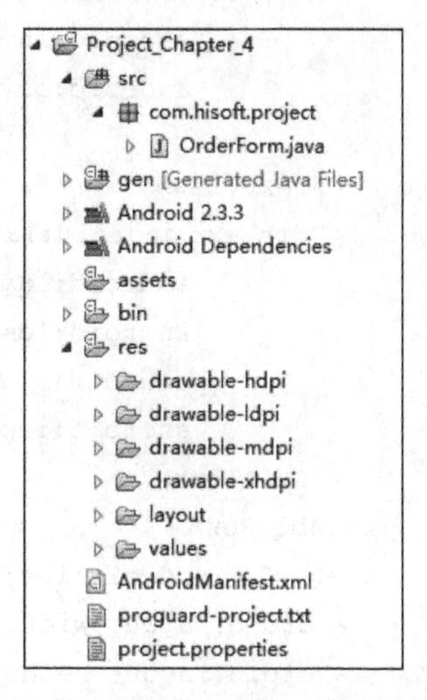

图 4-19 Project_Chapter_4 项目工程目录

拓展提示:在手机操作系统中,其他的手机操作系统,如 Windows Phone8 采用的框架与 Android 系统 4.1 采用的框架,它们的设计思路及设计模式,以及未来发展的设计趋势走向值得分析和对比思考。此外,不同的 Android UI 布局适应、适用于什么样的实际应用,进行归纳和总结规律。

4.5 项目案例

学习目标:学习 Android UI 布局的不同使用方法及应用,尤其是线性布局、表格布局的使用,以及它们的混合使用方法、属性的设置,修改 AndroidManifest. XML 文件、main. xml 文件的方法。

案例描述:使用 Android UI 布局中的线性布局,并在其中使用若干 TableRow,在每一个 TableRow 中添加 TextView、EditText 控件,然后设置每一个布局及控件的属性,实现订单界面。

案例要点:添加 TableRow 及属性设置,添加 TextView、EditText 控件及属性设置。

案例实施:

(1) 创建新的项目工程 Project_Chapter_4,选择目标平台 Android 2.3.3,如图 4-19

所示。

(2) 在 res 目录下的 layout 文件夹中创建 orderform. xml 文件,代码如下:

```
1.   <?xml version="1.0" encoding="utf-8"?>
2.   <LinearLayout xmlns:android="http://schemas.android.com/apk/res/android"
3.     android:orientation="vertical" android:layout_width="fill_parent"
4.     android:layout_height="fill_parent">
5.
6.     <TableRow android:id="@+id/TableRow01"
7.       android:layout_width="fill_parent"
8.       android:layout_height="wrap_content">
9.       <TextView android:text="@string/name"
10.                android:layout_width="70px"
11.                android:layout_height="40px"
12.                >
13.      </TextView>
14.      <EditText android:layout_width="fill_parent"
15.                android:layout_height="40px"
16.                android:id="@+id/ename"
17.                android:singleLine="true"
18.                android:inputType="textPersonName">
19.      </EditText>
20.    </TableRow>
21.    <TableRow android:id="@+id/TableRow02"
22.      android:layout_width="fill_parent"
23.      android:layout_height="wrap_content">
24.      <TextView android:text="@string/phone_number"
25.                android:layout_width="70px"
26.                android:layout_height="40px"
27.                >
28.      </TextView>
29.      <EditText android:layout_width="fill_parent"
30.                android:layout_height="40px"
31.                android:id="@+id/etel"
32.                android:singleLine="true"
33.                android:inputType="phone">
34.      </EditText>
35.    </TableRow>
36.    <TableRow android:id="@+id/TableRow03"
37.      android:layout_width="fill_parent"
38.      android:layout_height="wrap_content">
39.      <TextView android:text="@string/email"
40.                android:layout_width="70px"
41.                android:layout_height="40px"
42.                >
43.      </TextView>
```

```
44.       <EditText android:layout_width="fill_parent"
45.                 android:layout_height="40px"
46.                 android:id="@+id/email"
47.                 android:singleLine="true"
48.                 android:inputType="textEmailAddress">
49.       </EditText>
50.     </TableRow>
51.       <TableRow android:id="@+id/TableRow04"
52.       android:layout_width="fill_parent"
53.       android:layout_height="wrap_content">
54.       <TextView android:text="@string/addr"
55.                 android:layout_width="70px"
56.                 android:layout_height="40px"
57.                 >
58.       </TextView>
59.       <EditText android:layout_width="fill_parent"
60.                 android:layout_height="40px"
61.                 android:id="@+id/eaddress"
62.                 android:singleLine="true"
63.                 android:inputType="textPostalAddress">
64.       </EditText>
65.     </TableRow>
66.   </LinearLayout>
```

(3) 在 src 目录的 com. hisoft. project 包下创建 OrderForm. java 文件,代码如下:

```
1.     package com.hisoft.project;
2.
3.   import android.app.Activity;
4.   import android.os.Bundle;
5.
6.   /*********************************************************************
7.    * 程序名称:OrderForm.java                                          *
8.    * 功能:显示提交订单的窗口,输入用户的联系方式,提交表单             *
9.    * 作者:                                                            *
10.   * 日期:                                                            *
11.  *********************************************************************/
12.
13.
14.  public class OrderForm extends Activity {
15.
16.    /**
17.     * 创建订单页面
18.     */
19.    public void onCreate(Bundle savedInstanceState) {
20.
```

```
21.        super.onCreate(savedInstanceState);
22.        setContentView(R.layout.orderform);

23.    }

24.  }
```

(4) 修改 AndroidManifest. xml,代码如下:

```
1.         <?xml version="1.0" encoding="utf-8"?>
2.    <manifest xmlns:android="http://schemas.android.com/apk/res/android"
3.         package="com.hisoft.project"
4.         android:versionCode="1"
5.         android:versionName="1.0">
6.
7.         <uses-sdk android:minSdkVersion="10" />
8.
9.         <application
10.            android:icon="@drawable/ic_launcher"
11.            android:label="@string/app_name">
12.            <activity android:name=".OrderForm" android:label="我的订单">
13.              <intent-filter>
14.            <action android:name="android.intent.action.MAIN" />
15.               <category android:name="android.intent.category.LAUNCHER" />
16.               </intent-filter>
17.            </activity>
18.        </application>
19.
20.   </manifest>
```

(5) 部署运行 Project_Chapter_4 项目工程,项目运行效果如图 4-20 所示。

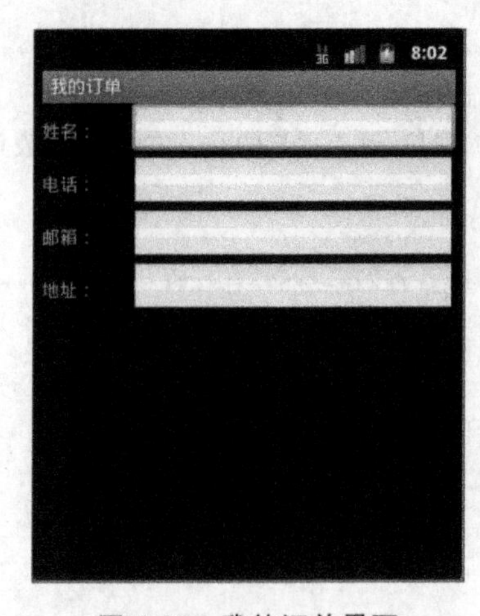

图 4-20　我的订单界面

习 题 4

1. 简答题

(1) Android UI 的设计原则包含哪些？描述 Android UI 设计当前不同版本的变化、未来发展的趋势是什么？

(2) 简述 Android UI 框架与 MVC 设计模式的关系，以及采用 MVC 设计有何优势？

(3) Android 中 View 和 ViewGroup 以及视图树模型有什么联系？并就控件分类。

(4) 简述 Android UI 不同界面布局适用于什么界面设计？它们常用的组合应用，在实际开发中有哪些？

(5) 什么情况下会采用绝对布局？采用它一般需要注意什么问题？

2. 完成下面的实训项目

要求：使用绝对布局和其他任意一种布局的组合完成我的订单界面设计，并要求在不同屏幕尺寸大小下(如 800、432、400 等)不发生变形且能正常显示。

第5章 Android UI系统控件基础

学习目标

本章主要介绍 Android UI 系统控件中文本控件 TextView 和 EditText、按钮控件 Button、图片按钮控件 ImageButton、单选按钮 RadioButton、复选按钮 CheckBox、时间选择器 TimePicker、日期选择器 DatePicker、图片控件 ImageView、模拟时钟 AnalogClock 和数字时钟 DiditalClock。使读者通过本章的学习，能够达到深入了解 Android UI 系统控件，掌握以下知识要点：

(1) 掌握文本控件 TextView、EditText 的类继承结构、属性及方法、应用。

(2) 掌握按钮控件 Button、图片按钮控件 ImageButton 的通常用法及属性。

(3) 掌握单选按钮 RadioButton、复选按钮 CheckBox 的常用方法、引用处理及属性方法设置。

(4) 掌握时间选择器 TimePicker、日期选择器 DatePicker 的使用方法及常用属性设置。

(5) 掌握图片按钮 ImageView 的常用方法、引用处理及属性方法设置。

(6) 熟悉模拟时钟 AnalogClock 和数字时钟 DiditalClock 的常用方法。

Android 系统提供了许多控件给开发者使用，开发者通过对这些控件编码与控件组合能够实现系统设想的模型和相应的功能。Android 系统的界面控件分为定制控件和系统控件，系统控件是 Android 系统提供给用户已经封装的界面控件，是在应用程序开发过程中常见的功能控件。系统控件更有利于帮助用户进行快速开发，同时能够使开发者在使用 Android 系统进行开发过程中保持应用程序的界面一致性。

在开发应用中，经常使用的系统控件有 TextView、EditText、Button、ImageButton、Checkbox、RadioButton、Spinner、ListView 和 TabHost 等。

Android UI 系统控件的使用，除了传统的程序代码中直接声明、创建之外，其最能体现其设计思想的是采用 XML 文件来描述控件，在 XML 中可以描述控件的宽度、长度、控件上的文本、控件的背景、控件的填充、设置源等。

5.1 文本控件简介

在 Android 系统中，文本控件包含 TextView 和 EditText 控件，它们都继承 android. view. View，在 android. widget 包中。本节就文本控件的属性及使用方法进行详述。

5.1.1 文本框

android. widget 包中的 TextView 是文本表示控件，一般用来文本展示，是一种用于显示字符串的控件。主要功能是向用户展示文本的内容，可以作为应用程序的标签或者邮件正文的显示，默认情况下不允许用户直接编辑。

在程序设计和开发中，使用 TextView 可以采用的方式有如下两种：

(1) 在程序中创建控件的对象方式来使用 TextView 控件。

例如 TextView 控件，可以通过编写如下代码完成控件使用。

```
TextView tv=new TextView(this);
        tv.setText("大家好");
        setContentView(tv);
```

(2) 使用 XML 描述控件，并在程序中引用和使用。

① 在 res/layout 文件下的 XML 文件中描述控件。

```
<TextView
Android:id="@+id/text_view"
Android:layout_width="fill_parent"
Android:layout_height="wrap_content"
Android:textSize="16sp"
Android:padding="10dip"
Android:background="#00f0d0"
Android:text="大家好,这里是 TextView"/>
```

② 在程序中引用 XML 描述的 TextView。

```
TextView text_view=(TextView) findViewById(R.id.text_view);
```

上述两种方式的使用各有优缺点，根据不同的需要采用相应的方法。相比而言，采用第二种方法更好，主要优势：一是方便代码的维护；二是编码的灵活；三是利于分工协作。

TextView 控件常用的方法：getText()、setText()。

TextView 控件有着与之相应的属性，通过选择不同的属性给予其值，能够实现不同的效果。TextView 控件属性的设置既可以在 XML 文件中通过属性名称进行设定赋值，也可以采用对应的方法在程序代码中设定。其常用属性及对应方法如表 5-1 所示。

表 5-1　TextView 控件常用 XML 属性及对应方法

属性名称	对应方法	说　明
android:text	setText(CharSequence)	设置 TextView 控件文字显示
android:autoLink	setAutoLinkMask(int)	设置是否当文本为 URL 链接/email/电话号码/map 时,文本显示为可点击的链接。可选值(none/web/email/phone/map/all)
android:hint	setHint(int)	当 TextView 中显示的内容为空时,显示该文本
android:textColor	setTextColor(ColorStateList)	设置字体颜色
android:textSize	setTextSize(float)	设置字体大小
android:typeface	setTypeface(Typeface)	设置文本字体,必须是以下常量值之一: normal 0, sans 1, serif 2, monospace(等宽字体) 3
android:ellipsize	setEllipsize(TextUtils.TruncateAt)	如果设置了该属性,当 TextView 中要显示的内容超过了 TextView 的长度时,会对内容进行省略。可取的值有 start、middle、end 和 marquee
android:gravity	setGravity(int)	定义 TextView 在 x 轴和 y 轴方向上的显示方式
android:height	setHeight(int)	设置文本区域的高度,支持度量单位: px(像素)/dp/sp/in/mm(毫米)
android:minHeight	setMinHeight(int)	设置文本区域的最小高度
android:maxHeight	setMaxHeight(int)	设置文本区域的最大高度
android:width	setWidth(int)	设置文本区域的宽度,支持度量单位: px(像素)/dp/sp/in/mm(毫米)
android:minWidth	setMinWidth(int)	设置文本区域的最小宽度
android:maxWidth	setMaxWidth(int)	设置文本区域的最大宽度

5.1.2　TextView 应用案例

本节在上述 TextView 控件讲解的基础之上,通过案例熟悉 TextView 控件的属性和用法,具体步骤如下:

(1) 创建一个新的 Android 工程,工程名为 TextViewDemo,目标 API 选择 10(即 Android 2.3.3 版本),应用程序名为 TextViewDemo,包名为 com.hisoft.activity,创建的 Activity 的名字为 TextViewActivity,最小 SDK 版本根据选择的目标 API 会自动添加为 10,创建后的项目工程如图 5-1 所示。

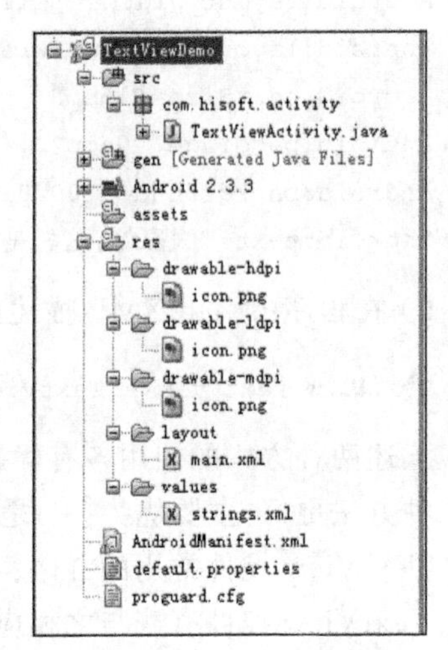

图 5-1　TextViewDemo 工程目录结构

(2) 打开项目工程中 res→layout 目录下的 main.xml 文件,设置线性布局,添加 4 个 TextView 控件,并设置属性,代码如下:

```
1.  <?xml version="1.0" encoding="utf-8"?>
2.  <?xml version="1.0" encoding="utf-8"?>
3.  <LinearLayout xmlns:android="http://schemas.android.com/apk/res/android"
4.    android:orientation="vertical"
5.    android:layout_width="fill_parent"
6.    android:layout_height="fill_parent"
7.    >

8.    <TextView
9.    android:layout_width="fill_parent"
10.   android:layout_height="wrap_content"
11.   android:text="字体大小为 14 的文本"
12.   android:textSize="14pt"
13.   />

14.   <TextView
15.   android:layout_width="fill_parent"
16.   android:layout_height="wrap_content"
17.   android:singleLine="true"
18.   android:text="TextView示例"
19.   android:ellipsize="middle"
20.   />

21.   <TextView
22.   android:layout_width="fill_parent"
23.   android:layout_height="wrap_content"
24.   android:singleLine="true"
25.   android:text="访问:http://www.zfjsjx.cn"
26.   android:autoLink="web"
27.   />

28.   <TextView
29.   android:layout_width="fill_parent"
30.   android:layout_height="wrap_content"
31.   android:text="红色并带阴影的文本"
32.   android:shadowColor="#0000ff"
33.   android:shadowDx="15.0"
34.   android:shadowDy="20.0"
35.   android:shadowRadius="45.0"
36.   android:textColor="#ff0000"
37.   android:textSize="20pt"
38.   />

39. </LinearLayout>
```

(3) src 目录下 com. hisoft. activity 包下的 TextViewActivity. java 文件和 res→values

目录下的 strings. xml 文件都暂不做修改。部署运行项目工程,项目运行效果如图 5-2 所示。

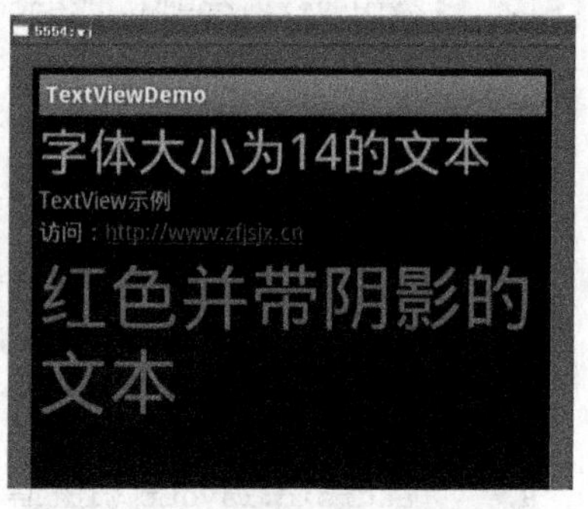

图 5-2　TextViewDemo 运行结果

5.1.3　编辑框

编辑框(EditText)控件继承自 android. widget. TextView,在 android. widget 包中。EditText 为输入框,是编辑文本控件,主要功能是让用户输入文本的内容,它是可以编辑的,是用来输入和编辑字符串的控件。

利用控件 EditText 不仅可以实现输入信息,还可以根据需要对输入信息进行限制约束。例如限制控件 EditText 输入信息:

```
<EditText
android:layout_width="fill_parent"
android:layout_height="wrap_content"
android:inputType="numeber"/>
```

与 5. 1. 2 节所讲述的 TextView 一样,EditText 控件的使用也有两种:一种是在程序中创建控件的对象方式来使用 EditText 控件;另外一种是在 res/layout 文件下的 XML 文件中描述控件,程序中使用 EditText 控件。

例如:

(1) 用 xml 描述一个 EditView。

```
<EditText Android:id="@+id/edit_text"
Android:layout_width="fill_parent"
Android:layout_height="wrap_content"
Android:text="这里可以输入文字" />
```

(2) 在程序中引用 xml 描述的 TextView。

```
EditText editText=(EditText) findViewById(R.id.editText);
```

EditText 的常用方法: getText()。它也有着与之相应的属性,通过选择不同的属性给予其值,能够实现其不同的效果。其常用属性及对应方法如表 5-2 所示。

表 5-2　EditText 控件常用 XML 属性及对应方法

属性名称	对应方法	说明
android:hint		输入框的提示文字
android:password	setTransformationMethod (TransformationMethod)	设置文本框中的内容是否显示为密码,当为 true 时,以小点“.”显示文本
android:phoneNumber	setKeyListener(KeyListner)	设置文本框中的内容只能是电话号码,当为 true 时,表示电话框
android:digits	setKeyListener(KeyListener),可使用此方法监听键盘来实现	设置允许输入哪些字符。如“1234567890.+－*/%\n()”

续表

属性名称	对应方法	说　明
android:numeric	setKeyListener(KeyListener),可使用此方法监听键盘来实现	设置只能输入数字,并且置顶可输入的数字格式,可选值有 integer| signed| decimal。integer 为正整数,signed 为整数(可带负号),decimal 为浮点数
android:singleLine	setTransformationMethod(TransformationMethod)	设置文本框的单行模式
android:maxLength	setFilters(InputFilter)	设置最大显示长度
android:cursorVisible	setCursorVisible(boolean)	设置光标是否可见,默认可见
android:lines	setLines(int)	通过设置固定的行数来决定 EditText 的高度
android:maxLines	setMaxLines(int)	设置最大的行数
android:minLines	setMinLines(int)	设置最小的行数
android:scrollHorizontally	setHorizontallyScrolling(boolean)	设置文本框是否可以进行水平滚动
android:selectAllOnFocus	setSelectAllOnFocus(boolean)	如果文本内容可选中,当文本框获得焦点时自动选中全部文本内容
android:shadowColor	setShadowLayer(float, float, float, int)	为文本框设置指定颜色的阴影,需要与 shadowRadius 一起使用
android:shadowDx	setShadowLayer(float, float, float, int)	设置阴影横向坐标开始位置,为浮点数
android:shadowDy	setShadowLayer(float, float, float, int)	设置阴影纵向坐标开始位置,为浮点数
android:shadowRadius	setShadowLayer(float, float, float, int)	为文本框设置阴影的半径,为浮点数

5.1.4 EditText 应用案例

在上述 EditText 控件讲解的基础之上,通过案例熟悉 EditText 控件的属性和用法,具体步骤如下:

(1) 创建一个新的 Android 工程,工程名为 EditTextDemo,目标 API 选择 10(即 Android 2.3.3 版本),应用程序名为 EditTextDemo,包名为 com. hisoft. activity,创建的 Activity 的名字为 EditTextActivity,最小 SDK 版本根据选择的目标 API 会自动添加为 10,创建后的项目工程如图 5-3 所示。

(2) 打开项目工程中 res→layout 目录下的 main. xml 文件,设置线性布局,添加一个 TextView 控件和一个 EditText 控件,并设置相关属性,代码如下:

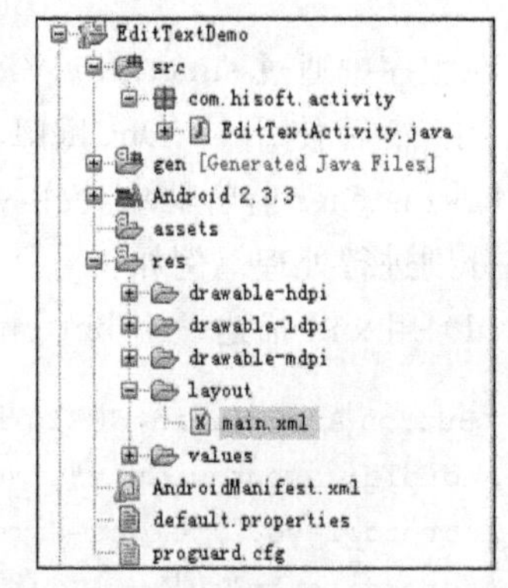

图 5-3　EditTextDemo 工程目录结构

```
1.   <?xml version="1.0" encoding="utf-8"?>
2.   <LinearLayout xmlns:android="http://schemas.android.com/apk/res/android"
3.     android:orientation="vertical"
4.     android:layout_width="fill_parent"
5.     android:layout_height="fill_parent"
6.     >
7.     <TextView android:text="请输入:"
8.         android:id="@+id/textView1"
9.         android:layout_width="wrap_content"
10.            android:layout_height="wrap_content">
11.      </TextView>
12.      <EditText android:layout_height="wrap_content"
13.            android:layout_width="match_parent"
14.            android:id="@+id/editText1"
15.            android:hint="这里键入输入内容">
16.          <requestFocus></requestFocus>
17.      </EditText>
```

(3) src 目录下 com. hisoft. activity 包下的 EditTextActivity. java 文件和 res→values 目录下的 strings. xml 文件都暂不做修改。部署运行项目工程，项目运行效果如图 5-4 所示。

图 5-4 EditTextDemo 工程运行结果

5.2 按钮控件简介

5.2.1 按钮

Button 是一种常用的按钮控件，继承自 android. widget. TextView，在 android. widget 包中。如图 5-5 所示，用户能够在该控件上单击，然后引发相应的事件处理函数。

```
java.lang.Object
  ↳android.view.View
      ↳android.widget.TextView
          ↳android.widget.Button
```

图 5-5 Button 类继承图

它的常用子类有 CheckBox、RadioButton 和 ToggleButton 等，在后续章节会讲到。

Button 按钮控件的通常用法是：

在程序中通过 super. findViewById(id)得到在 layout 中 XML 文件中声明的 Button 的引用，然后使用 setOnClickListener(View. OnClickListener)添加监听，再在 View. OnClickListener 监听器中使用 v. equals(View)方法判断是哪一个按钮被按下，调用不同方法分别进行处理。例如：

(1) 用 xml 描述一个 button。

```
<Button Android:id="@+id/button"
Android:layout_width="wrap_content"
Android:layout_height="wrap_content"
Android:text="这是一个 button" />
```

(2) 在程序代码中引用用 XML 描述的 button。

```
Button button= (Button) findViewById(R.id.button);
```

(3) 给 Button 设置事件响应。

```
button.setOnClickListener(button_listener);
```

(4) 生成一个按钮事件监听器。

```
private Button.OnClickListener button_listener=new
Button.OnClickListener() {
public void onClick(View v) {
switch(v.getId()){
      case R.id.Button:
        textView.setText("Button 按钮 1");
        return;
      case R.id.Button01:
        textView.setText("Button 按钮 2");
        return;
    }

}
};
```

此外,也可以采用在 layout 中的 XML 文件中声明分配一个方法给 Button 按钮,使用 android:onClick 属性,例如:

```
1.      <Button
2.       android:layout_height="wrap_content"
3.       android:layout_width="wrap_content"
4.       android:text="@string/self_destruct"
5.       android:onClick="selfDestruct" />
```

当用户单击 Button 按钮时,Android 系统会自动调用 activity 中的 selfDestruct (View)方法,但 selfDestruct(View)方法必须声明为 public,并只能接受 View 作为其唯一的参数。传递给这个方法的 View 是被单击的控件的一个引用,代码如下:

```
1.      public void selfDestruct(View view) {
2.        //Kabloey
3.    }
```

5.2.2 Button 应用案例

在开发应用中,Button 按钮的使用较为常见,下面通过一个单击 Button 按钮修改标题的案例来介绍 Button 的应用。具体步骤如下:

(1) 创建一个新的 Android 工程,工程名为 ButtonDemo,目标 API 选择 10(即 Android 2.3.3 版本),应用程序名为 ButtonDemo,包名为 com.hisoft.activity,创建的

Activity 的名字为 ButtonActivity，最小 SDK 版本根据选择的目标 API 会自动添加为 10，创建后的项目工程如图 5-6 所示。

(2) 打开项目工程中 res→layout 目录下的 main. xml 文件，设置线性布局，添加一个 Button 按钮控件，并设置相关属性，代码如下：

```
ButtonDemo
  src
    com.hisoft.activity
      ButtonActivity.java
  gen [Generated Java Files]
  Android 2.3.3
  assets
  res
    drawable-hdpi
      icon.png
    drawable-ldpi
    drawable-mdpi
    layout
      main.xml
    values
      strings.xml
  AndroidManifest.xml
  default.properties
  proguard.cfg
```

图 5-6　ButtonDemo 工程目录结构

```
1.   <?xml version="1.0" encoding="utf-8"?>
2.   <LinearLayout xmlns:android="http://
     schemas.android.com/apk/res/android"
3.     android:orientation="vertical"
4.     android:layout_width="fill_parent"
5.     android:layout_height="fill_parent"
6.     >
7.       <Button android:text="按钮 1"
8.           android:id="@+id/button1"
9.           android:layout_width="wrap_content"
10.              android:layout_height="wrap_content">
11.      </Button>

12.  </LinearLayout>
```

(3) 打开 src 目录下 com. hisoft. activity 包下的 ButtonActivity. java 文件，声明 Button 按钮，并获取引用，然后添加监听器，代码如下：

```
1.    package com.hisoft.activity;

2.   import android.app.Activity;
3.   import android.os.Bundle;
4.   import android.view.View;
5.   import android.view.View.OnClickListener;
6.   import android.widget.Button;
7.   public class ButtonActivity extends Activity
8.   {
9.       private Button button1;
10.
11.      @Override
12.      public void onCreate(Bundle savedInstanceState)
13.      {
14.          super.onCreate(savedInstanceState);
15.          setContentView(R.layout.main);
16.
17.          button1=(Button) this.findViewById(R.id.button1);
18.
19.          //给 button1 设置监听
20.          button1.setOnClickListener(new OnClickListener() {
21.
```

```
22.            public void onClick(View v) {
23.
24.              setTitle("按钮被点击了!!!");
25.            }
26.          });
27.       }
28.    }
```

(4) 部署运行 ButtonDemo 项目工程，项目运行效果如图 5-7 所示。单击“按钮 1”按钮后，效果如图 5-8 所示。

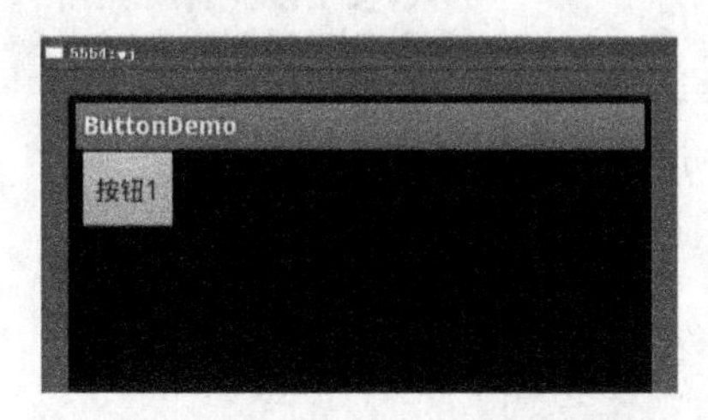

图 5-7　ButtonDemo 程序运行结果

图 5-8　单击“按钮 1”按钮的运行结果

5.2.3　图片按钮

ImageButton 继承自 ImageView 类，是用以实现能够显示图像功能的控件按钮，既可以显示图片，又可以作为 Button 使用。

ImageButton 与 Button 之间的区别：ImageButton 中没有 text 属性。

ImageButton 控件中设置按钮上显示的图片，可以通过 android：src 属性来设置，也可以通过 setImageResource(int)来设置。默认情况下，ImageButton 与 Button 具有一样的背景色，当按钮处于不同的状态时，背景色会发生变化，一般将 ImageButton 控件背景色设置为图片或者透明，以避免控件显示的图片不能完全覆盖背景色时影响显示效果。

下面通过例子说明使用 XML 描述 ImageButton 控件，并在程序中引用和使用的简要过程。

(1) 在 res/layout 文件下的 XML 文件中描述 ImageButton 控件。

```
<ImageButton android:id="@+id/ImageButton01"
  android:layout_width="wrap_content"
  android:layout_height="wrap_content">
</ImageButton>
```

(2) 在程序中引用 XML 描述的 ImageButton。

```
ImageButton imageButton=(ImageButton)findViewById(R.id.ImageButton01);
```

(3) 利用 setImageResource()将新加入的 png 文件 R.drawable.download 传递给 ImageButton。

```
imageButton.setImageResource(R.drawable.download);
```

5.2.4 ImageButton 应用案例

下面通过单击一个 Button 按钮显示 ImageButton 的案例来介绍 ImageButton 的应用。具体步骤如下：

(1) 创建一个新的 Android 工程，工程名为 ImageButtonDemo，目标 API 选择 10(即 Android 2.3.3 版本)，应用程序名为 ImageButtonDemo，包名为 com.hisoft.activity，创建的 Activity 的名字为 ImageButtonActivity，最小 SDK 版本根据选择的目标 API 会自动添加为 10，创建后的项目工程如图 5-9 所示。

```
ImageButtonDemo
  src
    com.hisoft.activity
      ImageButtonActivity.java
      MainActivity.java
  gen [Generated Java Files]
  Android 2.3.1
  assets
  res
    drawable-hdpi
      icon.png
    drawable-ldpi
      icon.png
    drawable-mdpi
    layout
      imagebutton.xml
      main.xml
    values
      strings.xml
  AndroidManifest.xml
  default.properties
  proguard.cfg
```

图 5-9 ImageButtonDemo 工程目录结构

(2) 打开项目工程中 res→layout 目录下的 main.xml 文件，设置线性布局，添加一个 Button 按钮控件，并设置相关属性，代码如下：

```
1.    <?xml version="1.0" encoding="utf-8"?>
2.  <LinearLayout xmlns:android="http://
schemas.android.com/apk/res/android"
3.    android:orientation="vertical"
4.    android:layout_width="fill_parent"
5.    android:layout_height="fill_parent"
6.  >
7.      <Button android:text="普通按钮"
8.          android:id="@+id/button1"
9.          android:layout_width="wrap_content"
10.         android:layout_height="wrap_content">
11.     </Button>

12. </LinearLayout>
```

(3) 在项目工程 res→layout 目录下创建 imagebutton.xml 文件，设置线性布局，添加一个 TextView 控件和一个 ImageButton 按钮，代码如下：

```
1.  <?xml version="1.0" encoding="utf-8"?>
2.  <LinearLayout xmlns:android="http://schemas.android.com/apk/res/android"
3.    android:orientation="vertical" android:layout_width="fill_parent"
4.    android:layout_height="wrap_content">

5.  <TextView
6.          android:layout_width="wrap_content"
7.          android:layout_height="wrap_content"
8.          android:text="图片按钮:" />
9.  <ImageButton android:src="@drawable/icon"
10.             android:layout_height="wrap_content"
11.             android:layout_width="wrap_content"
12.             android:id="@+id/imageButton1">
13. </ImageButton>
```

```
14.  </LinearLayout>
```

(4) 在 src 目录下的 com. hisoft. activity 包下打开 ImageButtonActivity. java 文件,设置界面显示 imagebutton. xml 文件内容。

```
1.   package com.hisoft.activity;

2.   import android.app.Activity;
3.   import android.os.Bundle;
4.   import android.view.View;
5.   import android.view.View.OnClickListener;
6.   import android.widget.Button;

7.   public class ImageButtonActivity extends Activity
8.   {
9.       private Button button1;

10.        @Override
11.        public void onCreate(Bundle savedInstanceState)
12.        {
13.            super.onCreate(savedInstanceState);
14.            setContentView(R.layout.imagebutton);

15.        }
16.    }
```

(5) 在 src 目录下的 com. hisoft. activity 包下新建 MainActivity. java 文件,代码如下:

```
1.   package com.hisoft.activity;

2.   import android.app.Activity;
3.   import android.content.Intent;
4.   import android.os.Bundle;
5.   import android.view.View;
6.   import android.view.View.OnClickListener;
7.   import android.widget.Button;

8.   public class MainActivity extends Activity {

9.     private Button button1;

10.    @Override
11.    protected void onCreate(Bundle savedInstanceState) {
12.      super.onCreate(savedInstanceState);
13.      setContentView(R.layout.main);

14.      button1=(Button) this.findViewById(R.id.button1);
```

```
15.        //给 button1 设置监听
16.        button1.setOnClickListener(new OnClickListener() {

17.          public void onClick(View v) {
18.            //通过 Intent 跳转到 ImageButtonActivity
19.            Intent intent=new Intent();
20.            intent.setClass(MainActivity.this, ImageButtonActivity.class);
21.            startActivity(intent);
22.          }
23.        });
24.    }

25.  }
```

(6) 部署运行 ImageButtonDemo 项目工程,项目运行效果如图 5-10 所示。单击"普通按钮"按钮后,效果如图 5-11 所示。

图 5-10　ImageButtonDemo 运行结果

图 5-11　图片按钮运行效果

5.3 单选与复选按钮简介

5.3.1 单选按钮

单选按钮(RadioButton)是仅可以选择一个选项的控件,继承自 android. widget. CompoundButton,在 android. widget 包中,如图 5-12 所示。

单选按钮要声明在 RadioGroup 中,RadioGroup 是 RadioButton 的承载体,程序运行时不可见,应用程序中可能包含一个或多个 RadioGroup,RadioGroup 是线性布局 LinearLayout 的子类。其类的继承结构如图 5-13 所示,一个 RadioGroup 包含多个 RadioButton,RadioGroup 用于对单选框进行分组,在每个 RadioGroup 中(相同组内的单选框),用户仅能够选择其中一个 RadioButton。

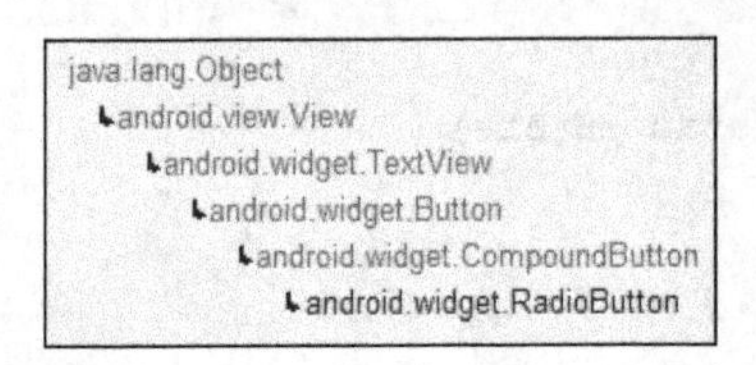

图 5-12　RadioButton 类继承图

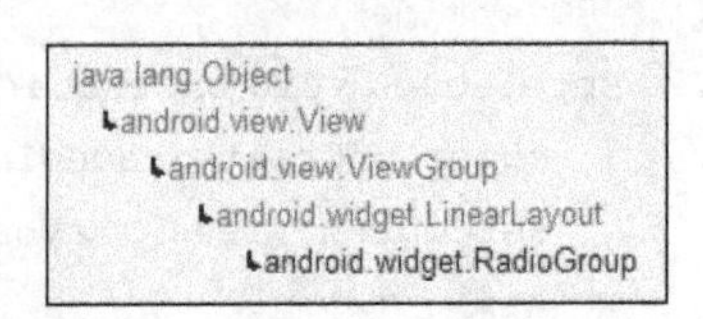

图 5-13　RadioGroup 的类继承图

单选按钮状态更改的监听是要给它的 RadioGroup 添加 setOnCheckedChangeListener (RadioGroup. OnCheckedChangeListener)监听器。注意,监听器类型和复选按钮(CheckBox)是不相同的。

单选按钮的通常用法:

(1) 用 xml 描述的 RadioGroup 和 RadioButton 应用的界面设计。

```
<?xml version="1.0" encoding="utf-8"?>
<LinearLayout xmlns:android="http://schemas.android.com/apk/res/android"
    android:orientation="vertical"
    android:layout_width="fill_parent"
    android:layout_height="fill_parent"
    >
<RadioGroup android:id="@+id/radioGroup"
  xmlns:android="http://schemas.android.com/apk/res/android"
  android:layout_width="wrap_content"
  android:layout_height="wrap_content">
<RadioButton android:id="@+id/java"
        android:layout_width="wrap_content"
        android:layout_height="wrap_content"
        android:text="java" />
    <RadioButton android:id="@+id/dotNet"
        android:layout_width="wrap_content"
        android:layout_height="wrap_content"
        android:text="dotNet" />
    <RadioButton android:id="@+id/php"
        android:layout_width="wrap_content"
        android:layout_height="wrap_content"
        android:text="PHP" />
</RadioGroup>
</LinearLayout>
```

(2) 引用处理程序。

```
public void onCreate(Bundle savedInstanceState) {
      …
        RadioGroup radioGroup= (RadioGroup) findViewById(R.id.radioGroup);
        radioGroup.setOnCheckedChangeListener(new RadioGroup.
        OnCheckedChangeListener() {
            public void onCheckedChanged(RadioGroup group, int checkedId) {
                RadioButton radioButton= (RadioButton) findViewById(checkedId);
                Log.i(TAG, String.valueOf(radioButton.getText()));
            }
        });
}
```

RadioButton 和 RadioGroup 常用的方法及说明如表 5-3 所示。

表 5-3 RadioButton 和 RadioGroup 常用的方法及描述

方法名称	描述
RadioGroup. check(int id)	通过传递的参数设置 RadioButton 单选框
RadioGroup. clearCheck()	清空选中的项
RadioGroup. setOnCheckedChangeListener()	处理单选框 RadioButton 被选择事件，把 RadioGroup. OnCheckedChangeListener 实例作为参数传入
RadioButton . getText()	获取单选框的值

如下代码：

```
RadioGroup.check(R.id.dotNet);         //将 id 名为 dotNet 的单选框设置成选中状态
(RadioButton) findViewById(radioGroup.getCheckedRadioButtonId());
                                       //获取被选中的单选框
RadioButton.getText();                 //获取单选框的值
```

5.3.2 复选按钮

复选按钮(CheckBox)是一个同时可以选择多个选项的控件，继承自 android. widget. CompoundButton，在 android. widget 包中，如图 5-14 所示。

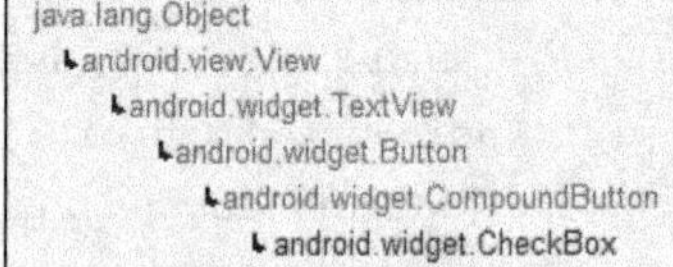

图 5-14 CheckBox 类继承结构

每个多选框都是独立的，可以通过迭代所有多选框，然后根据其状态是否被选中再获取其值。CheckBox 常用方法如表 5-4 所示。

表 5-4 CheckBox 常用方法及描述

方法名称	描述
isChecked()	检查是否被选中
setChecked(boolean)	如为 true，设置成选中状态
setOnCheckedChangeListener()	处理多选框 CheckBox 被选择事件，监听按钮状态是否更改，把 CompoundButton. OnCheckedChangeListener 实例作为参数传入
getText()	获取多选框的值

CheckBox 的通常用法：

(1) 用 xml 描述的 CheckBox 应用界面设计。

```
<?xml version="1.0" encoding="utf-8"?>
<LinearLayout
  xmlns:android="http://schemas.android.com/apk/res/android"
  android:layout_width="wrap_content"
  android:layout_height="fill_parent">
  <CheckBox android:id="@+id/checkboxjava"
    android:layout_width="wrap_content"
    android:layout_height="wrap_content"
```

```
    android:text="java" />
  <CheckBox android:id="@+id/checkboxdotNet"
    android:layout_width="wrap_content"
    android:layout_height="wrap_content"
    android:text="dotNet" />
  <CheckBox android:id="@+id/checkboxphp"
    android:layout_width="wrap_content"
    android:layout_height="wrap_content"
    android:text="PHP" />

    <Button android:id="@+id/checkboxButton"
    android:layout_width="fill_parent"
    android:layout_height="wrap_content"
    android:text="获取值" />
</LinearLayout>
```

(2) 引用 xml 描述的代码处理。

```
public class CheckBoxActivity extends Activity {
private static final String TAG="CheckBoxActivity";
private List<CheckBox>checkboxs=new ArrayList<CheckBox>();

    @Override
    public void onCreate(Bundle savedInstanceState) {
        super.onCreate(savedInstanceState);
        setContentView(R.layout.checkbox);
        checkboxs.add((CheckBox) findViewById(R.id.checkboxdotNet));
        checkboxs.add((CheckBox) findViewById(R.id.checkboxjava));
        checkboxs.add((CheckBox) findViewById(R.id.checkboxphp));
        checkboxs.get(1).setChecked(true);              //设置成选中状态
        for(CheckBox box : checkboxs){
          box.setOnCheckedChangeListener(listener);
        }
        Button button=(Button)findViewById(R.id.checkboxButton);
        button.setOnClickListener(new View.OnClickListener() {
  @Override
  public void onClick(View v) {
     List<String>values=new ArrayList<String>();
     for(CheckBox box : checkboxs){
      if(box.isChecked()){
          values.add(box.getText().toString());
      }
     }
     Toast.makeText(CheckBoxActivity.this, values.toString(), 1).show();
  }
  });
```

```
        }
          CompoundButton.OnCheckedChangeListener listener=
          new CompoundButton.OnCheckedChangeListener() {@Override
        public void onCheckedChanged(CompoundButton buttonView, boolean isChecked) {
        CheckBox checkBox= (CheckBox) buttonView;
        Log.i(TAG, "isChecked="+isChecked+",value="+checkBox.getText());
                                                              //输出单选框的值
        }
          };
    }
```

5.3.3 RadioButton 和 CheckBox 综合应用案例

通过上述的 RadioButton 和 CheckBox 基本介绍，下面通过一个案例应用加深读者对 RadioButton 和 CheckBox 的用法和应用的熟悉和掌握。具体步骤如下：

(1) 创建一个新的 Android 工程，工程名为 RadioButtonAndCheckboxDemo，目标 API 选择 10(即 Android 2.3.3 版本)，应用程序名为 RadioButtonAndCheckboxDemo，包名为 com.hisoft.activity，创建的 Activity 的名字为 RadioButtonAndCheckboxActivity，最小 SDK 版本根据选择的目标 API 会自动添加为 10，创建后的项目工程如图 5-15 所示。

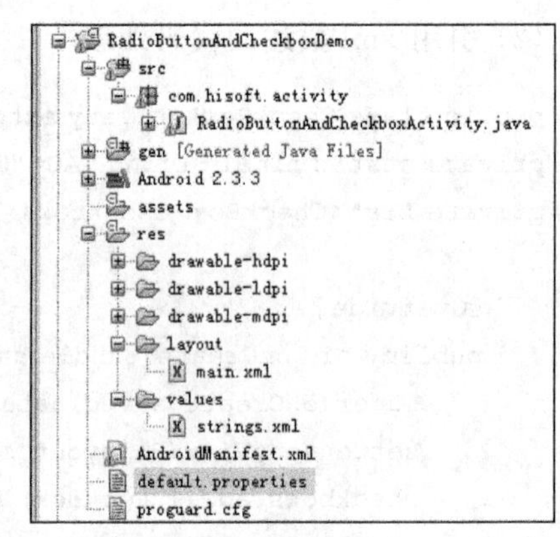

图 5-15 RadioButtonAndCheckboxDemo 工程目录结构

(2) 打开项目工程中 res→layout 目录下的 main.xml 文件，设置线性布局，添加两个 TextView 控件、一个 RadioGroup、两个 RadioButton 和三个 CheckBox，并设置相关属性，代码如下：

```
1.   <?xml version="1.0" encoding="utf-8"?>
2.   <LinearLayout xmlns:android="http://schemas.android.com/apk/res/android"
3.     android:orientation="vertical"
4.     android:layout_width="fill_parent"
5.     android:layout_height="fill_parent"
6.     >
7.       <TextView android:text="性别:"
8.             android:id="@+id/textView1"
9.             android:layout_width="wrap_content"
10.               android:layout_height="wrap_content">
11.      </TextView>
12.      <RadioGroup
13.          android:layout_width="fill_parent"
14.          android:layout_height="wrap_content"
15.          android:orientation="vertical"
16.          android:checkedButton="@+id/radioButton1"
```

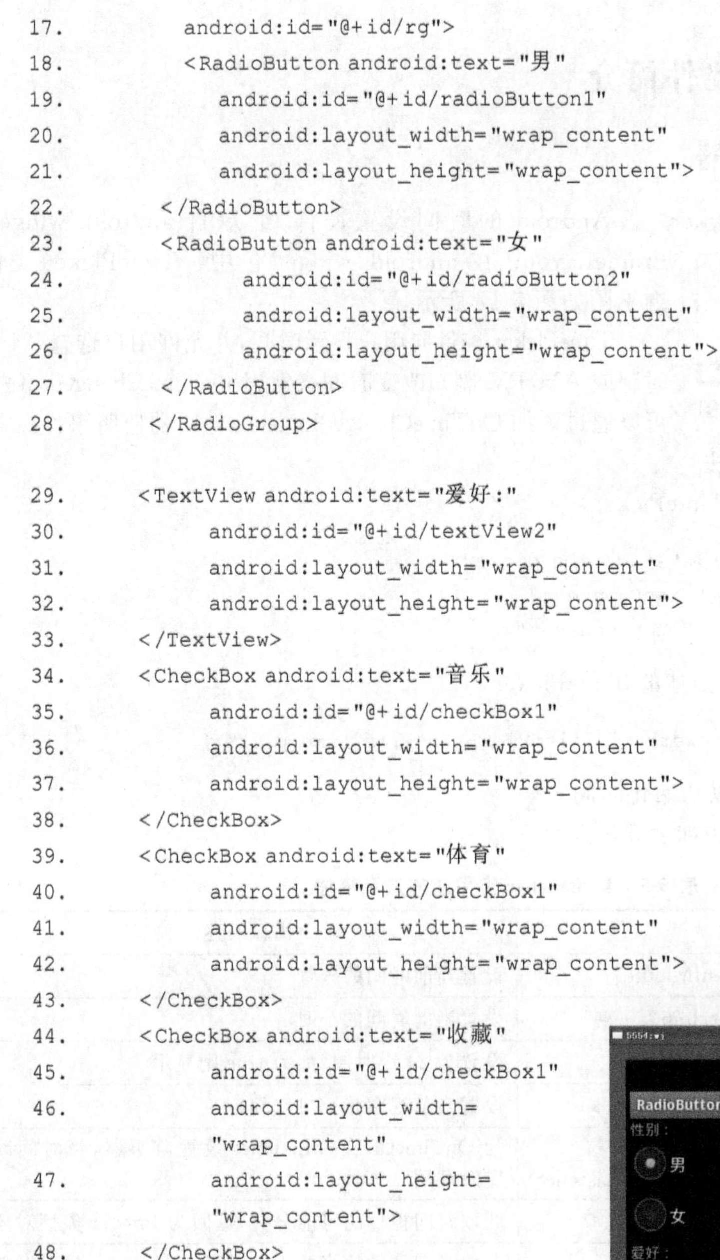

```
17.          android:id="@+id/rg">
18.          <RadioButton android:text="男"
19.             android:id="@+id/radioButton1"
20.             android:layout_width="wrap_content"
21.             android:layout_height="wrap_content">
22.        </RadioButton>
23.        <RadioButton android:text="女"
24.             android:id="@+id/radioButton2"
25.             android:layout_width="wrap_content"
26.             android:layout_height="wrap_content">
27.        </RadioButton>
28.      </RadioGroup>

29.      <TextView android:text="爱好:"
30.            android:id="@+id/textView2"
31.            android:layout_width="wrap_content"
32.            android:layout_height="wrap_content">
33.      </TextView>
34.      <CheckBox android:text="音乐"
35.            android:id="@+id/checkBox1"
36.            android:layout_width="wrap_content"
37.            android:layout_height="wrap_content">
38.      </CheckBox>
39.      <CheckBox android:text="体育"
40.            android:id="@+id/checkBox1"
41.            android:layout_width="wrap_content"
42.            android:layout_height="wrap_content">
43.      </CheckBox>
44.      <CheckBox android:text="收藏"
45.            android:id="@+id/checkBox1"
46.            android:layout_width=
               "wrap_content"
47.            android:layout_height=
               "wrap_content">
48.      </CheckBox>

49.  </LinearLayout>
```

(3) src 目录下 com. hisoft. activity 包下的 RadioButtonAndCheckboxActivity. java 文件和 res→values 目录下的 strings. xml 文件都暂不做修改。部署运行项目工程,项目运行效果如图 5-16 所示。

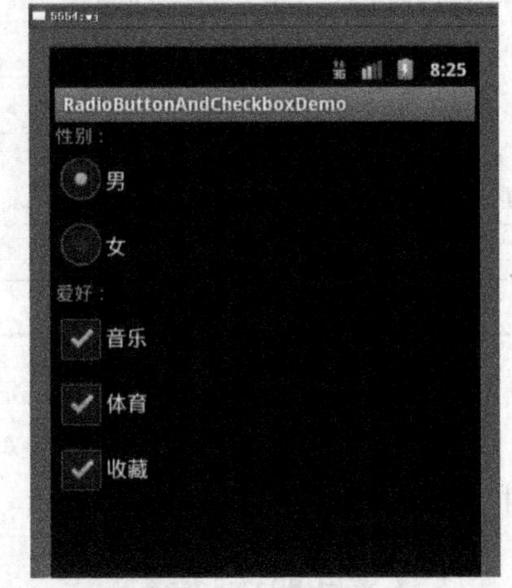

图 5-16 RadioButtonAndCheckboxDemo 运行结果

5.4 时间与日期控件简介

5.4.1 时间选择器

时间选择器(TimePicker)是 Android 的时间设置控件,继承自 android. widget. FrameLayout,在 android. widget 包中。TimePicker 类的继承图如图 5-17 所示。

java.lang.Object
↳android.view.View
↳android.view.ViewGroup
↳android.widget.FrameLayout
↳android.widget.TimePicker

图 5-17 TimePicker 类继承图

TimePicker 控件向用户显示时间,并允许用户选择(24 小时制或 AM/PM 制),改变时间会触发 OnTimeChanged 事件,可以通过添加 OnTimeChangedListener 监听器监听事件。

TimePicker 的通常用法:

(1) 用 xml 描述一个 TimePicker。

```
<TimePicker android:id="@+id/time_picker"
android:layout_width="wrap_content"
android:layout_height="wrap_content"/>
```

(2) 程序中引用 XML 描述的 TimePicker。

```
TimePicker tp= (TimePicker)this.findViewById(R.id.time_picker);
```

然后在使用的时候可以初始化时间。

TimePicker 常用的方法如表 5-5 所示。

表 5-5 TimePicker 常用的方法及描述

方法名称	描述
setCurrentMinute(Integer currentMinute)	设置当前时间的分钟
setCurrentHour(Integer currentHour)	设置当前时间的小时
setIs24HourView(boolean)	设置为 24 小时制,如为 true 则显示
setEnabled(boolean enabled)	设置当前视图是否可以编辑
setOnTimeChangedListener(TimePicker. OnTimeChangedListener onTimeChangedListener)	为 OnTimeChangedListener 设置监听器,当时间改变时调用
getCurrentMinute()	获取时间控件的当前分钟,返回为 Integer 类型对象
getCurrentHour()	获取时间控件的当前小时,返回为 Integer 类型对象

相关类包有 TimePickerDialog、DatePickerDialog,以对话框形式显示日期时间视图。

Calendar(日历)是设定年度日期对象和一个整数字段之间转换的抽象基类,如月,日,小时等。

5.4.2 日期选择器

日期选择器(DatePicker)是 Android 的日期设置控件,也继承自 android. widget. FrameLayout,在 android. widget 包中。DatePicker 类的继承图如图 5-18 所示。

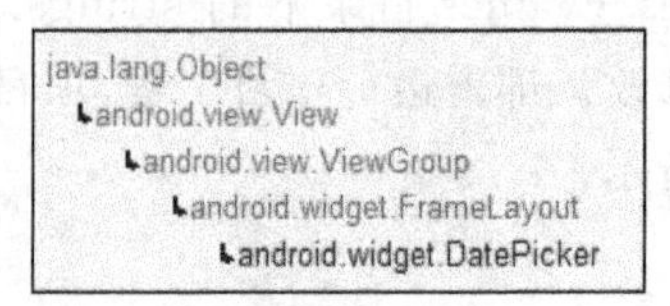

图 5-18 DatePicker 类继承图

DatePicker 控件提供年、月、日的日期数据，并允许用户进行选择。改变日期会触发 onDateChanged 事件，通过添加 onDateChangedListener 监听器可以监听捕获事件。

DatePicker 的通常用法：

(1) 用 xml 描述一个 DatePicker。

```
<DatePicker
android:id="@+id/date_picker"
android:layout_width="wrap_content"
android:layout_height="wrap_content" />
```

(2) 程序中引用 XML 描述的 DatePicker。

```
DatePicker dp= (DatePicker)this.findViewById(R.id.date_picker);
dp.init(2012, 8, 17, null);          //使用的时候可以初始化时间
```

DatePicker 常用的方法如表 5-6 所示。

表 5-6 DatePicker 常用的方法及说明

方法名称	描述
getDayOfMonth()	获取当前日
getMonth()	获取当前月
getYear()	获取当前年
updateDate(int year, int monthOfYear, int dayOfMonth)	更新日期
setEnabled(boolean enabled)	根据参数设置日期选择器控件是否可用或编辑
init(int year, int monthOfYear, int dayOfMonth, DatePicker.OnDateChangerdListener onDateChangedListener)	初始化日期选择器控件的属性，参数 onDateChangedListener 为监听器对象，监听日期数据变化

5.4.3 时间与日期控件综合应用案例

上面章节介绍了 TimePicker 和 DatePicker 的基本用法和方法，下面通过一个案例应用加深读者对 TimePicker 和 DatePicker 应用的熟悉和掌握。具体步骤如下：

(1) 创建一个新的 Android 工程，工程名为 TimeAndDatePickerDemo，目标 API 选择 10(即 Android 2.3.3 版本)，应用程序名为 TimeAndDatePickerDemo，包名为 com. hisoft. activity，创建的 Activity 的名字为 DatePickerActivity，最小 SDK 版本根据选择的目标 API 会自动添加为 10，创建后的项目工程图如图 5-19 所示。

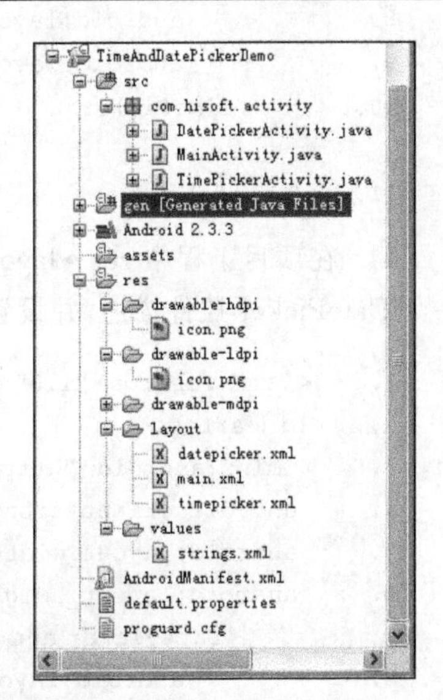

图 5-19 TimeAndDatePickerDemo 工程结构图

(2) 打开项目工程中 res→layout 目录下的 main.xml 文件，设置线性布局，添加两个 Button，并设置相关属性，代码如下：

```
1.  <?xml version="1.0" encoding="utf-8"?>
2.  <LinearLayout xmlns:android="http://schemas.android.com/apk/res/android"
3.    android:orientation="vertical"
4.    android:layout_width="fill_parent"
5.    android:layout_height="fill_parent"
6.    >
7.      <Button android:text="to TimePicker"
8.          android:id="@+id/bn_time"
9.          android:layout_width="wrap_content"
10.           android:layout_height="wrap_content">
11.     </Button>
12.     <Button android:text="to DatePicker"
13.         android:id="@+id/bn_date"
14.         android:layout_width="wrap_content"
15.         android:layout_height="wrap_content">
16.     </Button>

17. </LinearLayout>
```

(3) 在项目工程中 res→layout 目录下创建 timepicker.xml 文件,设置线性布局,添加一个 TimePicker 控件描述,并设置相关属性,代码如下:

```
1.      <?xml version="1.0" encoding="utf-8"?>
2.  <LinearLayout
3.    xmlns:android="http://schemas.android.com/apk/res/android"
4.    android:orientation="vertical"
5.    android:layout_width="match_parent"
6.    android:layout_height="match_parent">
7.      <TimePicker android:id="@+id/timePicker1"
8.            android:layout_width="wrap_content"
9.            android:layout_height="wrap_content">
10.     </TimePicker>

11. </LinearLayout>
```

(4) 在项目工程中 res→layout 目录下创建 datepicker.xml 文件,设置线性布局,添加一个 DatePicker 控件描述,并设置相关属性,代码如下:

```
1.   <?xml version="1.0" encoding="utf-8"?>
2.  <LinearLayout
3.    xmlns:android="http://schemas.android.com/apk/res/android"
4.    android:orientation="vertical"
5.    android:layout_width="match_parent"
6.    android:layout_height="match_parent">
7.      <DatePicker android:id="@+id/datePicker1"
8.            android:layout_width="wrap_content"
9.            android:layout_height="wrap_content">
10.     </DatePicker>
```

```
11.  </LinearLayout>
```

(5) 修改 src 目录中 com. hisoft. activity 包下的 DatePickerActivity. java 文件,代码如下:

```
1.   package com.hisoft.activity;
2.   import android.app.Activity;
3.   import android.os.Bundle;
4.   public class DatePickerActivity extends Activity {

5.     @Override
6.     protected void onCreate(Bundle savedInstanceState) {
7.       super.onCreate(savedInstanceState);
8.       this.setContentView(R.layout.datepicker);
9.     }

10.  }
```

(6) 在 src 目录中 com. hisoft. activity 包下创建 TimePickerActivity. java 文件,代码如下:

```
1.   package com.hisoft.activity;

2.   import android.app.Activity;
3.   import android.os.Bundle;

4.   public class TimePickerActivity extends Activity {

5.     @Override
6.     protected void onCreate(Bundle savedInstanceState) {
7.       super.onCreate(savedInstanceState);
8.       this.setContentView(R.layout.timepicker);
9.     }

10.  }
```

(7) 在 src 目录中 com. hisoft. activity 包下创建 MainActivity. java 文件,代码如下:

```
1.   package com.hisoft.activity;

2.   import android.app.Activity;
3.   import android.content.Intent;
4.   import android.os.Bundle;
5.   import android.view.View;
6.   import android.view.View.OnClickListener;
7.   import android.widget.Button;
8.   public class MainActivity extends Activity {
9.     private Button bn_time, bn_date;
```

```
10.    @Override
11.    protected void onCreate(Bundle savedInstanceState) {
12.      super.onCreate(savedInstanceState);
13.      setContentView(R.layout.main);

14.      this.bn_time=(Button) this.findViewById(R.id.bn_time);
15.      this.bn_date=(Button) this.findViewById(R.id.bn_date);

16.      MyListener ml=new MyListener();
17.      this.bn_time.setOnClickListener(ml);
18.      this.bn_date.setOnClickListener(ml);

19.    }

20.    class MyListener implements OnClickListener{

21.      private Intent intent=new Intent();
22.      public void onClick(View v) {
23.        if(v==bn_time){
24.          intent.setClass(MainActivity.this, TimePickerActivity.class);
25.        }
26.        if(v==bn_date){
27.          intent.setClass(MainActivity.this, DatePickerActivity.class);
28.        }
29.        startActivity(intent);
30.      }
31.    }
32.  }
```

(8) 部署运行 TimeAndDatePickerDemo 项目工程，项目运行效果如图 5-20 所示。

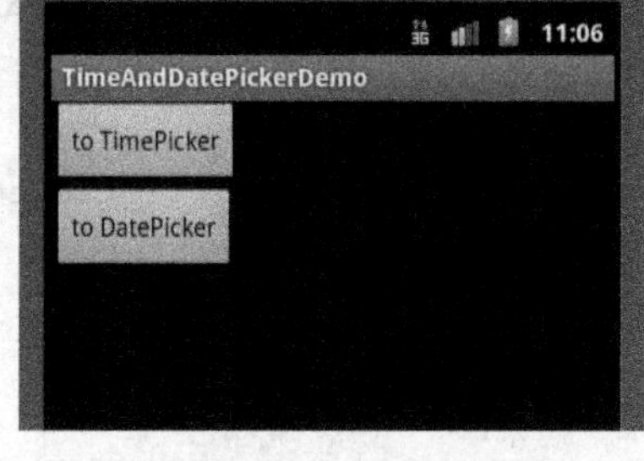

图 5-20 TimeAndDatePickerDemo 运行结果图

单击 to TimePicker 按钮，显示时间界面，如图 5-21 所示。

单击 to DatePicker 按钮，显示日期界面，如图 5-22 所示。

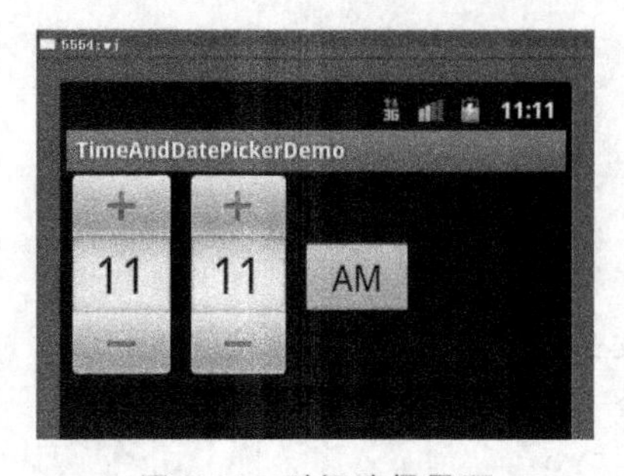

图 5-21 时间选择界面

图 5-22 日期选择界面

5.5 图片控件简介

5.5.1 图片控件

图片控件(ImageView)是最常用的组件之一,继承自 android. view. View,它的已知直接子类有 ImageButton、QuickContactBadge;已知间接子类有 ZoomButton。类图继承结构如图 5-23 所示。

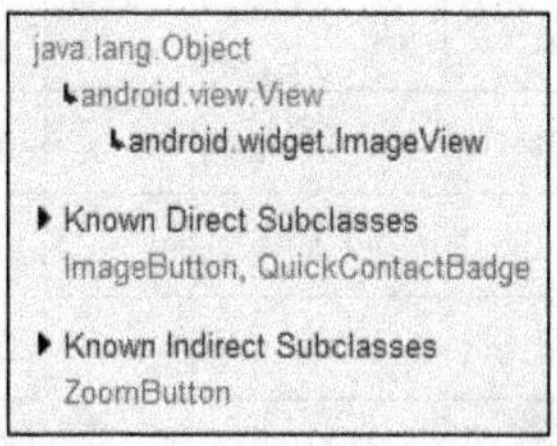

java.lang.Object
 ↳android.view.View
  ↳android.widget.ImageView

▶ Known Direct Subclasses
ImageButton, QuickContactBadge

▶ Known Indirect Subclasses
ZoomButton

图 5-23 ImageView类继承图

ImageView 控件显示任意图像,例如图标。ImageView 类可以加载各种来源的图片(如资源或图片库),其图片的来源可以是在资源文件中的 id,也可以是 Drawable 对象或者位图对象,还可以是 Content Provider 的 URI。需要计算图像的尺寸,以便可以在其他布局中使用,并提供例如缩放和着色(渲染)等各种显示选项。

ImageView 通常的用法:

(1) 用 xml 来描述 ImageView。

```
<ImageView
android:id="@+id/imagebutton"
android:src="@drawable/wjj "
android:layout_width="wrap_content"
android:layout_height="wrap_content"/>
```

(2) 在程序中引用 XML 描述的控件并处理。

```
ImageView  image1=(ImageView) findViewById(R.id.img1);
```

ImageView 设置图片、设置图片源的方法主要有三种:

(1) 设定图片相对路径。

```
ImageView iv;
String fileName="/data/com.test/aa.png;
Bitmap bm=BitmapFactory.decodeFile(fileName);
iv.setImageBitmap(bm);
```

(2) 通过传递 context,访问特定的图片源。

```
ImageView iv=new ImageView(context);
iv.setImageResource(iv[position]);
iv.setScaleType(ImageView.ScaleType.FIT_XY);
iv.setLayoutParams(new Gallery.LayoutParams(136,88));
```

(3) 通过获取 XML 描述中设定的图片或图片源。

```
mImageView=(ImageView)this.findViewById(R.id.myImageView1);
mImageView.setImageDrawable(getResources().getDrawable(R.drawable.right));
```

ImageView 常用的属性和方法如表 5-7 和表 5-8 所示。

表 5-7　ImageView 常用的属性及描述

属 性 名 称	描　　述
android：adjustViewBounds	是否保持宽高比。需要与 maxWidthMaxHeight 一起使用，否则单独使用没有效果
android：cropToPadding	是否截取指定区域用空白代替。单独设置无效果，需要与 scrollY 一起使用
android：tint	将图片渲染成指定的颜色
android：maxHeight	最大高度
android：maxWidth	最大宽度
android：src	图片路径
android：scaleType	调整或移动图片

表 5-8　ImageView 常用方法及对应 XML 属性和描述说明

方　　法	对应 XML 属性	描　　述
setAlpha(int)		设置 ImageView 透明度
setImageBitmap(Bitmap)		设置位图作为该 ImageView 的内容
setImageDrawable(Drawable)		设置 ImageView 所显示内容为 Drawable
setImageURI(Uri)		设置 ImageView 所显示内容为 Uri
setSelected(boolean)		设置 ImageView 的选择状态
setImageResource(int)	android：src	通过资源 ID 设置可绘制对象为该 ImageView 显示的内容
setBaselineAlignBottom (boolean aligned)	android：baselineAlignBottom	设置是否设置视图底部的视图基线。设置这个值覆盖 setBaseline()的所有调用
setAdjustViewBounds (boolean adjustViewBounds)	android：adjustViewBounds	当需要在 ImageView 调整边框时保持可绘制对象的比例时，将该值设为真
setScaleType (ImageView.ScaleType scaleType)	android：scaleType	控制图像应该如何缩放和移动，以使图像与 ImageView 一致
getScaleType ()	android：scaleType	返回当前 ImageView 使用的缩放类型

5.5.2　ImageView 应用案例

上面章节介绍了 ImageView 的基本用法、常用属性等知识，下面通过一个案例应用熟悉和掌握 ImageView 应用。具体步骤如下：

(1) 创建一个新的 Android 工程，工程名为 ImageViewDemo，目标 API 选择 10(即 Android 2.3.3 版本)，应用程序名为 ImageViewDemo，包名为 com.hisoft.activity，创建的 Activity 的名字为 ImageViewActivity，最小 SDK 版本根据选择的目标 API 会自动添加为 10，创建后的项目工程如图 5-24 所示。

(2) 打开项目工程中 res→layout 目录下的 main.xml 文件，设置线性布局，添加一个

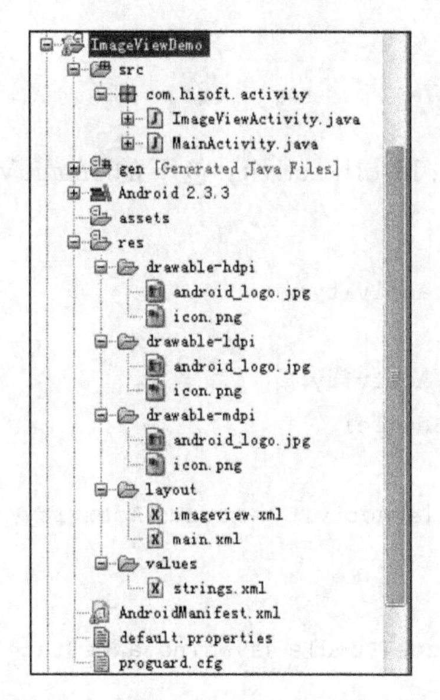

图 5-24 ImageViewDemo 工程结构

Button,并设置相关属性,代码如下:

```
1.  <?xml version="1.0" encoding="utf-8"?>
2.  <LinearLayout xmlns:android="http://schemas.android.com/apk/res/android"
3.    android:orientation="vertical"
4.    android:layout_width="fill_parent"
5.    android:layout_height="fill_parent"
6.    >
7.      <Button android:text="显示图片"
8.          android:id="@+id/button1"
9.          android:layout_width="wrap_content"
10.           android:layout_height="wrap_content">
11.     </Button>

12. </LinearLayout>
```

(3) 在项目工程中 res→layout 目录下创建 imageview.xml 文件,设置线性布局,添加一个 ImageView 控件描述,并设置相关属性,代码如下:

```
1.    <?xml version="1.0" encoding="utf-8"?>
2.  <LinearLayout
3.    xmlns:android="http://schemas.android.com/apk/res/android"
4.    android:orientation="vertical"
5.    android:layout_width="match_parent"
6.    android:layout_height="match_parent">
7.      <ImageView android:layout_height="wrap_content"
8.           android:id="@+id/imageView1"
9.           android:layout_width="wrap_content"
10.          android:src="@drawable/android_logo">
```

```
11.     </ImageView>

12. </LinearLayout>
```

(4) 修改 src 目录中 com. hisoft. activity 包下的 ImageViewActivity. java 文件，代码如下：

```
1.  package com.hisoft.activity;

2.  import android.app.Activity;
3.  import android.os.Bundle;

4.  public class ImageViewActivity extends Activity
5.  {
6.    @Override
7.    public void onCreate(Bundle savedInstanceState)
8.    {
9.      super.onCreate(savedInstanceState);
10.     setContentView(R.layout.imageview);
11.   }
12. }
```

(5) 在 src 目录中 com. hisoft. activity 包下创建 MainActivity. java 文件，代码如下：

```
1.  package com.hisoft.activity;

2.  import android.app.Activity;
3.  import android.content.Intent;
4.  import android.os.Bundle;
5.  import android.view.View;
6.  import android.view.View.OnClickListener;
7.  import android.widget.Button;
8.
9.  public class MainActivity extends Activity {

10.   private Button button1;

11.   @Override
12.   protected void onCreate(Bundle savedInstanceState) {
13.     super.onCreate(savedInstanceState);
14.     setContentView(R.layout.main);
15.
16.     button1=(Button) this.findViewById(R.id.button1);
17.
18.     //给 button1 设置监听
19.     button1.setOnClickListener(new OnClickListener() {
```

```
20.
21.         public void onClick(View v) {
22.           //通过 Intent 跳转到 ImageViewActivity
23.           Intent intent=new Intent();
24.           intent.setClass(MainActivity.this, ImageViewActivity.class);
25.           startActivity(intent);
26.         }
27.       });
28.     }
29.   }
```

(6) 把 android_logo.jpg 图片文件复制到资源 res 目录下的 drawable-hdpi、drawable-hdpi 和 drawable-hdpi 文件夹中。

(7) 部署运行 ImageViewDemo 项目工程，项目运行效果如图 5-25 所示。

单击“显示图片”按钮，显示结果如图 5-26 所示。

图 5-25 ImageViewDemo 运行结果

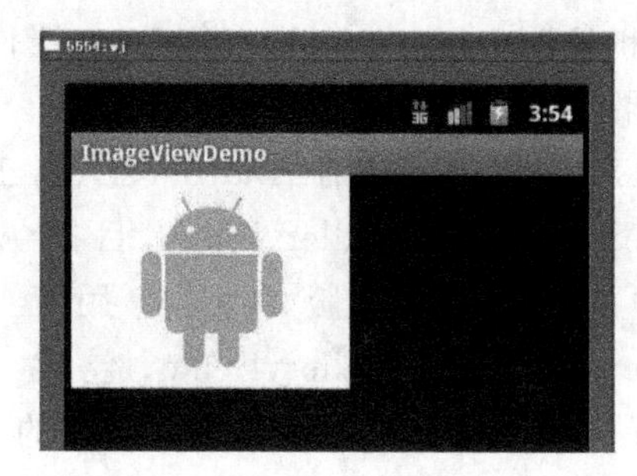

图 5-26 图片按钮显示

5.5.3 切换图片控件 ImageSwitcher、Gallery

ImageSwitcher 是 Android 中控制图片展示效果的一个控件，如幻灯片效果…，继承自 android. widget. ViewSwitcher。控件继承结构图如图 5-27 所示。

Gallery 控件是一个锁定中心条目并且拥有水平滚动列表的视图，它继承自 android. widget，其控件继承结构图如图 5-28 所示。这个控件目前已经被 Android 系统弃用，不再被长期系统支持，系统库支持的水平滚动部件有 HorizontalScrollView 和 ViewPager。

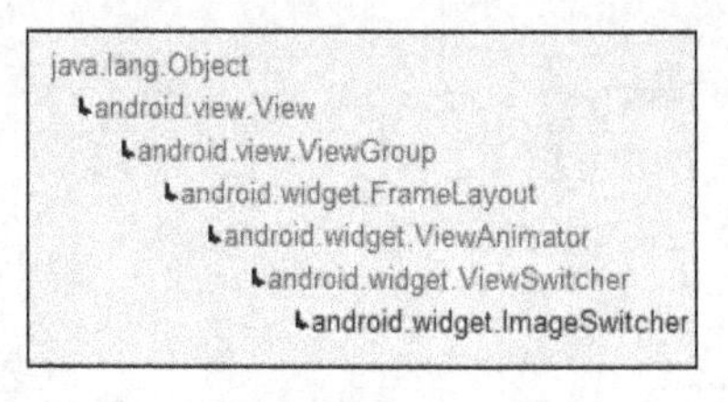

图 5-27 ImageSwitcher 类继承图

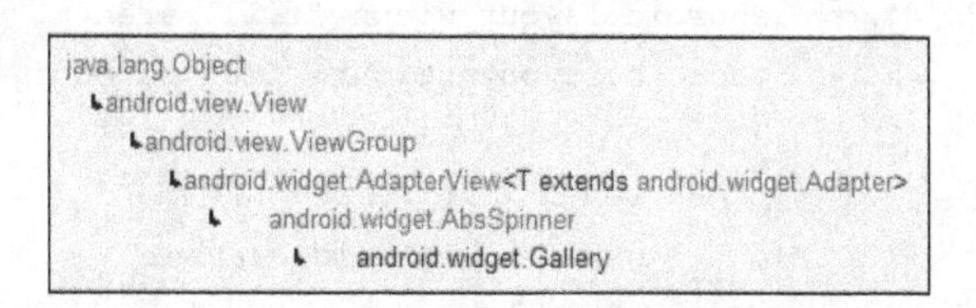

图 5-28 Gallery 类继承图

Gallery 使用 Theme_galleryItemBackground 作为 Gallery 适配器中各视图的默认参数。如果没有设置，就需要调整一些 Gallery(画廊)的属性，比如间距等。其常用的属性及对应方法如表 5-9 所示。

表 5-9 Gallery 常用的属性及对应方法、描述

属性名称	对应方法	描述
android:animationDuration	setAnimationDuration(int)	设置当子视图改变位置时动画转换时间。仅限于动画开始时生效
android:gravity	setGravity(int)	描述子视图的对齐方式
android:spacing	setSpacing(int)	设置 Gallery 中项的间距
android:unselectedAlpha	setUnselectedAlpha(float)	设置 Gallery 中未选中项的透明度(alpha)值

Gallery 中的视图应该使用 Gallery. LayoutParams 作为它们的布局参数类型。

注意：通常情况下，ImageSwitcher 组件和 Gallery 组件配合使用。

5.5.4 ImageSwitcher、Gallery 综合应用案例

上面章节介绍了 ImageSwitcher、Gallery 的常用属性和功能，下面通过一个案例应用熟悉和掌握 ImageSwitcher、Gallery 控件组合应用效果。具体步骤如下：

(1) 创建一个新的 Android 工程，工程名为 ImageSwitcherAndGalleryDemo，目标 API 选择 10(即 Android 2.3.3 版本)，应用程序名为 ImageSwitcherAndGalleryDemo，包名为 com. hisoft. activity，创建的 Activity 的名字为 ImageSwitcherAndGalleryActivity，最小 SDK 版本根据选择的目标 API 会自动添加为 10，创建后的项目工程如图 5-29 所示。

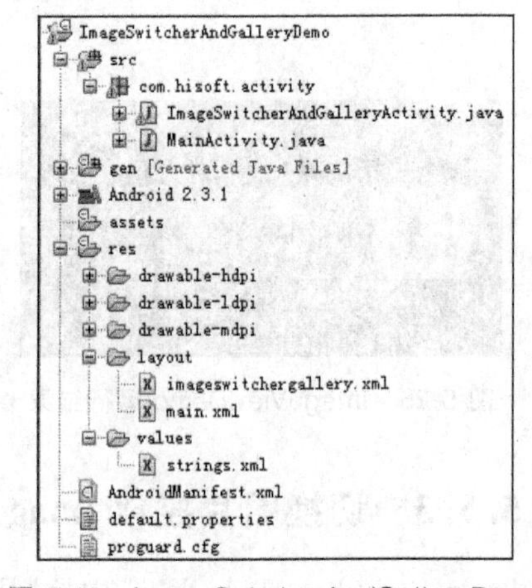

图 5-29 ImageSwitcherAndGalleryDemo 工程目录结构

(2) 打开项目工程中 res→layout 目录下的 main. xml 文件，设置线性布局，添加一个 Button，并设置相关属性，代码如下：

```
1.  <?xml version="1.0" encoding="utf-8"?>
2.  <LinearLayout xmlns:android="http://schemas.android.com/apk/res/android"
3.    android:orientation="vertical"
4.    android:layout_width="fill_parent"
5.    android:layout_height="fill_parent"
6.    >
7.     <Button android:text="浏览图片"
8.         android:id="@+id/button1"
9.         android:layout_width="wrap_content"
10.          android:layout_height="wrap_content">
11.     </Button>

12.  </LinearLayout>
```

(3) 在项目工程中 res→layout 目录下创建 imageswitchergallery. xml 文件，设置相对

布局,添加一个 Gallery 控件描述,并设置相关属性,代码如下:

```
1.    <?xml version="1.0" encoding="utf-8"?>
2.  <RelativeLayout
3.    xmlns:android="http://schemas.android.com/apk/res/android"
4.    android:layout_width="match_parent"
5.    android:layout_height="match_parent">
6.      <ImageSwitcher
7.          android:id="@+id/switcher"
8.                android:layout_width="fill_parent"
9.                android:layout_height="fill_parent"
10.                   android:layout_alignParentTop="true"
11.                   android:layout_alignParentLeft="true" />
12.
13.     <Gallery android:id="@+id/gallery"
14.                   android:background="#55000000"
15.                   android:layout_width="fill_parent"
16.                   android:layout_height="60dp"
17.                   android:layout_alignParentBottom="true"
18.                   android:layout_alignParentLeft="true"
19.                   android:gravity="center_vertical"
20.                   android:spacing="16dp" />
21. </RelativeLayout>
```

(4) 修改 src 目录中 com. hisoft. activity 包下的 ImageSwitcherAndGalleryActivity. java 文件,代码如下:

```
1.  package com.hisoft.activity;
2.
3.  import android.app.Activity;
4.  import android.content.Context;
5.  import android.os.Bundle;
6.  import android.view.View;
7.  import android.view.ViewGroup;
8.  import android.view.Window;
9.  import android.view.animation.AnimationUtils;
10. import android.widget.AdapterView;
11. import android.widget.BaseAdapter;
12. import android.widget.Button;
13. import android.widget.CheckBox;
14. import android.widget.EditText;
15. import android.widget.Gallery;
16. import android.widget.ImageSwitcher;
17. import android.widget.ImageView;
18. import android.widget.TextView;
19. import android.widget.ViewSwitcher;
```

```
20. import android.widget.Gallery.LayoutParams;
21.
22.
23. public class ImageSwitcherAndGalleryActivity  extends Activity implements
    AdapterView.OnItemSelectedListener, ViewSwitcher.ViewFactory {
24.
25.     @Override
26.     public void onCreate(Bundle savedInstanceState) {
27.         super.onCreate(savedInstanceState);
28.         requestWindowFeature(Window.FEATURE_NO_TITLE);
29.
30.         setContentView(R.layout.imageswitchergallery);
31.         setTitle("ImageShowActivity");
32.
33.         mSwitcher= (ImageSwitcher) findViewById(R.id.switcher);
34.         mSwitcher.setFactory(this);
35.         mSwitcher.setInAnimation(AnimationUtils.loadAnimation(this,
36.                 android.R.anim.fade_in));
37.         mSwitcher.setOutAnimation(AnimationUtils.loadAnimation(this,
38.                 android.R.anim.fade_out));
39.
40.
41.         Gallery g= (Gallery) findViewById(R.id.gallery);
42.         g.setAdapter(new ImageAdapter(this));
43.         g.setOnItemSelectedListener(this);
44.     }
45.
46.     public void onItemSelected(AdapterView parent, View v, int position,
    long id) {
47.         mSwitcher.setImageResource(mImageIds[position]);
48.     }
49.
50.     public void onNothingSelected(AdapterView parent) {
51.     }
52.
53.     public View makeView() {
54.         ImageView i=new ImageView(this);
55.         i.setBackgroundColor(0xFF000000);
56.         i.setScaleType(ImageView.ScaleType.FIT_CENTER);
57.         i.setLayoutParams(new ImageSwitcher.LayoutParams(LayoutParams.FILL
        _PARENT,
58.                 LayoutParams.FILL_PARENT));
59.         return i;
60.     }
61.
```

```
62.     private ImageSwitcher mSwitcher;
63.
64.     public class ImageAdapter extends BaseAdapter {
65.         public ImageAdapter(Context c) {
66.             mContext=c;
67.         }
68.
69.         public int getCount() {
70.             return mThumbIds.length;
71.         }
72.
73.         public Object getItem(int position) {
74.             return position;
75.         }
76.
77.         public long getItemId(int position) {
78.             return position;
79.         }
80.
81.         public View getView(int position, View convertView, ViewGroup parent) {
82.             ImageView i=new ImageView(mContext);
83.
84.             i.setImageResource(mThumbIds[position]);
85.             i.setAdjustViewBounds(true);
86.             i.setLayoutParams(new Gallery.LayoutParams(
87.                     LayoutParams.WRAP_CONTENT,
88.                     LayoutParams.WRAP_CONTENT));
89.             i.setBackgroundResource(R.drawable.picture_frame);
90.             return i;
91.         }
92.
93.         private Context mContext;
94.
95.     }
96.
97.    private Integer[] mThumbIds={
98.             R.drawable.sample_thumb_0,
99.             R.drawable.sample_thumb_1,
100.            R.drawable.sample_thumb_2,
101.            R.drawable.sample_thumb_3,
102.            R.drawable.sample_thumb_4,
103.            R.drawable.sample_thumb_5,
104.            R.drawable.sample_thumb_6,
105.            R.drawable.sample_thumb_7};
106.
```

```
107.       private Integer[] mImageIds={
108.               R.drawable.sample_0, R.drawable.sample_1,
109.               R.drawable.sample_2,
110.               R.drawable.sample_3, R.drawable.sample_4,
111.               R.drawable.sample_5,
112.               R.drawable.sample_6, R.drawable.sample_7};
113.   }
```

(5) 在 src 目录中 com. hisoft. activity 包下创建 MainActivity. java 文件，代码如下：

```
1.   package com.hisoft.activity;

2.   import android.app.Activity;
3.   import android.content.Intent;
4.   import android.os.Bundle;
5.   import android.view.View;
6.   import android.view.View.OnClickListener;
7.   import android.widget.Button;

8.   public class MainActivity extends Activity {

9.     private Button button1;

10.    @Override
11.    protected void onCreate(Bundle savedInstanceState) {
12.      super.onCreate(savedInstanceState);
13.      setContentView(R.layout.main);
14.
15.      button1=(Button) this.findViewById(R.id.button1);
16.
17.      //给 button1 设置监听
18.      button1.setOnClickListener(new OnClickListener() {
19.
20.        public void onClick(View v) {
21.          //通过 Intent 跳转到 ImageViewActivity
22.          Intent intent=new Intent();
23.          intent.setClass(MainActivity.this, ImageSwitcherAndGalleryActivity
             .class);
24.          startActivity(intent);
25.        }
26.      });
27.    }
28.    }
```

(6) 把图片资源文件复制到资源 res 目录下的 drawable-hdpi、drawable-hdpi 和 drawable-hdpi 文件夹中。

(7) 部署运行 ImageViewDemo 项目工程,项目运行效果如图 5-30 所示。

单击"浏览图片"按钮,显示结果如图 5-31 所示。

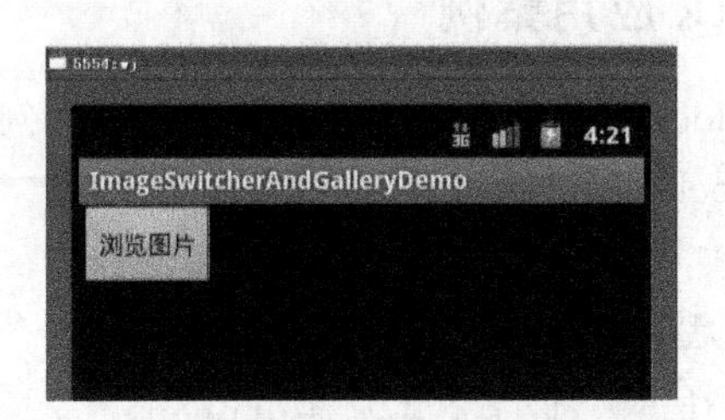

图 5-30 ImageSwitcherAndGalleryDemo 运行结果

图 5-31 图片相册效果

5.6 时钟控件简介

5.6.1 模拟时钟与数字时钟

时钟控件包括 AnalogClock 和 DigitalClock,它们都负责显示时钟,所不同的是 AnalogClock 控件显示模拟时钟,且只显示时针和分针;而 DigitalClock 显示数字时钟,可精确到秒。

AnalogClock 和 DigitalClock 控件的类继承结构不同,AnalogClock 控件继承自 android. view. View,AnalogClock 的类结构如图 5-32 所示。

DigitalClock 控件继承自 android. widget. TextView, DigitalClock 的类结构如图 5-33 所示。

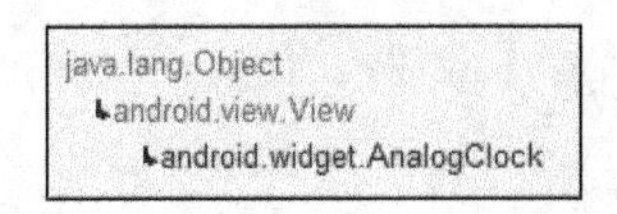

图 5-32 AnalogClock 类继承图

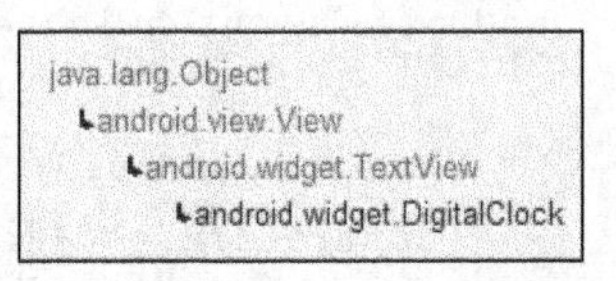

图 5-33 DigitalClock 类继承图

AnalogClock 和 DigitalClock 都不需要用户编写 Java 代码,只要在 res→layout 目录下的 xml 里插入以下代码即可自动调用显示时间。

(1) AnalogClock 控件在 XML 中添加的代码如下:

```
1.        <!--模拟时钟控件-->
2.     <AnalogClock android:id="@+id/analogClock"
3.          android:layout_width="wrap_content"
4.          android:layout_height="wrap_content"
```

```
5.        android:layout_gravity="center_horizontal"/>
```

(2) DigitalClock 控件在 XML 中添加的代码如下：

```
1.        <!--数字时钟控件-->
2.    <DigitalClock android:id="@+id/digitalClock"
3.        android:layout_width="wrap_content"
4.        android:layout_height="wrap_content"
5.        android:layout_gravity="center_horizontal"/>
```

5.6.2 AnalogClock 和 DigitalClock 应用案例

上面章节介绍了 AnalogClock 和 DigitalClock 的常用方法，下面通过一个案例时钟控制应用熟悉和掌握 AnalogClock 和 DigitalClock 控件应用效果。具体步骤如下：

(1) 创建一个新的 Android 工程，工程名为 AnalogAndDigitalClockDemo，目标 API 选择 10(即 Android 2.3.3 版本)，应用程序名为 AnalogAndDigitalClockDemo，包名为 com. hisoft. activity，创建的 Activity 的名字为 AnalogActivity，最小 SDK 版本根据选择的目标 API 会自动添加为 10，创建后的项目工程如图 5-34 所示。

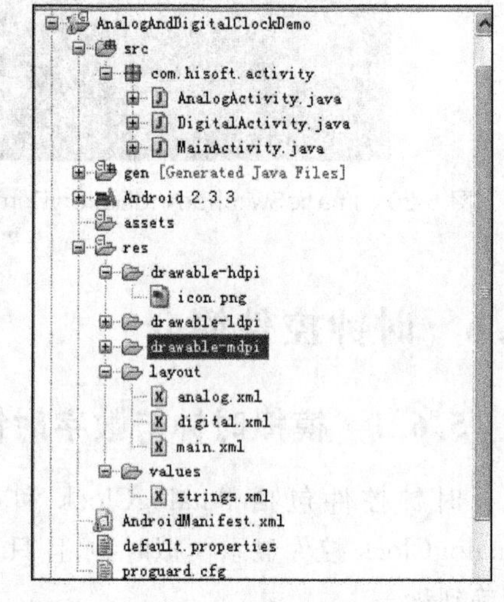

图 5-34　AnalogAndDigitalClockDemo 工程结构图

(2) 打开项目工程中 res→layout 目录下的 main.xml 文件，设置线性布局，添加两个 Button，并设置相关属性，代码如下：

```
1.    <?xml version="1.0" encoding=
      "utf-8"?>
2.    <LinearLayout xmlns:android="http://schemas.android.com/apk/res/android"
3.      android:orientation="vertical"
4.      android:layout_width="fill_parent"
5.      android:layout_height="fill_parent"
6.      >
7.         <Button android:text="to AnalogClock"
8.          android:id="@+id/bn_analog"
9.          android:layout_width="wrap_content"
10.         android:layout_height="wrap_content">
11.     </Button>
12.     <Button android:text="to DigitalClock"
13.         android:id="@+id/bn_digital"
14.         android:layout_width="wrap_content"
15.         android:layout_height="wrap_content">
16.     </Button>

17.   </LinearLayout>
```

(3) 在项目工程中的 res→layout 目录下创建 analog. xml 文件，设置相对布局，添加一个 AnalogClock 控件描述，并设置相关属性，代码如下：

```
1.  <?xml version="1.0" encoding="utf-8"?>
2.  <LinearLayout
3.    xmlns:android="http://schemas.android.com/apk/res/android"
4.    android:orientation="vertical"
5.    android:layout_width="match_parent"
6.    android:layout_height="match_parent">
7.      <AnalogClock android:id="@+id/analogClock1"
8.              android:layout_width="wrap_content"
9.              android:layout_height="wrap_content">
10.     </AnalogClock>
11. </LinearLayout>
```

(4) 在项目工程中的 res→layout 目录下创建 digital. xml 文件，设置相对布局，添加一个 DigitalClock 控件描述，并设置相关属性，代码如下：

```
1.  <?xml version="1.0" encoding="utf-8"?>
2.  <LinearLayout
3.    xmlns:android="http://schemas.android.com/apk/res/android"
4.    android:orientation="vertical"
5.    android:layout_width="match_parent"
6.    android:layout_height="match_parent">
7.      <DigitalClock android:text="DigitalClock"
8.               android:id="@+id/digitalClock1"
9.               android:layout_width="wrap_content"
10.              android:layout_height="wrap_content">
11.     </DigitalClock>

12. </LinearLayout>
```

(5) 修改 src 目录中 com. hisoft. activity 包下的 AnalogActivity. java 文件，代码如下：

```
1.  package com.hisoft.activity;
2.
3.  import android.app.Activity;
4.  import android.os.Bundle;
5.  public class AnalogActivity extends Activity {

6.    @Override
7.    protected void onCreate(Bundle savedInstanceState) {
8.      super.onCreate(savedInstanceState);
9.      this.setContentView(R.layout.analog);
10.   }
11. }
```

(6) 在 src 目录中的 com. hisoft. activity 包下创建 MainActivity. java 文件,代码如下:

```
1.  package com.hisoft.activity;
2.
3.  import android.app.Activity;
4.  import android.content.Intent;
5.  import android.os.Bundle;
6.  import android.view.View;
7.  import android.view.View.OnClickListener;
8.  import android.widget.Button;

9.  public class MainActivity extends Activity {

10.   private Button bn_analog, bn_digital;

11.   @Override
12.   protected void onCreate(Bundle savedInstanceState) {
13.     super.onCreate(savedInstanceState);
14.     setContentView(R.layout.main);

15.     this.bn_analog=(Button) this.findViewById(R.id.bn_analog);
16.     this.bn_digital=(Button) this.findViewById(R.id.bn_digital);

17.     MyListener ml=new MyListener();

18.     this.bn_analog.setOnClickListener(ml);
19.     this.bn_digital.setOnClickListener(ml);

20.   }

21.   class MyListener implements OnClickListener{
22.
23.     private Intent intent=new Intent();
24.
25.     public void onClick(View v) {
26.       if(v==bn_analog){
27.         intent.setClass(MainActivity.this, AnalogActivity.class);
28.       }
29.       if(v==bn_digital){
30.         intent.setClass(MainActivity.this, DigitalActivity.class);
31.       }
32.       startActivity(intent);
33.     }
34.   }
35. }
```

(7) 在 src 目录中的 com. hisoft. activity 包下创建 DigitalActivity. java 文件，代码如下：

```
1.   package com.hisoft.activity;
2.
3.   import android.app.Activity;
4.   import android.os.Bundle;
5.
6.   public class DigitalActivity extends Activity {
7.
8.     @Override
9.     protected void onCreate(Bundle savedInstanceState) {
10.      super.onCreate(savedInstanceState);
11.      this.setContentView(R.layout.digital);
12.    }
13.  }
```

(8) 部署运行 AnalogAndDigitalClockDemo 项目工程，项目运行效果如图 5-35 所示。

单击 to AnalogClock 按钮，如图 5-36 所示。

单击 to DigitalClock 按钮，如图 5-37 所示。

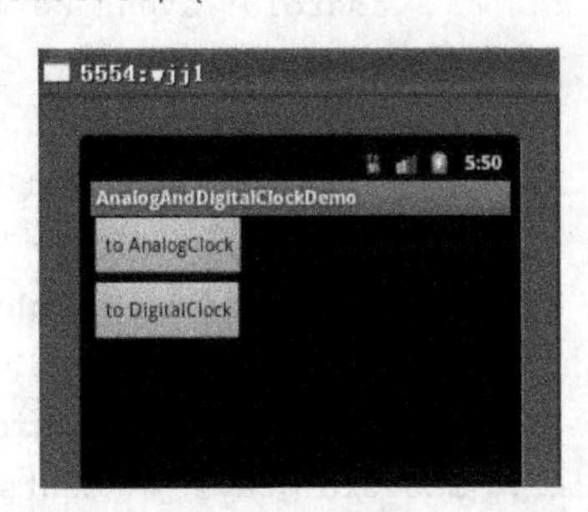

图 5-35　AnalogAndDigitalClockDemo 运行结果

图 5-36　模拟时钟

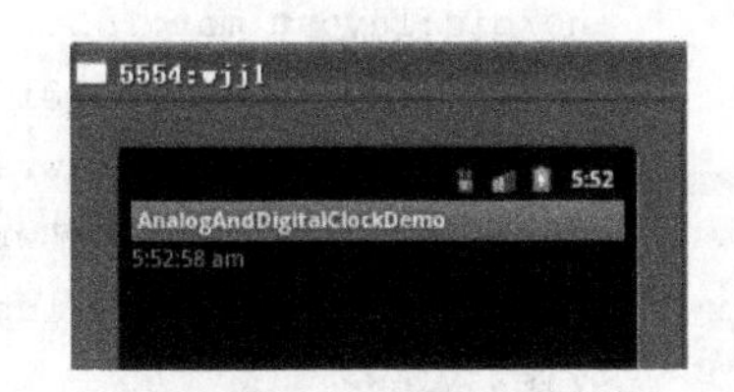

图 5-37　数字时钟

拓展提示：Android UI 系统控件应用的方式不同，对实际项目后期的开发及维护成本都会带来深远影响，选择使用 XML 描述并设置控件属性，还是在程序代码中直接使用方法设置，需要根据需要来评定，但不要忘记了 UI 的设计原则，通常松散耦合比紧密耦合更利于后续的工作开展。

5.7　项目案例

学习目标：学习 Android UI 系统控件的基本方法、属性的设置等应用。

案例描述：使用 RelativeLayout 相对布局、TextView 控件、EditText 控件、Button 按钮，并设置相对父控件的位置、控件之间相对位置的属性，实现 Ascent 移动版用户登录界面。

案例要点：RelativeLayout 相对布局、控件的属性设置，以及控件之间位置关系的属性设置。

案例实施：

(1) 创建工程 Project_Chapter_5，选择 Android 2.3.3 作为目标平台，如图 5-38 所示。

(2) 创建 login.xml 文件，将文件存放在 res/layout 下，代码如下：

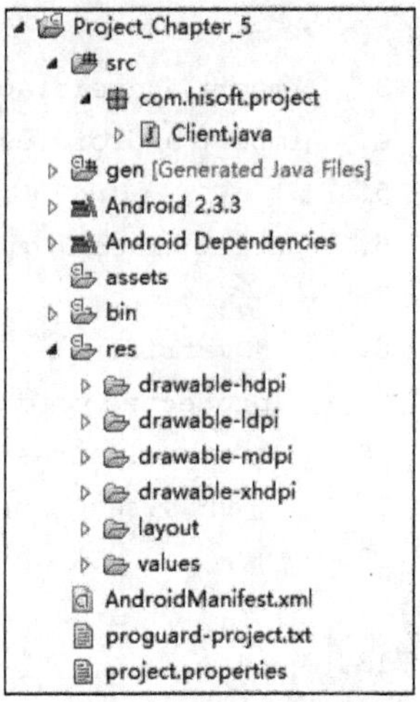

Project_Chapter_5
src
com.hisoft.project
Client.java
gen [Generated Java Files]
Android 2.3.3
Android Dependencies
assets
bin
res
drawable-hdpi
drawable-ldpi
drawable-mdpi
drawable-xhdpi
layout
values
AndroidManifest.xml
proguard-project.txt
project.properties

图 5-38　Project_Chapter_5 项目工程目录

```
1.   <?xml version="1.0" encoding="utf-8"?>
2.   <RelativeLayout xmlns:android=
     "http://schemas.android.com/apk/res/android"
3.       android:orientation="vertical"
4.       android:layout_width="fill_parent"
5.       android:layout_height="fill_parent"
6.       >
7.   <TextView
8.   android:id="@+id/TextView01"
9.   android:layout_width="fill_parent "
10.    android:layout_height="wrap_content"
11.    android:text="@string/login"
12.  android:textSize="30px"
13.  android:gravity="center_horizontal"
14.  />
15.  <TextView
16.    android:id="@+id/TextView02"
17.    android:layout_marginTop="12px"
18.    android:layout_marginLeft="5dip"
19.    android:layout_below="@id/TextView01"
20.    android:layout_width="wrap_content"
21.    android:layout_height="wrap_content"
22.    android:text="@string/input_username" android:textSize="20px">
23.    </TextView>
24.
25.  <EditText
26.    android:id="@+id/username"
27.    android:layout_alignTop="@id/TextView02"
28.    android:layout_toRightOf="@id/TextView02"
29.    android:layout_width="fill_parent"
30.    android:layout_height="wrap_content"
31.    android:singleLine="true">
32.    </EditText>
33.
34.  <TextView
35.    android:id="@+id/TextView03"
36.    android:layout_below="@id/username"
37.    android:layout_alignLeft="@id/TextView02"
38.    android:layout_width="wrap_content"
```

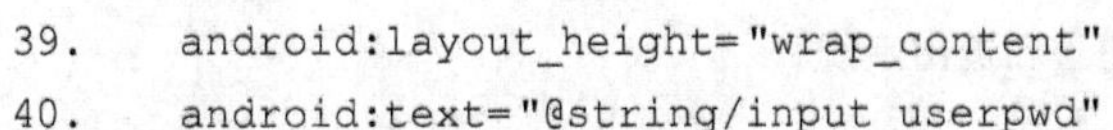

```
39.    android:layout_height="wrap_content"
40.    android:text="@string/input_userpwd" android:textSize="20px">
41.  </TextView>
42.
43.  <EditText
44.    android:id="@+id/password"
45.    android:layout_alignTop="@id/TextView03"
46.    android:layout_toRightOf="@id/TextView03"
47.    android:layout_alignLeft="@id/username"
48.    android:layout_width="fill_parent"
49.    android:layout_height="wrap_content" android:password="true"
       android:singleLine="true">
50.  </EditText>
51.
52.  <Button
53.    android:id="@+id/login"
54.    android:layout_marginTop="12px"
55.    android:layout_width="100px"
56.    android:layout_height="wrap_content"
57.    android:layout_below="@id/password"
58.    android:layout_alignLeft="@id/password"
59.    android:text="@string/bt_login"
60.    android:textSize="16px"
61.    >
62.  </Button>
63.
64.  <Button
65.    android:id="@+id/exit"
66.    android:layout_marginLeft="15px"
67.    android:layout_width="100px"
68.    android:layout_height="wrap_content"
69.    android:text="@string/bt_exit"
70.    android:layout_toRightOf="@id/login"
71.    android:layout_alignTop="@id/login"
72.    android:textSize="16px"
73.    >
74.  </Button>
75.  </RelativeLayout>
```

(3) 在 src 目录下的 com. hisoft. project 包下创建 Client. java，代码如下：

```
1.   package com.hisoft.project;
2.   import android.app.Activity;
3.   import android.os.Bundle;
4.   /*****************************************************************
5.    * 程序名称:Client.java                                              *
```

```
6.     * 功能:显示用户登录窗口,可登录和退出系统                              *
7.     * 作者:                                                             *
8.     * 日期:                                                             *
9.    /**************************************************************/
10.   public class Client extends Activity {
11.
12.     /**
13.      * 显示登录框页面
14.      */
15.     @Override
16.       public void onCreate(Bundle savedInstanceState) {
17.           super.onCreate(savedInstanceState);
18.           setContentView(R.layout.login);
19.
20.     }
21.
22.
23.   }
```

(4) 修改 res 目录下 values 文件夹中的 strings. xml 文件,代码如下:

```
1.    <?xml version="1.0" encoding="utf-8"?>
2.    <resources>
3.        <string name="app_name">Ascent 移动版医药商务系统</string>
4.        <string name="login">用户登录</string>
5.        <string name="input_username">用户名:</string>
6.        <string name="input_userpwd">密码:</string>
7.        <string name="bt_login">登录</string>
8.        <string name="bt_exit">退出</string>
9.    </resources>
```

(5) 部署项目工程,项目运行效果如图 5-39 所示。

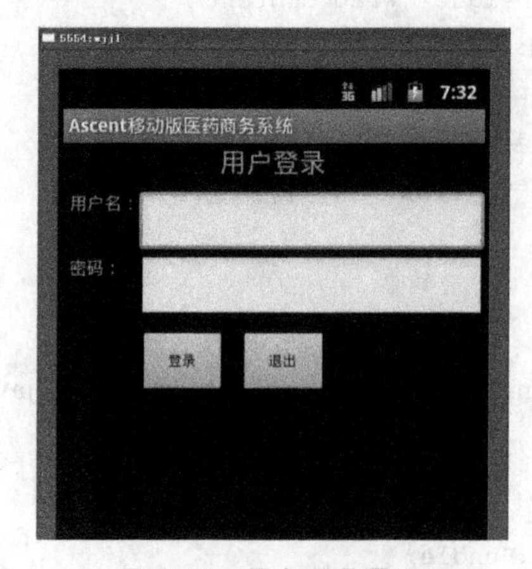

图 5-39　用户登录界面

1. 简答题

(1) 简述 TextView 和 EditText 控件的功能及用途，以及使用方法。

(2) Button 按钮和 ImageButton 按钮分别什么时候使用？它们之间的区别是什么？

(3) 单选按钮 RadioButton 和复选按钮 CheckBox 的常用用法是什么？它们的步骤包含哪些？

(4) ImageView 控件如何设置图片或图片源？

(5) 模拟时钟和数字时钟在用法上有什么不同？它们通用的方式是什么？

2. 完成下面的实训项目

要求：在本章项目案例的基础上完成用户界面文字及按钮上文字颜色的设置(自定义或红色)，以及“Ascent 移动版医药商务系统”由右到左的滚动。

第6章 Android UI系统控件进阶

学习目标

本章主要深入介绍 Android UI 系统控件中的列表控件 ListView、下拉列表控件 Spinner、进度条 ProgressBar、滑块控件 SeekBar、评分控件 RatingBar、自动完成文本控件 AutoCompleteTextView、Tabhost 控件、滚动视图控件 ScrollView、网格视图控件 GridView，以及 Android 事件处理监听器、事件处理的机制等。使读者通过本章的学习，能够深入熟悉 Android UI 系统控件，掌握以下知识要点：

(1) 列表控件 ListView 类继承关系、常用属性及设置描述、列表的适配器类型。

(2) 下拉列表控件 Spinner 的概念、属性设置及描述、常用的方法。

(3) 进度条 ProgressBar 和滑块控件 SeekBar 的属性及设置、常用方法、引用及事件处理。

(4) 评分控件 RatingBar、自动完成文本控件 AutoCompleteTextView 的属性描述及通常引用方法。

(5) Tabhost 控件的属性设置、通常用法及实现方式。

(6) 滚动视图控件 ScrollView、网格视图控件 GridView 的属性描述设置及通常用法。

(7) Android 事件类型、事件传递及处理原则、事件处理机制及事件处理常用方法。

6.1 列表控件简介

6.1.1 列表控件

ListView 是一种用于垂直显示的列表控件，其继承类结构图如图 6-1 所示。

ListView 是比较常用的组件，它以列表的形式展示具

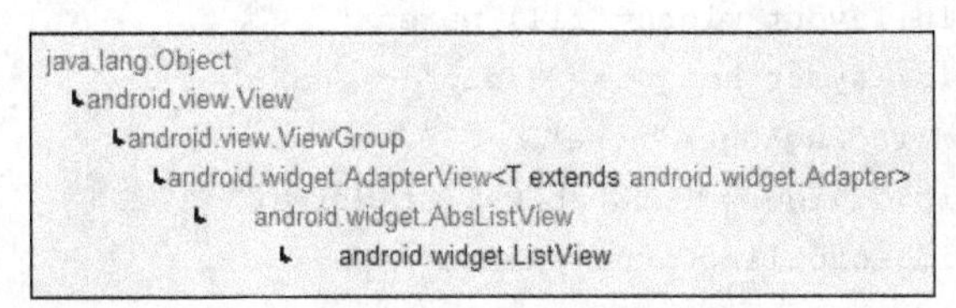

图 6-1 ListView类继承图

体内容。如果 ListView 控件显示内容过多，则会出现垂直滚动条，并且它能够根据数据的长度自适应显示。列表的显示需要三个元素：

(1) ListVeiw。用来展示列表的 View。

(2) 适配器。用来把数据映射到 ListView 上的中介。

(3) 数据。指被映射的字符串、图片或者基本组件。

根据列表的适配器类型，列表分为 ArrayAdapter、SimpleAdapter 和 SimpleCursorAdapter 三种。

ListView 能够通过适配器将数据和自身绑定，在有限的屏幕上提供大量内容供用户选择，所以是经常使用的用户界面控件。

其中以 ArrayAdapter 最为简单，只能展示一行字。SimpleAdapter 有最好的扩充性，可以自定义出各种效果。SimpleCursorAdapter 可以认为是 SimpleAdapter 对数据库的简单结合，可以方便地把数据库的内容以列表的形式展示出来。

ListView 支持单击事件处理，用户可以用少量的代码实现复杂的选择功能。ListView 常用的 XML 属性及描述如表 6-1 所示。

表 6-1 ListView 常用的 XML 属性及描述

属性名称	描述
android:dividerHeight	分隔符的高度。若没有指明高度，则用此分隔符固有的高度。必须为带单位的浮点数，如“14.5sp”。可用的单位如 px(pixel，像素)、dp(density-independent pixels，与密集度无关的像素)、sp(scaled pixels based on preferred font size，基于字体大小的固定比例的像素)、in(inches，英寸)和 mm(millimeters，毫米)
android:entries	引用一个将使用在此 ListView 里的数组。若数组是固定的，使用此属性将比在程序中写入更为简单
android:footerDividersEnabled	设成 flase 时，ListView 将不会在页脚视图前画分隔符。此属性缺省值为 true。属性值必须设置为 true 或 false
android:headerDividersEnabled	设成 flase 时，ListView 将不会在页眉视图后画分隔符。此属性缺省值为 true。属性值必须设置为 true 或 false
android:choiceMode	规定 ListView 所使用的选择模式。缺省状态下，list 没有选择模式。属性值必须设置为下列常量之一：none，值为 0，表示无选择模式；singleChoice，值为 1，表示最多可以有一项被选中；multipleChoice，值为 2，表示可以多项被选中

在布局文件中，用 XML 描述的 ListView 控件代码如下：

```
1.   <ListView
2.        android:id="@+id/myListView01"
```

```
3.         android:layout_width="fill_parent"
4.         android:layout_height="287dip"
5.        android:fadingEdge="none"
6.         android:divider="@drawable/list_driver"
7.         android:scrollingCache="false"
8.         android:background="@drawable/list">
9.     </ListView>
```

第 5 行是消除 listview 的上边和下边黑色的阴影。

第 7 行 listview 是消除在拖动的时候背景图片消失变成黑色背景。

第 8 行是 listview 的每一项之间设置一个图片做为间隔。其中@drawable/list_driver 是一个图片资源。

6.1.2 ListView 应用案例

上面章节介绍了 ListView 的常用方法和属性，下面通过一个 ListView 案例应用熟悉和掌握 ListView 控件应用效果。

ListView 控件编写程序的通常步骤如下：

(1) 在布局文件中声明 ListView 控件。

(2) 使用一维或多维动态数组保存 ListView 要显示的数据。

(3) 构建适配器 Adapter，将数据与显示数据的布局页面绑定。

(4) 通过 setAdapter()方法把适配器设置给 ListView。

案例具体步骤如下：

(1) 创建一个新的 Android 工程，工程名为 ListDemo，目标 API 选择 10(即 Android 2.3.3 版本)，应用程序名为 ListDemo，包名为 com.hisoft.activity，创建的 Activity 的名字为 ListDemoActivity，最小 SDK 版本根据选择的目标 API 会自动添加为 10，创建后的项目工程如图 6-2 所示。

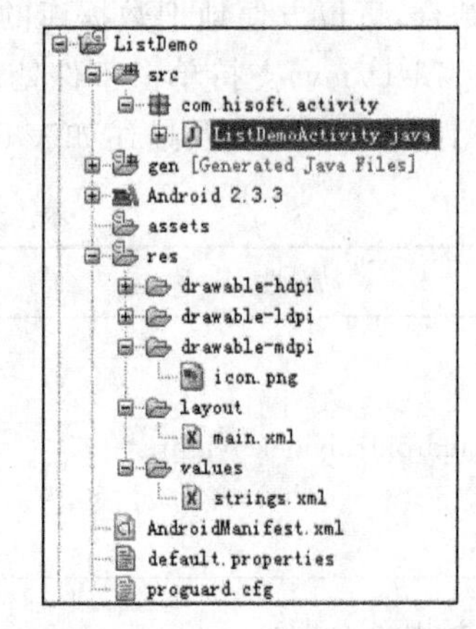

图 6-2　ListDemo 工程目录结构

(2) 修改布局文件 main.xml，添加三个 TextView 和 ListView 实现整体布局。具体代码如下：

```
1.   <?xml version="1.0" encoding="utf-8"?>
2.    <LinearLayout xmlns:android="http://schemas.android.com/apk/res/android"
3.        android:orientation="vertical"
4.        android:layout_width="fill_parent"
5.        android:layout_height="fill_parent"
6.        >
7.        <TextView android:layout_width="fill_parent"
8.               android:layout_height="wrap_content"
9.               android:text="@string/tv1"
10.              android:background="#FFF"
11.              android:textColor="#888"
12.              android:gravity="center"/>
```

```
13.         <ListView android:id="@+id/lvCheckedTextView"
14.                 android:layout_width="fill_parent"
15.                 android:layout_height="wrap_content"/>
16.
17.         <TextView android:layout_width="fill_parent"
18.                 android:layout_height="wrap_content"
19.                 android:text="@string/tv2"
20.                 android:background="#FFF"
21.                 android:textColor="#888"
22.                 android:gravity="center"/>
23.         <ListView android:id="@+id/lvRadioButton"
24.                 android:layout_width="fill_parent"
25.                 android:layout_height="wrap_content"/>
26.
27.         <TextView android:layout_width="fill_parent"
28.                 android:layout_height="wrap_content"
29.                 android:text="@string/tv3"
30.                 android:background="#FFF"
31.                 android:textColor="#888"
32.                 android:gravity="center"/>
33.         <ListView android:id="@+id/lvCheckedButton"
34.                 android:layout_width="fill_parent"
35.                 android:layout_height="wrap_content"/>
36.     </LinearLayout>
```

(3) 修改 strings. xml 文件，具体代码如下：

```
1.  <?xml version="1.0" encoding="utf-8"?>
2.   <resources>
3.       <string name="hello">Hello World, Listview02Activity!</string>
4.       <string name="app_name">ListDemo</string>
5.       <string name="tv1">单项选择应用</string>
6.       <string name="tv2">RadioButton 应用</string>
7.       <string name="tv3">CheckBox 应用</string>
8.   </resources>
```

(4) 修改 src 目录中 com. hisoft. activity 包下的 ListDemoActivity. java 文件，代码如下：

```
1.  package com.hisoft.activity;
2.  import android.app.Activity;
3.  import android.os.Bundle;
4.  import android.widget.ArrayAdapter;
5.  import android.widget.ListView;
6.  public class ListDemoActivity extends Activity {
7.      /** Called when the activity is first created. */
8.      @Override
```

```
9.        public void onCreate(Bundle savedInstanceState) {
10.           super.onCreate(savedInstanceState);
11.           setContentView(R.layout.main);
12.           //得到三个 listview
13.           ListView listview1= (ListView) findViewById(R.id.lvCheckedTextView);
14.           ListView listview2= (ListView) findViewById(R.id.lvRadioButton);
15.           ListView listview3= (ListView) findViewById(R.id.lvCheckedButton);
16.           //用 string 来保存 listview 要显示的数据
17.           String[] data=new String[]
18.             {"北京政法","海辉集团"};
19.           //构建适配器 Adapter,将数据与显示数据的布局页面绑定
20.           ArrayAdapter<String> lv1Adapter=new ArrayAdapter<String> (this,
              android.R.layout.simple_list_item_checked,data);
21.           //通过 setAdapter()方法把适配器设置给 ListView
22.           listview1.setAdapter(lv1Adapter);
23.           //listview 里的内容设置为单选
24.           listview1.setChoiceMode(ListView.CHOICE_MODE_SINGLE);
25.           ArrayAdapter<String> lv2Adapter=new ArrayAdapter<String> (this,
              android.R.layout.simple_list_item_single_choice,data);
26.           listview2.setAdapter(lv2Adapter);
27.           listview2.setChoiceMode(ListView.CHOICE_MODE_SINGLE);
28.
29.           ArrayAdapter<String> lv3Adapter=new ArrayAdapter<String> (this,
              android.R.layout.simple_list_item_multiple_choice,data);
30.           listview3.setAdapter(lv3Adapter);
31.           listview3.setChoiceMode(ListView.CHOICE_MODE_MULTIPLE);
32.           }
33.   }
```

(5) 部署运行 ListDemo 项目工程，项目运行效果如图 6-3 所示。

图 6-3　ListDemo 工程运行结果

6.1.3　下拉列表控件

下拉列表(Spinner)是 AdapterView 的子类，是一个每次只能选择所有项中一项的部件。它的项来自于与之相关联的适配器。类似于桌面程序的组合框(ComboBox)，但没有组合框的下拉菜单，而是使用浮动菜单为用户提供选择。Spinner 数据由 Adapter 提供，通过 Spinner.getItemAtPosition(Spinner.getSelectedItemPosition())；获取下拉列表框的值。

调用 setOnItemSelectedListener()方法，处理下拉列表框被选择事件，把 AdapterView.OnItemSelectedListener 实例作为参数传入。

Spinner 的类继承结构和常用的 XML 属性及对应方法

描述如图 6-4 和表 6-2 所示。

```
java.lang.Object
  ↳android.view.View
      ↳android.view.ViewGroup
          ↳android.widget.AdapterView<T extends android.widget.Adapter>
              ↳  android.widget.AbsSpinner
                  ↳  android.widget.Spinner
```

图 6-4 Spinner 的类继承关系

表 6-2 Spinner 常用的 XML 属性及对应方法、描述

属性名称	对应的方法	描述
android:prompt		该提示在下拉列表对话框或菜单显示时显示，如对话框的标题 setPrompt("选择颜色")；
android:dropDownHorizontalOffset	setDropDownHorizontalOffset(int)	在 spinnerMode 为下拉菜单(dropdown)时，设置下拉列表的水平偏移
android:dropDownVerticalOffset	setDropDownVerticalOffset(int)	设置下拉列表和文本框的垂直偏移
android:dropDownWidth	setDropDownWidth(int)	设置下拉列表的宽度
android:gravity	setGravity(int)	设置 listView 中当前选择的 item 位置
android:popupBackground	setPopupBackgroundResource(int)	在 spinnerMode 为下拉菜单(dropdown)时，设置下拉列表的背景

Spinner 控件应用的通常用法具体如下：

(1) 用 XML 描述的一个 Spinner 控件。

```
<?xml version="1.0" encoding="utf-8"?>
<LinearLayout
    xmlns:android="http://schemas.android.com/apk/res/android"
    android:layout_width="fill_parent"
    android:layout_height="wrap_content">
    <Spinner android:id="@+id/spinner"
        android:layout_height="wrap_content"
        android:layout_width="fill_parent"/>
</LinearLayout>
```

(2) 引用上述 XML 描述的 Spinner 控件，程序中调用：

```
public class SpinnerActivity extends Activity {
    private static final String TAG="SpinnerActivity";
    @Override
    public void onCreate(Bundle savedInstanceState) {
        super.onCreate(savedInstanceState);
        setContentView(R.layout.spinner);
```

```
        //第二个参数为下拉列表框每一项的界面样式,该界面样式由 Android 系统提供,当然也
        //可以自定义
        ArrayAdapter<String>adapter=new ArrayAdapter<String>(this,
        android.R.layout.simple_spinner_item);
        adapter.setDropDownViewResource(android.R.layout.simple_spinner_
        dropdown_item);
        adapter.add("java");
        adapter.add("dotNet");
        adapter.add("php");
        Spinner spinner=(Spinner) findViewById(R.id.spinner);
        spinner.setAdapter(adapter);
        spinner.setOnItemSelectedListener(new AdapterView.
        OnItemSelectedListener() {
    @Override
    public void onItemSelected(AdapterView<?>adapterView, View view,
    int position, long id) {
        Spinner spinner=(Spinner)adapterView;
        String itemContent=(String)adapterView.getItemAtPosition(position);
    }
    @Override
    public void onNothingSelected(AdapterView<?>view) {
        Log.i(TAG,  view.getClass().getName());
    }
      });
    }
}
```

6.1.4 Spinner 应用案例

上节介绍了 Spinner 的常用方法和属性,下面通过一个 ListView 案例应用熟悉和掌握 Spinner 控件应用效果。

创建一个 Spinner 控件应用的步骤如下:

(1) 在布局文件当中声明<Spinner>。

(2) 在 string.xml 当中声明一个数组<string-array>。

(3) 创建一个 ArrayAdapter(默认显示样式与下拉菜单中每个 item 的样式)。

(4) 得到 Spinner 对象,并设置数据:建立连接与提示。

案例具体步骤如下:

(1) 创建一个新的 Android 工程,工程名为 SpinnerDemo,目标 API 选择 10(即 Android 2.3.3 版本),应用程序名为 SpinnerDemo,包名为 com.hisoft.activity,创建的 Activity 的名字为 MainActivity,最小 SDK 版本根据选择的目标 API 会自动添加为 10,创建后的项目工程如图 6-5 所示。

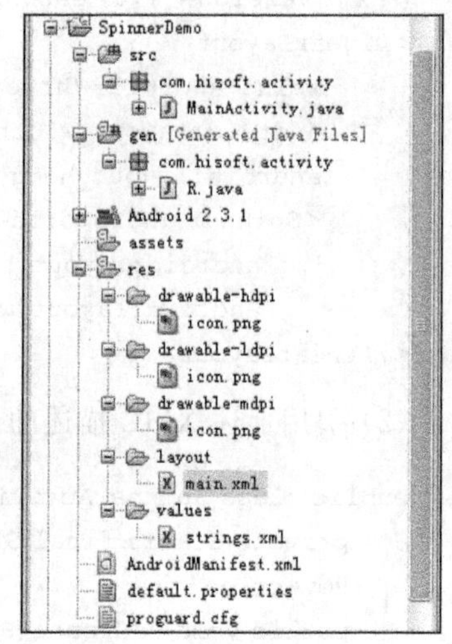

图 6-5 SpinnerDemo 工程目录结构

(2) 打开项目工程中 res→layout 目录下的 main. xml 文件,设置线性布局,添加一个 TextView 和一个 Spinner 控件,并设置相关属性,代码如下:

```
1.  <?xml version="1.0" encoding="utf-8"?>
2.  <LinearLayout xmlns:android="http://schemas.android.com/apk/res/android"
3.      android:layout_width="fill_parent"
4.      android:layout_height="fill_parent"
5.      android:orientation="horizontal">
6.      <TextView android:text="爱好:"
7.           android:id="@+id/textView1"
8.           android:layout_width="wrap_content"
9.           android:layout_height="wrap_content">
10.     </TextView>
11.     <Spinner android:id="@+id/favorite"
12.          android:layout_height="wrap_content"
13.          android:layout_width="wrap_content">
14.     </Spinner>
15. </LinearLayout>
```

(3) 修改 strings. xml 文件,添加<string-array>数组,具体代码如下:

```
1.  <?xml version="1.0" encoding="utf-8"?>
2.  <resources>
3.    <string name="hello">Hello World, MainActivity!</string>
4.    <string name="app_name">SpinnerDemo</string>
5.    <string-array name="favorite">
6.      <item>music</item>
7.      <item>sport</item>
8.      <item>programming</item>
9.      <item>watch TV</item>
10.         <item>shopping</item>
11.   </string-array>
12. </resources>
```

(4) 修改 src 目录中 com. hisoft. activity 包下的 MainActivity. java 文件,代码如下:

```
1.  package com.hisoft.activity;

2.  import android.app.Activity;
3.  import android.os.Bundle;
4.  import android.widget.ArrayAdapter;
5.  import android.widget.Spinner;

6.  public class MainActivity extends Activity {
7.
8.    private Spinner favorite;
9.
10.   /** Called when the activity is first created. */
11.   @Override
```

```
12.     public void onCreate(Bundle savedInstanceState) {
13.       super.onCreate(savedInstanceState);
14.       setContentView(R.layout.main);

15.       this.favorite= (Spinner) this.findViewById(R.id.favorite);
16.       ArrayAdapter<CharSequence>adapter=ArrayAdapter.createFromResource(
17.           this, R.array.favorite,
18.           android.R.layout.simple_spinner_item);
19.       adapter.setDropDownViewResource(android.R.layout.select_dialog_
          multichoice);
20.       this.favorite.setAdapter(adapter);

21.     }
22.   }
```

(5) 部署运行 SpinnerDemo 项目工程，项目运行效果如图 6-6 所示。

单击下拉按钮，弹出下拉列表菜单，如图 6-7 所示。

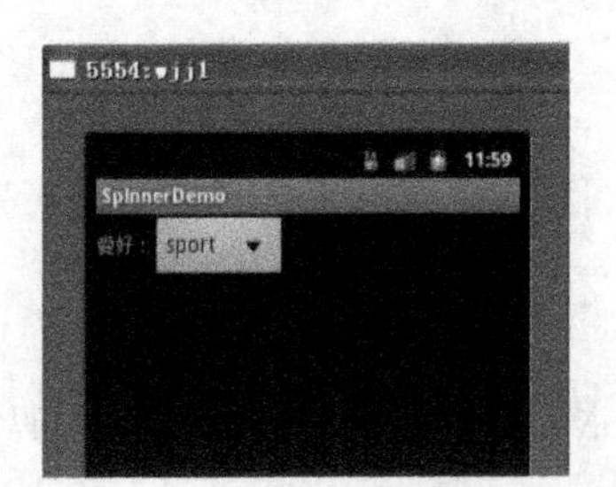

图 6-6　SpinnerDemo 项目运行结果

图 6-7　下拉列表

6.2　进度条与滑块控件简介

6.2.1　进度条

ProgressBar(进度条)控件就是一个表示运转的过程，例如发送短信、连接网络等，表示一个过程正在执行中。位于 android. widget 包下，类继承结构图如图 6-8 所示。

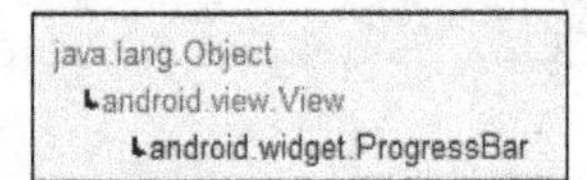

图 6-8　ProgressBar 控件类继承

ProgressBar 控件通用的方法具体如下：

(1) 在布局 xml 文件中添加进度条代码。

```
<ProgressBar android:layout_width="fill_parent"
    android:layout_height="20px"
    style="?android:attr/progressBarStyleHorizontal"
    android:id="@+id/downloadbar"/>
```

(2) 引用 XML 文件控件，在代码中操作进度条。

```
ProgressBar.setMax(100);      //设置总长度为 100
ProgressBar.setProgress(0);  //设置已经开启长度为 0,假设设置为 50,进度条将进行到一半
```

ProgressBar 常用的 XML 属性及描述如表 6-3 所示。

表 6-3 ProgressBar 常用的 XML 属性及描述

属性名称	描述
android:maxWidth	设置进度条的最大宽度
android:maxHeight	设置进度条的最大高度
android:max	设置进度条的最大值
android:progress	设置默认的进度值，取值为 0 到最大之间
android:progressDrawable	绘制进度模式
android:secondaryProgress	设置第二进度值。取值为 0 到最大之间

6.2.2 ProgressBar 应用案例

本节先对 ProgressBar 的应用和属性进行讲解，下面通过一个案例熟悉和掌握 ProgressBar 控件应用效果。具体步骤如下：

(1) 创建一个新的 Android 工程，工程名为 ProgressBarDemo，目标 API 选择 10(即 Android 2.3.3 版本)，应用程序名为 ProgressBarDemo，包名为 com.hisoft.activity，创建的 Activity 的名字为 MainActivity，最小 SDK 版本根据选择的目标 API 会自动添加为 10。

(2) 打开项目工程中 res→layout 目录下的 main.xml 文件，设置相对布局，添加一个 ProgressBar 和一个 Button 控件，并设置相关属性，代码如下：

```
1.   <?xml version="1.0" encoding="utf-8"?>
2.   <LinearLayout
3.     xmlns:android="http://schemas.android.com/apk/res/android"
4.     android:orientation="vertical"
5.     android:layout_width="match_parent"
6.     android:layout_height="match_parent">
7.       <ProgressBar android:layout_width="fill_parent"
8.              android:layout_height="wrap_content"
9.              style="?android:attr/progressBarStyleHorizontal"
10.                 android:id="@+id/progressBar1"
11.                 android:max="100"
```

```
12.                 android:progress="50"
13.                 android:secondaryProgress="70"
14.         >
15.         </ProgressBar>
16.
17.         <Button android:text="Click Button"
18.             android:id="@+id/button1"
19.             android:layout_width="wrap_content"
20.             android:layout_height="wrap_content">
21.         </Button>
22.
23.     </LinearLayout>
```

(3) 修改 src 目录中 com. hisoft. activity 包下的 MainActivity. java 文件,代码如下:

```
1.        package com.hisoft.activity;

2.    import android.app.Activity;
3.    import android.os.Bundle;
4.    import android.view.View;
5.    import android.view.View.OnClickListener;
6.    import android.widget.Button;
7.    import android.widget.ProgressBar;
8.
9.    public class MainActivity extends Activity {
10.
11.     private Button b1;
12.     private ProgressBar pb;
13.     private int currentValue=0;
14.
15.     @Override
16.     protected void onCreate(Bundle savedInstanceState) {
17.       //TODO Auto-generated method stub
18.       super.onCreate(savedInstanceState);
19.       this.setContentView(R.layout.main);
20.
21.       this.b1=(Button) this.findViewById(R.id.button1);
22.       this.pb=(ProgressBar) this.findViewById(R.id.progressBar1);
23.       this.b1.setOnClickListener(new OnClickListener() {
24.
25.         @Override
26.         public void onClick(View v) {
27.
28.           new Thread(new ProgressThread()).start();
29.
30.         }
```

```
31.       });
32.     }
33.
34.     class ProgressThread implements Runnable{
35.
36.       @Override
37.       public void run() {
38.         while(currentValue<=100){
39.           pb.setProgress(currentValue);
40.           try {
41.             Thread.sleep(100);
42.           } catch (InterruptedException e) {
43.             //TODO Auto-generated catch block
44.             e.printStackTrace();
45.           }
46.           currentValue+=10;
47.         }
48.         currentValue=0;
49.
50.       }
51.
52.     }
53.   }
```

(4) 部署运行 ProgressBarDemo 项目工程，项目运行效果如图 6-9 所示。

单击 Click Button 按钮，执行 ProgressBar，效果如图 6-10 所示。

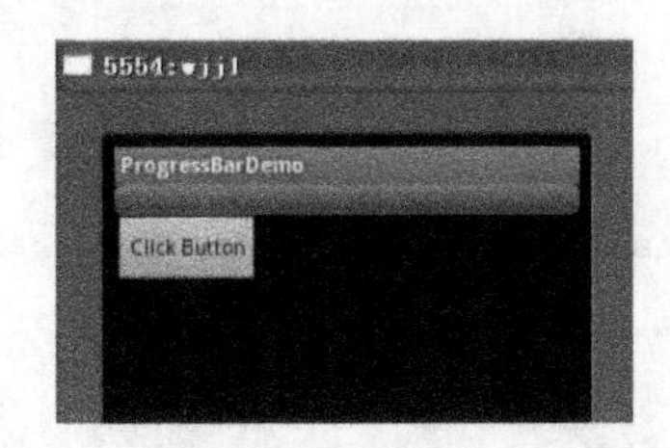

图 6-9 ProgressBarDemo 运行结果

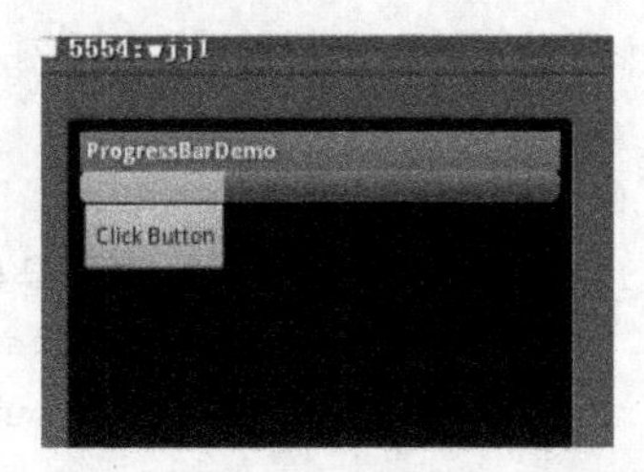

图 6-10 ProgressBar 效果

6.2.3 滑块

SeekBar(滑块)是 ProgressBar 的扩展，位于 android. widget 包中，在其基础上增加了一个可拖动的 thumb(就是那个可拖动的图标)。用户可以触摸 thumb 并向左或向右拖动，或者可以使用方向键设置当前的进度等级。不建议把可以获取焦点的 widget 放在 SeekBar 的左边或右边。SeekBar 的类继承结构图如图 6-11 所示。

SeekBar 可以附加一个 SeekBar. OnSeekBarChangeListener，以获得用户操作的通知。

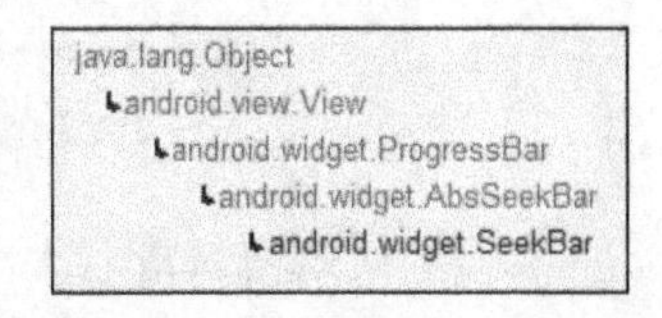

图 6-11 SeekBar 类继承关系

通过 SeekBar. getProgress()方法获取拖动条当前值。

通过调用 setOnSeekBarChangeListener()方法，处理拖动条值变化事件，把 SeekBar. OnSeekBarChangeListener 实例作为参数传入。

SeekBar 通用的方法及步骤如下：

(1) 用 XML 描述 SeekBar 控件。

```
<?xml version="1.0" encoding="utf-8"?>
<LinearLayout
  xmlns:android="http://schemas.android.com/apk/res/android"
  android:layout_width="fill_parent"
  android:layout_height="fill_parent"
  android:orientation="vertical">
  <SeekBar
    android:id="@+id/seekBar"
    android:layout_height="wrap_content"
    android:layout_width="fill_parent"/>

    <Button android:id="@+id/seekBarButton"
    android:layout_height="wrap_content"
    android:layout_width="wrap_content"
    android:text="获取值"
    />
</LinearLayout>
```

(2) 引用上述 XML 描述的控件，在程序中调用：

```
public class SeekBarActivity extends Activity {
    private SeekBar seekBar;
    @Override
    public void onCreate(Bundle savedInstanceState) {
        super.onCreate(savedInstanceState);
        setContentView(R.layout.seekbar);
        seekBar=(SeekBar) findViewById(R.id.seekBar);
        seekBar.setMax(100);                    //设置最大刻度
        seekBar.setProgress(30);                //设置当前刻度
        seekBar.setOnSeekBarChangeListener(new SeekBar.OnSeekBarChangeListener() {
            @Override
            public void onProgressChanged(SeekBar seekBar, int progress,
            boolean fromTouch) {
                Log.v("onProgressChanged()", String.valueOf(progress)+",
                "+String.valueOf(fromTouch));
            }
            @Override
            public void onStartTrackingTouch(SeekBar seekBar) {            //开始拖动
                Log.v("onStartTrackingTouch()", String.valueOf(seekBar.
```

```
            getProgress()));
        }
        @Override
        public void onStopTrackingTouch(SeekBar seekBar) {            //结束拖动
            Log.v("onStopTrackingTouch()", String.valueOf(seekBar.
            getProgress()));
        }
    });
    Button button= (Button)this.findViewById(R.id.seekBarButton);
    button.setOnClickListener(new View.OnClickListener() {
  @Override
  public void onClick(View v) {
  Toast.makeText(SeekBarActivity.this, String.valueOf(seekBar.
  getProgress()), 1).show();
  }
      });
  }
}
```

6.2.4 SeekBar 应用案例

在上节介绍 SeekBar 的常用方法和应用基础上，下面通过一个案例应用熟悉和掌握 SeekBar 控件应用效果。

(1) 创建一个新的 Android 工程，工程名为 SeekBarDemo，目标 API 选择 10(即 Android 2.3.3 版本)，应用程序名为 SeekBarDemo，包名为 com.hisoft.activity，创建的 Activity 的名字为 MainActivity，最小 SDK 版本根据选择的目标 API 会自动添加为 10，创建项目工程如图 6-12 所示。

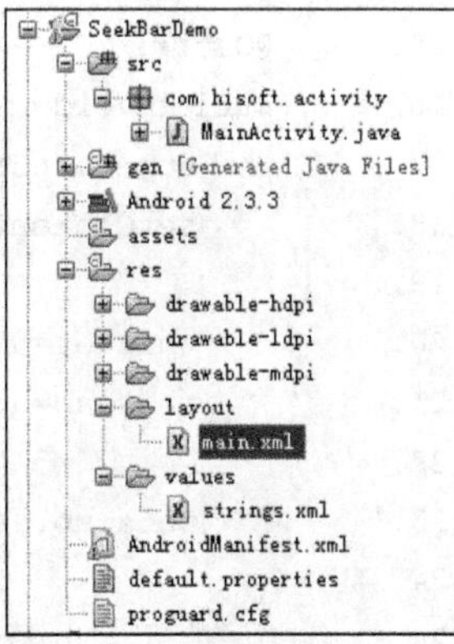

图 6-12　SeekBarDemo 工程目录结构

(2) 打开项目工程中 res→layout 目录下的 main.xml 文件，设置相对布局，添加一个 SeekBar 和一个 EditText 控件，并设置相关属性，代码如下：

```
1.   <?xml version="1.0" encoding="utf-8"?>
2.   <LinearLayout xmlns:android=
     "http://schemas.android.com/apk/res/android"
3.       android:orientation="vertical"
4.       android:layout_width="fill_parent"
5.       android:layout_height="fill_parent"
6.       >
7.       <EditText android:layout_height="wrap_content"
8.           android:layout_width="match_parent"
9.           android:id="@+id/editText1"
10.            android:text="当前值:">
```

```
11.         <requestFocus></requestFocus>
12.     </EditText>
13.     <SeekBar android:id="@+id/seekBar1"
14.         android:layout_height="wrap_content"
15.         android:layout_width="match_parent">
16.     </SeekBar>
17. </LinearLayout>
```

(3) 修改 src 目录中 com. hisoft. activity 包下的 MainActivity. java 文件，代码如下：

```
1.  package com.hisoft.activity;
2.
3.  import android.app.Activity;
4.  import android.os.Bundle;
5.  import android.widget.EditText;
6.  import android.widget.SeekBar;
7.  import android.widget.SeekBar.OnSeekBarChangeListener;
8.
9.  public class MainActivity extends Activity {
10.
11.   private EditText et;
12.   private SeekBar sb;
13.
14.     /** Called when the activity is first created. * /
15.     @Override
16.     public void onCreate(Bundle savedInstanceState) {
17.         super.onCreate(savedInstanceState);
18.         setContentView(R.layout.main);
19.
20.         this.et= (EditText) this.findViewById(R.id.editText1);
21.         this.sb= (SeekBar) this.findViewById(R.id.seekBar1);
22.
23.         this.sb.setOnSeekBarChangeListener(new OnSeekBarChangeListener() {
24.
25.             @Override
26.             public void onStopTrackingTouch(SeekBar seekBar) {
27.
28.
29.             }
30.
31.             @Override
32.             public void onStartTrackingTouch(SeekBar seekBar) {
33.
34.
35.             }
36.
```

```
37.            @Override
38.            public void onProgressChanged(SeekBar seekBar, int progress,
39.                boolean fromUser) {
40.
41.              et.setText("当前值:"+progress);
42.            }
43.        });
44.    }
45. }
```

(4) 部署运行 SeekBarDemo 项目工程,项目运行效果如图 6-13 所示。

图 6-13 SeekBar 运行效果

6.3 评分控件简介

6.3.1 评分控件

RatingBar(评分控件)位于 android. widget 包中,是基于 SeekBar 和 ProgressBar 的扩展,用星型来显示等级评定。使用 RatingBar 的默认大小时,用户可以触摸/拖动或使用键来设置评分。它有两种样式(小风格用 ratingBarStyleSmall,大风格用 ratingBarStyleIndicator),其中大的只适合指示,不适合于用户交互。RatingBar 的类继承结构如图 6-14 所示。

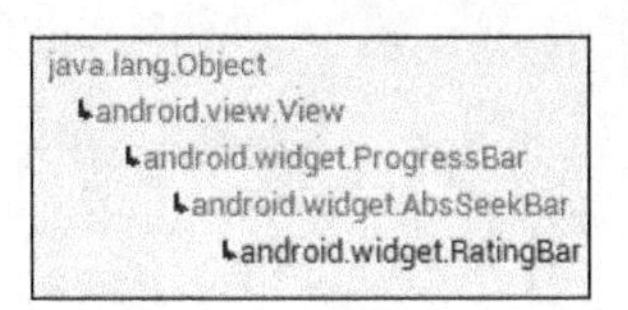

图 6-14 RatingBar 类继承关系

当使用可以支持用户交互的 RatingBar 时,无论将控件(widgets)放在它的左边还是右边都是不合适的。

只有当布局的宽被设置为 wrapcontent 时,设置的星星数量(通过函数 setNumStars(int)或者在 XML 的布局文件中定义)将显示出来(如果设置为另一种布局宽的话,后果无法预知)。

次级进度一般不应该被修改,因为它仅仅是被当作星型部分内部的填充背景。

RatingBar 控件的 XML 属性如表 6-4 所示。

表 6-4 RatingBar 控件的 XML 属性

属性名称	描述
android:isIndicator	RatingBar 是否是一个指示器(用户无法进行更改)
android:numStars	显示的星型数量,必须是一个整形值,如"100"
android:rating	默认的评分,必须是浮点类型,如"1.2"
android:stepSize	评分的步长,必须是浮点类型,如"1.2"

6.3.2 RatingBar 应用案例

上节介绍了 RatingBar 的基础知识和常用方法,下面通过一个案例应用熟悉和掌握 RatingBar 控件应用效果。

(1) 创建一个新的 Android 工程,工程名为 RatingBarDemo,目标 API 选择 10(即 Android 2.3.3 版本),应用程序名为 RatingBarDemo,包名为 com.hisoft.activity,创建的 Activity 的名字为 MainActivity,最小 SDK 版本根据选择的目标 API 会自动添加为 10,创建项目工程如图 6-15 所示。

图 6-15 RatingBarDemo 工程结构

(2) 打开项目工程中 res→layout 目录下的 main.xml 文件,设置相对布局,添加一个 RatingBar 和一个 EditText 控件,并设置相关属性,代码如下:

```
1.  <?xml version="1.0" encoding="utf-8"?>
2.  <LinearLayout xmlns:android=
    "http://schemas.android.com/apk/res/android"
3.      android:orientation="vertical"
4.      android:layout_width="fill_parent"
5.      android:layout_height="fill_parent"
6.      >
7.    <EditText android:layout_height="wrap_content"
8.          android:layout_width="match_parent"
9.          android:id="@+id/editText1"
10.           android:text="当前值:">
11.         <requestFocus></requestFocus>
12.     </EditText>
13.   <RatingBar android:id="@+id/ratingBar1"
14.         android:layout_width="wrap_content"
15.         android:layout_height="wrap_content"
16.         android:numStars="5"
17.         android:max="5">
18.   </RatingBar>

19. </LinearLayout>
```

(3) 修改 src 目录中 com.hisoft.activity 包下的 MainActivity.java 文件,代码如下:

```
1.  package com.hisoft.activity;
2.  import android.app.Activity;
3.  import android.os.Bundle;
4.  import android.widget.EditText;
5.  import android.widget.RatingBar;
6.  import android.widget.RatingBar.OnRatingBarChangeListener;

7.  public class MainActivity extends Activity {

8.    private EditText et;
9.    private RatingBar rb;
```

```
10.        /** Called when the activity is first created.* /
11.        @Override
12.        public void onCreate(Bundle savedInstanceState) {
13.            super.onCreate(savedInstanceState);
14.            setContentView(R.layout.main);
15.
16.            this.et= (EditText) this.findViewById(R.id.editText1);
17.            this.rb= (RatingBar) this.findViewById(R.id.ratingBar1);
18.
19.            this.rb.setOnRatingBarChangeListener(new OnRatingBarChangeListener() {
20.
21.                @Override
22.                public void onRatingChanged(RatingBar ratingBar, float rating,
23.                    boolean fromUser) {
24.
25.                  et.setText("当前值:"+rating);
26.
27.                }
28.            });
29.        }
30.    }
```

(4) 部署运行 RatingBarDemo 项目工程,项目运行效果如图 6-16 所示。

图 6-16　RatingBar 运行效果

6.4 自动完成文本控件简介

6.4.1 自动完成文本控件

AutoCompleteTextView(自动完成文本控件)继承于 EditText(编辑框),位于 android. widget 包下,能够完成自动提示功能。AutoCompleteTextView 类继承结构如图 6-17 所示。

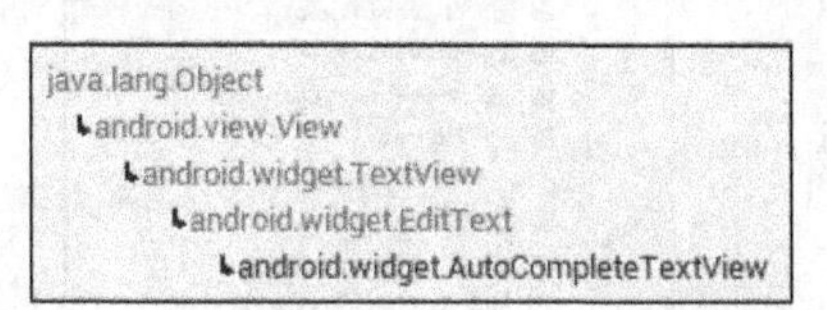

图 6-17　AutoCompleteTextView 类继承关系

AutoCompleteTextView 是一个用户输入时,能够通过显示一个下拉菜单自动提示一些与用户输入相关的文字提示信息,并可编辑的文本框。用户可以在下拉菜单列表中选择一项,简化输入。下拉菜单列表显示的数据一般是从一个数据适配器进行获取。AutoCompleteTextView 的 XML 属性及对应方法如表 6-5 所示。

AutoCompleteTextView 的通常用法及步骤如下:

(1) 用 xml 描述一个 AutoCompleteTextView。

```
<AutoCompleteTextView android:id="@+id/auto_complete"
android:layout_width="fill_parent"
android:layout_height="wrap_content"/>
```

表 6-5 AutoCompleteTextView 的 XML 属性及对应方法、描述

属性名称	对应方法	描述
android:completionHint	setCompletionHint(CharSequence)	提示信息可以直接显示在提示下拉框中
android:completionThreshold	setThreshold(int)	定义用户在下拉提示菜单出来之前需要输入的字符数，默认最多提示 20 条
android:dropDownAnchor	setDropDownAnchor(int)	设定 View 下弹出下拉菜单提示，它的值是 View 的 ID
android:dropDownHeight	setDropDownHeight(int)	设置下拉菜单的高度
android:dropDownWidth	setDropDownWidth(int)	设置下拉菜单的宽度

(2) 使用时需要设置一个 ArrayAdapter。

```
AutoCompleteTextView textView=(AutoCompleteTextView)findViewById(R.id.auto_
complete);
ArrayAdapter adapter=new ArrayAdapter(this,android.R.layout.simple_dropdown_
item_1line, COUNTRIES);
//定义匹配源的 adapter,COUNTRIES 是数据数组
textView.setAdapter(adapter);
```

6.4.2 AutoCompleteTextView 应用案例

在上述 AutoCompleteTextView 控件讲解的基础之上，通过案例熟悉 AutoCompleteTextView 控件的属性和用法，具体步骤如下：

(1) 创建一个新的 Android 工程，工程名为 AutoCompleteTextViewDemo，目标 API 选择 10(即 Android 2.3.3 版本)，应用程序名为 AutoCompleteTextViewDemo，包名为 com. hisoft. activity，创建的 Activity 的名字为 MainActivity，最小 SDK 版本根据选择的目标 API 会自动添加为 10，创建后的项目工程如图 6-18 所示。

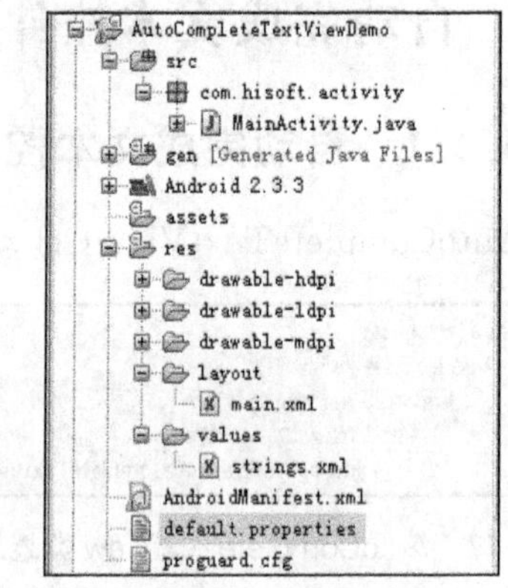

图 6-18 AutoCompleteTextViewDemo 工程目录

(2) 打开项目工程中 res→layout 目录下的 main.xml 文件，设置相对布局，添加一个 AutoCompleteTextView 和一个 TextView 控件，并设置相关属性，代码如下：

```
1.    <?xml version="1.0" encoding="utf-8"?>
2.    <LinearLayout xmlns:android="http://schemas.android.com/apk/res/android"
3.        android:orientation="vertical"
4.        android:layout_width="fill_parent"
5.        android:layout_height="fill_parent"
```

```
6.        >
7.        <TextView android:text="请输入查询信息:"
8.             android:id="@+id/textView1"
9.             android:layout_width="wrap_content"
10.              android:layout_height="wrap_content">
11.        </TextView>
12.
13.        <AutoCompleteTextView android:layout_height="wrap_content"
14.                  android:layout_width="match_parent"
15.                 android:id="@+id/autoCompleteTextView1"
16.                  android:text="">
17.           <requestFocus></requestFocus>
18.        </AutoCompleteTextView>
19.    </LinearLayout>
```

(3) 修改 src 目录中 com. hisoft. activity 包下的 MainActivity. java 文件,代码如下:

```
1.    package com.hisoft.activity;

2.    import android.app.Activity;
3.    import android.os.Bundle;
4.    import android.widget.ArrayAdapter;
5.    import android.widget.AutoCompleteTextView;

6.    public class MainActivity extends Activity {

7.      private AutoCompleteTextView actv;

8.        /** Called when the activity is first created. */
9.        @Override
10.       public void onCreate(Bundle savedInstanceState) {
11.           super.onCreate(savedInstanceState);
12.           setContentView(R.layout.main);

13.           this.actv=(AutoCompleteTextView) this.findViewById(R.id.
              autoCompleteTextView1);
14.
15.       String[] items={"android handy", "android pad", "android computer",
          "android tv"};
16.       ArrayAdapter<String>aa=new ArrayAdapter<String>(this,android.R.
          layout.simple_dropdown_item_1line,items);
17.       this.actv.setAdapter(aa);
18.       }
19.   }
```

(4) 部署运行 AutoCompleteTextViewDemo 项目工程,项目运行效果如图 6-19 所示。

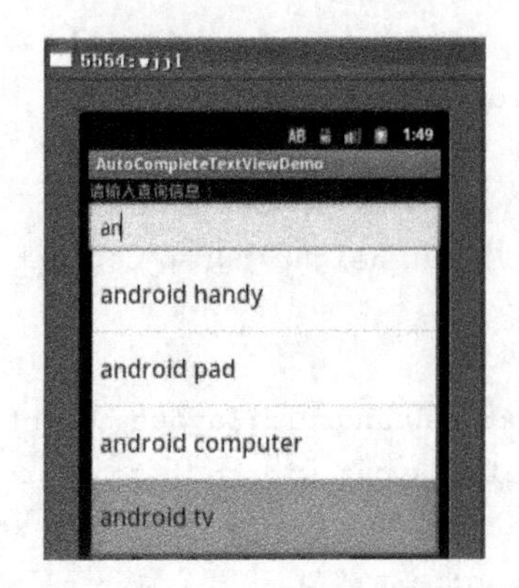

图 6-19　AutoCompleteTextView 运行效果

6.5　Tabhost 控件简介

6.5.1　Tabhost 控件

Tabhost 是提供选项卡(Tab 页)的窗口视图容器。此控件对象包含两个子对象：一组是用户可以选择指定 Tab 页的标签；另一组是 FrameLayout 用来显示该 Tab 页的内容。个别元素通常控制使用这个容器对象，而不是设置在子元素本身的值。Tabhost 是界面设计时经常使用的界面控件，可以实现多个分页之间的快速切换，每个分页可以显示不同的内容。Tabhost 的类继承结构图如图 6-20 所示。

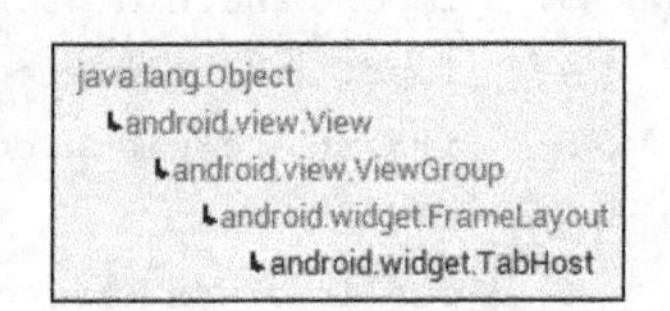

图 6-20　Tabhost 类继承关系

Tabhost 控件通常的用法及步骤如下：

(1) 设计所有分页的界面布局。

(2) 在分页设计完成后，使用代码建立 Tab 标签页，并给每个分页添加标识和标题。

(3) 确定每个分页所显示的界面布局。

每个分页建立一个 XML 文件，用以编辑和保存分页的界面布局，使用的方法与设计普通用户界面没有什么区别。

Tabhost 的实现方式有两种：

(1) 继承 TabActivity，从 TabActivity 中用 getTabHost()方法获取 TabHost。只要定义具体 Tab 内容布局即可。

(2) 不用继承 TabActivity，在布局文件中定义 TabHost 即可，但是 TabWidget 的 id 必须是@android:id/tabs，FrameLayout 的 id 必须是@android:id/tabcontent。TabHost 的 id 可以自定义。

注意：在使用 Tab 标签页时，可以将不同分页的界面布局保存在不同的 XML 文件中，也可以将所有分页的布局保存在同一个 XML 文件中。

第一种方法有利于在 Eclipse 开发环境中进行可视化设计，并且不同分页的界面布局在不同的文件中更加易于管理。

第二种方法则可以产生较少的 XML 文件，同时编码时的代码也会更加简洁。

6.5.2 Tabhost 应用案例

上节介绍了 Tabhost 的基础知识和常用方法，下面通过一个案例应用熟悉和掌握 Tabhost 控件应用效果。

(1) 创建一个新的 Android 工程，工程名为 TabhostDemo，目标 API 选择 10(即 Android 2.3.3 版本)，应用程序名为 TabhostDemo，包名为 com.hisoft.activity，创建的 Activity 的名字为 MainActivity，最小 SDK 版本根据选择的目标 API 会自动添加为 10，创建项目工程如图 6-21 所示。

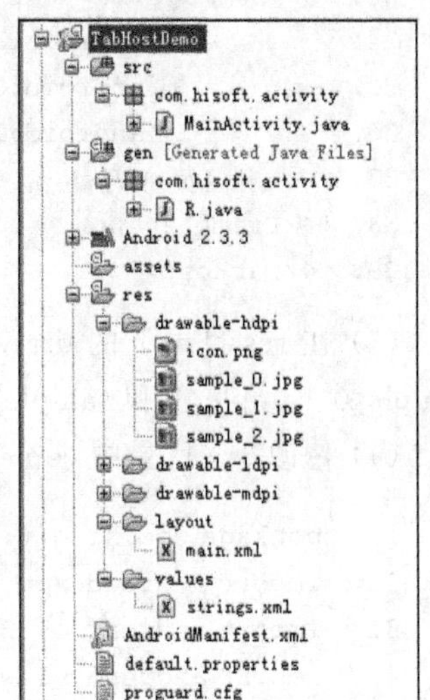

图 6-21 TabhostDemo 工程目录结构

(2) 打开项目工程中 res→layout 目录下的 main.xml 文件，设置 Tabhost 控件，并在其中设置三个帧布局，每一个帧布局中添加一个 ImageView 控件，并设置相关属性，代码如下：

```
1.   <?xml version="1.0" encoding="utf-8"?>
2.   <TabHost xmlns:android=
     "http://schemas.android.com/apk/res/android"
3.     android:layout_width="fill_parent"
4.     android:layout_height="fill_parent"
5.     >
6.   <!--定义第一个标签页的内容-->
7.   <FrameLayout
8.     android:id="@+id/frameLayout1"
9.     android:layout_width="fill_parent"
10.    android:layout_height="fill_parent">
11.      <ImageView android:src="@drawable/sample_0"
12.              android:id="@+id/imageView1"
13.              android:layout_width="wrap_content"
14.              android:layout_height="wrap_content">
15.      </ImageView>
16.  </FrameLayout>
17.  <!--定义第二个标签页的内容-->
18.  <FrameLayout
19.    android:id="@+id/frameLayout2"
20.    android:layout_width="fill_parent"
21.    android:layout_height="fill_parent">
22.      <ImageView android:src="@drawable/sample_1"
23.              android:id="@+id/imageView2"
24.              android:layout_width="wrap_content"
25.              android:layout_height="wrap_content">
26.      </ImageView>
27.  </FrameLayout>
28.  <!--定义第三个标签页的内容-->
29.  <FrameLayout
```

```
30.    android:id="@+id/frameLayout3"
31.    android:layout_width="fill_parent"
32.    android:layout_height="fill_parent">
33.      <ImageView android:src="@drawable/sample_2"
34.            android:id="@+id/imageView3"
35.            android:layout_width="wrap_content"
36.            android:layout_height="wrap_content">
37.      </ImageView>
38.  </FrameLayout>
39.  </TabHost>
```

（3）在 res 目录下的 drawable-hdpi、drawable-ldpi 和 drawable-mdpi 三个文件中添加 sample_0、sample_1 和 sample_2 三张图片。

（4）修改 src 目录中 com. hisoft. activity 包下的 MainActivity. java 文件，代码如下：

```
1.   package com.hisoft.activity;
2.   import android.app.TabActivity;
3.   import android.os.Bundle;
4.   import android.view.LayoutInflater;
5.   import android.widget.TabHost;

6.   public class MainActivity extends TabActivity {
7.       /** Called when the activity is first created. * /
8.       @Override
9.       public void onCreate(Bundle savedInstanceState) {
10.          super.onCreate(savedInstanceState);
11.
12.          TabHost tabHost=getTabHost();
13.      //设置使用 TabHost 布局
14.      LayoutInflater.from(this).inflate(R.layout.main,
15.          tabHost.getTabContentView(), true);
16.      //添加第一个标签页
17.      tabHost.addTab(tabHost.newTabSpec("tab1")
18.        .setIndicator("image1",this.getResources().getDrawable(R.drawable.
         sample_0))
19.        .setContent(R.id.frameLayout1));
20.      //添加第二个标签页
21.      tabHost.addTab(tabHost.newTabSpec("tab2")
22.        .setIndicator("image2",this.getResources().getDrawable(R.drawable.
         sample_1))
23.        .setContent(R.id.frameLayout2));
24.      //添加第三个标签页
25.      tabHost.addTab(tabHost.newTabSpec("tab3")
26.        .setIndicator("image3",this.getResources().getDrawable(R.drawable.
         sample_2))
27.          .setContent(R.id.frameLayout3));
28.      }
29.  }
```

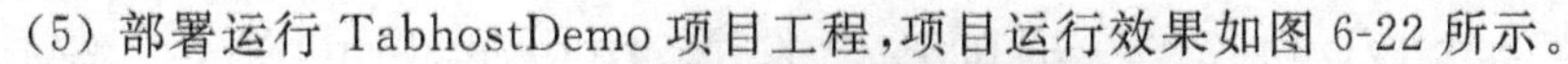

(5) 部署运行 TabhostDemo 项目工程,项目运行效果如图 6-22 所示。

图 6-22 TabhostDemo 运行效果

6.6 视图控件简介

6.6.1 滚动视图控件

ScrollView(滚动视图)控件是一种可供用户滚动的层次结构布局容器,位于 android.widget 包下,其类的继承关系如图 6-23 所示,允许显示比实际多的内容。ScrollView 是一种 FrameLayout,意味着需要在其上放置有自己滚动内容的子元素。子元素可以是一个复杂的对象的布局管理器。常用的子元素是垂直方向的 LinearLayout,显示在最上层的垂直方向可以让用户滚动的箭头。

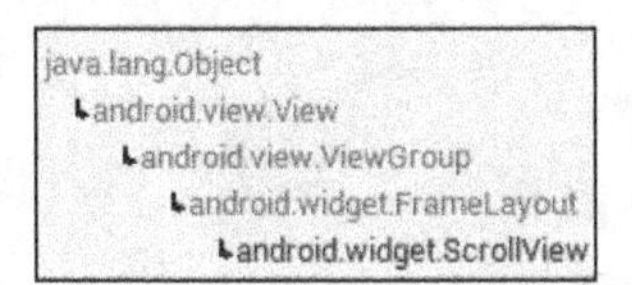

图 6-23 ScrollView 类继承关系

TextView 类也有自己的滚动功能,所以不需要使用 ScrollView,但是只有两个结合使用才能保证显示较多内容时候的效率。只有两者结合使用才可以实现在一个较大的容器中一个文本视图效果。

ScrollView 只支持垂直方向的滚动。

此类继承自 FrameLayout 类,因此内部要加入相应的布局控件才好用,否则都堆在一起了。如在 ScrollView 内部加入一个 LinearLayout 或是 RelativeLayout 等。

6.6.2 ScrollView 应用案例

前面介绍了 ScrollView 的常用方法和基础,下面通过一个案例应用熟悉和掌握 ScrollView 控件应用效果。

(1) 创建一个新的 Android 工程,工程名为 ScrollViewDemo,目标 API 选择 10(即 Android 2.3.3 版本),应用程序名为 ScrollViewDemo,包名为 com.hisoft.activity,创建的 Activity 的名字为 MainActivity,最小 SDK 版本根据选择的目标 API 会自动添加为 10,创建项目工程如图 6-24 所示。

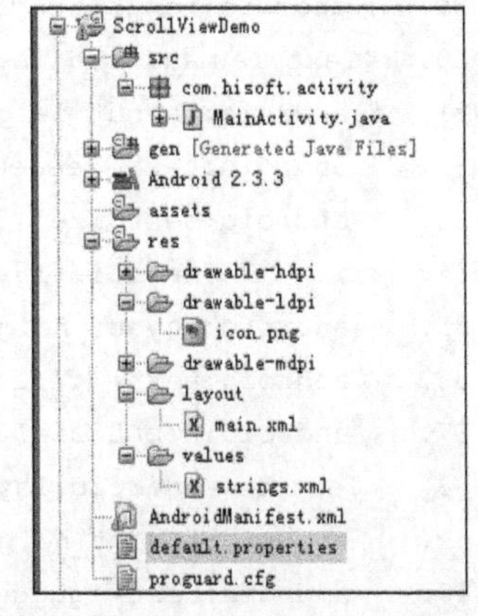

图 6-24 ScrollViewDemo 工程目录

(2) 打开项目工程中 res→layout 目录下的 main.

xml 文件，设置 ScrollView 控件，并在其中设置 HorizontalScrollView，然后在 HorizontalScrollView 控件中设置线性布局，接着在线性布局中添加多个 TextView 控件，并设置相关属性，代码如下：

```
1.  <?xml version="1.0" encoding="utf-8"?>
2.  <!--定义 ScrollView,为里面的组件添加垂直滚动条-->
3.  <ScrollView xmlns:android="http://schemas.android.com/apk/res/android"
4.    android:layout_width="fill_parent"
5.    android:layout_height="fill_parent"
6.    >
7.  <!--定义 HorizontalScrollView,为里面的组件添加水平滚动条-->
8.  <HorizontalScrollView
9.    android:layout_width="fill_parent"
10.   android:layout_height="wrap_content">
11. <LinearLayout android:orientation="vertical"
12.   android:layout_width="fill_parent"
13.   android:layout_height="fill_parent">
14. <TextView android:layout_width="wrap_content"
15.   android:layout_height="wrap_content"
16.   android:text="aaaaaaaaaaaa"
17.   android:textSize="30dp" />
18. <TextView android:layout_width="wrap_content"
19.   android:layout_height="wrap_content"
20.   android:text="bbbbbbbbbbbb"
21.   android:textSize="30dp" />
22. <TextView android:layout_width="wrap_content"
23.   android:layout_height="wrap_content"
24.   android:text="cccccccccccc"
25.   android:textSize="30dp" />
26. <TextView android:layout_width="wrap_content"
27.   android:layout_height="wrap_content"
28.   android:text="dddddddddddd"
29.   android:textSize="30dp" />
30. <TextView android:layout_width="wrap_content"
31.   android:layout_height="wrap_content"
32.   android:text="eeeeeeeeeeee"
33.   android:textSize="30dp" />
34. <TextView android:layout_width="wrap_content"
35.   android:layout_height="wrap_content"
36.   android:text="ffffffffffff"
37.   android:textSize="30dp" />
38. <TextView android:layout_width="wrap_content"
39.   android:layout_height="wrap_content"
40.   android:text="gggggggggggg"
41.   android:textSize="30dp" />
```

```
42.  <TextView android:layout_width="wrap_content"
43.    android:layout_height="wrap_content"
44.    android:text="hhhhhhhhhhhh"
45.    android:textSize="30dp" />
46.  <TextView android:layout_width="wrap_content"
47.    android:layout_height="wrap_content"
48.    android:text="iiiiiiiiiiii"
49.    android:textSize="30dp" />
50.  <TextView android:layout_width="wrap_content"
51.    android:layout_height="wrap_content"
52.    android:text="jjjjjjjjjjjj"
53.    android:textSize="30dp" />
54.  <TextView android:layout_width="wrap_content"
55.    android:layout_height="wrap_content"
56.    android:text="kkkkkkkkkkkk"
57.    android:textSize="30dp" />
58.  <TextView android:layout_width="wrap_content"
59.    android:layout_height="wrap_content"
60.    android:text="llllllllllll"
61.    android:textSize="30dp" />
62.  <TextView android:layout_width="wrap_content"
63.    android:layout_height="wrap_content"
64.    android:text="mmmmmmmmmmmm"
65.    android:textSize="30dp" />
66.  </LinearLayout>
67.  </HorizontalScrollView>
68.  </ScrollView>
```

(3) src 目录中 com. hisoft. activity 包下的 MainActivity. java 文件不做修改。

(4) 部署运行 ScrollViewDemo 项目工程，项目运行效果如图 6-25 所示。

图 6-25 ScrollViewDemo 运行效果

6.6.3 网格视图控件

GridView(网格视图)控件以二维滚动网格的格式显示其包含的子项控件，这些子项控件全部来自与视图相关的 ListAdapter 适配器。它位于 android. widget 包下，其类的继承结构图和其 XML 属性及对应方法如图 6-26 和表 6-6 所示。

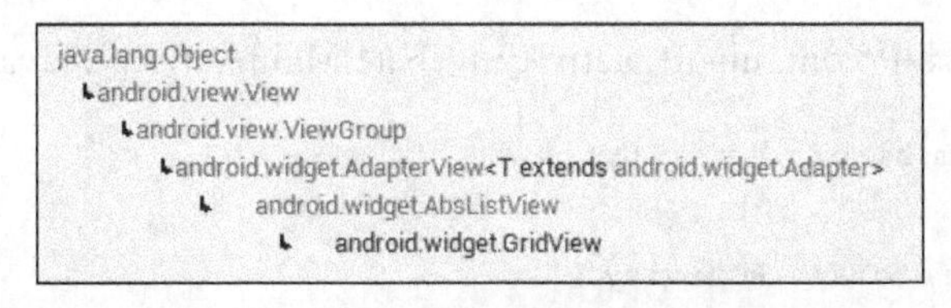

图 6-26 GridView 类继承关系

表 6-6　GridView 控件的 XML 属性及对应方法、描述

属性名称	对应方法	描　述
android:columnWidth	setColumnWidth(int)	设置列的宽度
android:gravity	setGravity(int)	设置元素的对齐方式
android:horizontalSpacing	setHorizontalSpacing(int)	设置列之间的水平间距
android:numColumns	setNumColumns(int)	设置列数
android:stretchMode	setStretchMode(int)	设置可自动填充空间的列数
android:verticalSpacing	setVerticalSpacing(int)	设置行之间默认的垂直间距

6.6.4　GridView 应用案例

上节介绍了 GridView 的基础知识和常用方法，下面通过一个案例应用熟悉和掌握 GridView 控件应用效果。

(1) 创建一个新的 Android 工程，工程名为 GridViewDemo，目标 API 选择 10(即 Android 2.3.3 版本)，应用程序名为 GridViewDemo，包名为 com.hisoft.activity，创建的 Activity 的名字为 MainActivity，最小 SDK 版本根据选择的目标 API 会自动添加为 10，创建项目工程如图 6-27 所示。

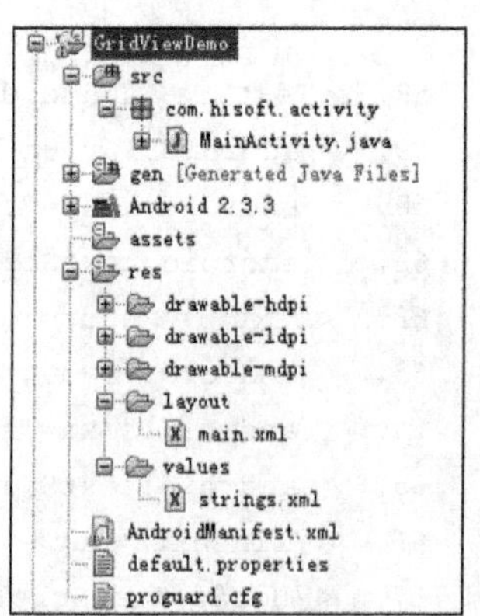

图 6-27　GridViewDemo 工程目录结构

(2) 打开项目工程中 res→layout 目录下的 main.xml 文件，设置相对布局，添加一个 GridView 控件，并设置相关属性，代码如下：

```
1.   <?xml version="1.0" encoding="utf-8"?>
2.   <GridView xmlns:android=
     "http://schemas.android.com/apk/res/android"
3.     android:id="@+id/gridview"
4.     android:layout_width="fill_parent"
5.     android:layout_height="fill_parent"
6.     android:columnWidth="90dp"
7.     android:numColumns="auto_fit"
8.     android:verticalSpacing="10dp"
9.     android:horizontalSpacing="10dp"
10.    android:stretchMode="columnWidth"
11.    android:gravity="center" />
```

(3) 修改 src 目录中 com.hisoft.activity 包下的 MainActivity.java 文件，代码如下：

```
1.  package com.hisoft.activity;

2.  import android.app.Activity;
3.  import android.os.Bundle;
4.  import android.view.View;
```

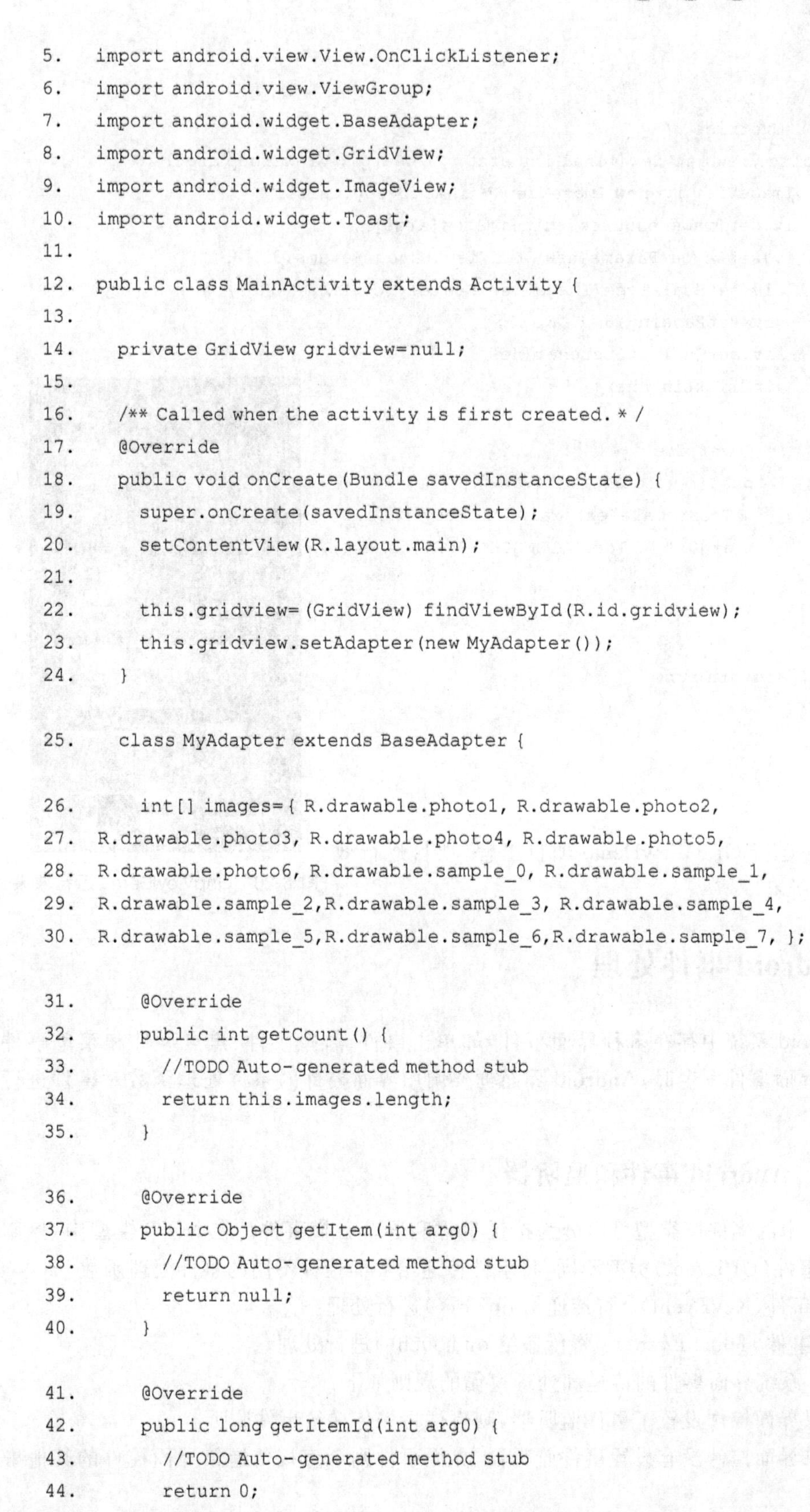

```
5.  import android.view.View.OnClickListener;
6.  import android.view.ViewGroup;
7.  import android.widget.BaseAdapter;
8.  import android.widget.GridView;
9.  import android.widget.ImageView;
10. import android.widget.Toast;
11.
12. public class MainActivity extends Activity {
13.
14.   private GridView gridview=null;
15.
16.   /** Called when the activity is first created. * /
17.   @Override
18.   public void onCreate(Bundle savedInstanceState) {
19.     super.onCreate(savedInstanceState);
20.     setContentView(R.layout.main);
21.
22.     this.gridview= (GridView) findViewById(R.id.gridview);
23.     this.gridview.setAdapter(new MyAdapter());
24.   }
25.   class MyAdapter extends BaseAdapter {
26.     int[] images={ R.drawable.photo1, R.drawable.photo2,
27. R.drawable.photo3, R.drawable.photo4, R.drawable.photo5,
28. R.drawable.photo6, R.drawable.sample_0, R.drawable.sample_1,
29. R.drawable.sample_2,R.drawable.sample_3, R.drawable.sample_4,
30. R.drawable.sample_5,R.drawable.sample_6,R.drawable.sample_7, };
31.     @Override
32.     public int getCount() {
33.       //TODO Auto-generated method stub
34.       return this.images.length;
35.     }
36.     @Override
37.     public Object getItem(int arg0) {
38.       //TODO Auto-generated method stub
39.       return null;
40.     }
41.     @Override
42.     public long getItemId(int arg0) {
43.       //TODO Auto-generated method stub
44.       return 0;
```

```
45.      }

46.      @Override
47.  public View getView(final int arg0, View arg1, ViewGroup arg2) {
48.      ImageView iv=new ImageView(MainActivity.this);
49.      iv.setImageResource(this.images[arg0]);
50.      iv.setLayoutParams(new GridView.LayoutParams(85, 85));
51.        iv.setScaleType(ImageView.ScaleType.CENTER_CROP);
52.        iv.setPadding(8, 8, 8, 8);
53.        iv.setOnClickListener(new
           OnClickListener() {

54.
55.          @Override
56.          public void onClick(View v) {
57.            Toast.makeText(MainActivity.this,
             arg0+" ", Toast.LENGTH_SHORT).show();
58.
59.          }
60.        });
61.        return iv;
62.      }
63.
64.    }
65.  }
```

图 6-28　GridViewDemo 运行效果

(4) 部署运行 GridViewDemo 项目工程，项目运行效果如图 6-28 所示。

6.7　Android 事件处理

在 Android 系统中存在多种界面事件，如单击事件、触摸事件、焦点事件和菜单事件等。在这些界面事件发生时，Android 界面框架调用界面控件的事件处理函数对事件进行处理。

6.7.1　Android 事件和监听器

Android 中的事件按类型可以分为按键事件和屏幕触摸事件。在 MVC 模型中，控制器根据界面事件(UI Event)类型不同，将事件传递给界面控件不同的事件处理函数。

- 按键事件(KeyEvent)：将传递给 onKey()进行处理。
- 触摸事件(TouchEvent)：将传递给 onTouch()进行处理。

Android 系统界面事件的传递和处理遵循的规则如下：

(1) 如果界面控件设置了事件监听器，则事件将先传递给事件监听器。

(2) 如果界面控件没有设置事件监听器，界面事件则会直接传递给界面控件的其他事件处理函数。

(3) 即使界面控件设置了事件监听器,界面事件也可以再次传递给其他事件处理函数。

(4) 是否继续传递事件给其他处理函数是由事件监听器处理函数的返回值决定的。

(5) 如果监听器处理函数的返回值为 true,表示该事件已经完成处理过程,不需要其他处理函数参与处理过程,这样事件就不会再继续进行传递。

(6) 如果监听器处理函数的返回值为 false,则表示该事件没有完成处理过程,或需要其他处理函数捕获到该事件,事件会被传递给其他的事件处理函数。

以 EditText 控件中的按键事件为例,说明 Android 系统界面事件传递和处理过程,假设 EditText 控件已经设置了按键事件监听器。

(1) 当用户按下键盘上的某个按键时,控制器将产生 KeyEvent 按键事件。

(2) Android 系统会首先判断 EditText 控件是否设置了按键事件监听器,因为 EditText 控件已经设置按键事件监听器 OnKeyListener,所以按键事件先传递到监听器的事件处理函数 onKey()中。

(3) 事件能够继续传递给 EditText 控件的其他事件处理函数,完全根据 onKey()的返回值来确定。

(4) 如果 onKey()返回 false,事件将继续传递,这样 EditText 控件就可以捕获到该事件,将按键的内容显示在 EditText 控件中。

(5) 如果 onKey()返回 true,将阻止按键事件的继续传递,这样 EditText 控件就不能够捕获到按键事件,也就不能够将按键内容显示在 EditText 控件中。

6.7.2 Android 事件处理机制

6.7.1 节讲述了 Android 事件的类型和事件处理的原则,按 Android 事件类别的处理可分为 Android 按键事件处理和屏幕触摸处理两类。不论是按键事件还是屏幕触摸处理,它们的 Android 的事件处理模型都分为基于监听接口的事件处理和基于回调机制的事件处理两类。此外,还有 Handler 消息传递机制,用于解决 Android 系统平台不允许新启动的线程访问该 Activity 里面的界面组件 Widget 问题,用户使用 Handler 可以完成 Activity 的界面组件与应用程序中线程之间的交互。

下面就 Android 的按键事件处理和屏幕触摸事件处理机制进行介绍,然后介绍 Android 事件处理机制中常用的方法。

1. Android 按键事件处理

Android 按键事件处理主要着重于 View 和 Activity 两个级别。

按键事件的处理如下:

(1) 默认情况下,如果没有 View 获得焦点,事件将传递给 Activity 处理。

(2) 如果 View 获得焦点,事件首先传递到 View 的回调方法中。View 的回调方法返回 false,事件继续传递到 Activity 处理。反之,事件不会继续传递。

使用 View. SetFocusable(true)设置可以获得焦点。

public boolean onKeyDown(int keyCode, KeyEvent msg)处理键盘按下事件。

public boolean onKeyUp(int keyCode, KeyEvent msg)处理键盘抬起事件。

注意：

(1) 要使按键可以被响应，需要在构造函数中调用 this. setFocusable(true)。

(2) 按键的 onKeyDown 和 onKeyUp 是相互独立的，不会相互影响。

(3) 无论是 View 还是 Activity 中，建议重写事件回调方法时，只对处理过的按键返回 true，没有处理的事件应该调用其父类方法。否则，其他未处理事件不会被传递到合适的目标组件中，例如 Back 按键失效问题。

下面就 Android 按键事件的监听及信息传递给处理函数举例如下：

为了处理 Android 控件的按键事件，需要先设置按键事件的监听器，并重载 onKey()。

```
1.   entryText.setOnKeyListener(new OnKeyListener(){
2.     @Override
3.   public boolean onKey(View view, int keyCode, KeyEvent keyEvent) {
4.       //过程代码…
5.       return true/false;
6.   }
```

第 1 行代码是设置控件的按键事件监听器。

第 3 行代码的 onKey()中的参数：

- 第 1 个参数 view 表示产生按键事件的界面控件。
- 第 2 个参数 keyCode 表示按键代码。
- 第 3 个参数 keyEvent 则包含了事件的详细信息，如按键的重复次数、硬件编码和按键标志等。

第 5 行代码是 onKey()的返回值：

- 返回 true，阻止事件传递。
- 返回 false，允许继续传递按键事件。

2. Android 屏幕触摸事件处理

在 Android 系统中，Touch 事件是屏幕触摸事件的基础事件。对于多层用户界面(UI)嵌套情况，如果用户单击的 UI 部分没有重叠，只是属于单独的某个 UI(如单击父 View 没有重叠的部分)，那么只有这个单独的 UI 能够捕获到 touch 事件。如果用户单击了 UI 重叠的部分，首先捕获到 touch 事件的是父类 View，然后再根据特定方法的返回值决定 Touch 事件的处理者。

在 Android 系统中，每个 View 的子类都有三个和 TouchEvent 处理密切相关的方法，分别是：

(1) public boolean dispatchTouchEvent(MotionEvent ev)； //用来分发 TouchEvent

(2) public boolean onInterceptTouchEvent(MotionEvent ev)； //用来拦截 TouchEvent

(3) public boolean onTouchEvent(MotionEvent ev)； //用来处理 TouchEvent

其中 onTouchEvent 方法定义在 View 类中，当 Touch 事件发生，首先传递到 View，由 View 处理时该方法将会被执行。

dispatchTouchEvent、onInterceptTouchEvnet 这两个方法定义在 ViewGroup 中，因为只有 ViewGroup 才会包含子 View 和子 ViewGroup，才需要在 UI 多层嵌套时，通过上述

的两个方法去决定是否监听处理连续 touch 动作和 touch 动作由谁去截获处理。

- dispatchTouchEvent 方法：默认返回值为 false。如果返回值为 false，表示捕获到一个 Touch 事件，View 便会调用 onInterceptTouchEvnet 方法进行处理，而忽略掉后面的事件。如果返回值为 true，View 将监听和处理一连串的事件。如用户单击 UI 会产生几次的 Touch 事件，如果该方法返回值为 false，View 将会处理第一次 Touch 事件，而忽略后续的 Touch 事件。如果返回值为 true，View 将处理所有的 Touch 事件。如果在第一个 Touch 事件的处理中，某个 View 的 onInterceptTouchEvnet 方法返回值为 true，把事件截获并处理。那么后续的 Touch 事件处理将不会调用该 View 的 onInterceptTouchEvnet 方法，而是直接调用该 View 的 onTouchEvent 方法。
- onInterceptTouchEvnet 方法：默认返回值为 false。如果返回值为 false，View 将不处理传递过来的 Touch 事件，而把事件传递给子 View。如果返回值为 true，View 将把事件截获并进行处理，不会把事件传递给子 View。因为 onInterceptTouchEvnet 方法的默认返回值为 false，所以在默认情况下，Touch 事件将由处于最里层的 View 的 onTouchEvent 方法去处理。如果有相邻 View 重叠，将由处于底下的 View 的 onTouchEvent 方法处理。父类 View 把事件传递给子类 View 后，子类 View 与父类 View 一样需要完成 dispatchTouchEvent、onInterceptTouchEvnet 的流程。如果 View 的 onInterceptTouchEvnet 方法返回 true，把 Touch 事件截获，将会调用自身的 onTouchEvent 事件进行处理。
- onTouchEvent 方法：默认返回值为 true。如果返回值为 true，表示事件处理完毕，将等待下一次事件。如果返回值为 false，则会返回调用重叠的处于上层的相邻 View 的 onTouchEvent 方法。如果没有重叠相邻 View，将返回调用父 View 的 onTouchEvent 方法。如果到了最外层的父 View 的 onTouchEvent 方法还是返回 false，则 Touch 事件消失。

如果为 View 设置了 OnTouchListener，而且 Touch 事件由该 View 进行处理时，监听器里面的 onTouch 方法将先于 View 自身的 onTouchEvent 方法的执行。如果 onTouch 方法返回 true，onTouchEvent 方法将不会执行。

当 TouchEvent 发生时，首先 Activity 将 TouchEvent 传递给最顶层的 View，TouchEvent 最先到达最顶层 View 的 dispatchTouchEvent，然后由 dispatchTouchEvent 方法进行分发，如果 dispatchTouchEvent 返回 true，则交给 View 的 onTouchEvent 处理；如果 dispatchTouchEvent 返回 false，则交给 View 的 interceptTouchEvent 方法来决定是否要拦截这个事件。如果 interceptTouchEvent 返回 true，也就是拦截掉了，则交给它的 onTouchEvent 来处理；如果 interceptTouchEvent 返回 false，那么就传递给子 View，由子 View 的 dispatchTouchEvent 再来重新开始这个事件的分发。如果事件传递到某一层的子 View 的 onTouchEvent 上了，这个方法返回了 false，那么这个事件会从这个 View 往上传递，都是 onTouchEvent 来接收。而如果传递到最上面的 onTouchEvent 也返回 false，则这个事件就会“消失”，系统认为事件处于阻塞状态，不再传递下一次事件。处理流程如图 6-29 所示。

上面已经介绍了 Android 系统的事件处理机制有基于回调机制的和基于监听接口的

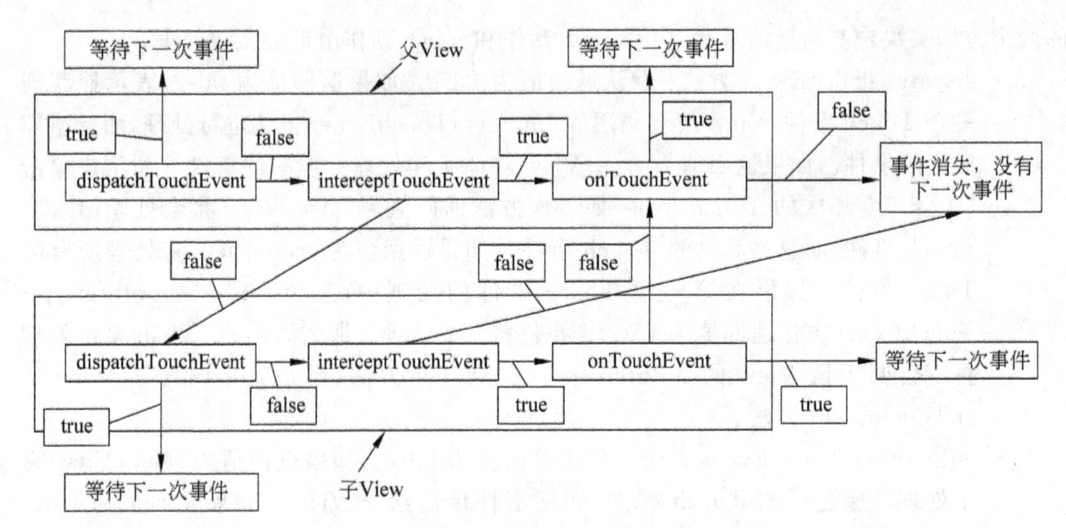

图 6-29 事件处理流程

两种，它们在事件处理中常用的方法如下所述。

(1) 基于回调机制的事件处理

Android 提供了 onKeyDown、onKeyUp、onTouchEvent、onTrackBallEvent 和 onFocusChanged 等回调方法供用户使用。

① onKeyDown：该方法是接口 KeyEvent.Callback 中的抽象方法，所有的 View 全部实现了该接口并重写了该方法。该方法用来捕捉手机键盘被按下的事件。

② onKeyUp：该方法也是接口 KeyEvent.Callback 中的一个抽象方法，并且所有的 View 同样全部实现了该接口并重写了该方法。onKeyUp 方法用来捕捉手机键盘按键抬起的事件。

③ onTouchEvent：该方法在 View 类中定义，并且所有的 View 子类全部重写了该方法，应用程序可以通过该方法处理手机屏幕的触摸事件。

④ onTrackBallEvent：该方法是手机中轨迹球的处理方法。所有的 View 同样全部实现了该方法。

⑤ onFocusChanged：该方法是焦点改变的回调方法。当某个控件重写了该方法后，当焦点发生变化时，会自动调用该方法来处理焦点改变的事件。

(2) 基于监听接口的事件处理

Android 提供的基于事件监听接口有 OnClickListener、OnLongClickListener、OnFocusChangeListener、OnKeyListener、OnTouchListener 和 OnCreateContextMenuListener 等。

① OnClickListener 接口：该接口处理的是单击事件。在触摸模式下，是在某个 View 上按下并抬起的组合动作；而在键盘模式下，是某个 View 获得焦点后单击确定键或者按下轨迹球事件。

② OnLongClickListener 接口：与上述 OnClickListener 接口的原理基本相同，只是该接口为 View 长按事件的捕捉接口，即当长时间按下某个 View 时触发的事件。

③ OnFocusChangeListener 接口：用来处理控件焦点发生改变的事件。如果注册了该接口，当某个控件失去焦点或者获得焦点时都会触发该接口中的回调方法。

④ OnKeyListener 接口：对手机键盘进行监听的接口，通过对某个 View 注册并监听，当 View 获得焦点并有键盘事件时便会触发该接口中的回调方法。

⑤ OnTouchListener 接口：用来处理手机屏幕事件的监听接口，当在 View 的范围内进行触摸按下、抬起或滑动等动作时都会触发该事件。

⑥ OnCreateContextMenuListener 接口：用来处理上下文菜单显示事件的监听接口。该方法是定义和注册上下文菜单的另一种方式。

6.7.3 Android 事件处理机制应用案例

上节讲述了 Android 事件处理机制的基础及详细原理，本节将通过一个案例帮助读者理解和掌握 Android 的事件处理机制，具体步骤如下：

(1) 创建一个新的 Android 工程，工程名为 TestTouchEventApp，目标 API 选择 10（即 Android 2.3.3 版本），应用程序名为 TestTouchEventApp，包名为 com. hisoft. activity，创建的 Activity 的名字为 TestTouchEventAppActivity，最小SDK 版本根据选择的目标 API 会自动添加为 10，创建项目工程如图 6-30 所示。

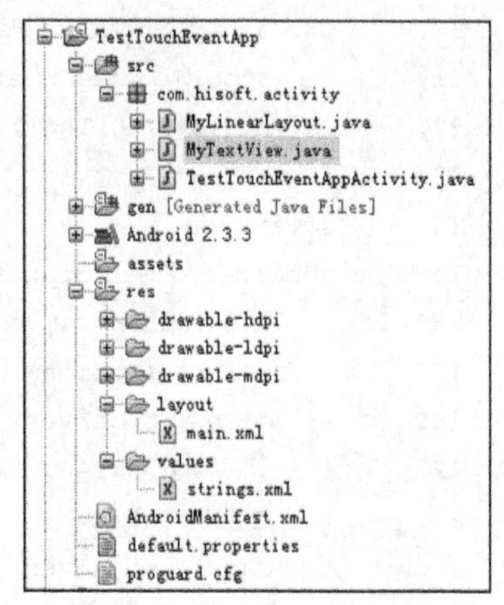

图 6-30 TestTouchEventApp 工程目录结构

(2) 打开项目工程中 res→layout 目录下的 main.xml 文件，设置自定义线性布局和自定义 TextView，并设置相关属性，代码如下：

```
1.  <?xml version="1.0" encoding="utf-8"?>
2.  <com.hisoft.activity.MyLinearLayout xmlns:
    android="http://schemas.android.com/apk/
    res/android"
3.      android:orientation="vertical"
4.      android:layout_width="fill_parent"
5.      android:layout_height="fill_parent"
6.      android:gravity="center"
7.      >
8.  <com.hisoft.activity.MyTextView
9.              android:layout_width="200px"
10.             android:layout_height="200px"
11.             android:id="@+id/tv"
12.             android:text="bjzf"
13.             android:textSize="40sp"
14.             android:textStyle="bold"
15.             android:background="#FFFFFF"
16.             android:textColor="#0000FF"
17.     />
18. </com.hisoft.activity.MyLinearLayout>
```

(3) 在 src 目录下包 com. hisoft. activity 下创建 MyLinearLayout. java 文件，代码如下：

```
1.    package com.hisoft.activity;

2.    import android.content.Context;
3.    import android.util.AttributeSet;
4.    import android.util.Log;
5.    import android.view.MotionEvent;
6.    import android.widget.LinearLayout;
7.
8.    public class MyLinearLayout extends LinearLayout {
9.      private final String TAG="MyLinearLayout";
10.     public MyLinearLayout(Context context, AttributeSet attrs) {
11.         super(context, attrs);
12.         Log.d(TAG, TAG);
13.     }
14.     @Override
15.     public boolean dispatchTouchEvent(MotionEvent ev) {
16.         int action=ev.getAction();
17.         switch (action) {
18.         case MotionEvent.ACTION_DOWN:
19.             Log.d(TAG, "dispatchTouchEvent action:ACTION_DOWN");
20.             break;
21.         case MotionEvent.ACTION_MOVE:
22.             Log.d(TAG, "dispatchTouchEvent action:ACTION_MOVE");
23.             break;
24.         case MotionEvent.ACTION_UP:
25.             Log.d(TAG, "dispatchTouchEvent action:ACTION_UP");
26.             break;
27.         case MotionEvent.ACTION_CANCEL:
28.             Log.d(TAG, "dispatchTouchEvent action:ACTION_CANCEL");
29.             break;
30.         }
31.         return super.dispatchTouchEvent(ev);
32.     }
33.     @Override
34.     public boolean onInterceptTouchEvent(MotionEvent ev) {
35.         int action=ev.getAction();
36.         switch (action) {
37.         case MotionEvent.ACTION_DOWN:
38.             Log.d(TAG, "onInterceptTouchEvent action:ACTION_DOWN");
39.             break;
40.         case MotionEvent.ACTION_MOVE:
41.             Log.d(TAG, "onInterceptTouchEvent action:ACTION_MOVE");
42.             break;
43.         case MotionEvent.ACTION_UP:
44.             Log.d(TAG, "onInterceptTouchEvent action:ACTION_UP");
```

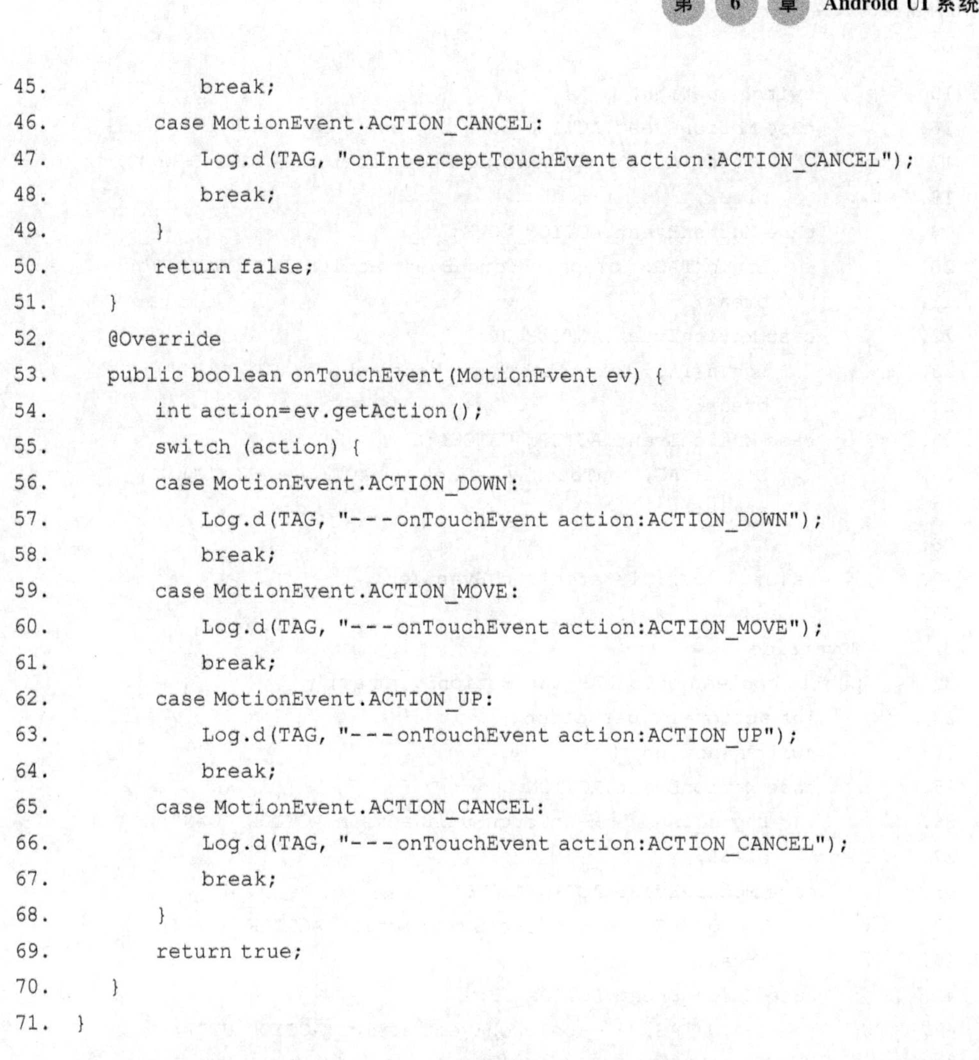

```
45.             break;
46.         case MotionEvent.ACTION_CANCEL:
47.             Log.d(TAG, "onInterceptTouchEvent action:ACTION_CANCEL");
48.             break;
49.         }
50.         return false;
51.     }
52.     @Override
53.     public boolean onTouchEvent(MotionEvent ev) {
54.         int action=ev.getAction();
55.         switch (action) {
56.         case MotionEvent.ACTION_DOWN:
57.             Log.d(TAG, "---onTouchEvent action:ACTION_DOWN");
58.             break;
59.         case MotionEvent.ACTION_MOVE:
60.             Log.d(TAG, "---onTouchEvent action:ACTION_MOVE");
61.             break;
62.         case MotionEvent.ACTION_UP:
63.             Log.d(TAG, "---onTouchEvent action:ACTION_UP");
64.             break;
65.         case MotionEvent.ACTION_CANCEL:
66.             Log.d(TAG, "---onTouchEvent action:ACTION_CANCEL");
67.             break;
68.         }
69.         return true;
70.     }
71. }
```

(4) 在 src 目录下的包 com. hisoft. activity 下创建 MyTextView. java 文件，代码如下：

```
1.     package com.hisoft.activity;

2.   import android.content.Context;
3.   import android.util.AttributeSet;
4.   import android.util.Log;
5.   import android.view.MotionEvent;
6.   import android.widget.TextView;

7.   public class MyTextView extends TextView {

8.       private final String TAG="MyTextView";
9.       public MyTextView(Context context, AttributeSet attrs) {
10.          super(context, attrs);
11.      }
12.      @Override
13.      public boolean dispatchTouchEvent(MotionEvent ev) {
14.          int action=ev.getAction();
```

```
15.         switch (action) {
16.         case MotionEvent.ACTION_DOWN:
17.             Log.d(TAG, "dispatchTouchEvent action:ACTION_DOWN");
18.             break;
19.         case MotionEvent.ACTION_MOVE:
20.             Log.d(TAG, "dispatchTouchEvent action:ACTION_MOVE");
21.             break;
22.         case MotionEvent.ACTION_UP:
23.             Log.d(TAG, "dispatchTouchEvent action:ACTION_UP");
24.             break;
25.         case MotionEvent.ACTION_CANCEL:
26.             Log.d(TAG, "onTouchEvent action:ACTION_CANCEL");
27.             break;
28.         }
29.         return super.dispatchTouchEvent(ev);
30.     }
31.     @Override
32.     public boolean onTouchEvent(MotionEvent ev) {
33.         int action=ev.getAction();
34.         switch (action) {
35.         case MotionEvent.ACTION_DOWN:
36.             Log.d(TAG, "---onTouchEvent action:ACTION_DOWN");
37.             break;
38.         case MotionEvent.ACTION_MOVE:
39.             Log.d(TAG, "---onTouchEvent action:ACTION_MOVE");
40.             break;
41.         case MotionEvent.ACTION_UP:
42.             Log.d(TAG, "---onTouchEvent action:ACTION_UP");
43.             break;
44.         case MotionEvent.ACTION_CANCEL:
45.             Log.d(TAG, "---onTouchEvent action:ACTION_CANCEL");
46.             break;
47.         }
48.         return true;
49.     }
50. }
```

(5) 部署运行 TestTouchEventApp 项目工程，项目运行效果如图 6-31 所示。

图 6-31　TestTouchEventApp 运行效果

(6) 在下面给定的条件下，通过程序运行时输出的 Log 来说明在不同条件下调用的时间顺序。

① 在 MyLinearLayout. onInterceptTouchEvent = false、MyLinearLayout. onTouchEvent = true、MyTextView. onTouchEvent=true 的条件下，输出的 log 如图 6-32 所示，表明 TouchEvent 完全由 TextView 处理。

② 在 MyLinearLayout. onInterceptTouchEvent =

Time		pid	Message
08-17 12:34:02.554	D	507	dispatchTouchEvent action:ACTION_DOWN
08-17 12:34:02.554	D	507	onInterceptTouchEvent action:ACTION_DOWN
08-17 12:34:02.614	D	507	dispatchTouchEvent action:ACTION_DOWN
08-17 12:34:02.624	D	507	---onTouchEvent action:ACTION_DOWN
08-17 12:34:02.664	D	507	dispatchTouchEvent action:ACTION_MOVE
08-17 12:34:02.664	D	507	onInterceptTouchEvent action:ACTION_MOVE
08-17 12:34:02.684	D	507	dispatchTouchEvent action:ACTION_MOVE
08-17 12:34:02.694	D	507	---onTouchEvent action:ACTION_MOVE
08-17 12:34:02.725	D	507	dispatchTouchEvent action:ACTION_MOVE
08-17 12:34:02.725	D	507	onInterceptTouchEvent action:ACTION_MOVE
08-17 12:34:02.758	D	507	dispatchTouchEvent action:ACTION_MOVE
08-17 12:34:02.764	D	507	---onTouchEvent action:ACTION_MOVE
08-17 12:34:02.865	D	507	dispatchTouchEvent action:ACTION_UP
08-17 12:34:02.865	D	507	onInterceptTouchEvent action:ACTION_UP
08-17 12:34:02.894	D	507	dispatchTouchEvent action:ACTION_UP
08-17 12:34:02.894	D	507	---onTouchEvent action:ACTION_UP

图 6-32 TouchEvent 完全由 TextView 处理

false、MyLinearLayout.onTouchEvent=true、MyTextView.onTouchEvent=false 的条件下，输出的 log 如图 6-33 所示，表明 TextView 只处理了 ACTION_DOWN 事件，LinearLayout 处理了所有的 TouchEvent。

Time		pid	Message
08-17 12:55:33.876	D	507	dispatchTouchEvent action:ACTION_DOWN
08-17 12:55:33.938	D	507	onInterceptTouchEvent action:ACTION_DOWN
08-17 12:55:33.944	D	507	---onTouchEvent action:ACTION_DOWN
08-17 12:55:34.154	D	507	dispatchTouchEvent action:ACTION_MOVE
08-17 12:55:34.154	D	507	---onTouchEvent action:ACTION_MOVE
08-17 12:55:34.204	D	507	dispatchTouchEvent action:ACTION_MOVE
08-17 12:55:34.204	D	507	---onTouchEvent action:ACTION_MOVE
08-17 12:55:34.244	D	507	dispatchTouchEvent action:ACTION_MOVE
08-17 12:55:34.244	D	507	---onTouchEvent action:ACTION_MOVE
08-17 12:55:35.219	D	507	dispatchTouchEvent action:ACTION_UP
08-17 12:55:35.224	D	507	---onTouchEvent action:ACTION_UP

图 6-33 TextView 只处理了 ACTION_DOWN 事件

③ 在 MyLinearLayout.onInterceptTouchEvent=true、MyLinearLayout.onTouchEvent=true 的条件下，输出的 log 如图 6-34 所示，表明 LinearLayout 处理了所有的 TouchEvent。

Time		pid	Message
08-17 13:00:36.454	D	507	dispatchTouchEvent action:ACTION_DOWN
08-17 13:00:36.464	D	507	onInterceptTouchEvent action:ACTION_DOWN
08-17 13:00:36.487	D	507	---onTouchEvent action:ACTION_DOWN
08-17 13:00:36.524	D	507	dispatchTouchEvent action:ACTION_MOVE
08-17 13:00:36.524	D	507	---onTouchEvent action:ACTION_MOVE
08-17 13:00:36.554	D	507	dispatchTouchEvent action:ACTION_MOVE
08-17 13:00:36.577	D	507	---onTouchEvent action:ACTION_MOVE
08-17 13:00:36.586	D	507	dispatchTouchEvent action:ACTION_MOVE
08-17 13:00:36.586	D	507	---onTouchEvent action:ACTION_MOVE
08-17 13:00:36.774	D	507	dispatchTouchEvent action:ACTION_UP
08-17 13:00:36.774	D	507	---onTouchEvent action:ACTION_UP

图 6-34 LinearLayout 处理了所有的 TouchEvent

④ 在 MyLinearLayout.onInterceptTouchEvent=true、MyLinearLayout.onTouchEvent=false 的条件下，输出的 log 如图 6-35 所示，表明 LinearLayout 只处理了 ACTION_DOWN 事件，其他的 TouchEvent 被 LinearLayout 最外层的 Activity 处理了。

Time		pid	Message
08-17 13:06:56.485	D	507	dispatchTouchEvent action:ACTION_DOWN
08-17 13:06:56.494	D	507	onInterceptTouchEvent action:ACTION_DOWN
08-17 13:06:56.516	D	507	---onTouchEvent action:ACTION_DOWN

图 6-35 其他的 TouchEvent 被 LinearLayout 最外层的 Activity 处理

6.7.4 按键事件应用案例

上节介绍了 Android 事件按键及处理机制的基础知识和常用方法，下面通过一个按钮移动红色小球案例应用熟悉和掌握 Android 事件处理流程和方法。

(1) 创建一个新的 Android 工程，工程名为 KeyEventDemo，目标 API 选择 10（即 Android 2.3.3 版本），应用程序名为 KeyEventDemo，包名为 com.hisoft.activity，创建的 Activity 的名字为 MainActivity，最小 SDK 版本根据选择的目标 API 会自动添加为 10，创建项目工程如图 6-36 所示。

图 6-36 KeyEventDemo 工程目录结构

(2) 打开项目工程中 res→layout 目录下的 main.xml 文件，设置线性布局，并设置相关属性，代码如下：

```
51.  <?xml version="1.0" encoding="utf-8"?>
52.  <LinearLayout xmlns:android=
     "http://schemas.android.com/apk/res/android"
53.      android:orientation="vertical"
54.      android:layout_width="fill_parent"
55.      android:layout_height="fill_parent"
56.      >
57.
58.  </LinearLayout>
```

(3) 修改 src 目录中 com.hisoft.activity 包下的 MainActivity.java 文件，代码如下：

```
1.   package com.hisoft.activity;
2.
3.   import android.app.Activity;
4.   import android.content.Context;
5.   import android.graphics.Canvas;
6.   import android.graphics.Color;
7.   import android.graphics.Paint;
8.   import android.os.Bundle;
9.   import android.view.Display;
10.      import android.view.KeyEvent;
11.      import android.view.View;
12.      import android.view.View.OnKeyListener;
13.      import android.view.WindowManager;
14.      import android.widget.Toast;
15.
16.      public class MainActivity extends Activity {
17.        /** Called when the activity is first created. * /
18.        @Override
19.        public void onCreate(Bundle savedInstanceState) {
```

```
20.             super.onCreate(savedInstanceState);
21.
22.         //获取窗口管理器
23.         WindowManager windowManager=getWindowManager();
24.         Display display=windowManager.getDefaultDisplay();
25.         //获得屏幕宽和高
26.         int screenWidth=display.getWidth();
27.         int screenHeight=display.getHeight();
28.         //设置小球的初始位置
29.         int radius=20;
30.         int x=screenWidth / 2;
31.         int y=screenHeight / 2-radius;
32.
33.         final BallView bv=new BallView(this, x, y, radius);
34.         this.setContentView(bv);
35.
36.         //监听上下左右键
37.         bv.setOnKeyListener(new OnKeyListener() {
38.
39.           @Override
40.           public boolean onKey(View v, int keyCode, KeyEvent event) {
41.             switch (keyCode) {
42.             case KeyEvent.KEYCODE_DPAD_DOWN:
43.               bv.y+=10;
44.               break;
45.             case KeyEvent.KEYCODE_DPAD_UP:
46.               bv.y-=10;
47.               break;
48.             case KeyEvent.KEYCODE_DPAD_LEFT:
49.               bv.x-=10;
50.               break;
51.             case KeyEvent.KEYCODE_DPAD_RIGHT:
52.               bv.x+=10;
53.               break;
54.
55.             }
56.
57.             bv.invalidate();                                  //重画
58.
59.             return true;
60.           }
61.         });
62.         }

63.     class BallView extends View {
```

```
64.
65.      private int x, y;                                        //代表圆心
66.      private int radius;                                      //半径
67.
68.      public BallView(Context context, int x, int y, int radius) {
69.        super(context);
70.        this.x=x;
71.        this.y=y;
72.        this.radius=radius;
73.        this.setFocusable(true);
74.        }

75.      @Override
76.      protected void onDraw(Canvas canvas) {
77.        //TODO Auto-generated method stub
78.        super.onDraw(canvas);
79.        canvas.drawColor(Color.WHITE);
80.        Paint p=new Paint();
81.        p.setStyle(Paint.Style.FILL);
82.        p.setColor(Color.RED);
83.        canvas.drawCircle(x, y, radius, p);
84.      }

85.    }
86.  }
```

图 6-37　小球运行效果

(4) 部署运行 KeyEventDemo 项目工程，项目运行效果如图 6-37 所示，通过键盘上的→、↓、←、↑键可以移动小球到指定位置。

6.7.5　触摸事件应用案例

前面章节已经介绍了 Android 触摸事件监听、处理机制及常用方法，本节通过屏幕触摸并获取触摸位置坐标案例来介绍触摸事件及其监听处理的通用方法，具体步骤如下：

(1) 创建一个新的 Android 工程，工程名为 TouchEventDemo，目标 API 选择 10(即 Android 2.3.3 版本)，应用程序名为 TouchEventDemo，包名为 com.hisoft.activity，创建的 Activity 的名字为 MainActivity，最小 SDK 版本根据选择的目标 API 会自动添加为 10，创建项目工程如图 6-38 所示。

(2) 打开项目工程中 res→layout 目录下的 main.xml 文件，设置线性布局，并设置相关属性，代码如下：

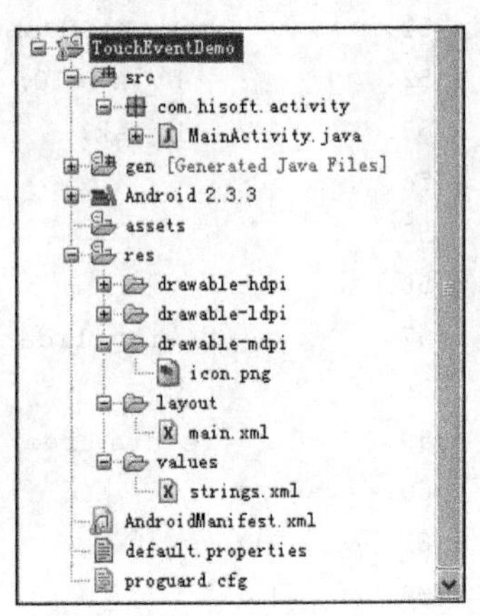

图 6-38　TouchEventDemo 工程目录结构

```
1.    <?xml version="1.0" encoding="utf-8"?>
2.    <LinearLayout xmlns:android=
```

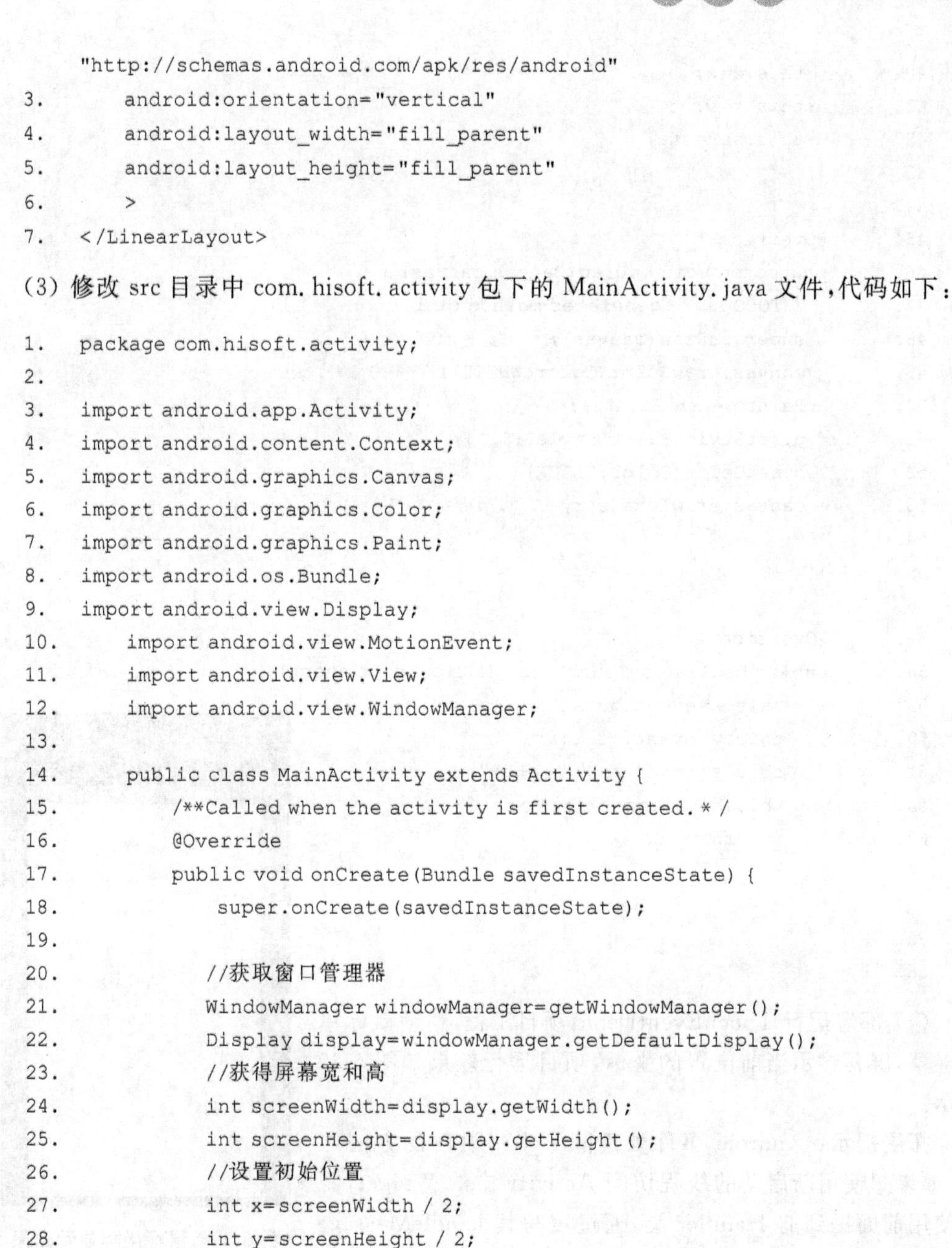

```
   "http://schemas.android.com/apk/res/android"
3.      android:orientation="vertical"
4.      android:layout_width="fill_parent"
5.      android:layout_height="fill_parent"
6.      >
7.   </LinearLayout>
```

(3) 修改 src 目录中 com. hisoft. activity 包下的 MainActivity. java 文件,代码如下:

```
1.   package com.hisoft.activity;
2.
3.   import android.app.Activity;
4.   import android.content.Context;
5.   import android.graphics.Canvas;
6.   import android.graphics.Color;
7.   import android.graphics.Paint;
8.   import android.os.Bundle;
9.   import android.view.Display;
10.      import android.view.MotionEvent;
11.      import android.view.View;
12.      import android.view.WindowManager;
13.
14.      public class MainActivity extends Activity {
15.          /**Called when the activity is first created. * /
16.          @Override
17.          public void onCreate(Bundle savedInstanceState) {
18.              super.onCreate(savedInstanceState);
19.
20.             //获取窗口管理器
21.             WindowManager windowManager=getWindowManager();
22.             Display display=windowManager.getDefaultDisplay();
23.             //获得屏幕宽和高
24.             int screenWidth=display.getWidth();
25.             int screenHeight=display.getHeight();
26.             //设置初始位置
27.             int x=screenWidth / 2;
28.             int y=screenHeight / 2;
29.
30.             setContentView(new TouchView(this, x, y, "( )"));
31.          }
32.
33.          class TouchView extends View {
34.
35.          private float x, y;                                  //初始位置坐标
36.          private String str;
37.
38.          public TouchView(Context context, float x, float y, String str) {
39.            super(context);
```

```
40.         this.x=x;
41.         this.y=y;
42.         this.str=str;
43.     }
44.
45.     @Override
46.     protected void onDraw(Canvas canvas) {
47.         //TODO Auto-generated method stub
48.         super.onDraw(canvas);
49.         canvas.drawColor(Color.WHITE);
50.         Paint p=new Paint();
51.         p.setStyle(Paint.Style.FILL);
52.         p.setColor(Color.BLACK);
53.         canvas.drawText(str, x, y, p);
54.
55.     }
56.
57.     @Override
58.     public boolean onTouchEvent(MotionEvent event) {
59.           this.x=event.getX();
60.           this.y=event.getY();
61.           this.str="("+x+","+y+")";
62.           this.invalidate();
63.           return true;
64.     }

65.   }
66. }
```

（4）部署运行 TouchEventDemo 项目工程，移动鼠标单击屏幕，屏幕显示当前位置的坐标，项目运行效果如图 6-39 所示。

图 6-39　鼠标坐标显示效果

拓展提示：Android 事件处理除了本书讲解的方法之外，如果想使用新启动的线程访问 Activity 中的 Widget，需要使用前面提到的 Handler 类，通过重写其 handleMessage 方法，在新线程中调用 sendEmptyMessage 方法向 Handler 发送消息，Handler 类调用 handleMessage 方法接收消息，然后根据消息的不同执行不同的操作，以完成 Activity 的 Widget 与应用程序中的线程交互。

6.8　项目案例

学习目标：学习 Android UI 系统控件的深入操作、属性的设置、事件处理等应用。

案例描述：使用 RelativeLayout 相对布局、TextView 控件、ListView 控件，并设置相对父控件的位置、控件之间相对位置的属性，实现医药药品的选择界面。

案例要点：ListView 控件、事件响应处理、ArrayAdapter。

案例实施：

(1) 创建工程 Project_Chapter_6，选择 Android 2.3.3 作为目标平台，项目工程目录结构如图 6-40 所示。

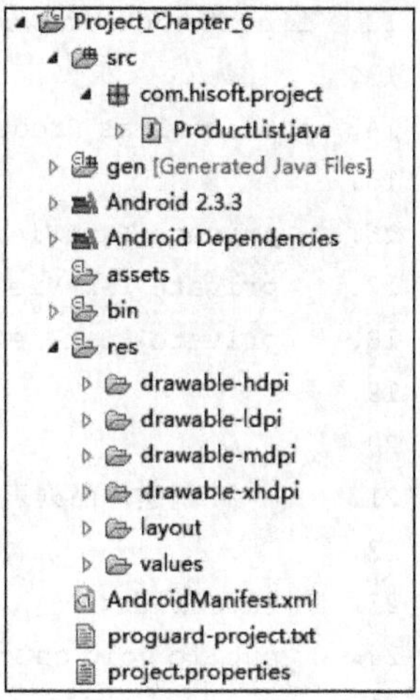

(2) 创建 productlist.xml 文件，将文件存放在 res/layout 目录下。

```
Project_Chapter_6
  src
    com.hisoft.project
      ProductList.java
  gen [Generated Java Files]
  Android 2.3.3
  Android Dependencies
  assets
  bin
  res
    drawable-hdpi
    drawable-ldpi
    drawable-mdpi
    drawable-xhdpi
    layout
    values
  AndroidManifest.xml
  proguard-project.txt
  project.properties
```

图 6-40 Project_Chapter_6 工程目录结构

```
1.   <?xml version="1.0" encoding="utf-8"?>
2.   <RelativeLayout
3.     xmlns:android="http://schemas.android.com/
     apk/res/android"
4.     android:orientation="vertical"
5.     android:layout_width="wrap_content"
6.     android:layout_height="wrap_content">
7.   <TextView
8.       android:text=""
9.       android:id="@+id/temp"
10.          android:layout_width="wrap_content"
11.          android:layout_height="wrap_content"
12.      ></TextView>
13.
14.     <ListView
15.      android:id="@+id/productlist"
16.      android:layout_below="@id/temp"
17.      android:layout_width="fill_parent"
18.      android:layout_height="wrap_content"
19.      android:focusable="true">
20.    </ListView>
21.    <TextView
22.      android:id="@+id/pageinfo"
23.      android:layout_alignParentRight="true"
24.      android:layout_width="wrap_content"
25.      android:layout_height="wrap_content"
26.      android:text=""
27.    ></TextView>
28.  </RelativeLayout>
```

(3) 在 src 目录下的包 com.hisoft.project 中编写 ProductList.java 文件，代码如下：

```
1.   package com.hisoft.project;
2.   /*****************************************************************
3.    * 程序名称:ProductList.java                                      *
4.    * 功能:显示所有的商品信息,可选择需要的商品添加到购物车             *
5.    * 作者:                                                          *
6.    * 日期:                                                          *
7.   *****************************************************************/
```

```
8.   import android.app.Activity;
9.   import android.os.Bundle;
10.  import android.widget.ArrayAdapter;
11.  import android.widget.ListView;
12.  import android.widget.TextView;
13.
14.  public class ProductList extends Activity{
15.
16.    private TextView temp;
17.    private TextView pageInfo;                    //页面显示信息
18.    private ListView productList;                 //商品列表展示控件
19.
20.       /**
21.        * 创建产品列表页面,初始化信息
22.        */
23.    @Override
24.    public void onCreate(Bundle savedInstanceState) {
25.      super.onCreate(savedInstanceState);
26.      setContentView(R.layout.productlist);
27.
28.      //拿到页面控件对象
29.      temp=(TextView) findViewById(R.id.temp);
30.      temp.setText("tempid:");
31.      pageInfo=(TextView) findViewById(R.id.pageinfo);
32.      //获取 ListView 对象
33.      productList=(ListView) findViewById(R.id.productlist);
34.
35.      productList.setItemsCanFocus(true);
36.      //设置商品可多选
37.      productList.setChoiceMode(ListView.CHOICE_MODE_MULTIPLE);
38.      productList.setTextFilterEnabled(true);
39.      //设置分页信息
40.      pageInfo.setText("共 6 件药品"+"\t"+"第 1 页"+"\t"+"共 1 页");
41.      //配置适配器
42.      String[] content={"1\t 药品-1","2\t 药品-2","3\t 药品-3","4\t 药品-4",
         "5\t 药品-5","6\t 药品-6"};
43.      ArrayAdapter<String>adapter=new ArrayAdapter<String>
         (ProductList.this,
44.         android.R.layout.simple_list_item_multiple_choice,
45.         content);
46.      productList.setAdapter(adapter);
47.    }
48.  }
```

(4) 部署运行 Project_Chapter_6 工程,运行效果如图 6-41 所示。

图 6-41 Project_Chapter_6 运行效果

1. 简答题

(1) 简述列表显示需要的元素有哪些?

(2) 下拉列表控件 Spinner 如何使用? 其步骤有哪些?

(3) 什么是自动完成文本控件 AutoCompleteTextView? 其应用步骤有哪些?

(4) Android 系统界面事件的传递和处理遵循的规则有哪些?

(5) 简述 Android 事件处理机制常用的方法有哪些?

2. 完成下面的实训项目

要求: 在本章项目案例的基础上完成用户登录、验证，如果登录成功，跳转到药品选择界面，进行药品的选择;如果登录失败，提示信息，不进行跳转。

第7章 Android UI菜单、对话框

学习目标

本章主要深入介绍 Android UI 菜单 Menu、选项菜单、子菜单、快捷菜单、对话框 Dialog、AlertDialog、日期选择对话框 DatePickerDialog、时间选择对话框 TimePickerDialog、ProgressDialog、Toast 控件、Notification 控件等。使读者通过本章的学习，能够深入熟悉 Android UI 菜单和对话框，并掌握以下知识要点：

(1) 菜单 Menu 的分类、onCreateOptionsMenu 创建方法。

(2) 选项菜单的分类、onPrepareOptionsMenu 等创建方法。

(3) 子菜单的添加方法及应用。

(4) 对话框控件 AlertDialog、日期选择对话框 DatePickerDialog、时间选择对话框 TimePickerDialog、ProgressDialog 的常用方法及属性设置。

(5) Toast 控件的属性设置、通常用法及实现方式。

(6) Notification 控件的属性描述设置及通常用法。

7.1 菜单控件 Menu

7.1.1 Menu 简介

菜单是应用程序中非常重要的组成部分，能够在不占用界面空间的前提下，为应用程序提供统一的功能和设置界面，并为程序开发人员提供易于使用的编程接口。在 Android 系统中，菜单和前面讲述的控件一样，不仅能够在代码中定义，而且可以像界面布局一样在 XML 文件中进行定义。使用 XML 文件定义界面菜单，将代码与界面设计分类，有助于简化代码的复杂程度，并且更有利于界面的可视化。

Android 系统支持三种菜单：

- 选项菜单(Option Menu)。
- 子菜单(Submenu)。
- 上下文菜单(Context Menu)。

在 Activity 中可以通过重写 onCreateOptionsMenu(Menu menu)方法创建选项菜单，然后在用户按下手机的 Menu 按钮时就会显示创建好的菜单，在 onCreateOptionsMenu (Menu menu)方法内部可以调用 Menu.add()方法实现菜单的添加。

如果处理选择事件，可以通过重写 Activity 的 onMenuItemSelected()方法，该方法常用于处理菜单被选择事件。

7.1.2 选项菜单

选项菜单是一种经常被使用的 Android 系统菜单。可以通过"菜单键(Menu)"打开浏览或选择。

选项菜单通常分为两类：图标菜单(Icon Menu)和扩展菜单(Expanded Menu)。

图标菜单是能够同时显示文字和图标的菜单，最多支持 6 个子项。此外，图标菜单不支持单选框和复选框控件，如图 7-1 所示。

扩展菜单是在图标菜单子项多于 6 个时才出现，通过单击图标菜单最后的子项 More 才能打开，如图 7-2 所示。

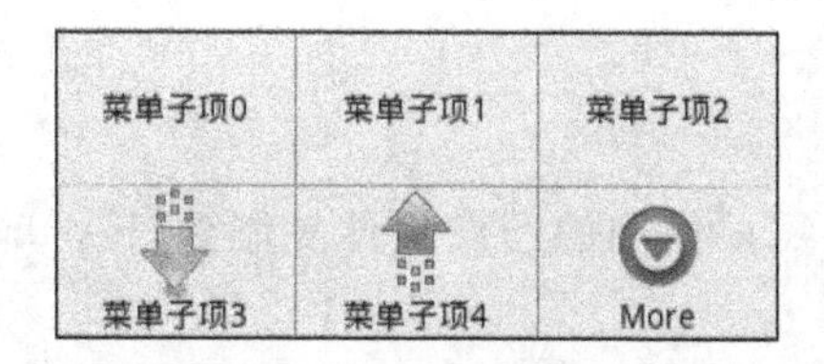

图 7-1 图标菜单

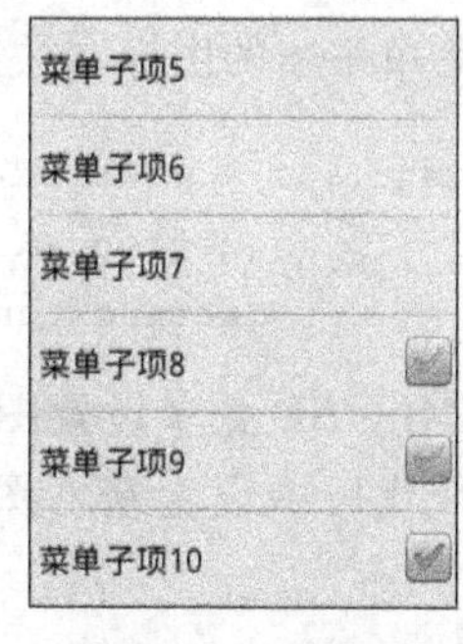

图 7-2 扩展菜单

扩展菜单是垂直的列表型菜单，它不支持显示图标，但支持单选框和复选框控件。

1. onCreateOptionsMenu()方法

只有在 Activity 中重载 onCreateOptionsMenu()方法才能够在 Android 应用程序中使用选项菜单。第一次使用选项菜单时会调用 onCreateOptionsMenu()方法，用来初始化菜单子项的相关内容(设置菜单子项自身的 ID 和组 ID、菜单子项显示的文字和图片等)。

```
1.  final static int DOWNLOAD=Menu.FIRST;
2.    final static int UPLOAD=Menu.FIRST+1;
3.      @Override
4.      public boolean onCreateOptionsMenu(Menu menu){
5.          menu.add(0,DOWNLOAD,0,"下载");
```

```
6.          menu.add(0,UPLOAD,1,"上传");
7.          return true;
8.    }
```

第 1 行和第 2 行代码将菜单子项 ID 定义成静态常量，并使用静态常量 Menu. FIRST（整数类型，值为 1）定义第一个菜单子项，后续的菜单子项仅需在 Menu. FIRST 增加相应的数值即可。

第 4 行 Menu 对象作为一个参数被传递到方法内部，因此在 onCreateOptionsMenu()方法中，用户可以使用 Menu 对象的 add()方法添加设置的菜单子项。

第 7 行代码是 onCreateOptionsMenu()方法返回值，返回 true 将显示在方法中设置的菜单，否则不能够显示菜单。

add()方法的语法如下：

```
MenuItem android.view.Menu.add(int groupId, int itemId, int order, CharSequence
title)
```

第 1 个参数 groupId 是组 ID，用以批量的对菜单子项进行处理和排序。

第 2 个参数 itemId 是子项 ID，是每一个菜单子项的唯一标识，通过子项 ID 使应用程序能够定位到用户所选择的菜单子项。

第 3 个参数 order 是定义菜单子项在选项菜单中的排列顺序。

第 4 个参数 title 是菜单子项所显示的标题。

添加菜单子项的图标和快捷键：使用 setIcon()方法和 setShortcut()方法。

```
1.   menu.add(0,DOWNLOAD,0,"下载")
2.          .setIcon(R.drawable.download);
3.          .setShortcut('2','d');
```

DOWNLOAD 菜单设置图标和快捷键。

第 2 行代码中设置新的图像资源，用户将需要使用的图像文件复制到/res/drawable 目录下。

setShortcut()方法的第 1 个参数是为数字键盘设定的快捷键；第 2 个参数是为全键盘设定的快捷键，且不区分字母的大小写。

2. onPrepareOptionsMenu()方法

重载 Activity 中的 onPrepareOptionsMenu()方法，能够实现动态的添加、删除菜单子项，或修改菜单的标题、图标和可见性等内容。

onPrepareOptionsMenu()方法的返回值的含义与 onCreateOptionsMenu()方法相同，即返回 true 显示菜单，返回 false 则不能够显示菜单。

7.1.3 选项菜单应用案例

7.1.2 节介绍了选项菜单 OptionsMenu 所涉及的 Menu、MenuItem 的基础知识，下面通过一个案例应用熟悉和掌握选项菜单 OptionsMenu 的应用开发过程。

(1) 创建一个新的 Android 工程，工程名为 OptionsMenuDemo，目标 API 选择 10(即 Android 2.3.3 版本)，应用程序名为 OptionsMenuDemo，包名为 com. hisoft. activity，创建

的 Activity 的名字为 MainActivity，最小 SDK 版本根据选择的目标 API 会自动添加为 10，创建项目工程如图 7-3 所示。

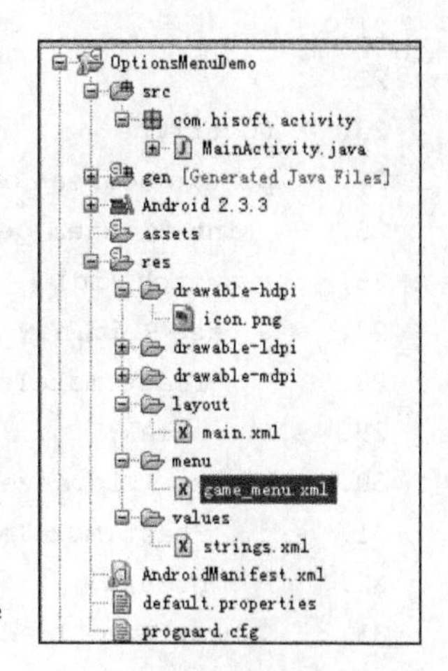

图 7-3　OptionsMenuDemo 工程目录结构

（2）在 res 目录下新建 menu 文件夹，然后创建 game_menu.xml 文件，添加 menu 和 item，并设置相关属性，代码如下：

```
4.    <?xml version="1.0" encoding="utf-8"?>
5.     <menu xmlns:android="http://schemas.android.
      com/apk/res/android">
6.         <item android:icon="@drawable/icon" android:
           title="新游戏" android:id="@+id/new_game"/>
7.        <item android:icon="@drawable/icon" android:title
          ="保存进度" android:id="@+id/save_game"/>
8.        <item android:icon="@drawable/icon" android:title
          ="载入进度" android:id="@+id/load_game"/>
9.         <item android:icon="@drawable/icon" android:
           title="退出游戏" android:id="@+id/exit_game"/>
10.     </menu>
```

（3）修改 src 目录中 com.hisoft.activity 包下的 MainActivity.java 文件，初始化菜单的操作主要通过在 onCreateOptionsMenu 方法中调用 MenuInflater 类自带的 inflate 方法绑定调用 game_menu.xml 文件和 menu 菜单，方法 onOptionsItemSelected 中放置菜单选项被选中时的处理程序代码，本案例使用了 Toast 控件及方法（后续章节会讲到），全部实现代码如下：

```
1.    package com.hisoft.activity;
2.    import android.app.Activity;
3.    import android.os.Bundle;
4.    import android.view.Menu;
5.    import android.view.MenuInflater;
6.    import android.view.MenuItem;
7.    import android.widget.Toast;
8.    public class MainActivity extends Activity {
9.      /**Called when the activity is first created.*/
10.     @Override
11.     public void onCreate(Bundle savedInstanceState) {
12.         super.onCreate(savedInstanceState);
13.         setContentView(R.layout.main);
14.     }
15.
16.     @Override
17.     public boolean onCreateOptionsMenu(Menu menu) {
18.       MenuInflater inflater=getMenuInflater();
19.       inflater.inflate(R.menu.game_menu, menu);
20.       return true;
```

```
21.     }
22.
23.     @Override
24.     public boolean onOptionsItemSelected(MenuItem item) {
25.       int id=item.getItemId();
26.       switch (id) {
27.       case R.id.new_game:
28.         Toast.makeText(this, item.getTitle(), Toast.LENGTH_LONG).show();
29.         break;
30.       case R.id.save_game:
31.         Toast.makeText(this, item.getTitle(), Toast.LENGTH_LONG).show();
32.         break;
33.       case R.id.load_game:
34.         Toast.makeText(this, item.getTitle(), Toast.LENGTH_LONG).show();
35.         break;
36.       case R.id.exit_game:
37.         Toast.makeText(this, item.getTitle(), Toast.LENGTH_LONG).show();
38.         break;
39.       default:
40.         break;
41.       }
42.       return super.onOptionsItemSelected(item);
43.     }

44.   }
```

(4) 部署运行 OptionsMenuDemo 项目工程,用鼠标单击移动设备上的 menu 按钮,程序运行显示如图 7-4 所示。然后单击"保存进度"按钮,程序运行如图 7-5 所示。

图 7-4　menu 菜单

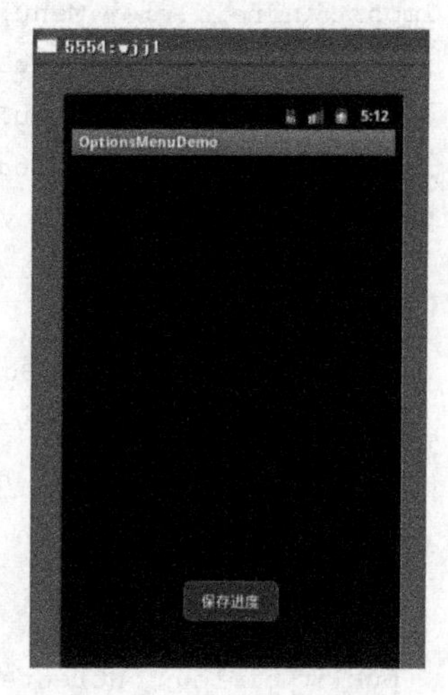

图 7-5　保存进度显示

7.1.4 子菜单

子菜单是指能够显示更加详细信息的菜单子项。在子菜单中,菜单子项使用浮动窗体的显示形式,更好地适应了小屏幕的显示。

Android 系统的子菜单使用非常灵活,可以在选项菜单或快捷菜单中使用子菜单,这样有利于将相同或相似的菜单子项组织在一起,便于显示和分类。此外,子菜单不支持嵌套,子菜单的添加是使用 addSubMenu()方法来实现。

```
1.   SubMenu uploadMenu= (SubMenu) menu.addSubMenu(0, UPLOAD,1,"上传").setIcon(R.
     drawable.upload);
2.   uploadMenu.setHeaderIcon(R.drawable.upload);
3.   uploadMenu.setHeaderTitle("上传");
4.   uploadMenu.add(0,SUB_UPLOAD_A,0,"上传参数 A");
5.   uploadMenu.add(0,SUB_UPLOAD_B,0,"上传参数 B");
```

第 1 行代码在上述的 onCreateOptionsMenu()方法传递的 menu 对象上调用 addSubMenu()方法,在选项菜单中添加一个菜单子项,用户单击后可以打开子菜单。

addSubMenu()方法与选项菜单中使用过的 add()方法支持相同的参数,同样可以指定菜单子项的 ID、组 ID 和标题等参数,并且能够通过 setIcon()方法设置菜单所显示的图标。

第 2 行代码调用 setHeaderIcon()方法,定义子菜单的图标。

第 3 行定义子菜单的标题,如果不设定子菜单的标题,子菜单将显示父菜单子项标题,即第 1 行代码中的"上传"。

第 4 行和第 5 行在子菜单中添加了两个菜单子项,菜单子项的更新方法和选择事件处理方法仍然使用 onPrepareOptionsMenu()方法和 onOptionsItemSelected()方法。

7.1.5 子菜单应用案例

7.1.4 节介绍了子菜单 SubMenu 的常用方法,下面通过一个接受用户菜单选项,然后弹出子菜单,选中子菜单项后显示所选项目的案例应用,熟悉和掌握菜单 SubMenu 的应用开发过程。

(1) 创建一个新的 Android 工程,工程名为 SubMenuDemo,目标 API 选择 10(即 Android 2.3.3 版本),应用程序名为 SubMenuDemo,包名为 com.hisoft.activity,创建的 Activity 的名字为 MainActivity,最小 SDK 版本根据选择的目标 API 会自动添加为 10,创建项目工程如图 7-6 所示。

(2) 在 res 目录下新建 menu 文件夹,然后创建 file_submenu.xml 文件,添加 menu 和 item,并设置相关属性,代码如下:

```
1.   <?xml version="1.0" encoding="utf-8"?>
2.   <menu xmlns:android="http://schemas.android.
     com/apk/res/android">
```

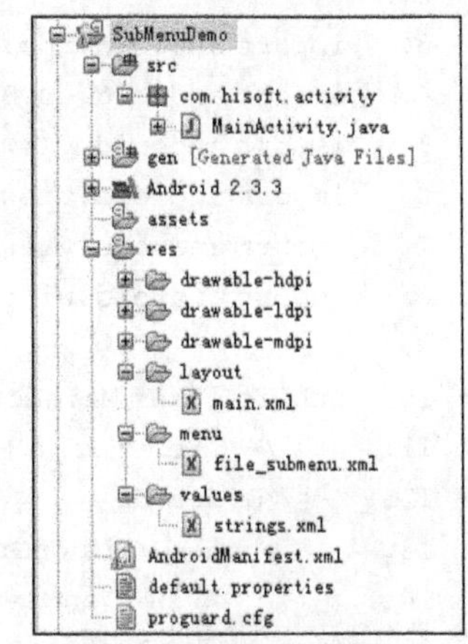

图 7-6 SubMenuDemo 工程目录结构

```
3.      <item android:id="@+id/file"
4.          android:icon="@drawable/icon"
5.          android:title="@string/file">
6.          <!--"file" submenu-->
7.          <menu>
8.              <item android:id="@+id/create_new"
9.                  android:title="@string/create_new"/>
10.             <item android:id="@+id/open"
11.                 android:title="@string/open"/>
12.             <item android:id="@+id/save"
13.                 android:title="@string/save"/>
14.             <item android:id="@+id/exit"
15.                 android:title="@string/exit"/>
16.         </menu>
17.     </item>
18. </menu>
```

(3) 修改 res 目录下 values 文件夹中的 strings. xml 文件，代码如下：

```
1.  <?xml version="1.0" encoding="utf-8"?>
2.  <resources>
3.      <string name="hello">Hello World, MainActivity!</string>
4.      <string name="app_name">SubMenuDemo</string>
5.      <string name="file">文件</string>
6.      <string name="create_new">新建</string>
7.      <string name="open">打开</string>
8.      <string name="save">保存</string>
9.      <string name="exit">退出</string>
10.     </resources>
```

(4) 修改 src 目录中 com. hisoft. activity 包下的 MainActivity. java 文件，代码如下：

```
1.  package com.hisoft.activity;
2.
3.  import android.app.Activity;
4.  import android.os.Bundle;
5.  import android.view.Menu;
6.  import android.view.MenuInflater;
7.  import android.view.MenuItem;
8.  import android.widget.Toast;
9.
10. public class MainActivity extends Activity {
11.     /**Called when the activity is first created.*/
12.     @Override
13.     public void onCreate(Bundle savedInstanceState) {
14.         super.onCreate(savedInstanceState);
15.         setContentView(R.layout.main);
16.     }
17.
```

```
18.      @Override
19.    public boolean onCreateOptionsMenu(Menu menu) {
20.      MenuInflater inflater=getMenuInflater();
21.      inflater.inflate(R.menu.file_submenu, menu);
22.      return true;
23.    }
24.
25.      @Override
26.    public boolean onOptionsItemSelected(MenuItem item) {
27.      int id=item.getItemId();
28.      switch (id) {
29.      case R.id.create_new:
30.  Toast.makeText(this,item.getTitle(),Toast.LENGTH_LONG).show();
31.        break;
32.      case R.id.open:
33.  Toast.makeText(this,item.getTitle(),Toast.LENGTH_LONG).show();
34.        break;
35.      case R.id.save:
36.  Toast.makeText(this,item.getTitle(),Toast.LENGTH_LONG).show();
37.        break;
38.      case R.id.exit:
39.  Toast.makeText(this,item.getTitle(),Toast.LENGTH_LONG).show();
40.        break;
41.      default:
42.        break;
43.      }
44.      return super.onOptionsItemSelected(item);
45.    }
46.  }
```

(5) 部署运行 OptionsMenuDemo 项目工程，用鼠标单击移动设备上的 menu 按钮，程序运行显示如图 7-7 所示。然后选择“文件”菜单，弹出子菜单，程序运行如图 7-8 所示。

图 7-7 menu 菜单显示

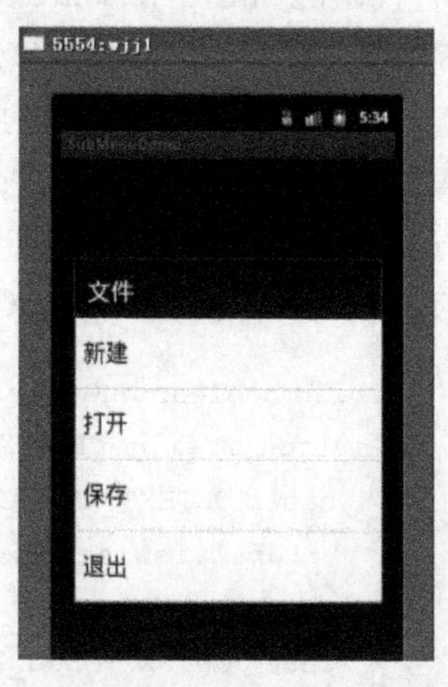

图 7-8 文件子菜单

选择“新建”子菜单，程序运行结果如图 7-9 所示。

图 7-9 新建子菜单显示

7.1.6 快捷菜单

快捷菜单同样采用了动窗体的显示方式，与子菜单的实现方式相同，但两种菜单的启动方式不同。快捷菜单类似于普通桌面程序中的“右键菜单”，当用户单击界面元素超过 2s 后，将启动注册到该界面元素的快捷菜单。

快捷菜单与使用选项菜单的方法大致相似，同样需要重载 onCreateContextMenu()方法和 onContextItemSelected()方法。onCreateContextMenu()方法主要用来添加快捷菜单所显示的标题、图标和菜单子项等内容。

选项菜单中的 onCreateOptionsMenu()方法仅在选项菜单第一次启动时被调用一次，而快捷菜单的 onCreateContextMenu()方法在每次启动时都会被调用一次。

```
1.   final static int CONTEXT_MENU_1=Menu.FIRST;
2.   final static int CONTEXT_MENU_2=Menu.FIRST+1;
3.   final static int CONTEXT_MENU_3=Menu.FIRST+2;
4.   @Override
5.   public void onCreateContextMenu(ContextMenu menu, View v, ContextMenuInfo
     menuInfo){
6.       menu.setHeaderTitle("快捷菜单标题");
7.       menu.add(0, CONTEXT_MENU_1, 0,"菜单子项 1");
8.       menu.add(0, CONTEXT_MENU_2, 1,"菜单子项 2");
9.       menu.add(0, CONTEXT_MENU_3, 2,"菜单子项 3");
10.      }
```

Android 系统中，ContextMenu 类支持 add()方法和 addSubMenu()方法，可以在快捷菜单中添加菜单子项和子菜单。

第 5 行代码的 onCreateContextMenu()方法中：

- 第 1 个参数 menu 是需要显示的快捷菜单。
- 第 2 个参数 v 是用户选择的界面元素。
- 第 3 个参数 menuInfo 是所选择界面元素的额外信息。

菜单选择事件的处理需要重载 onContextItemSelected()方法，该方法在用户选择快捷菜单中的菜单子项后被调用，与 onOptionsItemSelected()方法的使用方法基本相同。

```
1.     public boolean onContextItemSelected(MenuItem item){
2.       switch(item.getItemId()){
3.         case CONTEXT_MENU_1:
4.           LabelView.setText("子项 1");
5.           return true;
6.         case CONTEXT_MENU_2:
7.           LabelView.setText("子项 2");
```

```
8.             return true;
9.           case CONTEXT_MENU_3:
10.            LabelView.setText("子项 3");
11.        return true;
12.        }
13.        return false;
14.    }
```

```
1.     TextView LabelView=null;
2.     @Override
3.     public void onCreate(Bundle savedInstanceState) {
4.       super.onCreate(savedInstanceState);
5.       setContentView(R.layout.main);
6.       LabelView=(TextView)findViewById(R.id.label);
7.       registerForContextMenu(LabelView);
8.     }
```

第 7 行中使用 registerForContextMenu()方法，将快捷菜单注册到界面控件上。用户在长时间单击该界面控件时便会启动快捷菜单。

第 6 行中使用 TextView 是为了能够在界面上直接显示用户所选择快捷菜单的菜单子项。

第 5、8 和 11 行通过更改 TextView 的显示内容，显示用户所选择的菜单子项。

下方代码是/src/layout/main. xml 文件的部分内容，第 1 行声明了 TextView 的 ID 为 label。在上方代码的第 6 行中，通过 R. id. label 将 ID 传递给 findViewById()方法，这样用户便能够引用该界面元素，并能够修改该界面元素的显示内容。

```
1.   <TextView   android:id="@+id/label"
2.       android:layout_width="fill_parent"
3.       android:layout_height="fill_parent"
4.       android:text="@string/hello"
5.   />
```

需要注意的一点，上方代码的第 2 行将 android:layout_width 设置为 fill_parent，这样 TextView 将填充满父节点的所有剩余屏幕空间，用户单击屏幕 TextView 下方任何位置都可以启动快捷菜单。

如果将 android:layout_width 设置为 wrap_content，则用户必须准确单击 TextView 才能启动快捷菜单。

7.1.7 快捷菜单应用案例

7.1.6 节介绍了子菜单 ContextMenu 的常用方法，下面通过一个接受用户菜单选项，然后弹出子菜单，选中子菜单项后显示所选项目的案例应用，熟悉和掌握菜单 ContextMenuDemo 的应用开发过程。

(1) 创建一个新的 Android 工程，工程名为 ContextMenuDemo，目标 API 选择 10(即 Android 2.3.3 版本)，应用程序名为 ContextMenuDemo，包名为 com. hisoft. activity，创建

的 Activity 的名字为 MainActivity，最小 SDK 版本根据选择的目标 API 会自动添加为 10，创建项目工程如图 7-10 所示。

(2) 修改 res 目录下 layout 文件夹中的 main. xml 文件，添加一个 Button 按钮控件描述，并设置相关属性，代码如下：

```
1.   <?xml version="1.0" encoding="utf-8"?>
2.   <LinearLayout xmlns:android="http://schemas.
     android.com/apk/res/android"
3.       android:orientation="vertical"
4.       android:layout_width="fill_parent"
5.       android:layout_height="fill_parent"
6.       >
7.       <Button android:text="编辑按钮"
8.           android:id="@+id/btn_edit"
9.           android:layout_width="wrap_content"
10.          android:layout_height="wrap_content">
11.      </Button>
12.  </LinearLayout>
```

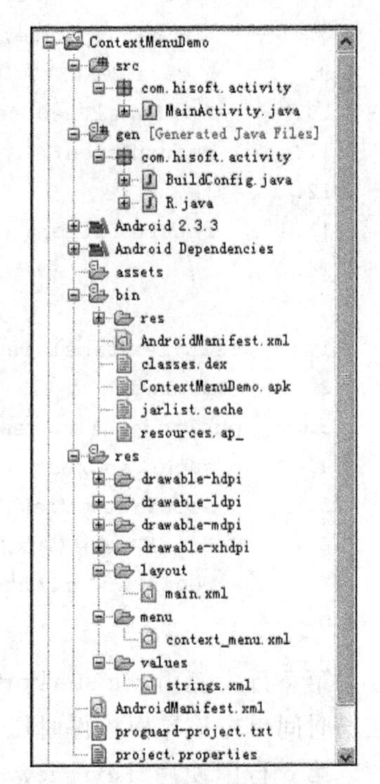

图 7-10 ContextMenuDemo 工程目录结构

(3) 在 res 目录下新建 menu 文件夹，然后创建 context_menu. xml 文件，添加 menu、group 和 item 描述，并设置相关属性，代码如下：

```
1.   <?xml version="1.0" encoding="utf-8"?>
2.   <menu xmlns:android="http://schemas.android.com/apk/res/android">
3.     <group android:checkableBehavior="single">
4.     <item android:icon="@drawable/ic_launcher" android:title="剪切" android:
       id="@+id/cut"/>
5.      <item android:icon="@drawable/ic_launcher" android:title="拷贝"
        android:id="@+id/copy"/>
6.       <item android:icon="@drawable/ic_launcher" android:title="粘贴"
         android:id="@+id/paste"/>
7.       </group>
8.   </menu>
```

(4) 修改 src 目录中 com. hisoft. activity 包下的 MainActivity. java 文件，代码如下：

```
1.   package com.hisoft.activity;
2.   import android.app.Activity;
3.   import android.os.Bundle;
4.   import android.view.ContextMenu;
5.   import android.view.ContextMenu.ContextMenuInfo;
6.   import android.view.MenuInflater;
7.   import android.view.MenuItem;
8.   import android.view.View;
```

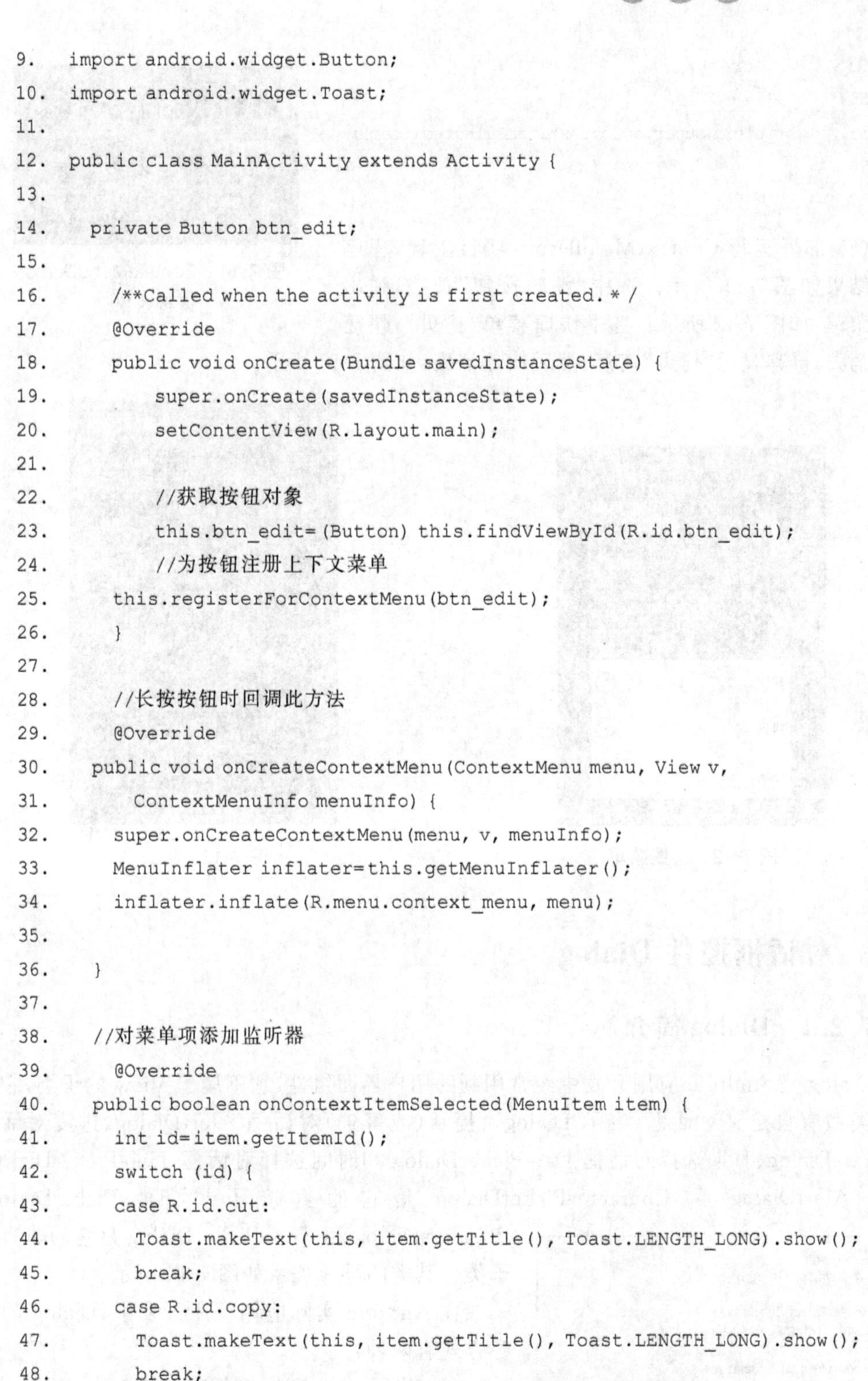

```
9.  import android.widget.Button;
10. import android.widget.Toast;
11.
12. public class MainActivity extends Activity {
13.
14.   private Button btn_edit;
15.
16.     /**Called when the activity is first created. * /
17.     @Override
18.     public void onCreate(Bundle savedInstanceState) {
19.         super.onCreate(savedInstanceState);
20.         setContentView(R.layout.main);
21.
22.         //获取按钮对象
23.         this.btn_edit=(Button) this.findViewById(R.id.btn_edit);
24.         //为按钮注册上下文菜单
25.     this.registerForContextMenu(btn_edit);
26.     }
27.
28.     //长按按钮时回调此方法
29.     @Override
30.   public void onCreateContextMenu(ContextMenu menu, View v,
31.       ContextMenuInfo menuInfo) {
32.     super.onCreateContextMenu(menu, v, menuInfo);
33.     MenuInflater inflater=this.getMenuInflater();
34.     inflater.inflate(R.menu.context_menu, menu);
35.
36.   }
37.
38.   //对菜单项添加监听器
39.     @Override
40.   public boolean onContextItemSelected(MenuItem item) {
41.     int id=item.getItemId();
42.     switch (id) {
43.     case R.id.cut:
44.       Toast.makeText(this, item.getTitle(), Toast.LENGTH_LONG).show();
45.       break;
46.     case R.id.copy:
47.       Toast.makeText(this, item.getTitle(), Toast.LENGTH_LONG).show();
48.       break;
49.     case R.id.paste:
50.       Toast.makeText(this, item.getTitle(), Toast.LENGTH_LONG).show();
51.       break;
52.     default:
```

```
53.            break;
54.        }
55.        return super.onContextItemSelected(item);
56.    }
57.  }
```

图7-11　ContextMenuDemo运行效果

(5) 部署运行 ContextMenuDemo 项目工程，程序运行结果如图 7-11 所示。选中"编辑按钮"长按，弹出快捷菜单，如图 7-12 所示。选中快捷菜单"拷贝"，快捷菜单消失，屏幕显示"拷贝"信息，然后快速消失，如图 7-13 所示。

图7-12　快捷菜单

图7-13　拷贝

7.2　对话框控件 Dialog

7.2.1　Dialog 简介

Dialog 是 Android 应用开发中经常用到的用户界面组件，它不属于 View 的子类，它包含的类型有自定义对话框（继承 Dialog）、提示（或警告）对话框 AlertDialog、进度对话框 ProgressDialog、日期选择对话框 DatePickerDialog 和时间选择对话框 TimePickerDialog。其中 AlertDialog 和 CharacterPickerDialog 是它的直接子类，DatePickerDialog、ProgressDialog 和 TimePickerDialog 是它的非直接子类。其类的继承关系如图 7-14 所示。

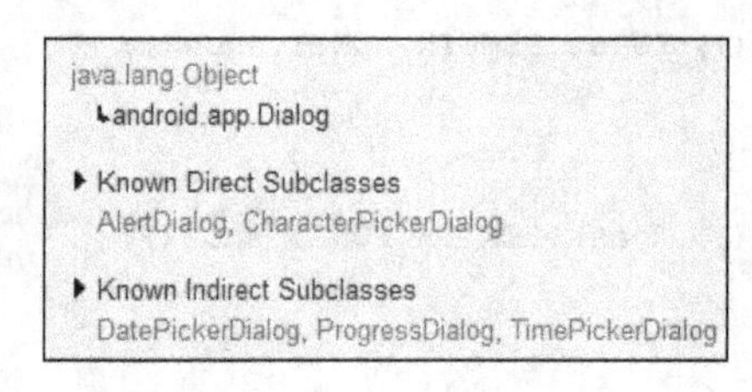

图7-14　Dialog类继承关系

在 Android 实际应用程序开发中，Dialog 的创建方式有两种：

(1) 使用 new 操作符创建一个新的 Dialog 对象，然后调用 Dialog 对象的 show 和 dismiss 方法来控制对话框的显示和隐藏。

(2) 在 Activity 的 onCreateDialog(int id)方法中创建 Dialog 对象并返回，然后调用

Activity 的 showDialog(int id)和 dismissDialog(int id)来显示和隐藏对话框，使用 getOwnerActivity()可以返回 Activity 并管理 Dialog。

上述两种方式的区别是：通过第二种方式创建的对话框会继承 Activity 的属性，比如获得 Activity 的 menu 事件等。下面分别进行介绍。

7.2.2 警告(提示)对话框 AlertDialog

AlertDialog 对话框是 Dialog 的子类，它有两个或者三个 Button 按钮，用 setMessage()方法可以在 AlertDialog 对话框上显示一个字符串。它继承自 Dialog，直接子类有 DatePickerDialog、ProgressDialog 和 TimePickerDialog，其类的继承结构图如图 7-15 所示。

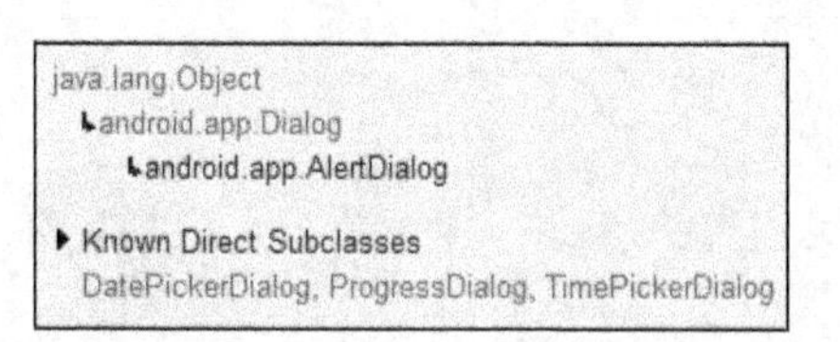

图 7-15 AlertDialog 类的继承关系

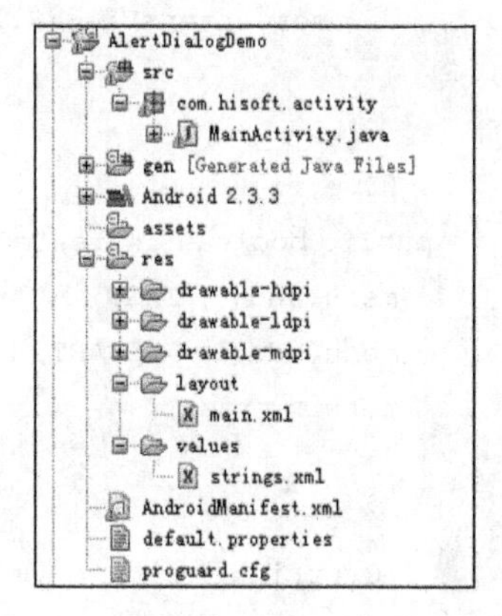

图 7-16 AlertDialogDemo 工程目录结构

7.2.3 AlertDialog 应用案例

7.2.2 节介绍了 Dialog 和 AlertDialog 的基础知识及类之间的关系，下面通过一个提示对话框的案例应用熟悉和掌握菜单 AlertDialog 的应用开发过程。

(1) 创建一个新的 Android 工程，工程名为 AlertDialogDemo，目标 API 选择 10(即 Android 2.3.3 版本)，应用程序名为 AlertDialogDemo，包名为 com.hisoft.activity，创建的 Activity 的名字为 MainActivity，最小 SDK 版本根据选择的目标 API 会自动添加为 10，创建项目工程如图 7-16 所示。

(2) 修改 src 目录中 com.hisoft.activity 包下的 MainActivity.java 文件，代码如下：

```
1.   package com.hisoft.activity;
2.
3.   import android.app.Activity;
4.   import android.app.AlertDialog;
5.   import android.app.ProgressDialog;
6.   import android.content.DialogInterface;
7.   import android.content.DialogInterface.OnClickListener;
8.   import android.os.Bundle;
9.   import android.view.Menu;
```

```
10.  import android.view.MenuItem;
11.
12.  public class MainActivity extends Activity {
13.
14.    private static final int EXIT=1;
15.    private static final int RESTART=2;
16.
17.      /**Called when the activity is first created. * /
18.      @Override
19.      public void onCreate(Bundle savedInstanceState) {
20.          super.onCreate(savedInstanceState);
21.          setContentView(R.layout.main);
22.      }
23.
24.      @Override
25.    public boolean onCreateOptionsMenu(Menu menu) {
26.      menu.add(1, EXIT, 1, "退出程序");
27.      menu.add(1, RESTART, 2, "重启应用");
28.      return true;
29.    }
30.
31.      @Override
32.    public boolean onOptionsItemSelected(MenuItem item) {
33.      if (item.getItemId()==EXIT) {
34.        showAlertDialog();
35.      }
36.
37.      return super.onOptionsItemSelected(item);
38.    }
39.
40.      public void showAlertDialog() {
41.      AlertDialog.Builder builder=new AlertDialog.Builder(this);
42.      builder.setTitle("退出");
43.      builder.setMessage("真的要退出程序吗?");
44.      builder.setPositiveButton("是", new OnClickListener() {
45.
46.        @Override
47.        public void onClick(DialogInterface dialog, int which) {
48.
49.          MainActivity.this.finish();
50.        }
51.      });
52.
53.      builder.setNegativeButton("否", new OnClickListener() {
```

```
54.
55.         @Override
56.         public void onClick(DialogInterface dialog, int which) {
57.
58.           dialog.cancel();
59.         }
60.     });
61.
62.     AlertDialog alert=builder.create();
63.     alert.show();
64.
65.   }
66. }
```

(3) 部署运行 AlertDialogDemo 项目工程，然后单击 menu 按钮，界面下方出现菜单选项，如图 7-17 所示。

单击"重启应用"按钮，程序重置到起始运行状态，菜单消失；单击"退出程序"按钮，弹出对话框，出现"是"或者"否"按钮，如图 7-18 所示。如单击"是"按钮，应用程序退出；单击"否"按钮，应用程序回到上一级状态。

图 7-17 菜单选项

图 7-18 对话框

7.2.4 日期选择对话框 DatePickerDialog

在 Android 应用中，日期控件有 DatePicker 和 DatePickerDialog，它们类的继承结构不同，所在的包也不一样。DatePicker 位于 android. widget 包下，继承自 android. widget. FrameLayout，类继承结构如图 7-19 所示；而 DatePickerDialog 位于 android. app 包下，继承自 android. app. AlertDialog，类继承结构如图 7-20 所示。在 DatePickerDialog 类中可以通过 getDatePicker()方法获取包含在这个对话框中的 DatePicker 对象。

java.lang.Object
 ↳android.app.Dialog
  ↳android.app.AlertDialog
   ↳android.app.DatePickerDialog

图 7-19　DatePickerDialog 类继承关系

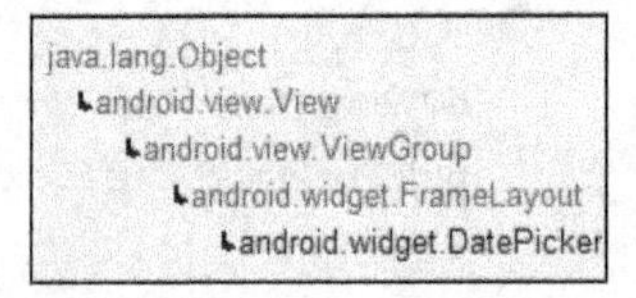

java.lang.Object
↳android.view.View
↳android.view.ViewGroup
↳android.widget.FrameLayout
↳android.widget.DatePicker

图 7-20　DatePicker 类继承关系

DatePickerDialog 的使用要复杂一些，它是以弹出式对话框形式出现的，并需要实现 DialogInterface. OnClickListener 和 DatePicker. OnDateChangedListener 接口。其主要是通过 DatePickerDialog 的 OnDateSetListener 方法实现。DatePicker 类主要是通过 OnDateChangedListener 方法实现用户选择的日期。

7.2.5　DatePickerDialog 应用案例

7.2.4 节介绍了 DatePicker 和 DatePickerDialog 的区别及联系，下面通过一个设置日期的案例应用讲解 DatePickerDialog 的应用，具体步骤如下：

(1) 创建一个新的 Android 工程，工程名为 DatePickerDialogDemo，目标 API 选择 10(即 Android 2.3.3 版本)，应用程序名为 DatePickerDialogDemo，包名为 com.hisoft.activity，创建的 Activity 的名字为 MainActivity，最小 SDK 版本根据选择的目标 API 会自动添加为 10，创建项目工程如图 7-21 所示。

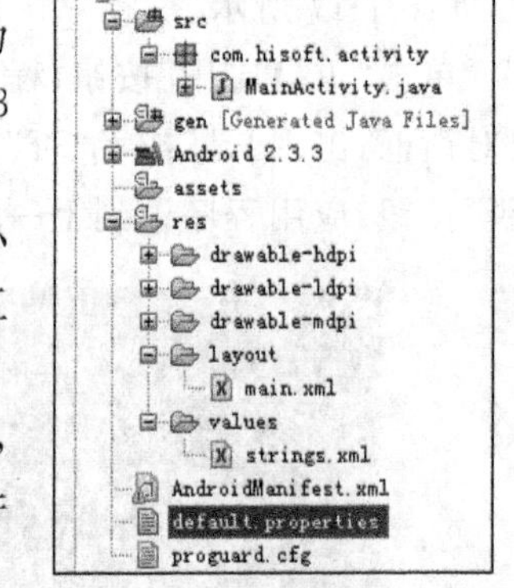

图 7-21　DatePickerDialogDemo 工程目录结构

(2) 修改 res 目录下 layout 文件夹中的 main.xml 文件，设置线性布局，添加一个 TextView 控件和 Button 按钮控件描述，并设置相关属性，代码如下：

```
1.  <?xml version="1.0" encoding="utf-8"?>
2.  <LinearLayout xmlns:android="http://schemas.
    android.com/apk/res/android"
3.      android:layout_width="wrap_content"
4.      android:layout_height="wrap_content"
5.      android:orientation="vertical">
6.      <TextView android:id="@+id/dateDisplay"
7.          android:layout_width="wrap_content"
8.          android:layout_height="wrap_content"
9.          android:text=""/>
10.         <Button android:id="@+id/pickDate"
11.           android:layout_width="wrap_content"
12.           android:layout_height="wrap_content"
13.           android:text="Change the date"/>
14.     </LinearLayout>
```

(3) 修改 src 目录中 com.hisoft.activity 包下的 MainActivity.java 文件，代码如下：

```
1.  package com.hisoft.activity;
```

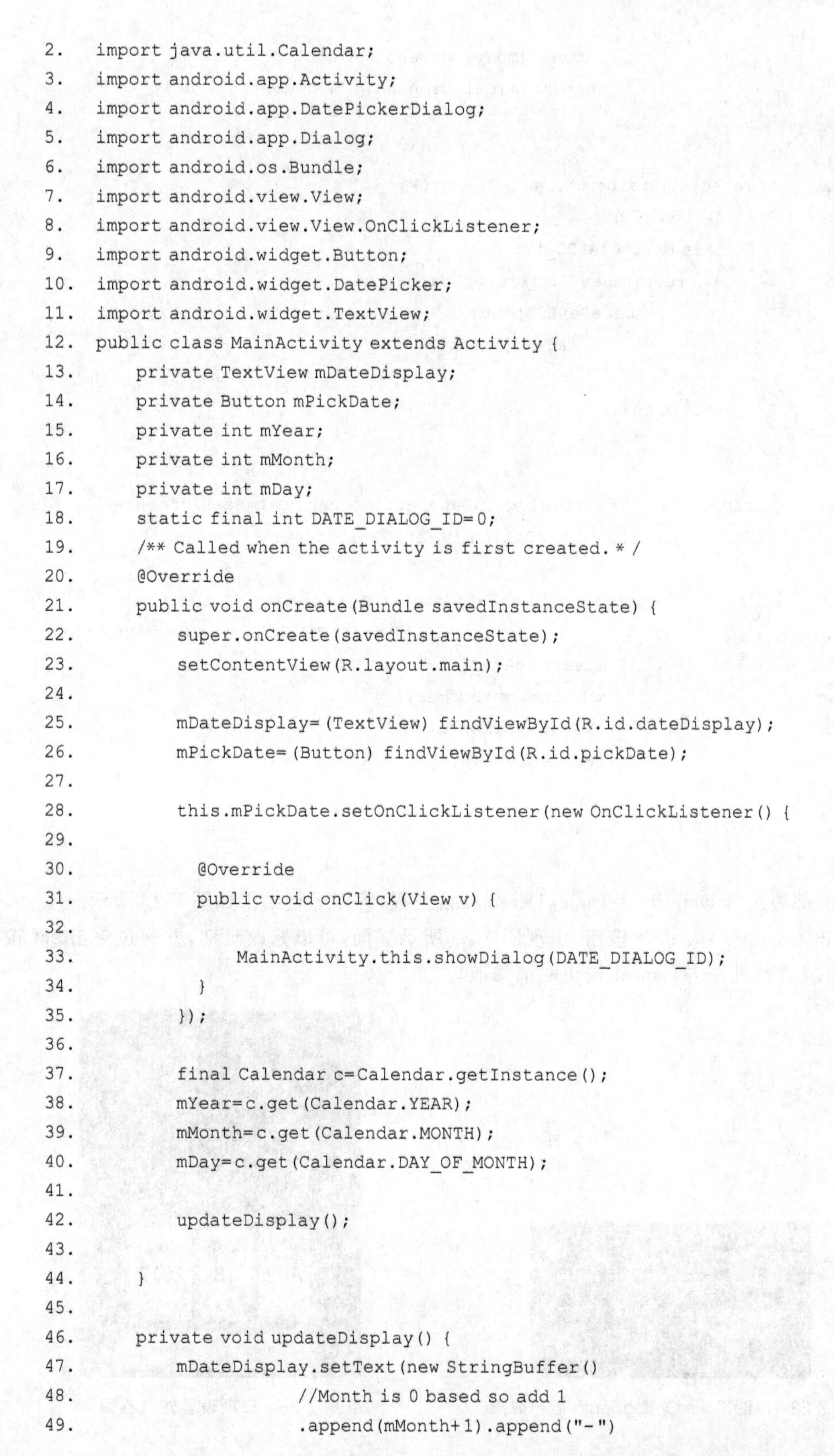

```
2.  import java.util.Calendar;
3.  import android.app.Activity;
4.  import android.app.DatePickerDialog;
5.  import android.app.Dialog;
6.  import android.os.Bundle;
7.  import android.view.View;
8.  import android.view.View.OnClickListener;
9.  import android.widget.Button;
10. import android.widget.DatePicker;
11. import android.widget.TextView;
12. public class MainActivity extends Activity {
13.     private TextView mDateDisplay;
14.     private Button mPickDate;
15.     private int mYear;
16.     private int mMonth;
17.     private int mDay;
18.     static final int DATE_DIALOG_ID=0;
19.     /** Called when the activity is first created. * /
20.     @Override
21.     public void onCreate(Bundle savedInstanceState) {
22.         super.onCreate(savedInstanceState);
23.         setContentView(R.layout.main);
24.
25.         mDateDisplay=(TextView) findViewById(R.id.dateDisplay);
26.         mPickDate=(Button) findViewById(R.id.pickDate);
27.
28.         this.mPickDate.setOnClickListener(new OnClickListener() {
29.
30.           @Override
31.           public void onClick(View v) {
32.
33.               MainActivity.this.showDialog(DATE_DIALOG_ID);
34.           }
35.         });
36.
37.         final Calendar c=Calendar.getInstance();
38.         mYear=c.get(Calendar.YEAR);
39.         mMonth=c.get(Calendar.MONTH);
40.         mDay=c.get(Calendar.DAY_OF_MONTH);
41.
42.         updateDisplay();
43.
44.     }
45.
46.     private void updateDisplay() {
47.         mDateDisplay.setText(new StringBuffer()
48.                 //Month is 0 based so add 1
49.                 .append(mMonth+1).append("-")
```

```
50.                     .append(mDay).append("-")
51.                     .append(mYear).append(" "));
52.     }
53.
54.     protected Dialog onCreateDialog(int id) {
55.         switch (id) {
56.         case DATE_DIALOG_ID:
57.             return new DatePickerDialog(this,
58.                 mDateSetListener,
59.                         mYear, mMonth, mDay);
60.         }
61.         return null;
62.     }
63.
64.     private DatePickerDialog.OnDateSetListener mDateSetListener=
65.             new DatePickerDialog.OnDateSetListener() {
66.
67.                 public void onDateSet(DatePicker view, int year,
68.                                         int monthOfYear, int dayOfMonth) {
69.                     mYear=year;
70.                     mMonth=monthOfYear;
71.                     mDay=dayOfMonth;
72.                     updateDisplay();
73.                 }
74.             };
75. }
```

(4) 部署运行 DatePickerDialogDemo 项目工程，程序运行结果如图 7-22 所示。

单击 Changer the date 按钮，出现图 7-23 所示界面，可以修改日期，并通过单击 set 按钮设定，或者通过单击 Cancel 按钮取消修改。

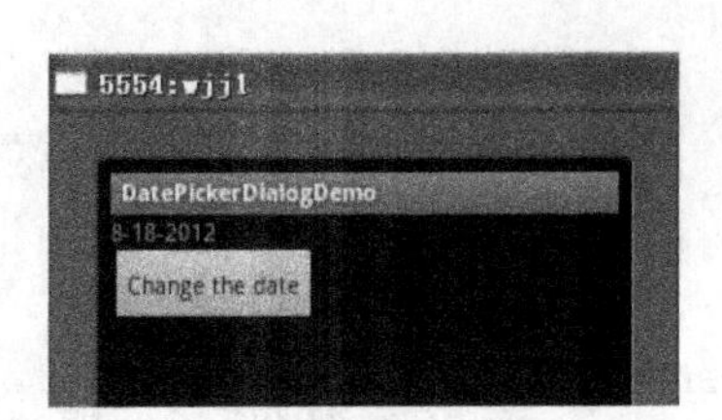

图 7-22 DatePickerDialogDemo 运行效果

图 7-23 日期设置修改界面

7.2.6 时间选择对话框 TimePickerDialog

同上述的日期选择对话框一样，时间选择对话框也有 TimePickerDialog 和 TimePicker 两个实现，TimePickerDialog 的类继承结构如图 7-24 所示，TimePicker 的类继承结构如图 7-25 所示。它们的类继承结构不同，所在的包也不同，但作用基本一样，用户在对话框中设置时间也是使用的 TimePicker 对象，主要区别是 TimePickerDialog 通过弹出对话框的方式调用 OnTimeSetListener 方法，设置选择时间(单击 Set 按钮)。一般用户选择对话框视图时采用它。

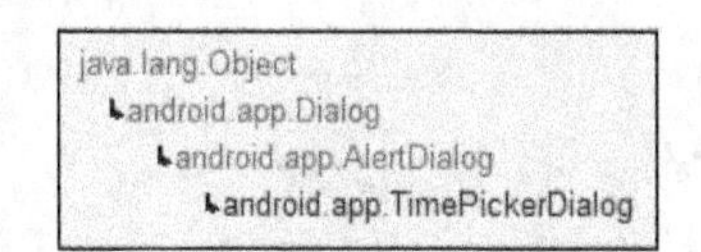

图 7-24 TimePickerDialog 类继承关系

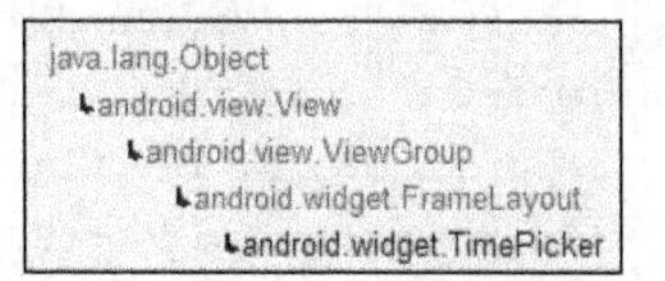

图 7-25 TimePicker 类继承关系

7.2.7 TimePickerDialog 应用案例

7.2.6 节简要介绍了 TimePickerDialog 和 TimePicker 两个类的不同与相同之处，下面通过一个 TimePickerDialog 应用案例介绍其应用。具体步骤如下：

(1) 创建一个新的 Android 工程，工程名为 TimePickerDialogDemo，目标 API 选择 10(即 Android 2.3.3 版本)，应用程序名为 TimePickerDialogDemo，包名为 com. hisoft. activity，创建的 Activity 的名字为 MainActivity，最小 SDK 版本根据选择的目标 API 会自动添加为 10，创建项目工程如图 7-26 所示。

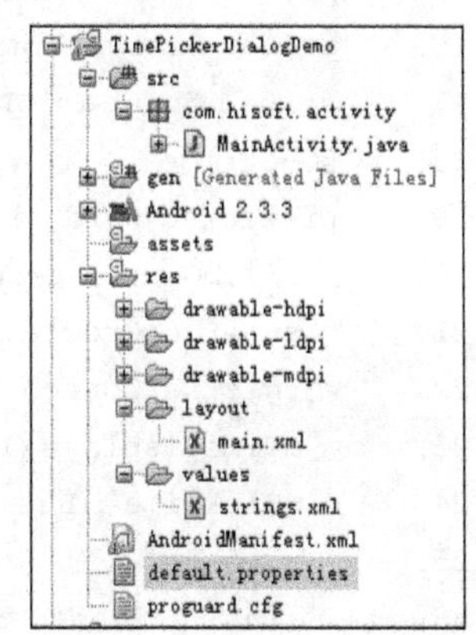

图 7-26 TimePickerDialogDemo 工程目录结构

(2) 修改 res 目录下 layout 文件夹中的 main. xml 文件，设置线性布局，添加一个 TextView 控件和 Button 按钮控件描述，并设置相关属性，代码如下：

```
1.  <?xml version="1.0" encoding="utf-8"?>
2.   <LinearLayout xmlns:android="http://
    schemas.android.com/apk/res/android"
3.    android:layout_width="wrap_content"
4.    android:layout_height="wrap_content"
5.    android:orientation="vertical">
6.    <TextView android:id="@+id/timeDisplay"
7.        android:layout_width="wrap_content"
8.        android:layout_height="wrap_content"
9.        android:text=""/>
10.   <Button android:id="@+id/pickTime"
11.       android:layout_width="wrap_content"
12.       android:layout_height="wrap_content"
```

```
13.          android:text="Change the time"/>
14.  </LinearLayout>
```

(3) 修改 src 目录中 com. hisoft. activity 包下的 MainActivity. java 文件，代码如下：

```
1.   package com.hisoft.activity;
2.   import java.util.Calendar;
3.   import android.app.Activity;
4.   import android.app.Dialog;
5.   import android.app.TimePickerDialog;
6.   import android.os.Bundle;
7.   import android.view.View;
8.   import android.view.View.OnClickListener;
9.   import android.widget.Button;
10.  import android.widget.TextView;
11.  import android.widget.TimePicker;
12.  public class MainActivity extends Activity {
13.    private TextView mTimeDisplay;
14.    private Button mPickTime;
15.    private int mHour;
16.    private int mMinute;
17.    static final int TIME_DIALOG_ID=0;
18.    @Override
19.    protected void onCreate(Bundle savedInstanceState) {
20.      //TODO Auto-generated method stub
21.      super.onCreate(savedInstanceState);
22.      this.setContentView(R.layout.main);
23.      mTimeDisplay=(TextView) findViewById(R.id.timeDisplay);
24.      mPickTime=(Button) findViewById(R.id.pickTime);
25.
26.      final Calendar c=Calendar.getInstance();
27.      mHour=c.get(Calendar.HOUR_OF_DAY);
28.      mMinute=c.get(Calendar.MINUTE);
29.
30.      updateDisplay();
31.
32.      this.mPickTime.setOnClickListener(new OnClickListener() {
33.
34.        @Override
35.        public void onClick(View v) {
36.
37.          MainActivity.this.showDialog(TIME_DIALOG_ID);
38.        }
39.      });
40.
41.    }
```

```
42.    private void updateDisplay() {
43.      mTimeDisplay.setText(new StringBuilder().append(pad(mHour)).append(":")
44.          .append(pad(mMinute)));
45.    }
46.
47.    private static String pad(int c) {
48.      if (c>=10)
49.        return String.valueOf(c);
50.      else
51.        return "0"+String.valueOf(c);
52.    }
53.
54.    protected Dialog onCreateDialog(int id) {
55.      switch (id) {
56.      case TIME_DIALOG_ID:
57.        return new TimePickerDialog(this, mTimeSetListener, mHour, mMinute,false);
58.      }
59.      return null;
60.    }
61.    private TimePickerDialog.OnTimeSetListener mTimeSetListener=
       new TimePickerDialog.OnTimeSetListener() {
62.      public void onTimeSet(TimePicker view, int hourOfDay, int minute) {
63.        mHour=hourOfDay;
64.        mMinute=minute;
65.        updateDisplay();
66.      }
67.    };
68.  }
```

(4) 部署运行 TimePickerDialogDemo 项目工程，程序运行结果如图 7-27 所示。

单击 Change the time 按钮，出现图 7-28 所示界面，可以修改时间，然后通过单击 set 按钮设定，或者通过单击 Cancel 按钮取消修改。

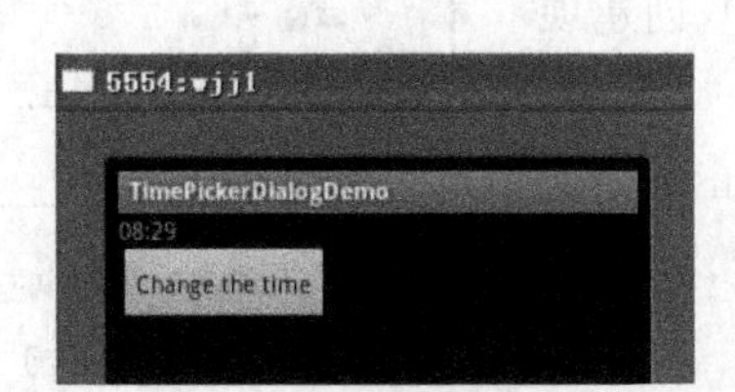

图 7-27　TimePickerDialogDemo 运行效果

图 7-28　时间设置修改界面

7.2.8　进度对话框 ProgressDialog

ProgressDialog 控件在 android. app 包中，继承自 android. app. AlertDialog，类继承结构如图 7-29 所示。它常用于显示载入进度、下载进度等。合理使用 ProgressDialog 能增加用户体验，让用户知道现在程序所处的状态。

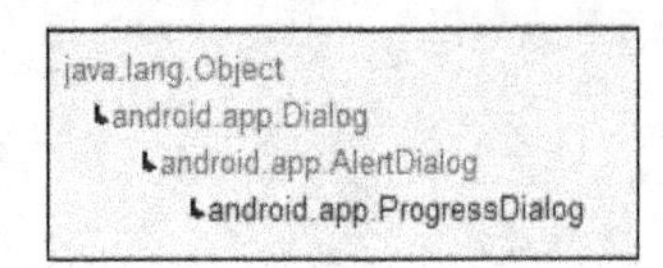

图 7-29　ProgressDialog 类继承关系

使用代码 ProgressDialog. show (ProgressDialogActivity. this,"请稍等","数据正在加载中...",true)；可以创建并显示一个进度对话框。调用 setProgressStyle()方法可以设置进度对话框风格。

ProgressDialog 的风格设置有两种，其对应的方法如下：

```
ProgressDialog.STYLE_SPINNER          旋体进度条风格（为默认风格）
ProgressDialog.STYLE_HORIZONTAL       横向进度条风格
```

ProgressDialog 控件的应用步骤主要有：

(1) 在布局上面加一个 Button，加入 OnClickListener。

(2) 把 ProgressDialog 声明成全局的，并在 Button 的 OnClickListener 中创建，然后使用的方法是 show(Context context, CharSequence title, CharSequence message, boolean indeterminate)方法，显示 ProgressDialog。第一个参数为当前运行 Activity 的 Context，第二个参数是标题，第三个参数是内容，最后一个参数可选。

(3) 在 Button 的 OnClickListener 中创建一个线程，让线程 run 的时候休眠 5s，然后使用 dismiss()方法，关闭刚才打开的 ProgressDialog 对话框。

7.2.9　ProgressDialog 应用案例

本节先对 ProgressDialog 的应用进行讲解，在应用中经常可以看见程序加载中的对话框，一般在程序运行前加载数据可以使用 ProgressDialog 对话框，当后台程序运行完毕，需要用 dismiss()方法来关闭获取焦点，以免不能关闭 ProgressDialog 对话框。

(1) 创建一个新的 Android 工程，工程名为 ProgressDialogDemo，目标 API 选择 10(即 Android 2. 3. 3 版本)，应用程序名为 ProgressDialogDemo，包名为 com. hisoft. activity，创建的 Activity 的名字为 MainActivity，最小 SDK 版本根据选择的目标 API 会自动添加为 10，创建项目工程如图 7-30 所示。

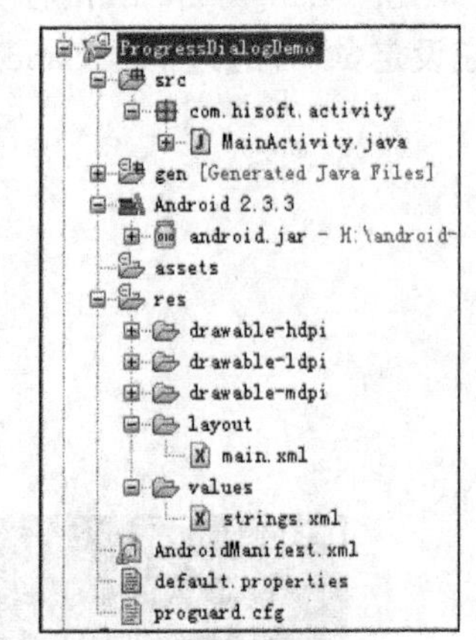

图 7-30　ProgressDialogDemo 工程目录结构

(2) 修改 res 目录下 layout 文件夹中的 main. xml 文件，设置线性布局，添加一个 Button 按钮控件描述，并设置相关属性，代码如下：

```
1.    <?xml version="1.0" encoding="utf-8"?>
```

```
2.  <LinearLayout xmlns:android="http://schemas.android.com/apk/res/android"
3.      android:orientation="vertical"
4.      android:layout_width="fill_parent"
5.      android:layout_height="fill_parent"
6.      >
7.      <Button android:text="启动程序"
8.          android:id="@+id/button1"
9.          android:layout_width="wrap_content"
10.         android:layout_height="wrap_content">
11.     </Button>
12. </LinearLayout>
```

(3) 修改 src 目录中 com. hisoft. activity 包下的 MainActivity. java 文件,代码如下:

```
1.  package com.hisoft.activity;
2.  import android.app.Activity;
3.  import android.app.ProgressDialog;
4.  import android.os.Bundle;
5.  import android.view.View;
6.  import android.view.View.OnClickListener;
7.  import android.widget.Button;
8.  public class MainActivity extends Activity {
9.    private Button b1;
10.   private ProgressDialog dialog;
11.   private int currentValue=0;
12.
13.     /**Called when the activity is first created. */
14.     @Override
15.     public void onCreate(Bundle savedInstanceState) {
16.         super.onCreate(savedInstanceState);
17.         setContentView(R.layout.main);
18.
19.         this.b1=(Button) this.findViewById(R.id.button1);
20.         this.b1.setOnClickListener(new OnClickListener() {
21.
22.           @Override
23.           public void onClick(View v) {
24.
25.           dialog=ProgressDialog.show(MainActivity.this, "",
26.                     "Loading. Please wait...", true);
27.           dialog.setMax(100);
28.           dialog.show();
29.           new Thread(new ProgressDialogThread()).start();
30.
```

```
31.
32.           }
33.       });
34.       }
35.
36.       class ProgressDialogThread implements Runnable{
37.
38.       @Override
39.       public void run() {
40.         while(currentValue<=100){
41.           try {
42.             Thread.sleep(100);
43.           } catch (InterruptedException e) {
44.             //TODO Auto-generated catch block
45.             e.printStackTrace();
46.           }
47.           currentValue+=2;
48.         }
49.
50.         dialog.dismiss();
51.       }
52.
53.     }
54. }
```

(4) 部署运行 ProgressDialogDemo 项目工程，程序运行结果如图 7-31 所示。单击“启动程序”按钮，程序运行后出现图 7-32 所示界面。

图 7-31　ProgressDialogDemo 运行效果

图 7-32　ProgressDialog 运行效果

7.3 信息提示控件

7.3.1 Toast 控件简介

Toast 控件位于 android. widget 包中，其类的继承结构图如图 7-33 所示。Toast 是一种提供给用户快速、简短信息的视图。借助 Toast 类可以创建和显示该信息。

java.lang.Object
 └android.widget.Toast

图 7-33 Toast 类继承结构

Toast 视图以浮于应用程序之上的视图形式呈现给用户。因为它并不获得焦点，即使用户正在输入什么也不会受到影响。它的目标是尽可能以不显眼的方式使用户看到提供的信息。音量控制和信息设置、保存成功这两个例子就是使用 Toast。

使用该类最简单的方法就是调用它的一个静态方法 makeText 方法，让它来构造需要的一切并返回一个新的 Toast 对象。即生成一个从资源中取得的包含文本视图的标准 Toast 对象。

7.3.2 Toast 应用案例

上面简述了 Toast 的功能和创建方法，下面通过一个 Toast 应用案例，简要介绍 Toast 显示快速、简短消息，以及显示图片的方法，具体步骤如下：

(1) 创建一个新的 Android 工程，工程名为 ToastDemo，目标 API 选择 10(即 Android 2.3.3 版本)，应用程序名为 ToastDemo，包名为 com. hisoft. activity，创建的 Activity 的名字为 MainActivity，最小 SDK 版本根据选择的目标 API 会自动添加为 10，创建项目工程如图 7-34 所示。

(2) 修改 res 目录下 layout 文件夹中的 main. xml 文件，设置线性布局，添加一个 Button 按钮控件描述，并设置相关属性，代码如下：

```
1.   <?xml version="1.0" encoding="utf-8"?>
2.   <LinearLayout xmlns:android="http://schemas.android.com/apk/res/android"
3.       android:orientation="vertical"
4.       android:layout_width="fill_parent"
5.       android:layout_height="fill_parent"
6.       >
7.       <Button android:text="弹出 Toast"
8.           android:id="@+id/button1"
9.           android:layout_width="wrap_content"
10.          android:layout_height="wrap_content">
11.      </Button>
12.  </LinearLayout>
```

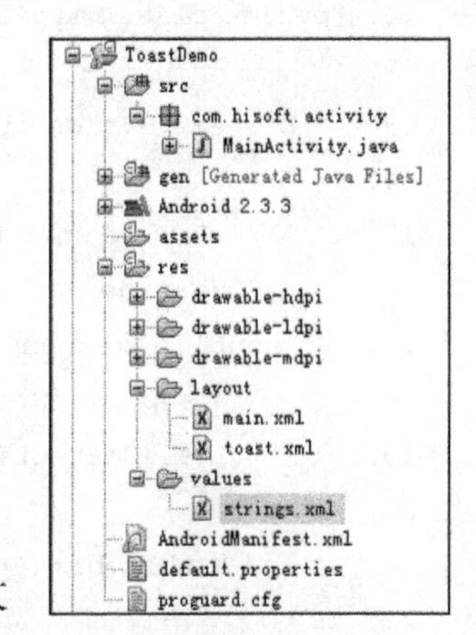

图 7-34 ToastDemo 工程目录结构

(3) 在 res 目录下的 layout 文件夹中新建 toast. xml 文件，设置线性布局，添加一个 ImageView 控件和 TextView 控件描述，并设置相关属性，代码如下：

```
1.  <?xml version="1.0" encoding="utf-8"?>
2.  <LinearLayout
3.    xmlns:android="http://schemas.android.com/apk/res/android"
4.    android:layout_width="match_parent"
5.    android:layout_height="match_parent" android:orientation="horizontal">
6.      <ImageView android:src="@drawable/icon"
7.            android:layout_height="wrap_content"
8.            android:layout_width="wrap_content"
9.            android:id="@+id/imageView1">
10.     </ImageView>
11.     <TextView android:id="@+id/textView1"
12.           android:text="@string/info"
13.           android:layout_height="wrap_content"
14.           android:layout_width="wrap_content"
15.     android:textAppearance="?android:attr/textAppearanceMedium">
16.     </TextView>
17. </LinearLayout>
```

(4) 修改 src 目录中 com.hisoft.activity 包下的 MainActivity.java 文件,代码如下:

```
1.  package com.hisoft.activity;
2.
3.  import android.app.Activity;
4.  import android.os.Bundle;
5.  import android.view.LayoutInflater;
6.  import android.view.View;
7.  import android.view.View.OnClickListener;
8.  import android.widget.Button;
9.  import android.widget.Toast;
10.
11. public class MainActivity extends Activity {
12.
13.   private Button b1;
14.
15.     /**Called when the activity is first created. */
16.     @Override
17.     public void onCreate(Bundle savedInstanceState) {
18.        super.onCreate(savedInstanceState);
19.        setContentView(R.layout.main);
20.
21.        this.b1=(Button) this.findViewById(R.id.button1);
22.
23.        this.b1.setOnClickListener(new OnClickListener() {
24.
```

```
25.         @Override
26.         public void onClick(View v) {
27.
28.           Toast t=Toast.makeText(MainActivity.this, "", Toast.LENGTH_SHORT);
29.           LayoutInflater inflater=LayoutInflater.from(MainActivity.this);
30.           View view=inflater.inflate(R.layout.toast, null);
31.           t.setView(view);
32.           t.show();
33.
34.         }
35.       });
36.
37.     }
38. }
```

(5) 修改 res 目录下 values 文件夹中的 strings.xml 文件，代码如下：

```
1.  <?xml version="1.0" encoding="utf-8"?>
2.  <resources>
3.      <string name="hello">Hello World, MainActivity!</string>
4.      <string name="app_name">ToastDemo</string>
5.      <string name="info">这是一条提示信息</string>
6.  </resources>
```

(6) 部署运行 ToastDemo 项目工程，程序运行结果如图 7-35 所示。

单击“弹出 Toast”按钮，出现 Toast 提示信息，如图 7-36 所示。

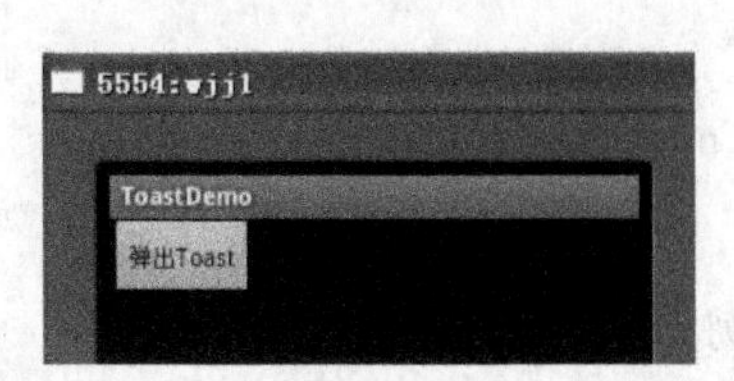

图 7-35　ToastDemo 运行效果

图 7-36　Toast 提示信息

7.3.3 Notification 控件简介

Notification 控件位于 android. app 包下，类的继承结构如图 7-37 所示。Notification 也是 Android 系统中给用户消息提示的方式，位于屏幕最顶部的状态栏，通知的同时可以播放声音，以及振动提示用户，单击通知还可以返回指定的 Activity。通常它用来在状态栏中显示电池电量、信息强度等信息。用鼠标按住状态栏，然后往下拖或者拉，可以打开状态栏并查看系统的提示信息。

```
java.lang.Object
  android.app.Notification
```

图 7-37 Notification 类继承关系

通知的设置等操作相对比较简单，就是新建一个 Notification 对象，然后设置好通知的各项参数，使用系统后台运行的 NotificationManager 服务将通知发出来。可以通过 Notification. Builder 快速地创建构造 Notification 对象。

Notification 的通常用法及步骤如下：

(1) 得到 NotificationManager。

```
String ns=Context.NOTIFICATION_SERVICE;
NotificationManager mNotificationManager= (NotificationManager) getSystemService(ns);
```

(2) 创建一个新的 Notification 对象。

```
Notification notification=new Notification();
notification.icon=R.drawable.notification_icon; //或者复杂一些的方式创建 Notification
int icon=R.drawable.notification_icon;          //通知图标
CharSequence tickerText="Hello";                //状态栏(Status Bar)显示的通知文本提示
long when=System.currentTimeMillis();           //通知产生的时间,会在通知信息里显示
Notification notification=new Notification(icon, tickerText, when);
```

(3) 填充 Notification 的各个属性。

```
Context context=getApplicationContext();
CharSequence contentTitle="My notification";
CharSequence contentText="Hello World!";
Intent notificationIntent=new Intent(this, MyClass.class);
PendingIntent contentIntent=PendingIntent.getActivity(this, 0, notificationIntent, 0);
notification.setLatestEventInfo(context, contentTitle, contentText, contentIntent);
```

(4) 发送通知。

```
private static final int ID_NOTIFICATION=1;
mNotificationManager.notify(ID_NOTIFICATION, notification);
```

Notification 的手机提示方式有 4 种，具体如下：

① 在状态栏(Status Bar)显示的通知文本提示，例如：

```
notification.tickerText="hello";
```

② 发出提示音，例如：

```
notification.defaults|=Notification.DEFAULT_SOUND;
```

```
notification.sound=Uri.parse("file:///sdcard/notification/ringer.mp3");
notification.sound=Uri.withAppendedPath(Audio.Media.INTERNAL_CONTENT_URI, "6");
```

③ 手机振动，例如：

```
notification.defaults|=Notification.DEFAULT_VIBRATE;
long[] vibrate={0,100,200,300};
notification.vibrate=vibrate;
```

④ LED 灯闪烁，例如：

```
notification.defaults|=Notification.DEFAULT_LIGHTS;
notification.ledARGB=0xff00ff00;
notification.ledOnMS=300;
notification.ledOffMS=1000;
notification.flags|=Notification.FLAG_SHOW_LIGHTS;
```

Notification 如果需要更新一个通知，只需要在设置好 notification 之后再调用 setLatestEventInfo，然后重新发送一次通知即可。

7.3.4 Notification 应用案例

7.3.3 节讲述了 Notification 的通常用法及提示方式等基础知识，下面通过 Notification 案例讲解如何向状态栏添加信息以及图片。

(1) 创建一个新的 Android 工程，工程名为 NotificationDemo，目标 API 选择 10(即 Android 2.3.3 版本)，应用程序名为 NotificationDemo，包名为 com. hisoft. activity，创建的 Activity 的名字为 MainActivity，最小 SDK 版本根据选择的目标 API 会自动添加为 10，创建项目工程如图 7-38 所示。

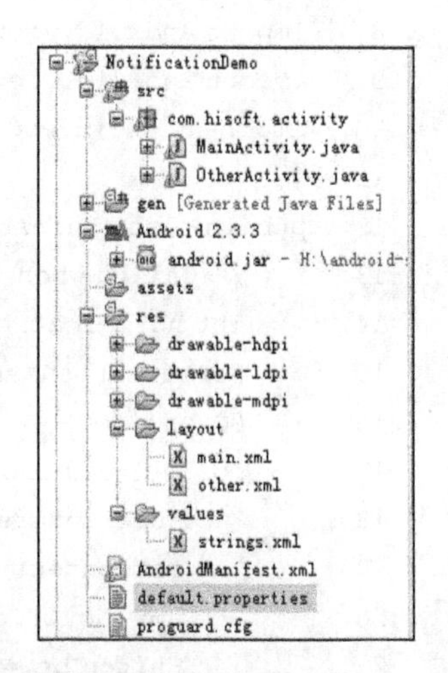

图 7-38 NotificationDemo 工程目录结构

(2) 修改 res 目录下 layout 文件夹中的 main. xml 文件，设置线性布局，添加两个 Button 按钮控件描述，并设置相关属性，代码如下：

```
1.  <?xml version="1.0" encoding="utf-8"?>
2.  <LinearLayout xmlns:android="http://schemas.android.com/apk/res/android"
3.      android:orientation="vertical"
4.      android:layout_width="fill_parent"
5.      android:layout_height="fill_parent"
6.      >
7.  <Button android:id="@+id/bt1"
8.      android:layout_height="wrap_content"
9.      android:layout_width="fill_parent"
10.     android:text="测试 Notification"
11. />
12. <Button android:id="@+id/bt2"
```

```
13.     android:layout_height="wrap_content"
14.     android:layout_width="fill_parent"
15.     android:text="清除 Notification"
16. />
17. </LinearLayout>
```

(3) 修改 src 目录中 com. hisoft. activity 包下的 MainActivity. java 文件，代码如下：

```
1.  package com.hisoft.activity;
2.  import android.app.Activity;
3.  import android.app.Notification;
4.  import android.app.NotificationManager;
5.  import android.app.PendingIntent;
6.  import android.content.Intent;
7.  import android.os.Bundle;
8.  import android.view.View;
9.  import android.view.View.OnClickListener;
10. import android.widget.Button;
11.
12. public class MainActivity extends Activity {
13.     /**Called when the activity is first created. * /
14.     int notification_id=19172439;
15.     NotificationManager nm;
16.     @Override
17.     public void onCreate(Bundle savedInstanceState) {
18.         super.onCreate(savedInstanceState);
19.         setContentView(R.layout.main);
20.         nm= (NotificationManager)getSystemService(NOTIFICATION_SERVICE);
21.         Button bt1= (Button)findViewById(R.id.bt1);
22.         bt1.setOnClickListener(bt1lis);
23.         Button bt2= (Button)findViewById(R.id.bt2);
24.         bt2.setOnClickListener(bt2lis);
25.
26.     }
27.     OnClickListener bt1lis=new OnClickListener(){
28.
29.     @Override
30.     public void onClick(View v) {
31.       //TODO Auto-generated method stub
32.       showNotification(R.drawable.ic_launcher_home,"测试信息","短信","北京
          政法与海辉测试内容");
33.     }
34.
35.     };
36.     OnClickListener bt2lis=new OnClickListener(){
37.
```

```
38.     @Override
39.     public void onClick(View v) {
40.       //TODO Auto-generated method stub
41.       //showNotification(R.drawable.home,"测试信息","短信","北京政法测试内容");
42.       nm.cancel(notification_id);
43.     }
44.
45. };
46.     public void showNotification(int icon,String tickertext,String title,
        String content){
47.       //设置一个唯一的 ID,随便设置
48.
49.       //Notification 管理器
50.        Notification notification=new Notification(icon,tickertext,System.
           currentTimeMillis());
51.       //后面的参数分别是显示在顶部通知栏的小图标,小图标旁的文字(短暂显示,自动消
          //失)是系统当前时间
52.       notification.defaults=Notification.DEFAULT_ALL;
53.       //设置通知是否同时播放声音或振动,声音为 Notification.DEFAULT_SOUND
54.       //振动为 Notification.DEFAULT_VIBRATE;
55.       //Light 为 Notification.DEFAULT_LIGHTS
56.       //全部为 Notification.DEFAULT_ALL
57.       //如果是振动或者全部,必须在 AndroidManifest.xml 加入振动权限
58.       PendingIntent pt=PendingIntent.getActivity(this, 0, new Intent(this,
          MainActivity.class), 0);
59.       //单击通知后的动作,这里是转回 MainActicity
60.       notification.setLatestEventInfo(this,title,content,pt);
61. //如果需要更新一个通知,在设置好 notification 之后再调用 setLatestEventInfo,然
    //后重新发送一次通知即可
62.       nm.notify(notification_id, notification);
63.
64.     }
65. }
```

(4) 在 AndroidMainfest.xml 文件中添加振动器的权限,代码如下:

```
1.  <uses-permission android:name="android.permission.VIBRATE"/>
```

(5) 部署运行 NotificationDemo 项目工程,程序运行结果如图 7-39 所示。

单击"测试 Notification"按钮,状态栏显示如图 7-40 所示。

然后用鼠标按住状态栏往下拖动,显示信息标题及内容,如图 7-41 所示。

如果需要更新一个通知,单击"测试 Notification"按钮,然后即可重新发送信息。

拓展提示:实际开发中,Android UI 菜单、对话框经常需要和基本控件结合应用。此外,菜单、对话框的美观程度对整个系统的推广具有非常重要的影响,值得注意实现效果。

图 7-39　NotificationDemo 运行效果

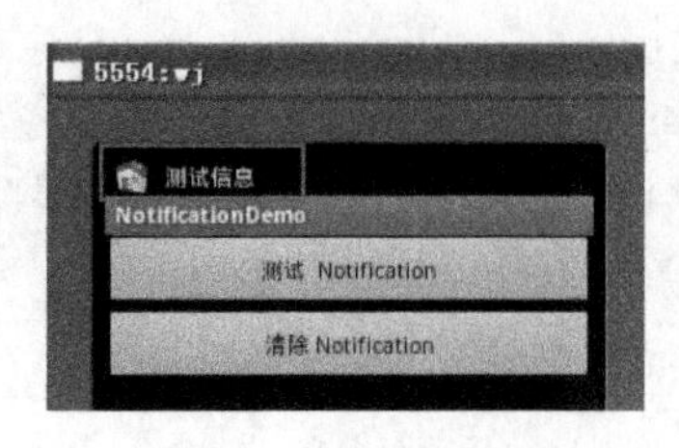

图 7-40　Notification 状态栏显示

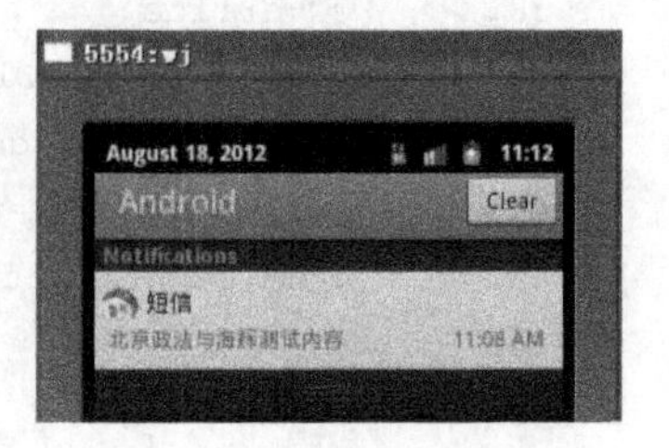

图 7-41　信息标题及内容

7.4　项目案例

学习目标：学习 Android UI 菜单、对话框、信息提示控件分类、方法、属性的设置等应用。

案例描述：使用 Android UI 菜单、对话框、信息提示控件、RelativeLayout 相对布局、TextView 控件、ListView 控件，并设置相对父控件的位置、控件之间相对位置的属性，实现 Ascent 医药菜单选择界面。

案例要点：Menu、onOptionsItemSelected 方法、Toast 控件。

案例实施：

(1) 创建工程 Project_Chapter_7，选择 Android 2.3.3 作为目标平台，如图 7-42 所示。

```
Project_Chapter_7
  src
    com.hisoft.project
      ProductList.java
  gen [Generated Java Files]
  Android 2.3.3
  Android Dependencies
  assets
  bin
  res
    drawable-hdpi
    drawable-ldpi
    drawable-mdpi
    drawable-xhdpi
    layout
      productlist.xml
    values
  AndroidManifest.xml
  proguard-project.txt
  project.properties
```

图 7-42　Project_Chapter_7 工程目录结构

(2) 创建 productlist.xml 文件，将文件存放在 res/layout 下，代码如下：

```
1.   <?xml version="1.0" encoding="utf-8"?>
2.   <RelativeLayout
3.     xmlns:android="http://schemas.android.com/apk/res/android"
4.     android:orientation="vertical"
5.     android:layout_width="wrap_content"
6.     android:layout_height="wrap_content">
7.     <TextView
8.       android:text=""
9.       android:id="@+id/temp"
10.      android:layout_width="wrap_content"
11.      android:layout_height="wrap_content"
12.    ></TextView>
13.     <ListView
14.      android:id="@+id/productlist"
15.      android:layout_below="@id/temp"
16.      android:layout_width="fill_parent"
17.      android:layout_height="wrap_content"
```

```
18.      android:focusable="true">
19.    </ListView>
20.  <TextView
21.      android:id="@+id/pageinfo"
22.      android:layout_alignParentRight="true"
23.      android:layout_width="wrap_content"
24.      android:layout_height="wrap_content"
25.      android:text=""
26.    ></TextView>
27.  </RelativeLayout>
```

(3) 在 src 目录下的 com. hisoft. project 包下编写 ProductList. java,代码如下:

```
1.   package com.hisoft.project;
2.
3.   /**********************************************************************
4.     * 程序名称:ProductList.java                                          *
5.     * 功能:显示所有的商品信息,可选择需要的商品添加到购物车                 *
6.     * 作者:                                                              *
7.     * 日期:                                                              *
8.     **********************************************************************/
9.   import android.app.Activity;
10.  import android.os.Bundle;
11.  import android.view.Menu;
12.  import android.view.MenuItem;
13.  import android.widget.ArrayAdapter;
14.  import android.widget.ListView;
15.  import android.widget.TextView;
16.  import android.widget.Toast;
17.
18.  public class ProductList extends Activity {
19.
20.    private TextView temp;
21.    private TextView pageInfo;                          //页面显示信息
22.    private ListView productList;                       //商品列表展示控件
23.
24.    //menu 菜单
25.    public Menu menu;
26.    public static final int HANDLE=0;                   //添加到购物车
27.    public static final int PRE=1;                      //上一页
28.    public static final int NEXT=2;                     //下一页
29.    public static final int CART=3;                     //查看购物车
30.
31.    /**
32.      * 创建产品列表页面,初始化信息
33.      */
34.    @Override
35.    public void onCreate(Bundle savedInstanceState) {
```

```
36.      super.onCreate(savedInstanceState);
37.      setContentView(R.layout.productlist);
38.
39.      //拿到页面控件对象
40.      temp=(TextView) findViewById(R.id.temp);
41.      temp.setText("tempid:");
42.      pageInfo=(TextView) findViewById(R.id.pageinfo);
43.      //获取 ListView 对象
44.      productList=(ListView) findViewById(R.id.productlist);
45.
46.      productList.setItemsCanFocus(true);
47.      //设置商品可多选
48.      productList.setChoiceMode(ListView.CHOICE_MODE_MULTIPLE);
49.      productList.setTextFilterEnabled(true);
50.      //设置分页信息
51.      pageInfo.setText("共 6 件药品"+"\t"+"第 1 页"+"\t"+"共 1 页");
52.      //配置适配器
53.      String[] content={ "1\t 药品-1", "2\t 药品-2", "3\t 药品-3", "4\t 药品-4",
54.         "5\t 药品-5", "6\t 药品-6" };
55.      ArrayAdapter<String>adapter=new ArrayAdapter<String>(
56.         ProductList.this,
57.         android.R.layout.simple_list_item_multiple_choice, content);
58.      productList.setAdapter(adapter);
59.    }
60.
61.    /**
62.       * 创建 MENU 菜单
63.       */
64.    public boolean onCreateOptionsMenu(Menu menu) {
65.
66.
67.      menu.add(0, HANDLE, 0, "添加到购物车");             //添加到购物车 0
68.
69.      menu.add(0, PRE, 0, "上一页");                      //1 翻到上一页
70.
71.      menu.add(0, NEXT, 0, "下一页");                     //2 翻到下一页
72.
73.      menu.add(0, CART, 0, "我的购物车");                 //3 查看购物车
74.
75.      return true;
76.
77.    }
78.
79.    /**
80.     * MENU 菜单的选择事件
81.     */
82.    @Override
```

```
83.     public boolean onOptionsItemSelected(MenuItem item) {
84.       switch (item.getItemId()) {
85.       case HANDLE:
86.         //添加到购物车
87.          Toast.makeText(ProductList.this, "所选择的是:"+item.getTitle(),
             Toast.LENGTH_SHORT).show();
88.         return true;
89.       case PRE:
90.         //上一页
91.          Toast.makeText(ProductList.this, "所选择的是:"+item.getTitle(),
             Toast.LENGTH_SHORT).show();
92.         return true;
93.       case NEXT:
94.         //下一页
95.          Toast.makeText(ProductList.this, "所选择的是:"+item.getTitle(),
             Toast.LENGTH_SHORT).show();
96.         return true;
97.
98.       case CART:
99.         //查看购物车
100.          Toast.makeText(ProductList.this, "所选择的是:"+item.getTitle(),
              Toast.LENGTH_SHORT).show();
101.         return true;
102.       }
103.       return false;                                  //should never happen
104.     }
105. }
```

(4) 部署 Project_Chapter_7 工程,运行效果如图 7-43 所示。

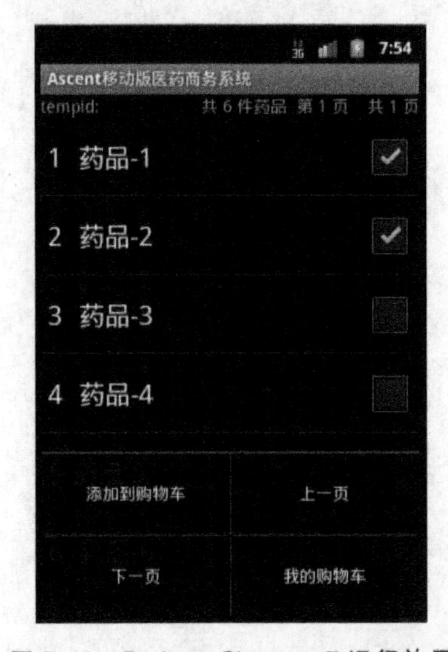

图 7-43 Project_Chapter_7 运行效果

1. 简答题

(1) Android 系统支持的菜单有哪些？如何在程序中创建菜单？

(2) 选项菜单可以分为哪些类别？它们各自的创建方法是什么？

(3) 简述创建子菜单的创建过程。

(4) 快捷菜单和选项菜单有什么区别？它们的创建有何不同？

(5) 什么是 Notification 控件？其主要作用是什么？

2. 完成下面的实训项目

要求：

(1) 单击“注册”按钮跳转到用户注册界面,包含用户名、密码、确认密码、性别(男、女)、个人爱好(多选)、邮箱、电话等基本信息。

(2) 单击 menu 按钮弹出菜单选项,如提交、编辑、返回和删除等。

第8章 Android组件广播消息与服务

学习目标

本章主要介绍 Android 组件 Intent、Intent 对象包含的信息、使用 Intent 进行组件通信、Intent 广播消息、BroadcastReceiver 监听广播消息、Service 组件服务、Service 与 Activity 通信等。使读者通过本章的学习，能够深入了解 Android 组件之间的消息传递机制及应用，掌握以下知识要点：

(1) Intent 分类、消息机制及启动方式。

(2) Activity、Service 和 BroadcastReceiver 及 Intent 进行通信。

(3) Intent 启动 Activity 的方法、获取 Activity 返回值。

(4) Intent 解析原理、机制、匹配规则。

(5) Intent 广播消息常用的方法。

(6) BroadcastReceiver 监听广播消息过程及方法。

(7) Service 类的继承关系及组件服务应用。

8.1 Intent 消息通信

8.1.1 Intent 简介

Intent 提供了一种通用的消息系统，它允许在用户的应用程序与其他的应用程序间传递 Intent 来执行动作和产生事件。

Intent 负责对应用中一次操作的动作、动作涉及数据、附加数据进行描述，Android 则根据此 Intent 的描述，负责找到对应的组件，将 Intent 传递给调用的组件，并完成组件的调用。使用 Intent 可以激活 Android 应用的三个核心组件：活动、服务和广播接收器。

在 Android 系统中，Intent 的用途主要有三个：

(1) 启动 Activity。

(2) 启动 Service。

(3) 在 Android 系统上发布广播消息(广播消息可以是接收到特定数据或消息，也可以是手机的信号变化或电池的电量过低等信息)。

通常 Intent 分为显式和隐式两类。显式 Intent 就是指定了组件名字，是由程序指定具体的目标组件来处理，即在构造 Intent 对象时就指定接收者，指定了一个明确的组件(setComponent 或 setClass)来使用处理 Intent。

```
Intent intent=new Intent(
    getApplicationContext(),
    Test.class
);
startActivity(intent);
```

特别注意：被启动的 Activity 需要在 AndroidManifest. xml 中进行定义。

隐式 Intent 就是没有指定 Intent 的组件名字，没有制定明确的组件来处理该 Intent，使用这种方式时，需要让 Intent 与应用中的 IntentFilter 描述表相匹配。需要 Android 根据 Intent 中的 Action、data 和 Category 等来解析匹配。由系统接受调用并决定如何处理，即 Intent 的发送者在构造 Intent 对象时并不知道也不关心接收者是谁，有利于降低发送者和接收者之间的耦合。如 startActivity(new Intent(Intent. ACTION_DIAL));。

```
Intent intent=new Intent();
intent.setAction("test.intent.IntentTest");
startActivity(intent);
```

目标组件(Activity、Service、Broadcast Receiver)是通过设置它们的 Intent Filter 来界定其处理的 Intent。如果一个组件没有定义 Intent Filter，那么它只能接受处理显式的 Intent，只有定义了 Intent Filter 的组件才能同时处理隐式和显式的 Intent。

一个 Intent 对象包含了很多数据的信息，由 6 个部分组成：

(1) Action：要执行的动作。

(2) Data：执行动作要操作的数据。

(3) Category：被执行动作的附加信息。

(4) Extras：其他所有附加信息的集合。

(5) Type：显式指定 Intent 的数据类型(MIME)。

(6) Component：指定 Intent 的目标组件的类名称，比如要执行的动作、类别、数据和附加信息等。

下面就一个 Intent 中包含的信息进行简要介绍。

1) Action

一个 Intent 的 Action 在很大程度上说明这个 Intent 要做什么，是查看(View)、删除(Delete)、编辑(Edit)等。Action 一个字符串命名的动作，Android 中预定义了很多 Action，可以参考 Intent 类查看。表 8-1 是 Android 文档中的几个动作。

表 8-1 Action

Constant	Target component	Action
ACTION_CALL	activity	Initiate a phone call
ACTION_EDIT	activity	Display data for the user to edit
ACTION_MAIN	activity	Start up as the initial activity of a task, with no data input and no returned output
ACTION_SYNC	activity	Synchronize data on a server with data on the mobile device
ACTION_BATTERY_LOW	broadcast receiver	A warning that the battery is low
ACTION_HEADSET_PLUG	broadcast receiver	A headset has been plugged into the device, or unplugged from it
ACTION_SCREEN_ON	broadcast receiver	The screen has been turned on
ACTION_TIMEZONE_CHANGED	broadcast receiver	The setting for the time zone has changed

此外,用户也可以自定义 Action,比如 com. flysnow. intent. ACTION_ADD。定义的 Action 最好能表明其所表示的意义,要做什么,这样 Intent 中的数据才好填充。Intent 对象的 getAction()可以获取动作,使用 setAction()可以设置动作。

2) Data

Data 实质上是一个 URI,用于执行一个 Action 时所用到的数据的 URI 和 MIME。不同的 Action 有不同的数据规格,比如 ACTION_EDIT 动作,数据就可以包含一个用于编辑文档的 URI;如果是一个 ACTION_CALL 动作,那么数据就是一个包含了 tel:6546541 的数据字段,所以上面提到的自定义 Action 时要规范命名。数据的 URI 和类型对于 Intent 的匹配是很重要的,Android 往往根据数据的 URI 和 MIME 找到能处理该 Intent 的最佳目标组件。

3) Component(组件)

Component 指定 Intent 的目标组件的类名称。通常 Android 会根据 Intent 中包含的其他属性的信息,比如 action、data/type 和 category 进行查找,最终找到一个与之匹配的目标组件。

如果设置了 Intent 目标组件的名字,那么这个 Intent 就会被传递给特定的组件,而不再执行上述查找过程。指定了这个属性以后,Intent 的其他所有属性都是可选的,也就是我们说的显式 Intent。如果不设置,则是隐式的 Intent,Android 系统将根据 Intent Filter 中的信息进行匹配。

4) Category

Category 指定了用于处理 Intent 的组件的类型信息,一个 Intent 可以添加多个 Category,使用 addCategory()方法即可,使用 removeCategory()删除一个已经添加的类别。Android 的 Intent 类里定义了很多常用的类别,可以参考使用。

5) Extras

Extras 用于处理 Intent 的目标组件需要一些额外的信息时,通过 Intent 的 put()方法把额外的信息塞入到 Intent 对象中,用于目标组件的使用,一个附件信息就是一个 key-value 的键值对。Intent 有一系列的 put 和 get 方法用于处理附加信息的塞入和取出。

8.1.2 使用 Intent 进行组件通信

8.1.1 节已经讲述了 Intent 的作用、分类及其包含的信息，从上述内容可以得知，Intent 就是一个动作的完整描述，包含了动作的产生组件、接收组件和传递的数据信息。Intent 也可称为一个在不同组件之间传递的消息，这个消息在到达接收组件后，接收组件会执行相关的动作。Intent 为 Activity、Service 和 BroadcastReceiver 等组件提供了交互的能力，如图 8-1 所示。

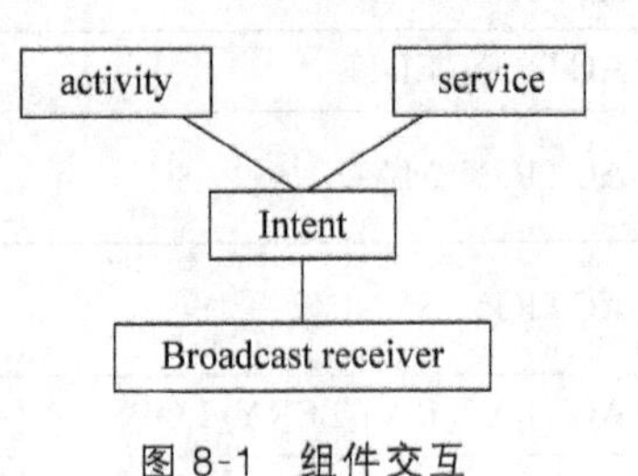

图 8-1 组件交互

对于 Activity、Service 和 BroadcastReceiver 这三个组件，它们都有自己独立的传递 Intent 的机制。

- Activity：对于 Activity 来说，它主要是通过 Context. startActivity()或 Activity. startActivityForRestult()来启动一个存在的 Activity 做一些事情。当使用 Activity. startActivityForResult()启动一个 Activity 时，可以使用 Activity. setResult()返回一些结果信息，可以在 Activity. onActivityResult()中得到返回的结果。
- Service：对于 Service 来说，主要是通过 Context. startService()初始化一个 Service 或者传递消息给正在运行的 Service。同样，也可以通过 Context. bindService()建立一个调用组件和目标服务之间的连接。
- BroadcastReceiver：可以通过 Context. sendBroadcast()、Context. sendOrderedBroadcast()以及 Context. sendStickyBroadcast()这些方法传递 Intent 给感兴趣的广播。

消息之间的传递是没有重叠的，比如调用 startActivity()传播一个 Intent，只会传递给 Activity，而不会传递给 Service 和 BroadcastReceiver，反过来也是这样。

8.1.3 使用 Intent 启动 Activity

在 Android 系统中，应用程序一般都有多个 Activity，Intent 可以实现不同 Activity 之间的切换和数据传递。

使用 Intent 启动 Activity 方式主要有两种：显式启动和隐式启动。如前面章节所述一样，显式启动必须在 Intent 中指明启动的 Activity 所在的类。而隐式启动，Android 系统根据 Intent 的动作和数据来决定启动哪一个 Activity，也就是说在隐式启动时，Intent 中只包含需要执行的动作和所包含的数据，并没有指明具体启动的 Activity，而是由 Android 系统和最终用户来决定。下面就显式和隐式启动 Activity 的通常用法进行介绍。

1. 显式启动 Activity 的通常用法

(1) 新建一个 Intent。

(2) 指定当前的应用程序上下文以及要启动的 Activity。

(3) 把新建好的这个 Intent 作为参数传递给 startActivity()方法。

```
1.    Intent intent=new Intent(IntentTestDemo.this, NewActivity.class);
2.    startActivity(intent);
```

上述包含了两个 Activity 类，分别是 IntentTestDemo 和 NewActivity，程序默认启动的是 IntentTestDemo。具体步骤如下：

(1) 依照前面案例创建的步骤，新创建一个工程名为 IntentTestDemo 的工程，然后打开工程中的 AndroidManifest. xml 文件，在<application>根节点下添加<activity>标签，注册新添加的 activity，嵌套在<application>根节点标签下，添加代码如下：

```
1.    <activity android:name=".NewActivity"
2.    android:label="@string/app_name"
3.    /activity>
```

在 Android 应用程序中，用户使用的每个组件都必须在 AndroidManifest. xml 文件中的<application>节点内定义。<application>节点下共有两个<activity>节点：分别代表应用程序中所使用的两个 Activity，即 IntentTestDemo（创建工程时自动生成）和 NewActivity。

(2) 修改 res 目录下 layout 文件夹中的 main. xml 文件，设置线性布局，添加一个 Button 按钮控件描述，并设置相关属性，代码如下：

```
1.    <?xml version="1.0" encoding="utf-8"?>
2.    <LinearLayout xmlns:android="http://schemas.android.com/apk/res/android"
3.        android:orientation="vertical"
4.        android:layout_width="fill_parent"
5.        android:layout_height="fill_parent"
6.        >
7.    <Button android:id="@+id/bt1"
8.        android:layout_height="wrap_content"
9.        android:layout_width="fill_parent"
10.       android:text="测试显式 Intent"
11.   />
12.   </LinearLayout>
```

(3) 修改 src 目录下 com. hisoft. activity 包下的 IntentTestDemoActivity. java 文件，添加显示使用 Intent 启动 Activity 的核心代码，代码如下：

```
1.    Button button=(Button)findViewById(R.id.bt1);
2.    button.setOnClickListener(new OnClickListener(){
3.      public void onClick(View view){
4.        Intent intent=new Intent(IntentTestDemoActivity.this, NewActivity.class);
5.        startActivity(intent);
6.      }
7.    });
```

在单击事件的处理函数中，Intent 构造函数的第 1 个参数是应用程序上下文，程序中的应用程序上下文就是 IntentTestDemo；第 2 个参数是接收 Intent 的目标组件，使用的是显式启动方式，直接指明了需要启动的 Activity。

(4) 在 src 目录下的 com. hisoft. activity 包下创建新的 NewActivity，在 res 目录下的 layout 文件夹中创建 new_main. xml 文件，在 values 文件夹下的 strings. xml 文件中添加

text 引用，让 NewActivity 界面显示 NewActivity application。

(5) 部署运行程序，程序运行效果如图 8-2 所示。单击"测试显式 Intent"按钮，程序运行如图 8-3 所示。

图 8-2 IntentTestDemo 运行效果

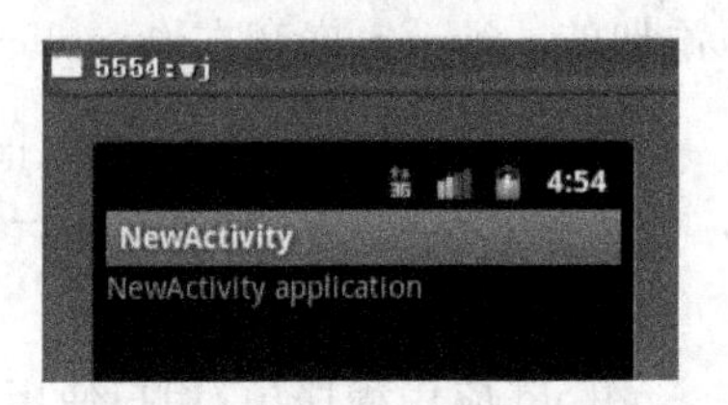

图 8-3 测试显式 Intent 效果

2. 隐式启动 Activity 的通常用法

隐式启动 Activity 时，Android 系统在应用程序运行时解析 Intent，并根据一定的规则对 Intent 和 Activity 进行匹配，使 Intent 上的动作、数据与 Activity 完全匹配。

(1) 在 AndroidManifest. xml 中注册声明需要匹配 Activity。

(2) 程序代码中创建新的 Intent(可以向 Intent 中添加运行 Activity 所需要的附加信息)。

(3) 将 Intent 传递给 startActivity()。

创建 Intent 时，在缺省情况下 Android 系统会调用内置的 Web 浏览器，如：

```
Intent intent=new Intent(Intent.ACTION_VIEW, Uri.parse("http://www.google.com"));
startActivity(intent);
```

上述代码中 Intent 的动作是 Intent. ACTION_VIEW，根据 URI 的数据类型来匹配动作。数据部分的 URI 是 Web 地址，使用 Uri. parse(urlString)方法可以简单地把一个字符串解释成 Uri 对象。

创建 Intent 对象的语法如下：

```
Intent intent=new Intent(Intent.ACTION_VIEW, Uri.parse(urlString));
```

Intent 构造函数的第 1 个参数是 Intent 需要执行的动作；第 2 个参数是 URI，表示需要传递的数据。

Android 系统支持的常见动作字符串常量如表 8-2 所示。

表 8-2 Android 系统支持的常见动作字符串常量

动　作	说　明
ACTION_ANSWER	打开接听电话的 Activity，默认为 Android 内置的拨号盘界面
ACTION_CALL	打开拨号盘界面并拨打电话，使用 Uri 中的数字部分作为电话号码
ACTION_DELETE	打开一个 Activity，对所提供的数据进行删除操作
ACTION_DIAL	打开内置拨号盘界面，显示 Uri 中提供的电话号码
ACTION_EDIT	打开一个 Activity，对所提供的数据进行编辑操作
ACTION_INSERT	打开一个 Activity，在提供数据的当前位置插入新项

续表

动　作	说　明
ACTION_PICK	启动一个 Activity，从提供的数据列表中选取一项
ACTION_SEARCH	启动一个 Activity，执行搜索动作
ACTION_SENDTO	启动一个 Activity，向数据提供的联系人发送信息
ACTION_SEND	启动一个可以发送数据的 Activity
ACTION_VIEW	最常用的动作，对以 Uri 方式传送的数据，根据 Uri 协议部分以最佳方式启动相应的 Activity 进行处理。对于 http:address，将打开浏览器查看；对于 tel:address，将打开拨号呼叫指定的电话号码
ACTION_WEB_SEARCH	打开一个 Activity，对提供的数据进行 Web 搜索

隐式 Intent 应用的具体步骤如下：

(1) 同显式 Intent 一样，新创建一个工程，然后打开工程中的 AndroidManifest. xml 文件，在＜application＞根节点下添加＜activity＞标签，注册新添加的 activity，嵌套在＜application＞根节点标签下，添加代码如下：

```
1.    <activity
2.            android:name=".FirstActivity"
3.            android:label="First Activity">
4.           <intent-filter>
5.               <action android:name="com.android.activity.Me_Action"/>
6.              <category android:name="android.intent.category.DEFAULT"/>
7.            </intent-filter>
8.        </activity>
```

(2) 修改 res 目录下 layout 文件夹中的 main. xml 文件，设置线性布局，添加一个 Button 按钮控件描述，并设置相关属性，代码如下：

```
1.  <?xml version="1.0" encoding="utf-8"?>
2.  <LinearLayout xmlns:android="http://schemas.android.com/apk/res/android"
3.      android:orientation="vertical"
4.      android:layout_width="fill_parent"
5.      android:layout_height="fill_parent"
6.      >
7.  <Button android:id="@+id/bt1"
8.    android:layout_height="wrap_content"
9.    android:layout_width="fill_parent"
10.    android:text="测试隐式 Intent"
11.  />
12.  </LinearLayout>
```

(3) 修改 src 目录下 com. hisoft. activity 包下的 IntentTestDemoActivity. java 文件，添加显示使用 Intent 启动 Activity 的核心代码，代码如下：

```
1.   Button button=(Button)findViewById(R.id.bt1);
```

```
2.          button.setOnClickListener(new OnClickListener(){
3.
4.
5.          @Override
6.          public void onClick(View view){
7.              Intent intent=new Intent();
8.              intent.setAction("com.android.activity.Me_Action");
9.              startActivity(intent);
10.         }
11.         });
```

(4) 在 src 目录下 com. hisoft. activity 包中的 FirstActivity. java 文件，代码如下：

```
1.   public class FirstActivity extends Activity {
2.       @Override
3.       protected void onCreate(Bundle savedInstanceState) {
4.           super.onCreate(savedInstanceState);
5.           setContentView(R.layout.second);
6.           Intent intent=new Intent(Intent.ACTION_VIEW, Uri.parse("http://
             www.google.com"));
7.           startActivity(intent);
8.       }
9.   }
```

(5) 修改 res 目录下 layout 文件夹中新创建的 second. xml 文件，并设置相关属性，代码如下：

```
1.   <?xml version="1.0" encoding="utf-8"?>
2.   <LinearLayout xmlns:android="http://schemas.android.com/apk/res/android"
3.       android:orientation="vertical"
4.       android:layout_width="fill_parent"
5.       android:layout_height="fill_parent"
6.       >
7.   <TextView
8.       android:layout_width="fill_parent"
9.       android:layout_height="wrap_content"
10.      android:text="@string/start"
11.      />
12.  </LinearLayout>
```

(6) 修改 res 目录下 values 文件夹中的 strings. xml 文件，并设置相关属性，代码如下：

```
1.   <?xml version="1.0" encoding="utf-8"?>
2.   <resources>
3.       <string name="hello">Hello World, IntentTestDemoActivity!</string>
4.       <string name="app_name">IntentTestDemo</string>
5.       <string name="start">NewActivity application</string>
6.       <string name="app">NewActivity</string>
```

```
7.    </resources>
```

(7) 部署运行程序，程序运行效果如图 8-4 所示。单击"测试隐式 Intent"按钮，程序根据设定的网址生成一个 Intent，并以隐式启动的方式调用 Android 内置的 Web 浏览器，并打开指定的 google 网站运行，如图 8-5 所示。

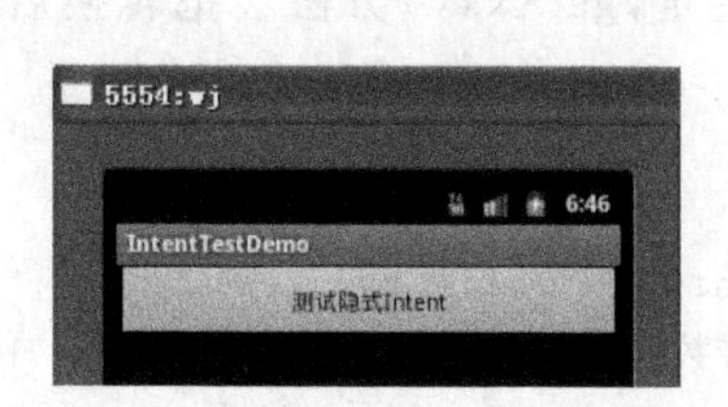

图 8-4 隐式 Intent 运行结果

图 8-5 Web 浏览

注意：Android 本地的应用程序组件和第三方应用程序一样，都是 Intent 解析过程中的一部分。它们没有更高的优先度，可以被新的 Activity 完全代替，这些新的 Activity 宣告自己的 Intent Filter 能响应相同的动作请求。

隐式 Intent 与显式 Intent 相比更有优势，它不需要指明需要启动哪一个 Activity，而由 Android 系统来决定，有利于使用第三方组件。此外，匹配的 Activity 可以是应用程序本身的，也可以是 Android 系统内置的，还可以是第三方应用程序提供的。因此，这种方式更加强调了 Android 应用程序中组件的可复用性。

在一个 Activity 中可以使用系统提供的 startActivity(Intent intent)方法打开新的 Activity。在打开新的 Activity 前，可以决定是否为新的 Activity 传递参数：

```
startActivity(new Intent(MainActivity.this, NewActivity.class));
```

Bundle 类用作携带数据，它类似于 Map，用于存放 key-value 名值对形式的值。相对于 Map，它提供了各种常用类型的 putXxx()/getXxx()方法，如 putString()/getString()和 putInt()/getInt()，putXxx()用于往 Bundle 对象放入数据，getXxx()用于从 Bundle 对象里获取数据。Bundle 的内部实际上是使用了 HashMap<String,Object>类型的变量来存放 putXxx()方法放入的值。

启动 Activity 并传递数据：

```
1.    public final class Bundle implements Parcelable, Cloneable {
2.         ...
3.     Map<String, Object>mMap;
4.     public Bundle() {
5.          mMap=new HashMap<String, Object>();
6.          ...
7.     }
8.     public void putString(String key, String value) {
```

```
9.          mMap.put(key, value);
10.    }
11.  public String getString(String key) {
12.         Object o=mMap.get(key);
13.          return (String) o;
14.  }
15.  }
```

在调用 Bundle 对象的 getXxx()方法时，方法内部会从该变量中获取数据，然后对数据进行类型转换，转换成什么类型由方法的 Xxx 决定，getXxx()方法会把转换后的值返回。

打开新的 Activity，并传递若干个参数给它：

```
1.   Intent intent=new Intent(MainActivity.this, NewActivity.class)
2.   Bundle bundle=new Bundle();   //该类用作携带数据 bundle.putString("name", "lee");
3.   bundle.putInt("age", 4);
4.   intent.putExtras(bundle);  //附带上额外的数据
5.   startActivity(intent);
```

在新的 Activity 中接收前面 Activity 传递过来的参数：

```
1.   public class NewActivity extends Activity
2.   {
3.               @Override
4.      protected void onCreate(Bundle savedInstanceState)
5.     {
6.          ...
7.          Bundle bundle=this.getIntent().getExtras();
8.          String name=bundle.getString("name");
9.                   int age=bundle.getInt("age");
10.         }
11.  }
```

8.1.4 获取 Activity 返回值

在 Activity 中得到新打开的 Activity 关闭后返回的数据，则需要完成以下方面：

(1) 在 Activity 中使用系统提供的 startActivityForResult(Intent intent, int requestCode)方法打开新的 Activity。

(2) 在 Activity 中重写 onActivityResult(int requestCode, int resultCode, Intent data)方法。

当新的 Activity 关闭后，新的 Activity 返回的数据通过 Intent 进行传递，Android 平台会调用前面 Activity 的 onActivityResult()方法把存放了返回数据的 Intent 作为第三个输入参数传入，这样在 onActivityResult()方法中使用第三个输入参数可以取出新 Activity 返回的数据。代码如下：

```
1.     public class MainActivity extends Activity {
2.        @Override
3.        protected void onCreate(Bundle savedInstanceState) {
4.     ...
5.     Button button=(Button) this.findViewById(R.id.button);
6.       button.setOnClickListener(new View.OnClickListener(){
                                          //单击该按钮会打开一个新的 Activity
7.       public void onClick(View v) {
8.       //第二个参数为请求码,可以根据需求自己编号
9.       startActivityForResult (new Intent(MainActivity.this, NewActivity.class),1);
10.      }});
11.         }
12.      //第一个参数为请求码,即调用 startActivityForResult()传递过去的值
13.      //第二个参数为结果码,结果码用于标识返回数据来自哪一个新 Activity
14.     @Override
15.     protected void onActivityResult(int requestCode, int resultCode, Intent data) {
16.       String result=data.getExtras().getString("result"));
                                          //得到新 Activity 关闭后返回的数据
17.       }
18.  }
```

上面讲述了使用 startActivityForResult(Intent intent,int requestCode)方法打开新的 Activity,新 Activity 关闭前需要向前面的 Activity 返回数据,需要使用系统提供的 setResult(int resultCode,Intent data)方法实现,代码如下:

```
1.   public class NewActivity extends Activity {
2.     @Override protected void onCreate(Bundle savedInstanceState) {
3.      ...
4.             button.setOnClickListener(new View.OnClickListener(){
5.       public void onClick(View v) {
6.         Intent intent=new Intent();                  //数据是使用 Intent 返回
7.         intent.putExtra("result", "返回的数据!");     //把返回数据存入 Intent
8.          NewActivity.this.setResult(RESULT_ CANCELED, intent);   //设置返回数据
9.          NewActivity.this.finish();                           //关闭 Activity
10.       }});
11.     }
12.  }
```

setResult()方法的第一个参数值可以根据需要自己定义。上面代码中使用到的 RESULT_是 CANCELED 系统 Activity 类定义的一个常量,值为 0,代码片段如下:

```
1.   public class android.app.Activity extends ...{
2.     public static final int RESULT_CANCELED=0;
3.     public static final int RESULT_OK=-1;
4.     public static final int RESULT_FIRST_USER=1;
5.   }
```

上述代码中请求码的作用主要在于：使用 startActivityForResult(Intent intent,int requestCode)方法打开新的 Activity，需要为 startActivityForResult()方法传入一个请求码(第二个参数)。请求码的值是根据业务需要由自己设定，用于标识请求来源。

例如，一个 Activity 有两个 button 按钮，单击这两个按钮都会打开同一个 Activity。不管是 button1 还是 button2 按钮打开新 Activity，当这个新 Activity 关闭后，系统都会调用前面 Activity 的 onActivityResult(int requestCode,int resultCode,Intent data)方法。在 onActivityResult()方法中，如果需要知道新 Activity 是由哪个按钮打开的，并且要做出相应的业务处理，则参考代码如下：

```
1.   public void onCreate(Bundle savedInstanceState) {
2.           …
3.           button1.setOnClickListener(new View.OnClickListener(){
4.       public void onClick(View v) {
5.            startActivityForResult (new Intent (MainActivity.this, NewActivity.
              class), 1);
6.         }});
7.         button2.setOnClickListener(new View.OnClickListener(){
8.       public void onClick(View v) {
9.           startActivityForResult (new Intent(MainActivity.this, NewActivity.
             class), 2);
10.        }});
11.       @Override
12.  protected void onActivityResult(int requestCode, int resultCode, Intent data){
13.               switch(requestCode){
14.                   case 1:
15.                        //来自按钮 1 的请求,作相应处理
16.                   case 2:
17.                 //来自按钮 2 的请求,作相应处理
18.                 }
19.          }
20.  }
```

同样，上述结果码的主要作用是：在一个 Activity 中，可能会使用 startActivityForResult()方法打开多个不同的 Activity 处理不同的业务，当这些新 Activity 关闭后，系统都会调用前面 Activity 的 onActivityResult(int requestCode,int resultCode,Intent data)方法。为了知道返回的数据来自于哪个新 Activity，在 onActivityResult()方法(假设 ResultActivity 和 NewActivity 为要打开的新 Activity)中处理代码参考如下：

```
1.     public class ResultActivity extends Activity {
2.          …
3.          ResultActivity.this.setResult(1, intent);
4.          ResultActivity.this.finish();
5.   }
6.   public class NewActivity extends Activity {
```

```
7.          ...
8.          NewActivity.this.setResult(2, intent);
9.          NewActivity.this.finish();
10. }
11. public class MainActivity extends Activity {
                        //在该 Activity 会打开 ResultActivity 和 NewActivity
12.      @Override
13. protected void onActivityResult(int requestCode, int resultCode, Intent data) {
14.              switch(resultCode){
15.                  case 1:
16.                      //ResultActivity 的返回数据
17.                  case 2:
18.                //NewActivity 的返回数据
19.                  }
20.          }
21. }
```

8.1.5 Intent Filter 原理与匹配机制

Intent Filter(Intent 过滤器)是一种根据 Intent 中的动作(Action)、类别(Categorie)和数据(Data)等内容，对适合接收该 Intent 的组件进行匹配和筛选的机制。

Intent 过滤器可以匹配数据类型、路径和协议，还包括可以用来确定多个匹配项顺序的优先级(Priority)。

应用程序的 Activity 组件、Service 组件和 BroadcastReceiver 都可以注册 Intent 过滤器，则这些组件在特定的数据格式上就可以产生相应的动作。

1. 注册 Intent Filter

(1) 在 AndroidManifest.xml 文件的各个组件的节点下定义<intent-filter>节点，然后在<intent-filter>节点中声明该组件所支持的动作、执行的环境和数据格式等信息。

(2) 在程序代码中动态地为组件设置 Intent 过滤器。

在上述(1)中，定义的<intent-filter>节点包含的标签有<action>标签、<category>标签和<data>标签。

- <action>标签定义 Intent Filter 的“动作”。
- <category>标签定义 Intent Filter 的“类别”。
- <data>标签定义 Intent Filter 的“数据”。

<intent-filter>节点支持的标签和属性如表 8-3 所示。

<category>标签用来指定 Intent Filter 的服务方式，每个 Intent Filter 可以定义多个<category>标签，开发者可使用自定义的类别，或使用 Android 系统提供的类别。如表 8-4 所示。

表 8-3 <intent-filter>节点支持的标签和属性

标签	属性	说　明
<action>	android:name	指定组件所能响应的动作,用字符串表示,通常使用 Java 类名和包的完全限定名构成
<category>	android:category	指定以何种方式去服务 Intent 请求的动作
<data>	android:host	指定一个有效的主机名
	android:mimetype	指定组件能处理的数据类型
	android:path	有效的 URI 路径名
	android:port	主机的有效端口号
	android:scheme	所需要的特定的协议

表 8-4 Android 系统提供的类别

常　量　值	描　　述
ALTERNATIVE	Intent 数据默认动作的一个可替换的执行方法
SELECTED_ALTERNATIVE	和 ALTERNATIVE 类似,但替换的执行方法不是指定的,而是被解析出来的
BROWSABLE	声明 Activity 可以由浏览器启动
DEFAULT	为 Intent 过滤器中定义的数据提供默认动作
HOME	设备启动后显示的第一个 Activity
LAUNCHER	在应用程序启动时首先被显示

AndroidManifest.xml 文件中的每个组件的<intent-filter>都被解析成一个 Intent Filter 对象。当应用程序安装到 Android 系统时,所有的组件和 Intent Filter 都会注册到 Android 系统中。这样,Android 系统便知道了如何将任意一个 Intent 请求通过 Intent Filter 映射到相应的组件上。

2. Intent 解析机制

当使用 startActivity 时,隐式 Intent 解析到一个单一的 Activity。如果存在多个 Activity 都有能够匹配在特定的数据上执行给定的动作,Android 会从这些中选择最好的一个进行启动。决定哪个 Activity 来运行的过程称为 Intent 解析,即 Intent 到 Intent Filter 的映射过程。

Intent 解析机制主要是通过查找已注册在 AndroidManifest.xml 中的所有 IntentFilter 及其中定义的 Intent,最终找到一个可以与请求的 Intent 达成最佳匹配的 Intent Filter。

Intent 解析的匹配规则:

- Android 系统把所有应用程序包中的 Intent 过滤器集合在一起,形成一个完整的 Intent 过滤器列表。
- 在 Intent 与 Intent 过滤器进行匹配时,Android 系统会将列表中所有 Intent 过滤器的"动作"和"类别"与 Intent 进行匹配,任何不匹配的 Intent 过滤器都将被过滤掉。

没有指定“动作”的 Intent 过滤器可以匹配任何的 Intent，但是没有指定“类别”的 Intent 过滤器只能匹配没有“类别”的 Intent。

- 把 Intent 数据 Uri 的每个子部与 Intent 过滤器的<data>标签中的属性进行匹配，如果<data>标签指定了协议、主机名、路径名或 MIME 类型，那么这些属性都要与 Intent 的 Uri 数据部分进行匹配，任何不匹配的 Intent 过滤器均被过滤掉。
- 如果 Intent 过滤器的匹配结果多于一个，则可以根据在<intent-filter>标签中定义的优先级标签来对 Intent 过滤器进行排序，优先级最高的 Intent 过滤器将被选择。

在根据 Intent 解析匹配规则解析的过程中，Android 是通过 Intent 的 action、category 和 data 这三个属性进行判断的，判断方法如下：

(1) 如果 Intent 指定了 action，则目标组件的 IntentFilter 的 action 列表中就必须包含有这个 action，否则不能匹配。

(2) 如果 Intent 没有提供 mimetype，系统将从 data 中得到数据类型。和 action 一样，目标组件的数据类型列表中必须包含 Intent 的数据类型，否则不能匹配。

(3) 如果 Intent 中的数据不是 content：类型的 URI，而且 Intent 也没有明确指定它的 type，将根据 Intent 中数据的 scheme（比如 http：或者 mailto:）进行匹配。同上，Intent 的 scheme 必须出现在目标组件的 scheme 列表中。

(4) 如果 Intent 指定了一个或多个 category，这些类别必须全部出现在组建的类别列表中。比如 Intent 中包含了两个类别：LAUNCHER_CATEGORY 和 ALTERNATIVE_CATEGORY，解析得到的目标组件必须至少包含这两个类别。

一个 Intent 对象只能指定一个 action，而一个 Intent Filter 可以指定多个 action。action 的列表不能为空，否则它将组织所有的 Intent。

一个 Intent 对象的 action 必须和 intent filter 中的某一个 action 匹配才能通过测试。如果 intent filter 的 action 列表为空，则不通过。如果 intent 对象不指定 action，并且 intentfilter 的 action 列表不为空，则通过测试。

下面针对 Intent 和 Intent Filter 中包含的子元素 Action（动作）、Data（数据）以及 Category(类别)进行比较检查的具体规则详细介绍。

1. 动作匹配测试

动作匹配指 Intent Filter 包含特定的动作或没有指定的动作。一个 Intent Filter 有一个或多个定义的动作，如果没有任何一个能与 Intent 指定的动作匹配的话，这个 Intent Filter 算作是动作匹配检查失败。

<intent-filter>元素中可以包括子元素<action>，比如：

```
<intent-filter>
<action android:name="com.example.project.SHOW_CURRENT"/>
<action android:name="com.example.project.SHOW_RECENT"/>
<action android:name="com.example.project.SHOW_PENDING"/>
</intent-filter>
```

一条<intent-filter>元素至少应该包含一个<action>，否则任何 Intent 请求都不能和该<intent-filter>匹配。如果 Intent 请求的 Action 和<intent-filter>中某一条<action>匹配,那么该 Intent 就通过了这条<intent-filter>的动作测试。如果 Intent 请求或<intent-filter>中没有说明具体的 Action 类型,那么会出现下面两种情况:

(1) 如果<intent-filter>中没有包含任何 Action 类型,那么无论什么 Intent 请求都无法和这条<intent-filter>匹配。

(2) 反之,如果 Intent 请求中没有设定 Action 类型,那么只要<intent-filter>中包含有 Action 类型,这个 Intent 请求就将顺利地通过<intent-filter>的行为测试。

2. 类别匹配测试

Intent Filter 必须包含所有在解析的 Intent 中定义的种类。一个没有特定种类的 Intent Filter 只能与没有种类的 Intent 匹配。

<intent-filter>元素可以包含<category>子元素,例如:

```
<intent-filter…>
<category android:name="android.Intent.Category.DEFAULT"/>
<category android:name="android.Intent.Category.BROWSABLE"/>
</intent-filter>
```

只有当 Intent 请求中所有的 Category 与组件中某一个 IntentFilter 的<category>完全匹配时,才会让该 Intent 请求通过测试,IntentFilter 中多余的<category>声明并不会导致匹配失败。一个没有指定任何类别测试的 IntentFilter 仅仅只会匹配没有设置类别的 Intent 请求。

3. 数据匹配测试

Intent 的数据 URI 中的部分会与 Intent Filter 中的 data 标签比较。如果 Intent Filter 定义 scheme、host/authority、path 或 mimetype,这些值都会与 Intent 的 URI 比较。任何不匹配都会导致 Intent Filter 从列表中删除。

没有指定 data 值的 Intent Filter 会和所有的 Intent 数据匹配。

数据在<intent-filter>中的描述如下:

```
<intent-filter…>
<data android:type="video/mpeg" android:scheme="http"…/>
<data android:type="audio/mpeg" android:scheme="http"…/>
</intent-filter>
```

<data>元素指定了希望接受的 Intent 请求的数据 URI 和数据类型,URI 被分成三部分来进行匹配:scheme、authority 和 path。其中,用 setData()设定的 Inteat 请求的 URI 数据类型和 scheme 必须与 IntentFilter 中所指定的一致。scheme 是 URI 部分的协议,例如 http:、mailto:和 tel:。

若 IntentFilter 中还指定了 authority 或 path,它们也需要相匹配才会通过测试。如下:

mimetype 是正在匹配的数据的数据类型。当匹配数据类型时,可以使用通配符来匹配子类型(如 bjzfs/ *)。如果 Intent Filter 指定一个数据类型,它必须与 Intent 匹配;没有指定数据的话全部匹配。

host-name 是介于 URI 中 scheme 和 path 之间的部分(如 www. google. com)。匹配主机名时,Intent Filter 的 scheme 也必须通过匹配。

path 紧接在 host-name 的后面(如/ig)。path 只在 scheme 和 host-name 部分都匹配的情况下才匹配。

如果这个过程中多于一个组件解析出来的话,它们会以优先度来排序,可以在 Intent Filter 的节点里添加一个可选的标签。最高等级的组件会返回。

具体实例见隐式 Intent 启动 Activity。

8.2 Intent 广播消息

前面章节已经讲述了 Intent 的用途,其中一个重要用途是发送广播消息,广播消息的内容可以是与应用程序密切相关的数据信息,也可以是 Android 的系统信息,例如网络连接变化、电池电量变化、接收到短信和系统设置变化等,应用程序和 Android 系统都可以使用 Intent 发送广播消息。如果应用程序注册了 BroadcastReceiver,则可以接收到指定的广播消息。

8.2.1 广播消息

使用 Intent 广播消息常用的方法:

(1) 创建一个 Intent,在构造 Intent 时必须用一个全局唯一的字符串标识其要执行的动作,通常使用应用程序包的名称。

(2) 调用 sendBroadcast()方法就可把 Intent 携带的消息广播出去。如果要在 Intent 传递额外数据,可以用 Intent 的 putExtra()方法。

如果利用 Intent 发送广播消息,并添加了额外的数据,然后调用 sendBroadcast()发送广播消息,代码如下:

```
1.  String UNIQUE_STRING="com.hisoft.BroadcastReceiverDemo";
2.  Intent intent=new Intent(UNIQUE_STRING);
3.  intent.putExtra("key1", "testValue1");
4.  intent.putExtra("key2", "testValue2");
5.  sendBroadcast(intent);
```

8.2.2 BroadcastReceiver 监听广播消息

BroadcastReceiver(广播接收者)位于 android. content 包下,其类的继承结构如图 8-6 所示,是用于接收 sendBroadcast()广播的 Intent,广播 Intent 的发送是通过调用 Context. sendBroadcast()、Context. sendOrderedBroadcast()来实现的。通常一个广播 Intent 可以被订阅了此 Intent 的多个广播接收者所接收。

```
java.lang.Object
  ↳android.content.BroadcastReceiver

▶ Known Direct Subclasses
  AppWidgetProvider, DeviceAdminReceiver
```

图 8-6 BroadcastReceiver 类继承关系

广播是一种广泛运用的在应用程序之间传输信息的机制。而 BroadcastReceiver 是对发送出来的广播进行过滤接收并响应的一类组件。

BroadcastReceiver 自身并不实现图形用户界面，但是当它收到某个通知后，BroadcastReceiver 可以启动 Activity 作为响应，或者通过 NotificationMananger 提醒用户，或者启动 Service 等。

BroadcastReceiver 为广播接收器，它和事件处理机制类似，只不过事件的处理机制是程序组件级别的，广播处理机制是系统级别的。它用于接收并处理广播通知，如由系统发起的地域变换、电量不足、来电来信等，或者程序播放的广播。

BroadcastReceiver 通知用户的方式有多种，如启动 activity、使用 NotificationManager、开启背景灯、振动设备、播放声音等。最典型的是在状态栏显示一个图标，用户通过单击它打开浏览通知内容。

1. Broadcast Receiver 组件监听过程

使用 Broadcast Receiver 组件监听过滤接收的过程是：首先在需要发送信息的地方把要发送的信息和用于过滤的信息(如 action、category)封装入一个 Intent 对象，然后通过调用 sendBroadcast()方法把 Intent 对象以广播方式发送出去。当 Intent 发送以后，所有已在 AndroidManifest. xml 中或代码中注册的 BroadcastReceiver 会检查注册时的 IntentFilter 是否与发送的 Intent 相匹配，若匹配就会调用 BroadcastReceiver 的 onReceive()方法。所以在定义一个 BroadcastReceiver 时需继承 BroadcastReceiver 类，并重载 onReceive()方法。代码如下：

```
1.  public class TestMeBroadcastReceiver extends BroadcastReceiver {
2.    @Override
3.    public void onReceive(Context context, Intent intent) {
4.      …
5.    }
6.  }
```

注册 BroadcastReceiver 的应用程序不需要一直运行，当 Android 系统接收到与之匹配的广播消息时，系统会自动启动此 BroadcastReceiver，在 BroadcastReceiver 接收到与之匹配的广播消息后，onReceive()方法会被调用。onReceive()方法必须要在 5s 执行完毕，否则 Android 系统会认为该组件失去响应，并提示用户强行关闭该组件。

由于它的典型特征，BroadcastReceiver 通常适合用于做一些资源管理的工作。

2. BroadcastReceiver 用于监听实现方式

BroadcastReceiver 用于监听广播的 Intent，Broadcast Receiver 监听的运用，可以有两种方式来实现：

第一种方式是在 AndroidManifest. xml 文件中注册一个 BroadcastReceiver，并在其中使用 Intent Filter 指定要处理的广播消息 Intent。这是一种推荐的方法，因为它不需要手动注销广播(如果广播未注销，程序退出时可能会出错)。

第二种方式是直接在代码中实现，但需要手动注册注销。

它们通常的开发步骤包含：

(1) 继承 BroadcastReceiver 类，实现自己的类，重写父类 BroadcastReceiver 中的

onReceive()方法。

(2) 在 AndroidManifest. xml 文件中为应用程序添加需要的权限。

(3) 在 AndroidManifest. xml 文件中或者程序代码中注册 BroadcastReceiver 对象。

(4) 等待接收广播,然后匹配。

第一种方式在 AndroidManifest. xml 文件中注册的实现过程如下:

(1) 在 AndroidManifest. xml 文件中注册一个 BroadcastReceiver,在<application>根节点标签下添加<receiver>标签和为应用程序添加需要的权限。

```
1.  <receiver android:name=".MyBroadcastReceiver">
2.          <intent-filter android:priority="1000">
3.              <action android:name=" android.provider.Telephony.SMS_RECEIVED"/>
4.          </ intent-filter>
5.      </ receiver>
6.  <uses-permission android:name="android.permission.RECEIVE_SMS"/>
        //添加权限
7.  <uses-permission android:name="android.permission.SEND_SMS"/>
```

(2) 在程序代码中调用 BroadcastReceiver 的 onReceive()方法。

```
1.   public class MyBroadcastReceiver extends BroadcastReceiver {
2.     //action 名称
3.     String SMS_RECEIVED="android.provider.Telephony.SMS_RECEIVED";
4.       public void onReceive(Context context, Intent intent) {
5.        if (intent.getAction().equals(SMS_RECEIVED)) {
6.            //相关处理:地域变换、电量不足、来电来信
7.        }
8.     }
9.  }
```

第二种方式在代码中注册的实现过程如下:

(1) 在程序代码中使用 registerReceiver 方法注册。

```
1.   IntentFilter intentFilter=new IntentFilter("android.provider.Telephony.
     SMS_RECEIVED ");
2.   registerReceiver(mBatteryInfoReceiver,intentFilter);
```

(2) 在程序代码中调用 BroadcastReceiver 重写的 onReceive()方法。

```
1.  private BroadcastReceiver myBroadcastReceiver=new BroadcastReceiver() {
2.          @Override
3.          public void onReceive(Context context, Intent intent) {
4.          //相关处理,如收短信,监听电量变化信息
5.          }
6.      };
```

(3) 广播注销

```
1.          //代码中注销广播
2.     unregisterReceiver(mBatteryInfoReceiver);
```

注意：在 Activity 中代码注销广播通常在 onPause()中注销。不在 Activity. onSaveInstanceState()中注销是因为这个方法是用来保存 Intent 状态的。

另外，因为 BroadcastReceiver 的生命周期很短，如果需要完成一项比较耗时的工作，应该通过发送 Intent 给 Service，由 Service 来完成。

3. BroadcastReceiver 广播发送方式

BroadcastReceiver 广播的发送方式有三种，分别是普通广播（Normal broadcasts）、异步广播（sendStickyBroadcast(intent)）和有序广播（Ordered broadcasts）。

- 普通广播：发送一个广播，所以监听该广播的广播接收者都可以监听到该广播。
- 异步广播：处理完之后的 Intent 依然存在，这时 registerReceiver(BroadcastReceiver, IntentFilter) 还能收到它的值，直到把它去掉。不能将处理结果传给下一个接收者，无法终止广播。
- 有序广播（Ordered broadcasts）：按照接收者的优先级顺序接收广播，优先级别在 intent-filter 中的 priority 中声明，范围在－1000～1000 之间，值越大，优先级越高。可以终止广播意图的继续传播，接收者可以修改其内容。

由于篇幅原因，此处不再详述，可参考相关文档详细了解它的收发及应用。

8.2.3 Broadcast Receiver 应用案例

上面介绍了 BroadcastReceiver 监听方式和使用方法，下面通过一个 SMS(短信，Short Message Service)案例，通过 Emulator Control 向模拟器发送短信，模拟器收到短信将会提示，详细介绍 BroadcastReceiver 的应用。

（1）创建一个新工程名为 NotificationDemo 的 Android 工程，目标 API 选择 10(即 Android 2. 3. 3 版本)，应用程序名为 NotificationDemo，包名为 com. hisoft. activity，创建的 Activity 的名字为 MainActivity，最小 SDK 版本根据选择的目标 API 会自动添加为 10，创建项目工程如图 8-7 所示。

（2）修改 res 目录下 layout 文件夹中的 main. xml 文件，设置线性布局，添加一个 TextView 控件描述，并设置相关属性，代码如下：

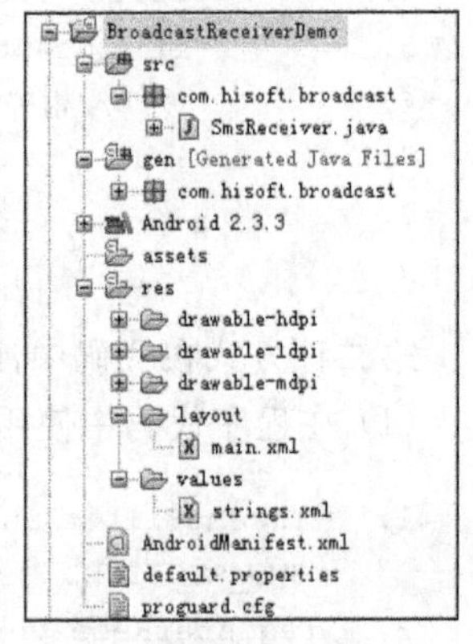

图 8-7 NotificationDemo 工程目录结构

```
1.   <?xml version="1.0" encoding="utf-8"?>
2.   <LinearLayout xmlns:android="http://schemas.android.com/apk/res/android"
3.       android:orientation="vertical"
4.       android:layout_width="fill_parent"
5.       android:layout_height="fill_parent"
6.       >
7.   <TextView
8.       android:layout_width="fill_parent"
9.       android:layout_height="wrap_content"
10.      android:text="@string/hello"
11.      />
```

```
12.  </LinearLayout>
```

(3) 修改 src 目录中 com.hisoft.activity 包下的 MainActivity.java 文件,代码如下:

```
1.   package com.hisoft.broadcast;
2.
3.   import android.content.BroadcastReceiver;
4.   import android.content.Context;
5.   import android.content.Intent;
6.   import android.os.Bundle;
7.   import android.telephony.SmsMessage;
8.   import android.widget.Toast;
9.
10.
11.  public class SmsReceiver extends BroadcastReceiver
12.  {
13.    //当接收到短信时被触发
14.     @Override
15.     public void onReceive(Context context, Intent intent)
16.     {
17.       //如果是接收到短信
18.       if (intent.getAction().equals(
19.         "android.provider.Telephony.SMS_RECEIVED"))
20.       {
21.         //abortBroadcast()方法是取消广播,将会让系统收不到短信(如果不写或者注释
          //掉,系统状态栏会有收到短信息提示,短信息收件箱会收到发送的短信息)
22.         //abortBroadcast();
23.         StringBuilder sb=new StringBuilder();
24.         //接收由 SMS 传过来的数据
25.         Bundle bundle=intent.getExtras();
26.         //判断是否有数据
27.         if (bundle !=null)
28.         {
29.             //通过 pdus 可以获得接收到的所有短信消息
30.             Object[] pdus= (Object[]) bundle.get("pdus");
31.             //构建短信对象 array,并依据收到的对象长度来创建 array 的大小
32.             SmsMessage[] messages=new SmsMessage[pdus.length];
33.             for (int i=0; i<pdus.length; i++)
34.             {
35.               messages[i]=SmsMessage
36.                 .createFromPdu((byte[]) pdus[i]);
37.             }
38.             //将送来的短信合并自定义信息于 StringBuilder 当中
39.             for (SmsMessage message: messages)
40.             {
41.               sb.append("短信来源:");
42.               //获得接收短信的电话号码
43.               sb.append(message.getDisplayOriginatingAddress());
```

```
44.                sb.append("\n------短信内容------\n");
45.                //获得短信的内容
46.                sb.append(message.getDisplayMessageBody());
47.              }
48.          }
49.          Toast.makeText(context, sb.toString()
50.              , Toast.LENGTH_LONG).show();
51.      }
52.    }
53.  }
```

(4) 修改 AndroidManifest. xml 文件，在 application 根节点下添加配置 SMSReceiver 类，代码如下：

```
1.    <receiver android:name=".SmsReceiver">
2.      <intent-filter android:priority="800">
3.        <action android:name="android.provider.Telephony.SMS_RECEIVED"/>
4.      </intent-filter>
5.    </receiver>
```

同时，在 AndroidManifest. xml 文件中的 manifest 根节点下添加设置应用程序接收短信的权限，以使应用程序可以成功地接收 SMS_RECEIVED 广播，代码如下：

```
1.    <uses-permission android:name="android.permission.RECEIVE_SMS"/>
```

(5) 测试发送短信息。打开 DDMS 视图，在 Emulator Control 面板中的 Telephone Actions 选项区域中选择 SMS 单选按钮，然后在 Incoming number 文本框中输入接收短信息的手机号码，在 Message 列表框中输入内容，最后单击 Send 按钮发送短信息，如图 8-8 所示。

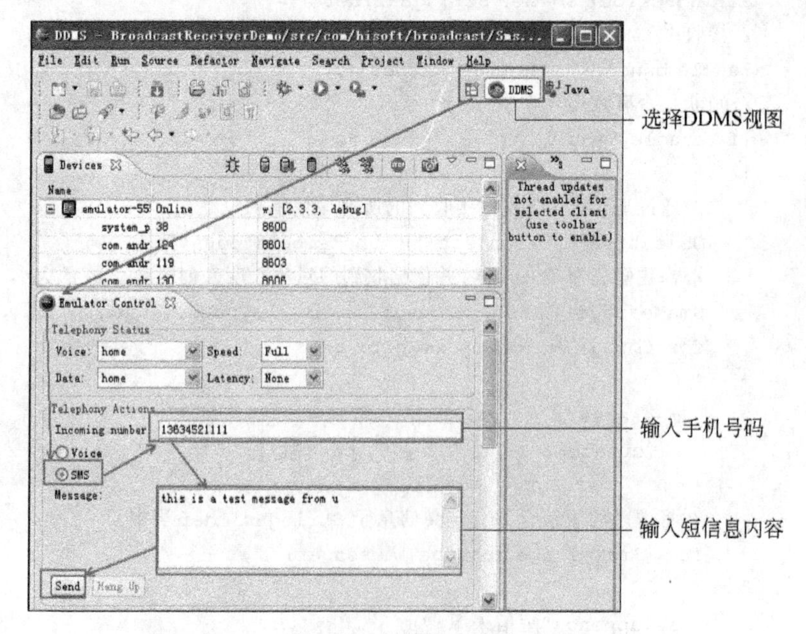

图 8-8　测试信息界面

(6) 发送短信息后，短信发送号码、内容信息出现，如图 8-9 所示。然后在界面顶部状态栏出现有新短信息提示，鼠标按住往下拖动，显示如图 8-10 所示。

(7) 选择 messaging，信息显示如图 8-11 所示。可以选中信息进行查看，也可以单击 menu 按钮，在菜单中选择“删除”等操作。

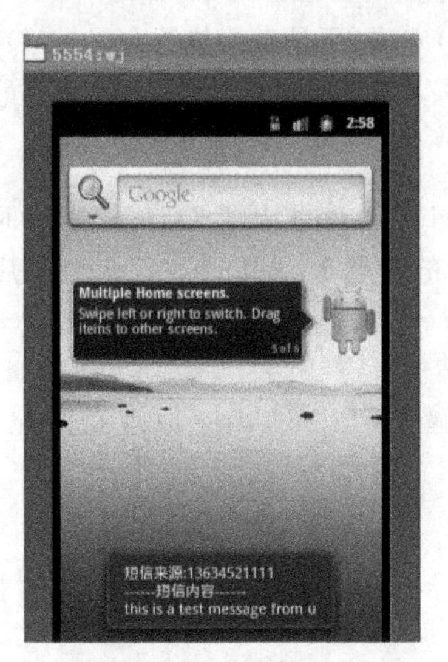

图 8-9 发送显示

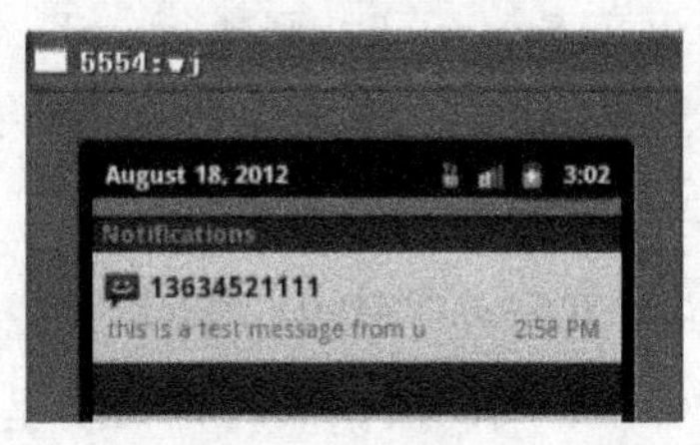

图 8-10 信息内容

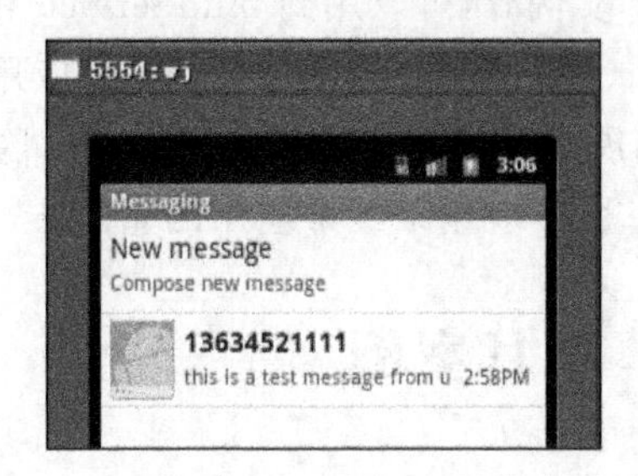

图 8-11 查看信息内容

在编写 SMSReceiver 类时需要注意如下 4 点：

(1) 接收短信的 Broadcast Action 是 android. provider. Telephony. SMS_RECEIVED，因此要在 onReceiver 方法的开始部分判断接收到的是否是接收短信的 Broadcast Action。

(2) 需要通过 Bundle. get("pdus")来获得接收到的短信消息。这个方法返回了一个表示短信内容的数组。每一个数组元素表示一条短信，这就意味着通过 Bundle. get("pdus")可以返回多条系统接收到的短信内容。

(3) 通过 Bundle. get("pdus")返回的数组一般不能直接使用，需要使用 SmsMessage. createFromPdu 方法将这些数组元素转换成 SmsMessage 对象才可以使用。每一个 SmsMessage 对象表示一条短信。

(4) 通过 SmsMessage 类的 getDisplayOriginatingAddress 方法可以获得发送短信的电话号码。通过 getDisplayMessageBody 方法可以获得短信的内容。

8.3 Service 组件服务

Service 位于 android. app 包下，其类的继承结构如图 8-12 所示。它是一个不能与用户交互的，不能自己启动的，长期运行在后台的应用组件。每一个 Service 都必须在 AndroidMainfest. xml 文件中使用 <service>标签在<application>标签根节点下进行声明。Service 的启动可以通过 Context. startService()或 Context. bindService()。

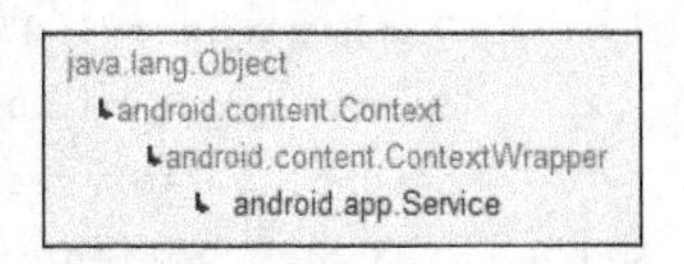

图 8-12 Service 类继承关系

1. Service 的生命周期

Android Service 的生命周期相对简单，它只继承了 onCreate()、onStart()和 onDestroy()三个方法，当第一次启动 Service 时，先后调用了 onCreate()、onStart()这两个方法；当停止 Service 时，则执行 onDestroy()方法。这里需要注意的是，如果 Service 已经启动了，当再次启动 Service 时不会再执行 onCreate()方法，而是直接执行 onStart()方法，具体的应用见后续的案例。

2. Service 与 Activity 通信

Service 后端的数据最终是需要呈现在前端 Activity 之上的，因为启动 Service 时系统会重新开启一个新的进程，这就涉及不同进程间通信的问题了(AIDL)。当想获取启动的 Service 实例时，可以用到 bindService 和 onBindService 方法，它们分别执行了 Service 中的 IBinder()和 onUnbind()方法，具体应用见后续项目案例。

拓展提示：在复杂应用程序中，隐式 Intent 的解析、IntentFilter 原理及匹配机制、Service 服务应用是经常应用的重点。

8.4 项目案例

学习目标：Android 组件 Intent 对象信息、IntentFilter 匹配机制、Service 服务等方法及应用。

案例描述：使用线性布局 LinearLayout，添加 Button 按钮设置界面布局，通过 Handler 类、handleMessage、ServiceConnection、startService、stopService、bindService、unbindService、onBind 和 onUnbind 等方法实现 Service 服务的启动、停止、绑定、解除绑定操作。

案例要点：startService、stopService、bindService、unbindService、onBind 和 onUnbind 等相关方法。

(1) 创建一个新的 Android 工程，工程名为 ServiceDemo，目标 API 选择 10(即 Android 2.3.3 版本)，应用程序名为 ServiceDemo，包名为 com.hisoft.activity，创建的 Activity 的名字为 MainActivity，最小 SDK 版本根据选择的目标 API 会自动添加为 10，创建项目工程如图 8-13 所示。

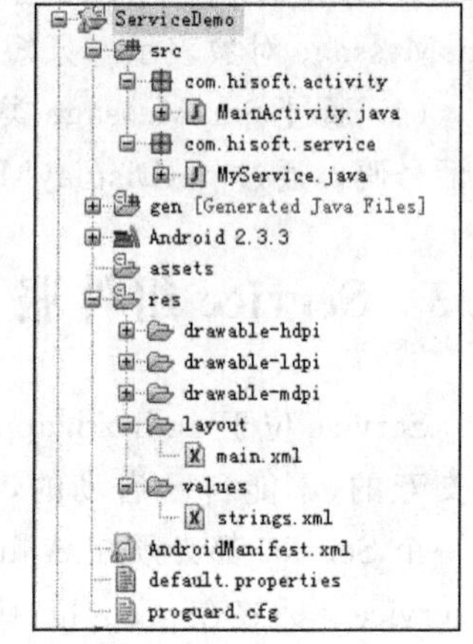

图 8-13 ServiceDemo 工程目录结构

(2) 修改 res 目录下 layout 文件夹中的 main.xml 文件，设置线性布局，添加 4 个 Button 按钮控件描述，并设置相关属性，代码如下：

```
1.    <?xml version="1.0" encoding="utf-8"?>
2.    <LinearLayout xmlns:android="http://schemas.
      android.com/apk/res/android"
3.        android:orientation="vertical"
4.        android:layout_width="fill_parent"
5.        android:layout_height="fill_parent"
```

```
6.       >
7.     <Button
8.         android:text="启动 Service"
9.         android:id="@+id/Button01"
10.        android:layout_width="fill_parent"
11.        android:layout_height="wrap_content">
12.    </Button>
13.    <Button
14.        android:text="停止 Service"
15.        android:id="@+id/Button02"
16.        android:layout_width="fill_parent"
17.        android:layout_height="wrap_content">
18.      </Button>
19.    <Button
20.        android:text="绑定 Service"
21.        android:id="@+id/Button03"
22.        android:layout_width="fill_parent"
23.        android:layout_height="wrap_content">
24.    </Button>
25.    <Button
26.        android:text="解除 Service"
27.        android:id="@+id/Button04"
28.        android:layout_width="fill_parent"
29.        android:layout_height="wrap_content">
30.      </Button>
31.  </LinearLayout>
```

(3) 修改 src 目录中 com. hisoft. activity 包下的 MainActivity. java 文件，代码如下：

```
1.   package com.hisoft.broadcast;
2.   import android.app.Activity;
3.   import android.content.ComponentName;
4.   import android.content.Intent;
5.   import android.content.ServiceConnection;
6.   import android.os.Bundle;
7.   import android.os.Handler;
8.   import android.os.IBinder;
9.   import android.os.Message;
10.  import android.view.View;
11.  import android.view.View.OnClickListener;
12.  import android.widget.Toast;
13.
14.  import com.hisoft.service.MyService;
15.
16.  public class MainActivity extends Activity {
17.    ServiceConnection sc;
```

```
18.        OnClickListener listener;
19.      Handler hd=new Handler()
20.      {
21.        @Override
22.        public void handleMessage(Message msg)
23.        {
24.          switch(msg.what)
25.          {
26.            case 0:
27.              Toast.makeText(
28.                  MainActivity.this,
29.                  "调用 Service 的 onCreate 和 onBind 方法",
30.                  Toast.LENGTH_SHORT).show();
31.            break;
32.            case 1:
33.              Toast.makeText(
34.                  MainActivity.this,
35.                  "调用 Service 的 onUnbind 和 onDestroy 方法",
36.                  Toast.LENGTH_SHORT).show();
37.            break;
38.            case 2:
39.              Toast.makeText(
40.                  MainActivity.this,
41.                  "调用 Service 的 onDestroy 方法",
42.                  Toast.LENGTH_SHORT).show();
43.            break;
44.            case 3:
45.              Toast.makeText(
46.                  MainActivity.this,
47.                  "调用 Service 的 onCreate 方法",
48.                  Toast.LENGTH_SHORT).show();
49.            break;
50.          }
51.        }
52.      };
53.        @Override
54.        public void onCreate(Bundle savedInstanceState) {
55.            super.onCreate(savedInstanceState);
56.            setContentView(R.layout.main);
57.            sc=new ServiceConnection()
58.            {
59.              @Override
60.              public void onServiceConnected(ComponentName name, IBinder service) {
61.                //TODO Auto-generated method stub
```

```
62.
63.       }
64.
65.       @Override
66.       public void onServiceDisconnected(ComponentName name) {
67.         //TODO Auto-generated method stub
68.
69.         }
70.       };
71.
72.     listener=new OnClickListener()
73.     {
74.       @Override
75.       public void onClick(View v) {
76.         Intent intent=new Intent(MainActivity.this,MyService.class);
77.         switch(v.getId())
78.         {
79.           case R.id.Button01:         //Start Service
80.             startService(intent);
81.             hd.sendEmptyMessage(3);
82.           break;
83.           case R.id.Button02://Stop Service
84.             stopService(intent);
85.             hd.sendEmptyMessage(2);
86.           break;
87.           case R.id.Button03://Bind Service
88.             bindService(intent,sc,BIND_AUTO_CREATE);
89.             hd.sendEmptyMessage(0);
90.           break;
91.           case R.id.Button04://Unbind Service
92.             unbindService(sc);
93.             hd.sendEmptyMessage(1);
94.           break;
95.         }
96.       }
97.     };
98.
99.         this.findViewById(R.id.Button01).setOnClickListener(listener);
100.        this.findViewById(R.id.Button02).setOnClickListener(listener);
101.        this.findViewById(R.id.Button03).setOnClickListener(listener);
102.        this.findViewById(R.id.Button04).setOnClickListener(listener);
103.    }
104. }
```

(4) 在 src 目录下创建 com. hisoft. service 包的 MyService. java 文件,代码如下:

```
1.  package com.hisoft.service;
2.  import android.app.Service;
3.  import android.content.Intent;
4.  import android.os.IBinder;
5.  import android.util.Log;
6.  public class MyService extends Service{
7.
8.    @Override
9.    public IBinder onBind(Intent arg0) {
10.
11.     Log.d("MyService", "=========onBind=========");
12.     return null;
13.   }
14.
15.   @Override
16.   public boolean onUnbind(Intent arg0)
17.   {
18.     Log.d("MyService", "=========onUnbind=========");
19.     return super.onUnbind(arg0);
20.   }
21.
22.   @Override
23.   public void onRebind(Intent arg0)
24.   {
25.     super.onRebind(arg0);
26.     Log.d("MyService", "=========onRebind=========");
27.   }
28.
29.   @Override
30.   public void onCreate()
31.   {
32.     super.onCreate();
33.     Log.d("MyService", "=========onCreate=========");
34.   }
35.
36.   @Override
37.   public void onDestroy()
38.   {
39.     super.onDestroy();
40.     Log.d("MyService", "=========onDestroy=========");
41.   }
42.
43. }
```

（5）部署运行结果如图 8-14 所示。

单击“启动 Service”按钮，程序运行结果如图 8-15 所示。Toast 显示“调用 Service 的 onCreate 方法”，并在后台日志中显示图 8-16 所示内容。

图 8-14　ServiceDemo 运行效果

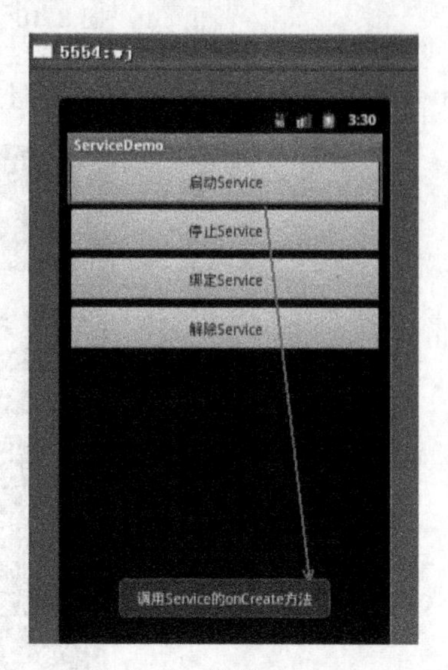

图 8-15　启动 Service

图 8-16　Service 启动后台调用显示

单击“停止 Service”按钮，程序运行结果如图 8-17 所示。Toast 显示“调用 Service 的 onDestroy 方法”，并在后台日志中显示图 8-18 所示内容。

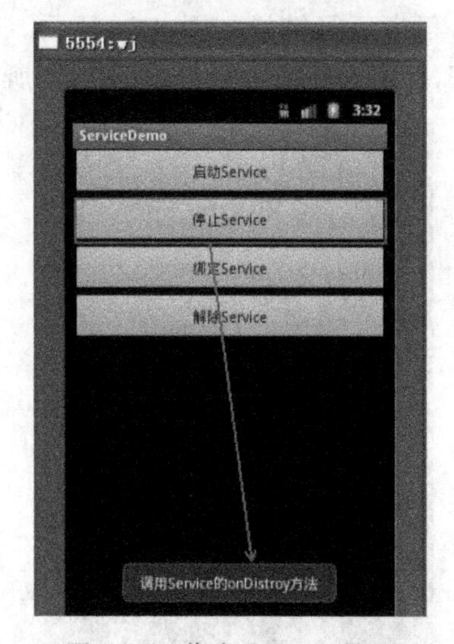

图 8-17　停止 Service 显示

单击“绑定 Service”按钮，程序运行结果如图 8-19 所示。Toast 显示“调用 Service 的

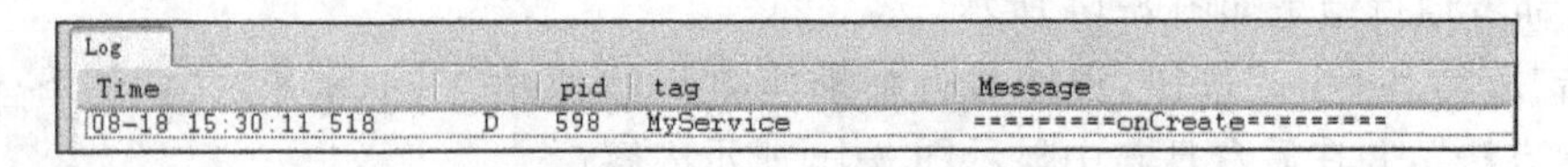

Time		pid	tag	Message
08-18 15:30:11.518	D	598	MyService	=========onCreate=========

图 8-18　停止 Service 后台显示

onCreate 和 onBind 方法”,并在后台日志中显示图 8-20 所示内容。

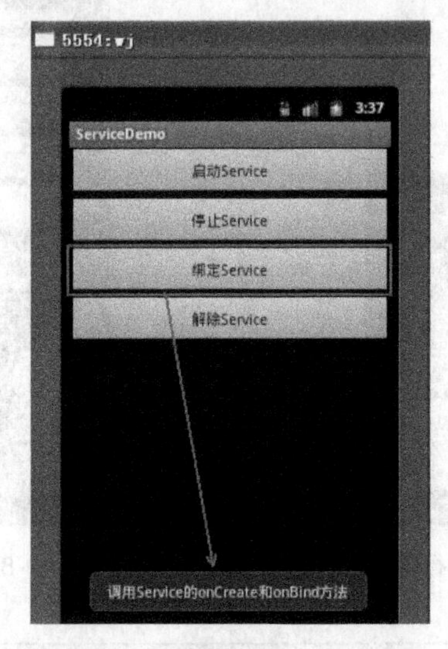

图 8-19　绑定 Service 显示

Time		pid	tag	Message
08-18 15:37:26.660	D	598	MyService	=========onCreate=========
08-18 15:37:26.660	D	598	MyService	=========onBind=========

图 8-20　绑定 Service 后台显示

单击“解除 Service”按钮,程序运行结果如图 8-21 所示。Toast 显示“调用 Service 的 onUnbind 和 onDestroy 方法”,并在后台日志中显示图 8-22 所示内容。

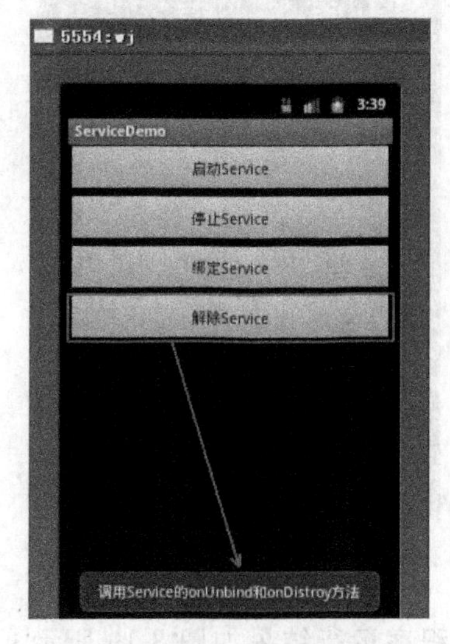

图 8-21　解除 Service 显示

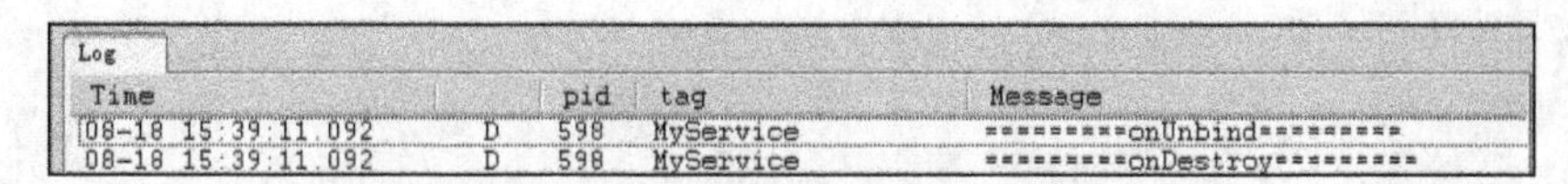

图 8-22 解除 Service 后台显示

1. 简答题

(1) Intent 是什么？在 Android 中,其主要用途有哪些？

(2) 一个 Intent 对象由几个部分组成？它们的作用分别是什么？

(3) 显式启动 Activity 的通常步骤包含哪些？

(4) 简述隐式启动 Activity 的通常用法及过程。

(5) 简述 Intent Filter 原理与匹配机制过程及步骤。

2. 完成下面的实训项目

要求：

(1) 使用 Service 实现音乐播放盒,当用户单击“播放”按钮,即使用户退出本操作界面,进行其他操作,音乐也可以继续在后台播放。当用户单击“停止”按钮,停止音乐的播放。当用户单击“暂停”按钮,暂停音乐播放。

(2) 用户可以通过菜单选项选择退出音乐播放,并在退出前给出提示信息对话框,以便用户进行确认是否退出音乐播放。

第9章 Android数据存储与访问

学习目标

本章主要介绍 Android 数据存储与访问的 SharedPreferences、文件存储 openFileOutput 和 openFileInput、SD 卡存储与访问方法、SQLite 数据库(创建、操作、管理及应用)、数据共享(Uri、UriMatcher 和 ContentUris、ContentResolver 操作数据)、网络存储应用等。使读者通过本章的学习,能够深入熟悉 Android 数据存储与访问的常用方法及途径,能够掌握以下知识要点:

(1) SharedPreferences 访问模式、访问本程序数据的通常用法。

(2) 文件存储 openFileOutput 和 openFileInput 及属性、文件操作模式设置。

(3) 创建、访问 SD 卡及应用。

(4) SQLite 数据库体系结构组成。

(5) SQLite 数据库创建、操作、管理及应用。

(6) Uri、UriMatcher 和 ContentUris。

(7) 创建 ContentProvider、ContentResolver 操作数据。

(8) 网络存储应用。

在 Android 系统中,数据存储和使用与通常的数据操作有很大的不同。首先,Android 中所有的应用程序数据都为自己应用程序所有,其他应用程序如果共享、访问别的应用程序数据,必须通过 Android 系统提供的方式才能访问或者暴露自己的私有数据供其他应用程序使用。Android 平台中实现数据存储的方式有 5 种,分别是使用 SharedPreferences 存储数据、文件存储数据、SQLite 数据库存储数据、使用 ContentProvider 存储数据和网络存储数据。

(1) SharedPreferences。

SharedPreferences 的功能类似于 Windows 系统上的 ini 配置文件,主要用于系统的配置信息的保存,比如保留界面设置的颜色、保留登录用户名等,以便下次登录时使用。

(2) 文件(Files)存储。

Android 移动操作系统是基于 Linux 核心,文件也是 Linux 形式的文件系统。文件保存在设备的内部存储器上,在 Linux 系统下的/data/data/<package name>/files 目录中。

(3) 数据库(SQLite Databases)。

在 Andriod 系统中,数据存储、管理使用的数据库是轻便型的数据库 SQLite。SQLite 是一个开源的嵌入式关系数据库,与普通关系型数据库一样,也具有 ACID 的特性。

(4) ContentProvider(数据提供者)。

ContentProvider 是在应用程序间共享数据的一种接口机制。ContentProvider 提供了更为高级的数据共享方法,应用程序可以指定需要共享的数据,而其他应用程序则可以在不知数据来源、路径的情况下,对共享数据进行查询、添加、删除和更新等操作。

(5) 网络存储。

前面介绍的几种存储都是将数据存储在本地设备上,除此之外,还有一种存储(获取)数据的方式,通过网络来实现数据的存储和获取。通过调用 WebService 返回的数据或是解析 HTTP 协议实现网络数据交互。

在 Android 系统中,按数据的共享方式可以分为应用程序内自用和数据被其他应用程序共享两种。

(1) 本应用程序内使用。

通常应用中需要的数据一般都是只能为本应用程序使用。使用 SharedPreferences、文件存储、SQLite 数据库存储方式创建的应用程序默认为本程序使用,其他程序无法获取数据操作。

在 Android 中,可以在控制台使用 adb 命令查看本程序使用的数据:

```
adb shell                        //进入手机的文件系统
cd  /data/data                   //进入目录
```

ls 查看,可以发现在系统中安装的每个应用程序在那个目录下都有一个文件夹,再次进入应用程序后,使用 ls 命令查看,会出现 shared_prefs、files、databases 几个目录,它们其实就是存放应用程序内自用的数据,内容分别由 SharedPreferences、文件存储、SQLite 数据库存储这三种方式创建。当然,如果没有创建过,这个目录可能不存在。

(2) 应用程序数据共享。

这类数据通常是一些共用数据,很多程序都会来调用,比如电话簿数据等。在 Android 系统中,由文件存储、数据库存储、SharedPreferences 创建的数据都可以通过系统提供的特定方式访问,实现数据共享。

9.1 SharedPreferences

9.1.1 SharedPreferences 简介

前面已经讲述了在 Android 系统中可以使用一个 SharedPreferences 类来保存一些系

统的配置信息、窗口的状态等。SharedPreferences 接口位于 android. content 包下,它是一个轻量级的存储类,特别适合用于保存软件配置参数。

使用 SharedPreferences 保存数据,最终是以 xml 文件存放数据,是基于 xml 文件存储键值对(Name/Value Pair,NVP)数据。xml 处理时 Dalvik 会通过自带底层的本地 XML Parser 解析,比如 XMLpull 方式。SharedPreferences 保存数据的文件存放在目录/data/data/<package name>/shared_prefs 下。SharedPreferences 不仅能够保存数据,还能够实现不同应用程序间的数据共享。

由于 SharedPreferences 完全对用户屏蔽对文件系统的操作过程,在开发中 SharedPreferences 对象本身只能获取数据而不支持存储和修改,存储修改是通过 Editor 对象实现的。

SharedPreferences 支持各种基本数据类型,包括整型、布尔型、浮点型和长型等。

1. SharedPreferences 访问模式

在 Android 系统中,SharedPreferences 分为许多权限,其支持的访问模式有三种:私有、全局读和全局写。

- 私有(Context. MODE_PRIVATE):为默认操作模式,代表该文件是私有数据,只能被应用本身访问,在该模式下写入的内容会覆盖原文件的内容。
- 全局读(Context. MODE_WORLD_READABLE):不仅创建程序可以对其进行读取或写入,其他应用程序也读取操作的权限,但没有写入操作的权限。
- 全局写(Context. MODE_WORLD_WRITEABLE):创建程序和其他程序都可以对其进行写入操作,但没有读取的权限。

2. 使用 SharedPreferences 实现存储,访问自身程序数据的通常用法如下:

(1) 定义 SharedPreferences 的访问模式。

在使用 SharedPreferences 前,先定义 SharedPreferences 的访问模式。

如将访问模式定义为私有模式:

```
public static int MODE=Context.MODE_PRIVATE;
```

也可以将 SharedPreferences 的访问模式设定为即可以全局读,也可以全局写。设定如下:

```
public static int MODE=Context.MODE_WORLD_READABLE+Context.MODE_WORLD_WRITEABLE;
```

(2) 定义 SharedPreferences 的名称。

SharedPreferences 的名称与在 Android 文件系统中保存的文件同名。因此,只要具有相同的 SharedPreferences 名称的 NVP 内容都会保存在同一个文件中,例如

```
public static final String  PR_NAME="SaveFile";
```

(3) 获取 SharedPreferences 对象。

使用 SharedPreferences,需要将上述定义的访问模式和 SharedPreferences 名称作为参数,传递到 getSharedPreferences 方法并获取到 SharedPreferences 对象。

```
SharedPreferences sharedPreferences=getSharedPreferences(PR_NAME, MODE);
```

(4) 利用 edit()方法获取 Editor 对象。

在获取到 SharedPreferences 对象后，可以通过 SharedPreferences. Editor 类对 SharedPreferences 进行修改。

```
Editor editor=sharedPreferences.edit();
```

(5) 通过 Editor 对象存储 key-value 键值对数据。

```
editor.putString("Name", "John");
editor.putInt("Age",28);
editor.putFloat("Height", 1.77);
```

(6) 通过 commit()方法提交数据。

```
editor.commit();
```

完成上述步骤后，如果需要从已经保存的 SharedPreferences 中读取数据，同样是调用 getSharedPreferences()方法，并在方法的第 1 个参数中指明需要访问的 SharedPreferences 名称，然后通过 get<Type>()方法获取保存在 SharedPreferences 中的 NVP。

```
1.    SharedPreferences sharedPreferences=getSharedPreferences(PR_NAME, MODE);
2.    String name=sharedPreferences.getString("Name","name");
3.    int age=sharedPreferences.getInt("Age", 20);
4.    float height=sharedPreferences.getFloat("Height",);
```

上述代码中，get<Type>()方法中的第 1 个参数是 NVP 的名称。

第 2 个参数是在无法获取到数值的时候使用的缺省值。如 getFloat()的第 2 个参数为缺省值，如果 preference 中不存在该 key，将返回缺省值。

3. 访问其他应用程序数据的 SharedPreferences

如果需要创建访问其他应用程序数据的 SharedPreferences，其前提条件是：

在 SharedPreferences 对象创建时，为其指定 Context. MODE_WORLD_READABLE 或者 Context. MODE_WORLD_WRITEABLE 权限。

```
Context otherApps = createPackageContext ( " com. hisoft. sharedpreferences ",
Context.CONTEXT_IGNORE_SECURITY);
SharedPreferences sharedPreferences = otherApps. getSharedPreferences ( " testApp ",
Context.MODE_WORLD_READABLE);
String name=sharedPreferences.getString("name", "");
int age=sharedPreferences.getInt("age", 1);
```

如果想采用读取 xml 文件方式，直接访问其他应用 SharedPreferences 对应的 xml 文件，代码如下：

```
File sfx=new File("/data/data/<package name>/shared_prefs/mypreferences.xml");
//<package name>应替换成应用的包名
```

4. 访问资源文件

(1) 访问存储在 res 目录下的文件，如 res/raw 目录下：

```
InputStream ismp3=getResources().openRawResource(R.raw.testVideo);  //存放声音文件
```

(2) 访问存储在 assets 目录下的文件：

```
InputStream anyFile=getAssets().open(name);                          //存放数据文件
```

注意：*存储文件的大小有限制。*

SharedPreferences 对象与后续讲解的 SQLite 数据库相比，省略了创建数据库、创建表、写 SQL 语句等诸多操作，相对而言更加方便、简洁。但是 SharedPreferences 也有其自身缺陷，比如其只能存储 boolean、int、float、long 和 String 这 5 种简单的数据类型，无法进行条件查询等。所以不论 SharedPreferences 的数据存储操作是如何简单，它也只能是存储方式的一种补充，而无法完全替代如 SQLite 数据库这样的其他数据存储方式。

9.1.2 读取应用程序数据案例

上面简单介绍了 SharedPreferences 的基础知识和存储访问应用方法，下面通过一个案例详细介绍 SharedPreferences 访问本程序数据的应用。

(1) 创建一个新的 Android 工程，工程名为 SharedPreferencesDemo，目标API 选择 10(即 Android 2.3.3 版本)，应用程序名为 SharedPreferencesDemo，包名为 com. hisoft. sharedpreferences，创建的 Activity 的名字为 MainActivity，最小 SDK 版本根据选择的目标 API 会自动添加为 10，创建项目工程如图 9-1 所示。

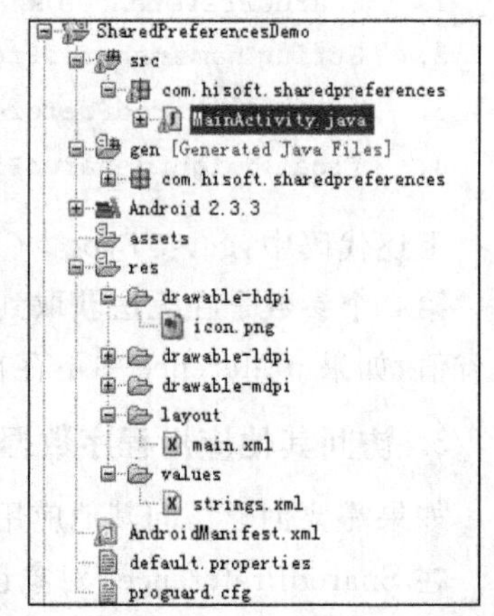

图 9-1 SharedPreferencesDemo 工程目录结构

(2) 修改 res 目录下 layout 文件夹中的 main. xml 文件，设置线性布局，添加一个 TextView 控件和两个 EditText 控件描述，并设置相关属性，代码如下：

```
1.    <?xml version="1.0" encoding="utf-8"?>
2.    <LinearLayout xmlns:android="http://schemas.
      android.com/apk/res/android"
3.        android:orientation="vertical"
4.        android:layout_width="fill_parent"
5.        android:layout_height="fill_parent"
6.        >
7.    <TextView
8.        android:layout_width="fill_parent"
9.        android:layout_height="wrap_content"
10.       android:text="@string/inputname"
11.       />
12.   <EditText android:layout_width="match_parent"
13.         android:layout_height="wrap_content"
14.         android:id="@+id/username">
```

```
15.         <requestFocus></requestFocus>
16.     </EditText>
17.  </LinearLayout>
```

(3) 修改 res 目录下 values 文件夹中的 strings. xml 文件,代码如下:

```
1.  <?xml version="1.0" encoding="utf-8"?>
2.  <resources>
3.      <string name="hello">Hello World, MainActivity!</string>
4.      <string name="app_name">SharedPreferencesDemo</string>
5.      <string name="inputname">请输入用户名:</string>
6.  </resources>
```

(4) 修改 src 目录中 com. hisoft. sharedpreferences 包下的 MainActivity. java 文件,代码如下:

```
1.   package com.hisoft.sharedpreferences;
2.   import com.hisoft.sharedpreferences.R;
3.   import android.app.Activity;
4.   import android.content.SharedPreferences;
5.   import android.content.SharedPreferences.Editor;
6.   import android.os.Bundle;
7.   import android.view.Menu;
8.   import android.view.MenuItem;
9.   import android.widget.EditText;
10.
11.  public class MainActivity extends Activity {
12.
13.     private EditText et_name;
14.     private static final String NAME="name";
15.     private static final int EXIT=1;
16.
17.       /** Called when the activity is first created. */
18.       @Override
19.       public void onCreate(Bundle savedInstanceState) {
20.           super.onCreate(savedInstanceState);
21.           setContentView(R.layout.main);
22.
23.           this.et_name=(EditText) this.findViewById(R.id.username);
24.
25.           SharedPreferences sp=this.getSharedPreferences("mypreference",
              MODE_WORLD_READABLE);
26.           String username=sp.getString(NAME, "");
27.
28.           this.et_name.setText(username);
29.
30.       }
```

```
31.
32.        @Override
33.        protected void onDestroy() {
34.          super.onDestroy();
35.          SharedPreferences sp=this.getSharedPreferences("mypreference", MODE_
             WORLD_READABLE);
36.          SharedPreferences.Editor edit=sp.edit().putString(NAME, this.et_
             name.getText().toString());
37.          edit.commit();
38.        }
39.
40.      @Override
41.      public boolean onCreateOptionsMenu(Menu menu) {
42.        menu.add(0, EXIT, 0, "退出程序");
43.        return true;
44.      }
45.
46.      @Override
47.      public boolean onOptionsItemSelected(MenuItem item) {
48.
49.        if(item.getItemId()==EXIT){
50.          this.finish();
51.        }

52.        return super.onOptionsItemSelected(item);
53.      }
54.    }
```

(5) 部署运行程序,SharedPreferencesDemo 工程运行结果如图 9-2 所示。

输入用户名“张学友”,下次程序启动后会自动读取用户名在编辑框,如图 9-3 所示。

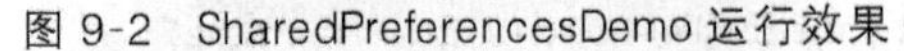

图 9-2 SharedPreferencesDemo 运行效果

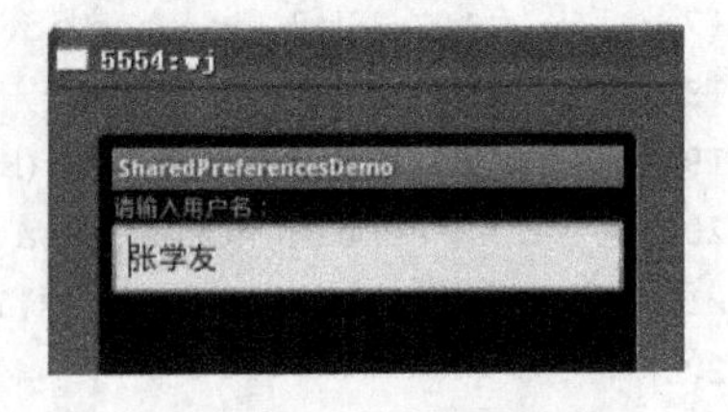

图 9-3 自动读取用户名

数据存储在路径 data/data/com. hisoft. sharepreferences/shared_prefs/目录下,通过选择 Eclipse 菜单中的 Window→Show View→Other 菜单项,在对话框中展开 android 文件夹,选择下面的 File Explorer 视图,然后在 File Explorer 视图中展开,如图 9-4 所示,名称为 mypreferences. xml 文件,单击 Pull a file from a device 按钮导出文件,文件内容如下面代码所示。

com.hisoft.sharedpreferences		2012-08-22 03:30	drwxr-x--x
lib		2012-08-22 03:30	drwxr-xr-x
shared_prefs		2012-08-22 03:44	drwxrwx--x
mypreference.xml	109	2012-08-22 03:44	-rw-rw-r--

图 9-4　数据存储路径

```
1.   <?xml version='1.0' encoding='utf-8' standalone='yes' ?>
2.   <map>
3.   <string name="name">张学友</string>
4.   </map>
```

9.1.3　读取其他应用程序数据案例

上述简单介绍了 SharedPreferences 访问其他应用程序数据的条件和方法，下面通过一个案例详细介绍 SharedPreferences 访问其他应用程序数据的应用。

(1) 创建一个新的 Android 工程，工程名为 OtherSharedPreferencesDemo，目标 API 选择 10（即 Android 2.3.3 版本），应用程序名为 OtherSharedPreferencesDemo，包名为 com.hisoft.activity，创建的 Activity 的名字为 MainActivity，最小 SDK 版本根据选择的目标 API 会自动添加为 10，创建项目工程如图 9-5 所示。

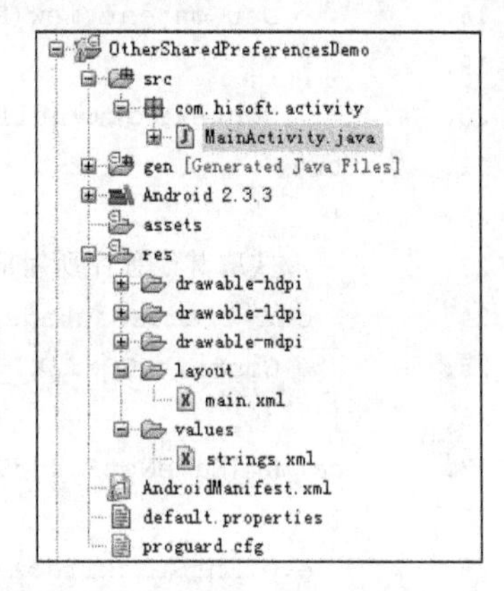

图 9-5　OtherSharedPreferencesDemo 工程目录结构

(2) 修改 res 目录下 layout 文件夹中的 main.xml 文件，设置线性布局，添加一个 TextView 控件描述，并设置相关属性，代码如下：

```
1.   <?xml version="1.0" encoding="utf-8"?>
2.   <LinearLayout xmlns:android="http://schemas.android.com/apk/res/android"
3.       android:orientation="vertical"
4.       android:layout_width="fill_parent"
5.       android:layout_height="fill_parent"
6.       >
7.   <TextView  android:id="@+id/textview1"
8.       android:layout_width="fill_parent"
9.       android:layout_height="wrap_content"
10.      android:text=""
11.      />
12.  </LinearLayout>
```

(3) 修改 src 目录中 com.hisoft.activity 包下的 MainActivity.java 文件，代码如下：

```
1.   package com.hisoft.activity;
2.
3.   import android.app.Activity;
4.   import android.content.Context;
```

```
5.  import android.content.SharedPreferences;
6.  import android.content.pm.PackageManager.NameNotFoundException;
7.  import android.os.Bundle;
8.  import android.widget.TextView;
9.
10. public class MainActivity extends Activity {
11.
12.   private TextView tv;
13.
14.     /**Called when the activity is first created. * /
15.     @Override
16.     public void onCreate(Bundle savedInstanceState) {
17.         super.onCreate(savedInstanceState);
18.         setContentView(R.layout.main);
19.
20.         Context ctx=null;
21.     try
22.     {
23.       //获取其他程序所对应的 Context
24.       ctx=createPackageContext("com.hisoft.sharedpreferences",
25.         Context.CONTEXT_IGNORE_SECURITY);
26.     }
27.     catch (NameNotFoundException e)
28.     {
29.       e.printStackTrace();
30.     }
31.     //使用其他程序的 Context 获取对应的 SharedPreferences
32.     SharedPreferences prefs=ctx.getSharedPreferences("mypreference",
33.       Context.MODE_WORLD_READABLE);
34.     //读取数据
35.     String name=prefs.getString("name", "");
36.     this.tv=(TextView) findViewById(R.id.textview1);
37.     //显示读取的数据内容
38.     this.tv.setText("被其他应用程序写入的 name
        的值:"+name);
39.     }
40. }
```

(4) 部署运行 OtherSharedPreferencesDemo 工程，读取上一案例存储的 name 值，程序运行结果如图 9-6 所示。

图 9-6　程序运行结果

9.2　文件存储

Android 系统使用的是基于 Linux 的文件系统，应用程序开发人员可以建立和访问程序自身的私有文件，也可以访问保存在资源目录中的原始文件和 XML 文件。此外，还可以在 SD 卡等外部存储设备中保存文件信息等。

9.2.1 文件存储简介

在 Android 系统中，允许应用程序创建仅能够自身访问的私有文件，文件保存在设备的内部存储器上，文件默认保存路径位置是/data/data/<package name>/ files/目录下。

Android 系统不仅支持标准 Java 的 IO 类和方法，还提供了能够简化读写流式文件过程的方法。关于文件存储，Activity 提供了 openFileOutput()方法和 openFileInput()方法。

openFileOutput()方法可以用于把数据输出到文件中。

openFileInput()方法为打开应用程序私有文件读取数据。

具体的实现过程与在 J2SE 环境中保存数据到文件中是一样的。文件可用来存放大量数据，如文本、图片和音频等。

下述是 openFileOutput()方法和 openFileInput()方法的用法。

(1) openFileOutput()方法的用法。

openFileOutput()方法为打开应用程序私有文件写入数据，如果指定的文件不存在，则创建一个新的文件。

openFileOutput()方法的语法声明：

```
public FileOutputStream openFileOutput(String name,int mode)
```

第 1 个参数是文件名称，这个参数不能包含路径分隔符“/”。

第 2 个参数是文件操作模式。

使用 openFileOutput()方法创建新文件，代码如下：

```
1.   String  NAME="test.txt";        //定义了创建文件的名称 test.txt
2.   FileOutputStream fos=openFileOutput(NAME,Context.MODE_PRIVATE);
                                     //使用 openFileOutput()方法以私有模式建立文件
3.   String ts="This is a test data";
4.   fos.write(ts.getBytes());       //将数据写入文件
5.   fos.flush();                    //将缓存中所有剩余的数据写入文件
6.   fos.close();                    //关闭流
```

(2) openFileInput()方法的用法。

如果要打开存放在/data/data/<package name>/files 目录下的应用程序私有的文件，可以使用 openFileInput()方法。

openFileInput()方法的语法声明：

```
public FileInputStream openFileInput (String name)
```

方法参数也是文件名称，字符串中不能包含路径分隔符“/”。

使用 openFileInput ()方法打开已有文件，代码如下：

```
7.   FileInputStream inStream=this.getContext().openFileInput("test.txt");
8.   try {
9.     ByteArrayOutputStream outStream=
```

```
10.         new ByteArrayOutputStream();
11.      byte[] buffer=new byte[1024];
12.      int length=-1;
13.      while((length=inStream.read(buffer)) !=-1){
14.         outStream.write(buffer, 0, length);
15.      }
16.      outStream.close();
17.      inStream.close();
18.      return outStream.toString();
19.   } catch (IOException e) {
20.      Log.i("FileTest", e.getMessage());
21.   }
```

如果想直接使用文件的绝对路径,可以使用如下代码:

```
1.    File file=new File("/data/data/ com.hisoft.activity/files/test.txt");
2.    FileInputStream inStream=new FileInputStream(file);
```

上面第 1 行文件路径中的 com. hisoft. activity 为应用所在包。

对于私有文件只能被创建该文件的应用访问。如果希望文件能被其他应用读和写,可以在创建文件时指定 Context. MODE_WORLD_READABLE 和 Context. MODE_WORLD_WRITEABLE 权限。

Activity 还提供了 getCacheDir()和 getFilesDir()方法。

getCacheDir()方法用于获取/data/data/<package name>/cache 目录。

getFilesDir()方法用于获取/data/data/<package name>/files 目录。

注意:

(1) openFileOutput()和 openFileInput()方法使用时必须使用 try{} catch{}捕获异常。

(2) 创建的文件保存在/data/data/<package name>/files 目录下,如/data/data/com. hisoft. activity/files/test. txt。

同上节讲述的一样,通过 File Explorer 视图,在 File Explorer 视图中展开/data/data/<package name>/files 目录即可看到该文件。Android 系统支持 4 种文件操作模式,如表 9-1 所示。

表 9-1　Android 系统支持的 4 种文件操作模式

文件操作模式	值	描　述
MODE_PRIVATE	0	私有模式,缺陷模式,文件仅能够被文件创建程序访问,或具有相同 UID 的程序访问。为默认操作模式,代表该文件是私有数据,只能被应用本身访问。在该模式下写入的内容会覆盖原文件的内容,如果想把新写入的内容追加到原文件中,可以使用 Context. MODE_APPEND
MODE_APPEND	32 768	追加模式,模式会检查文件是否存在,存在就往文件追加内容,否则就创建新文件
MODE_WORLD_READABLE	1	全局读模式,允许任何程序读取私有文件
MODE_WORLD_WRITEABLE	2	全局写模式,允许任何程序写入私有文件

注意：在使用上述模式时，可以用“＋”来选择多种模式，比如 openFileOutput(FILENAME, Context.MODE_PRIVATE ＋ MODE_WORLD_READABLE)；。

9.2.2 文件存储应用案例

上节介绍了文件存储访问方式及访问方法，下面通过一个文件存储案例详细介绍访问 File 的应用。

(1) 创建一个新的 Android 工程，工程名为 FileWriteAndReadDemo，目标 API 选择 10（即 Android 2.3.3 版本），应用程序名为 FileWriteAndReadDemo，包名为 com.hisoft，创建的 Activity 的名字为 MainActivity，最小 SDK 版本根据选择的目标 API 会自动添加为 10，创建项目工程如图 9-7 所示。

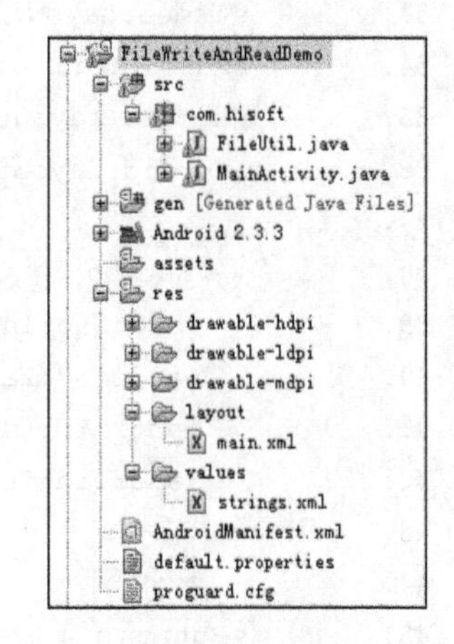

图 9-7 FileWriteAndReadDemo 工程目录结构

(2) 修改 res 目录下 layout 文件夹中的 main.xml 文件，添加 EditText、TextView 和 Button 控件，代码如下：

```
1.   <?xml version="1.0" encoding="utf-8"?>
2.     <LinearLayout xmlns:android="http://schemas.
       android.com/apk/res/android"
3.         android:orientation="vertical"
4.         android:layout_width="fill_parent"
5.         android:layout_height="fill_parent"
6.         >
7.
8.      <TextView
9.           android:layout_width="fill_parent"
10.          android:layout_height="wrap_content"
11.          android:text="@string/filename"
12.        />
13.
14.        <EditText
15.          android:layout_width="fill_parent"
16.          android:layout_height="wrap_content"
17.          android:id="@+id/filename"
18.        />
19.
20.        <TextView
21.          android:layout_width="fill_parent"
22.          android:layout_height="wrap_content"
23.          android:text="@string/content"
24.        />
25.
26.        <EditText
27.          android:layout_width="fill_parent"
28.          android:layout_height="wrap_content"
```

```
29.        android:minLines="3"
30.        android:id="@+id/content"
31.      />
32.
33.      <LinearLayout
34.      android:orientation="horizontal"
35.      android:layout_width="fill_parent"
36.      android:layout_height="fill_parent">
37.
38.        <Button
39.          android:layout_width="wrap_content"
40.          android:layout_height="wrap_content"
41.          android:id="@+id/button"
42.          android:text="@string/save"
43.        />
44.
45.      <Button
46.          android:layout_width="wrap_content"
47.          android:layout_height="wrap_content"
48.          android:id="@+id/read"
49.          android:text="@string/read"
50.        />
51.      </LinearLayout>
52.    </LinearLayout>
```

(3) 在 src 目录中的 com. hisoft 包下创建 FileUtil. java 文件，代码如下：

```
1.   package com.hisoft;
2.   import java.io.ByteArrayOutputStream;
3.   import java.io.FileInputStream;
4.   import java.io.FileOutputStream;
5.
6.   import android.content.Context;
7.   import android.util.Log;
8.
9.   /**
10.    * 文件保存与读取功能实现类
11.    * @author Administrator
12.    */
13.  public class FileUtil{
14.
15.    public static final String TAG="FileService";
16.    private Context context;
17.
18.    //得到传入的上下文对象的引用
19.    public FileUtil(Context context) {
```

```
20.        this.context=context;
21.    }
22.    public FileUtil(){
23.
24.    }
25.
26.    /**
27.      * 保存文件
28.      *
29.      * @param fileName 文件名
30.      * @param content   文件内容
31.      * @throws Exception
32.      * /
33.    public void save(String fileName, String content) throws Exception {
34.
35.      //由于页面输入的都是文本信息,因此当文件名不是以.txt后缀名结尾时,自动加
         //上.txt后缀
36.      if (!fileName.endsWith(".txt")) {
37.        fileName=fileName+".txt";
38.      }
39.
40.      byte[] buf=fileName.getBytes("iso8859-1");
41.
42.      Log.e(TAG, new String(buf,"utf-8"));
43.
44.      fileName=new String(buf,"utf-8");
45.
46.      Log.e(TAG, fileName);
47.
48.      //如果希望文件被其他应用读和写,可以传入
49.      //openFileOutput("output.txt", Context.MODE_WORLD_READABLE+Context.
           MODE_WORLD_WRITEABLE);
50.
51.       FileOutputStream fos=context.openFileOutput(fileName, context.MODE_
          PRIVATE);
52.      fos.write(content.getBytes());
53.      fos.close();
54.    }
55.
56.    /**
57.      * 读取文件内容
58.      *
59.      * @param fileName 文件名
60.      * @return 文件内容
61.      * @throws Exception
```

```
62.       * /
63.     public String read(String fileName) throws Exception {
64.
65.   //由于页面输入的都是文本信息,因此当文件名不是以.txt后缀名结尾时,自动加上.txt
      //后缀
66.       if (!fileName.endsWith(".txt")) {
67.         fileName=fileName+".txt";
68.       }
69.
70.       FileInputStream fis=context.openFileInput(fileName);
71.       ByteArrayOutputStream baos=new ByteArrayOutputStream();
72.
73.       byte[] buf=new byte[1024];
74.       int len=0;
75.
76.       //将读取后的数据放置在内存中——ByteArrayOutputStream
77.       while ((len=fis.read(buf)) !=-1) {
78.         baos.write(buf, 0, len);
79.       }
80.
81.       fis.close();
82.       baos.close();
83.
84.       //返回内存中存储的数据
85.       return baos.toString();
86.
87.     }
88.
89. }
```

(4) 修改 src 目录下包 com. hisoft 中的 MainActivity. java 文件,代码如下:

```
1.  package com.hisoft;
2.  import android.app.Activity;
3.  import android.os.Bundle;
4.  import android.util.Log;
5.  import android.view.View;
6.  import android.widget.Button;
7.  import android.widget.EditText;
8.  import android.widget.Toast;
9.  public class MainActivity extends Activity {
10.   /**Called when the activity is first created. * /
11.
12.   //得到 FileUtil 对象
13.   private FileUtil fileService=new FileUtil(this);
14.   //定义视图中的 filename 输入框对象
```

```
15.     private EditText fileNameText;
16.     //定义视图中的 contentText 输入框对象
17.     private EditText contentText;
18.     //定义一个 Toast 提示对象
19.     private Toast toast;
20.       @Override
21.       public void onCreate(Bundle savedInstanceState) {
22.     super.onCreate(savedInstanceState);
23.     setContentView(R.layout.main);
24.
25.     //得到视图中的两个输入框和两个按钮的对象引用
26.     Button button=(Button)this.findViewById(R.id.button);
27.     Button read=(Button)this.findViewById(R.id.read);
28.     fileNameText=(EditText) this.findViewById(R.id.filename);
29.     contentText=(EditText) this.findViewById(R.id.content);
30.
31.     //为保存按钮添加保存事件
32.     button.setOnClickListener(new View.OnClickListener() {
33.         @Override
34.         public void onClick(View v) {
35.
36.           String fileName=fileNameText.getText().toString();
37.           String content=contentText.getText().toString();
38.
39.           //当文件名为空的时候,提示用户文件名为空,并记录日志
40.           if(isEmpty(fileName)) {
41.             toast=Toast.makeText(MainActivity.this, R.string.empty_filename,
                Toast.LENGTH_LONG);
42.             toast.setMargin(RESULT_CANCELED, 0.345f);
43.             toast.show();
44.             Log.w(fileService.TAG, "The file name is empty");
45.             return;
46.           }
47.
48.           //当文件内容为空的时候,提示用户文件内容为空,并记录日志
49.           if(isEmpty(content)) {
50.             toast=Toast.makeText(MainActivity.this, R.string.empty_content,
                Toast.LENGTH_LONG);
51.             toast.setMargin(RESULT_CANCELED, 0.345f);
52.             toast.show();
53.             Log.w(fileService.TAG, "The file content is empty");
54.             return;
55.           }
56.
57.           //当文件名和内容都不为空的时候,调用 fileService 的 save 方法
```

```
58.            //当成功执行的时候,提示用户保存成功,并记录日志
59.            //当出现异常的时候,提示用户保存失败,并记录日志
60.            try {
61.              fileService.save(fileName, content);
62.              toast=Toast.makeText(MainActivity.this, R.string.success, Toast.
               LENGTH_LONG);
63.              toast.setMargin(RESULT_CANCELED, 0.345f);
64.              toast.show();
65.              Log.i(fileService.TAG, "The file save successful");
66.            } catch (Exception e) {
67.               toast= Toast.makeText(MainActivity.this, R.string.fail, Toast.
                LENGTH_LONG);
68.              toast.setMargin(RESULT_CANCELED, 0.345f);
69.              toast.show();
70.              Log.e(fileService.TAG, "The file save failed");
71.            }
72.
73.        }
74.    });
75.
76.
77.    //为读取按钮添加读取事件
78.    read.setOnClickListener(new View.OnClickListener() {
79.        @Override
80.        public void onClick(View v) {
81.
82.            //得到文件名输入框中的值
83.            String fileName=fileNameText.getText().toString();
84.
85.            //如果文件名为空,则提示用户输入文件名,并记录日志
86.            if(isEmpty(fileName)) {
87.              toast=Toast.makeText(MainActivity.this, R.string.empty_filename,
               Toast.LENGTH_LONG);
88.              toast.setMargin(RESULT_CANCELED, 0.345f);
89.              toast.show();
90.              Log.w(fileService.TAG, "The file name is empty");
91.              return;
92.            }
93.
94.    //调用 fileService 的 read 方法,并将读取出来的内容放入到文本内容输入框里面
95.        //如果成功执行,提示用户读取成功,并记录日志
96.        //如果出现异常信息(例如文件不存在),提示用户读取失败,并记录日志
97.            try {
98.              contentText.setText(fileService.read(fileName));
```

```
99.             toast=Toast.makeText(MainActivity.this, R.string.read_success,
                Toast.LENGTH_LONG);
100.          toast.setMargin(RESULT_CANCELED, 0.345f);
101.          toast.show();
102.          Log.i(fileService.TAG, "The file read successful");
103.        } catch (Exception e) {
104.          toast=Toast.makeText(MainActivity.this, R.string.read_fail, Toast.
              LENGTH_LONG);
105.          toast.setMargin(RESULT_CANCELED, 0.345f);
106.          toast.show();
107.          Log.e(fileService.TAG, "The file read failed");
108.        }
109.      }
110.    });
111.
112.
113.    }
114.
115.      //isEmpty方法,判断字符串是否为空
116.      private boolean isEmpty(String s) {
117.      if(s==null||"".equals(s.trim())) {
118.          return true;
119.      }
120.      return false;
121.      }
122.
123. }
```

(5) 修改 res 目录下 values 文件下的 strings. xml 文件,代码如下:

```
1.   <?xml version="1.0" encoding="utf-8"?>
2.   <resources>
3.       <string name="hello">Hello World, MainActivity!</string>
4.       <string name="app_name">FileWriteAndReadDemo</string>
5.       <string name="filename">文件名</string>
6.       <string name="read">读文件</string>
7.       <string name="save">保存文件</string>
8.       <string name="content">文件内容</string>
9.       <string name="success">保存成功</string>
10.      <string name="fail">保存失败</string>
11.      <string name="empty_filename">空文件名</string>
12.      <string name="read_success">读取成功</string>
13.      <string name="read_fail">读取失败</string>
14.      <string name="empty_content">空文件内容</string>
15.    </resources>
```

（6）如需存入 SD 卡，需要在 AndroidManifest.xml 文件中的<manifest>中添加读写文件的权限，具体代码见 9.2.3 节的案例。

（7）部署工程 FileWriteAndReadDemo，程序运行效果如图 9-8 所示。

输入存储的文件名及文件内容，单击“保存文件”按钮，如果保存成功，Toast 会显示保存成功提示信息，如图 9-9 和图 9-10 所示。文件存储到目录/data/data/com.hisoft/files/下，打开 DDMS 视图下的 File Explorer 面板进行查看，如图 9-11 所示。

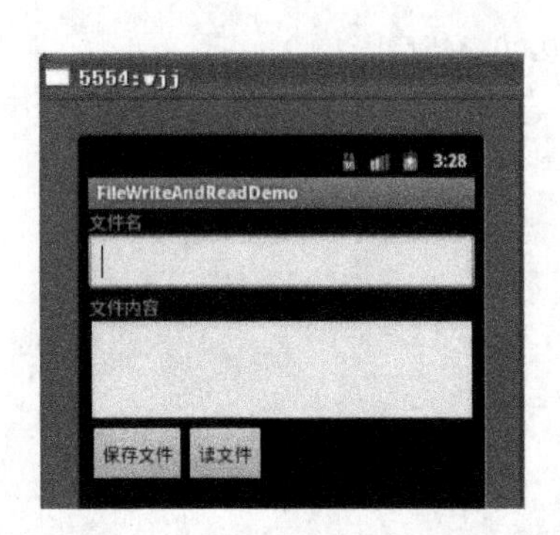

图 9-8　FileWriteAndReadDemo 运行效果

图 9-9　保存文件成功

图 9-10　保存成功日志

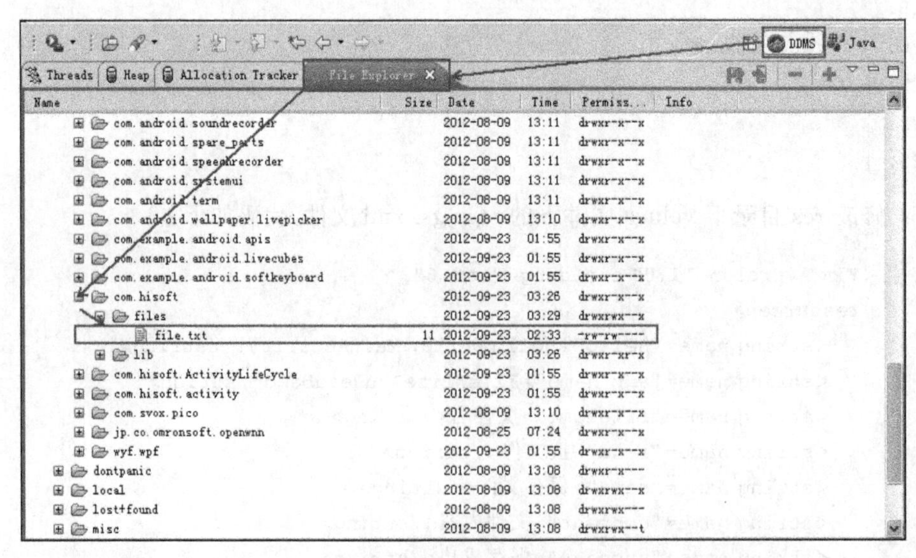

图 9-11　文件保存路径

下次启动后，在“文件名”文本框中输入文件名，单击“读文件”按钮，文件内容会自动读取出来。然后 Toast 会显示提示信息，logcat 显示读取成功，如图 9-12 和图 9-13 所示。如果没有填写“文件名”，单击“读文件”按钮，Toast 信息会提示“空文件名”，Logcat 会显示“The file name is empty”。

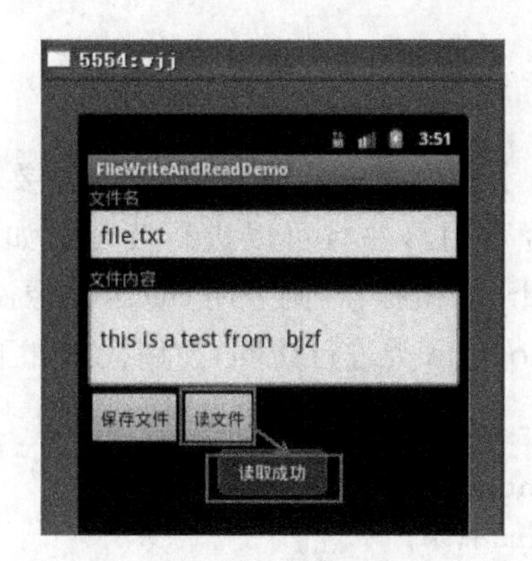

图 9-12 文件读取成功

```
09-23 03:51:14.739  I  7728  FileService        The file read successful
```

图 9-13 logcat 显示文件读取成功

9.2.3 SDCard 存储简介

SD 卡(Secure Digital Memory Card)是 Android 的外部存储设备，广泛使用于数码设备上，Android 系统提供了对 SD 卡的便捷的访问方法。

上节讲述了使用 Activity 的 openFileOutput()方法保存文件，文件是存放在手机自身空间内，一般手机的自身存储空间不大，如果要存放像视频这样的大文件，人们通常把它放置在外部的存储设备 SDCard 上。

SD 卡适用于保存大尺寸的文件或者是一些无需设置访问权限的文件，可以保存录制的大容量的视频文件和音频文件等。

SD 卡使用的是 FAT(File Allocation Table)的文件系统，不支持访问模式和权限控制，但可以通过 Linux 文件系统的文件访问权限的控制保证文件的私密性。

1. SD 卡创建方式

Android 模拟器支持 SD 卡，但模拟器中没有默认的 SD 卡，应用程序开发人员必须在模拟器中手工添加 SD 卡的映像文件。

创建 SD 卡有两种方式：一种是通常在 Eclipse 创建模拟器时创建 SD 卡；另外一种是使用＜Android SDK＞/tools 目录下的 mksdcard 工具创建 SD 卡映像文件。

在控制台窗口中进入 android SDK 安装路径的 tools 目录下，使用 mksdcard 工具，命令如下：

```
mksdcard  -l  SDCa 1024M d:\android\sdcard_file
```

第 1 个参数-l 表示后面的字符串是 SD 卡的标签，这个新建立的 SD 卡的标签是 SDCa。

第 2 个参数 1024M 表示 SD 卡的容量是 1GB。

第 3 个参数表示 SD 卡映像文件的保存位置，上面的命令将映像保存在 d:\android 目录下的 sdcard_file 文件中。在 CMD 中执行该命令后，可在所指定的目录中找到生产的

SD 卡映像文件。

2. 访问 SD 卡

在编程访问 SD 卡，往 SD 卡存放文件之前，首先程序需要先判断手机是否装有 SD 卡（检测系统的/sdcard 目录是否可用），并且可以进行读写。如果不可用，则说明设备中的 SD 卡已经被移除（如用在 Android 模拟器，则表明 SD 卡映像没有被正确加载）；如果可用，则直接通过使用标准的 Java.io.File 类进行访问，使用代码如下：

```
16.  if(Environment.getExternalStorageState().
17.    equals(Environment.MEDIA_MOUNTED)){
18.           //获取 SDCard 目录
19.           File sdCardDir=Environment.getExternalStorageDirectory();
20.           File saveFile=new File(sdCardDir, "test.txt");
21.           FileOutputStream outStream=new FileOutputStream(saveFile);
22.           outStream.write("How are you!".getBytes());
23.           outStream.close();
24.  }
```

上述代码中 Environment.getExternalStorageState()方法用于获取 SDCard 的状态，如果手机装有 SDCard，并且可以进行读写，那么方法返回的状态等于 Environment.MEDIA_MOUNTED。

第 4 行的 Environment.getExternalStorageDirectory()方法用于获取 SDCard 的目录。或使用下面的代码完成：

```
1.  File saveFile=new File("/sdcard/test.txt");
2.  FileOutputStream outStream=new FileOutputStream(saveFile);
3.  outStream.write("How are you!".getBytes());
4.  outStream.close();
```

注意：在处理中文字符时需要注意编码问题，发送和接收、保存和读取都采用相同的字符编码，一般采用 utf-8 编码，以防出现乱码。

9.2.4 SD 卡存储应用案例

上述介绍了 SDCard 创建方式及访问方法，下面通过一个案例详细介绍访问 SDCard 的应用。

（1）创建一个新的 Android 工程，工程名为 SDCardDemo，目标 API 选择 10（即 Android 2.3.3 版本），应用程序名为 SDCardDemo，包名为 com.hisoft.activity，创建的 Activity 的名字为 MainActivity，最小 SDK 版本根据选择的目标 API 会自动添加为 10，创建项目工程如图 9-14 所示。

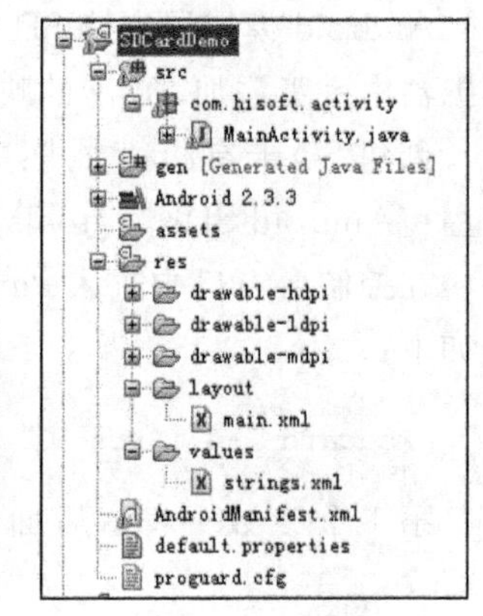

图 9-14 SDCardDemo 工程目录结构

（2）修改 res 目录下 layout 文件夹中的 main.xml 文件，设置线性布局，添加两个 EditText 控件和两个 Button 控件描述，并设置相关属性，代码如下：

```
1.  <?xml version="1.0" encoding="utf-8"?>
2.  <LinearLayout xmlns:android="http://schemas.android.com/apk/res/android"
3.    android:orientation="vertical"
4.    android:layout_width="fill_parent"
5.    android:layout_height="fill_parent">
6.  <EditText android:id="@+id/edit1"
7.    android:layout_width="fill_parent"
8.    android:layout_height="wrap_content"
9.    android:lines="4"/>
10. <Button android:id="@+id/write"
11.   android:layout_width="wrap_content"
12.   android:layout_height="wrap_content"
13.   android:text="@string/write"/>
14. <EditText android:id="@+id/edit2"
15.   android:layout_width="fill_parent"
16.   android:layout_height="wrap_content"
17.   android:editable="false"
18.   android:cursorVisible="false"
19.   android:lines="4"/>
20. <Button android:id="@+id/read"
21.   android:layout_width="wrap_content"
22.   android:layout_height="wrap_content"
23.   android:text="@string/read"/>
24. </LinearLayout>
```

(3) 修改 res 目录下 values 文件夹中的 strings.xml 文件,代码如下:

```
1.  <?xml version="1.0" encoding="utf-8"?>
2.  <resources>
3.      <string name="hello">Hello World, MainActivity!</string>
4.      <string name="app_name">SDCardDemo</string>
5.      <string name="read">从 SD 卡读取</string>
6.      <string name="write">写入 SD 卡</string>
7.  </resources>
```

(4) 修改 src 目录中 com.hisoft.activity 包下的 MainActivity.java 文件,代码如下:

```
1.  package com.hisoft.activity;
2.  import java.io.BufferedReader;
3.  import java.io.File;
4.  import java.io.FileInputStream;
5.  import java.io.FileOutputStream;
6.  import java.io.FileWriter;
7.  import java.io.IOException;
8.  import java.io.InputStreamReader;
9.  import java.io.OutputStreamWriter;
10. import java.io.PrintWriter;
```

```
11.  import android.app.Activity;
12.  import android.os.Bundle;
13.  import android.os.Environment;
14.  import android.view.View;
15.  import android.view.View.OnClickListener;
16.  import android.widget.Button;
17.  import android.widget.EditText;
18.
19.  public class MainActivity extends Activity {
20.
21.    private Button btn_read, btn_write;
22.    final String FILE_NAME="/myfile.txt";
23.
24.    @Override
25.    public void onCreate(Bundle savedInstanceState)
26.    {
27.      super.onCreate(savedInstanceState);
28.      setContentView(R.layout.main);
29.      //获取两个按钮
30.      this.btn_read=(Button) findViewById(R.id.read);
31.      this.btn_write=(Button) findViewById(R.id.write);
32.      //获取两个文本框
33.      final EditText edit1=(EditText) findViewById(R.id.edit1);
34.      final EditText edit2=(EditText) findViewById(R.id.edit2);
35.      //为 write 按钮绑定事件监听器
36.      this.btn_write.setOnClickListener(new OnClickListener()
37.      {
38.        @Override
39.        public void onClick(View source)
40.        {
41.          //将 edit1 中的内容写入文件中
42.          write(edit1.getText().toString());
43.          edit1.setText("");
44.        }
45.      });
46.
47.      this.btn_read.setOnClickListener(new OnClickListener()
48.      {
49.        @Override
50.        public void onClick(View v)
51.        {
52.          //读取指定文件中的内容,并显示出来
53.          edit2.setText(read());
54.        }
55.      });
```

```
56.     }
57.
58.     private String read()
59.     {
60.       BufferedReader br=null;
61.       try
62.       {
63.         //如果手机插入了 SD 卡,而且应用程序具有访问 SD 的权限
64.         if (Environment.getExternalStorageState()
65.           .equals(Environment.MEDIA_MOUNTED))
66.         {
67.           //获取 SD 卡对应的存储目录
68.           File sdCardDir=Environment.getExternalStorageDirectory();
69.           //获取指定文件对应的输入流
70.           FileInputStream fis=new FileInputStream(sdCardDir
71.             .getCanonicalPath()  +FILE_NAME);
72.           //构造 BufferedReader 从文件中读取
73.           br=new BufferedReader(new
74.             InputStreamReader(fis));
75.           StringBuilder sb=new StringBuilder("");
76.           String line=null;
77.           while((line=br.readLine()) !=null)
78.           {
79.             sb.append(line);
80.           }
81.           return sb.toString();
82.         }
83.       }
84.       catch (Exception e)
85.       {
86.         e.printStackTrace();
87.       }
88.       finally{
89.         if(br !=null){
90.           try {
91.             br.close();
92.           } catch (IOException e) {
93.             //TODO Auto-generated catch block
94.             e.printStackTrace();
95.           }
96.           br=null;
97.         }
98.       }
99.       return null;
100.    }
```

```
101.
102.    private void write(String content)
103.    {
104.      PrintWriter pw=null;
105.      try
106.      {
107.        //如果手机插入了SD卡,而且应用程序具有访问SD的权限
108.        if (Environment.getExternalStorageState()
109.          .equals(Environment.MEDIA_MOUNTED))
110.        {
111.          //获取SD卡的目录
112.          File sdCardDir=Environment.getExternalStorageDirectory();
113.          File targetFile=new File(sdCardDir.getCanonicalPath()
114.            +FILE_NAME);
115.          //构造PrintWriter对象向文件中写入
116.          FileOutputStream fos=new FileOutputStream(targetFile);
117.          pw=new PrintWriter(new OutputStreamWriter(fos));
118.          pw.write(content);
119.          pw.flush();
120.
121.        }
122.      }
123.      catch (Exception e)
124.      {
125.        e.printStackTrace();
126.      }
127.      finally{
128.        if(pw !=null){
129.          pw.close();
130.          pw=null;
131.        }
132.      }
133.    }
134.  }
```

(5) 在AndroidManifest.xml文件中,<manifest>根节点下添加在SD卡中创建、删除、写入数据的权限,代码如下:

```
1.  <!--在SD卡中创建与删除文件权限-->
2.  <uses-permission
    android:name="android.permission.MOUNT_UNMOUNT_FILESYSTEMS"/>
3.  <!--向SD卡写入数据权限-->
4.  <uses-permission android:name="android.permission.WRITE_EXTERNAL_
    STORAGE"/>
```

（6）部署运行 SDCardDemo 工程，然后在“写入 SD 卡”按钮上方的文本框中输入“this is a sdcard data app”，单击“写入 SD 卡”按钮，如图 9-15 所示。数据写入到 mnt/sdcard/myfile. txt 文件中，如图 9-16 所示。在 myfile. txt 文件中存储着刚才写入的内容，如图 9-17 所示。具体导出文件步骤及操作详见 9.3.5 节中的案例，此处不再赘述。

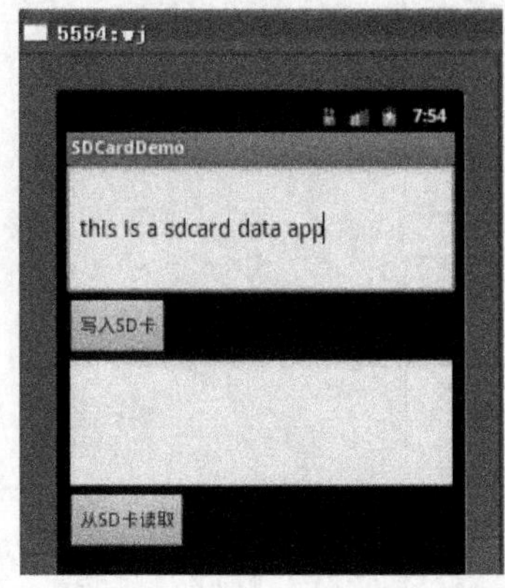

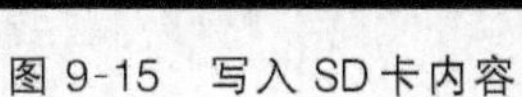

图 9-15　写入 SD 卡内容

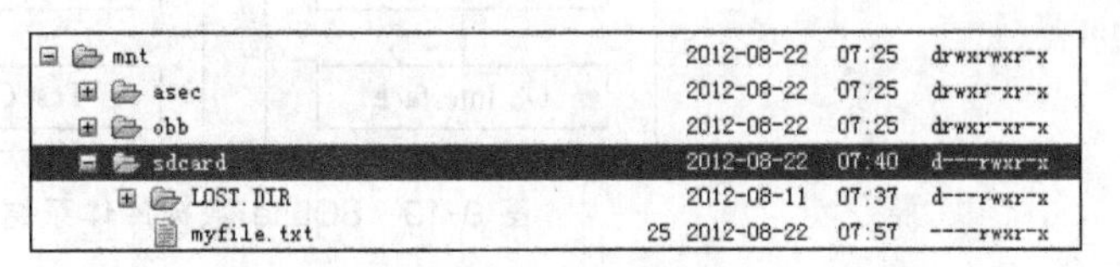

图 9-16　存储路径

单击“从 SD 卡读取”按钮，运行结果显示如图 9-18 所示。

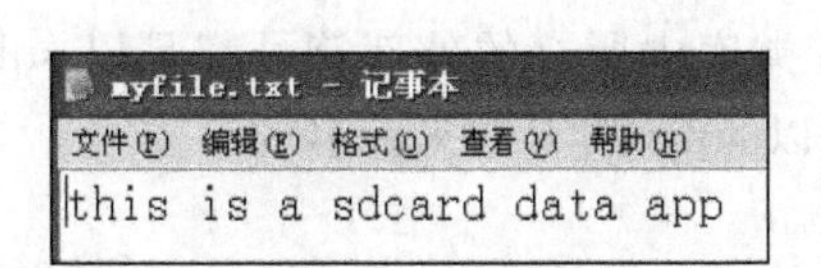

图 9-17　文件内容

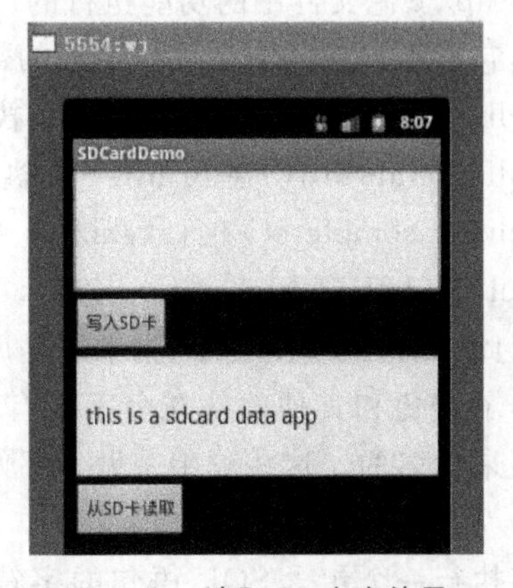

图 9-18　读取 SD 内容效果

9.3　SQLite 数据库存储

9.3.1　SQLite 数据库简介

SQLite 是在 2000 年由 D. Richard Hipp 发布的轻量级嵌入式关系型数据库，它支持 SQL 语言，是开源的项目，在 Android 系统平台中集成了嵌入式关系型数据库(SQLite)。

1. SQLite 数据库体系结构

SQLite 数据库由 SQL 编译器、内核、后端以及附件 4 部分组成。SQLite 通过利用虚拟机和虚拟数据库引擎（VDBE），使调试、修改和扩展 SQLite 的内核变得更加方便。SQLite 数据库体系结构如图 9-19 所示。

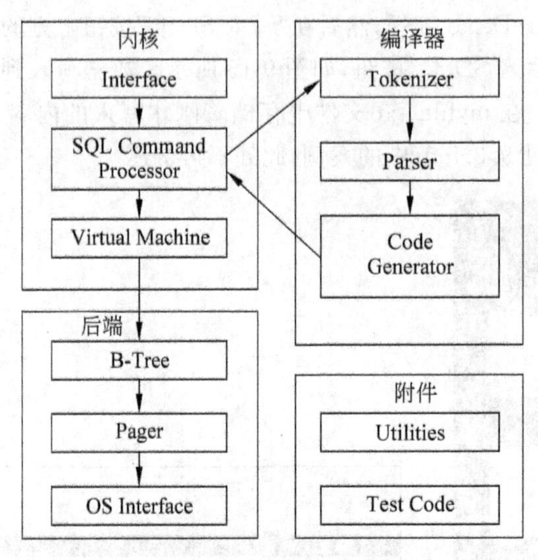

图 9-19 SQLite 数据库体系结构

1) Interface

接口由 SQLite C API 组成，SQLite 类库大部分的公共接口程序是由 main. c，legacy. c 和 vdbeapi. c 源文件中的功能执行的。但有些程序是分散在其他文件夹的，因为在其他文件夹里它们可以访问有文件作用域的数据结构。例如：

sqlite3_get_table()在 table. c 中执行。

sqlite3_mprintf()在 printf. c 中执行。

sqlite3_complete()在 tokenize. c 中执行。

Tcl 接口程序用 tclsqlite. c 执行。

因此，无论是应用程序、脚本还是库文件，最终都是通过接口与 SQLite 交互。

为了避免和其他软件在名字上有冲突，SQLite 类库中所有的外部符号都是以 sqlite3 为前缀来命名的。这些被用来做外部使用的符号是以 sqlite3_开头来命名的。

2) Tokenizer

当执行一个包含 SQL 语句的字符串时，接口程序要把这个字符串传递给 tokenizer。Tokenizer 的任务是把原有字符串分成一个个标示符，并把这些标示符传递给剖析器。Tokenizer 是在 C 文件夹 tokenize. c 中用手编译的。

在这个设计中需要注意的一点是 tokenizer 调用 parser。即用 tokenizer 调用 parser 会使程序运行的更顺利。

3) Parser

Parser 是 Lemon(LALR(1)文法分析器生成工具)生成的分析器的核心例程，在分析器调用 ParserAlloc 后，分词器就可以把切分的词传递给 Parser 进行语法分析。Tokenizer 和 Parser 对 SQL 语句进行语法检查，然后把 SQL 语句转化为底层能更方便处理的分层的数据结构，这种分层的数据结构称为“语法树”，把语法树传给 Code Generator 进行处理。

4) Code Generator

在 Parser 收集完符号并转换成完全的 SQL 语句时，它调用 Code Generator 来产生虚拟的机器代码，这些机器代码将按照 SQL 语句的要求来工作。在代码产生器中有许多文件，如

attach. c、auth. c、build. c、delete. c、expr. c、insert. c、pragma. c、select. c、trigger. c、update. c、vacuum. c 和 where. c。在这些文件中，expr. c 处理表达式代码的生成。where. c 处理 SELECT、UPDATE 和 DELETE 语句中 WHERE 子句的代码的生成。文件 attach. c、delete. c、insert. c、select. c、trigger. c update. c 和 vacuum. c 处理 SQL 语句中具有同样名字的语句的代码的生成（每个文件调用 expr. c 和 where. c 中的程序）。所有 SQL 的其他语句的代码是由 build. c 生成的。文件 auth. c 执行 sqlite3_set_authorizer()的功能。

5）Virtual Machine

由 Code Generator（代码生成器）产生的程序由 Virtual Machine（虚拟机器）来运行。总而言之，虚拟机器主要用来执行一个为操作数据库而设计的抽象的计算引擎。机器有一个 用来存储中间数据的存储栈。每个指令包含一个操作代码和三个额外的操作数。

虚拟机器本身是被包含在一个单独的文件 vdbe. c 中的。虚拟机器也有它自己的标题文件：vdbe. h 在虚拟机器和剩下的 SQLite 类库之间定义了一个接口程序，vdbeInt. h 定义了虚拟机器的结构。文件 vdbeaux. c 包含了虚拟机器所使用的实用程序和一些被其他类库用来建立 VM 程序的接口程序模块。文件 vdbeapi. c 包含虚拟机器的外部接口，比如 sqlite3_bind_... 类的函数。单独的值（字符串，整数，浮动点数值，BLOBS）被存储在一个叫 Mem 的内部目标程序里，Mem 是由 vdbemem. c 执行的。

6）B-Tree

SQLite 数据库在磁盘里维护，使用源文件 btree. c 中的 B-Tree（B-树）执行。数据库中的每个表格和目录使用一个单独的 B-tree。所有的 B-trees 被存储在同样的磁盘文件里。文件格式的细节被记录在 btree. c 开头的备注里。B-tree 子系统的接口程序被标题文件 btree. h 所定义。主要功能就是索引，它维护着各个页面之间复杂的关系，便于快速找到所需数据。

7）Pager

B-tree 模块要求信息来源于磁盘上固定规模的程序块。默认程序块的大小是 1024 个字节，但是可以在 512～65 536 个字节间变化。Pager（页面高速缓存）负责读、写和高速缓存这些程序块。页面高速缓存还提供重新运算和提交抽象命令，它还管理关闭数据库文件夹。B-tree 驱动器要求页面高速缓存器中的特别的页，当它想修改页或重新运行改变的时候会通报页面高速缓存。为了保证所有的需求被快速、安全和有效的处理，页面高速缓存处理所有微小的细节。运行页面高速缓存的代码在专门的 C 源文件 pager. c 中。页面高速缓存的子系统的接口程序被目标文件 pager. h 所定义。页缓存的主要作用就是通过操作系统接口在 B-树和磁盘之间传递页面。

8）OS Interface

为了在 POSIX 和 Win32 之间提供一些可移植性，SQLite 操作系统的接口程序使用一个提取层。OS 提取层的接口程序被定义在 os. h。每个支持的操作系统有自己的执行文件：UNIX 使用 os_unix. c，Windows 使用 os_win. c。每个具体的操作器具有自己的标题文件，如 os_unix. h、os_win. h 等。

9）Utilities

内存分配和字符串比较程序位于 util. c。Parser 使用的表格符号被 hash. c 中的无用信息表格维护。源文件 utf. c 包含 UNICODE 转换子程序。SQLite 有自己的执行文件

printf()(有一些扩展)。在 printf.c 中,还有它自己随机数量产生器在 random.c。

10) Test Code

如果计算回归测试脚本,多于一半的 SQLite 代码数据库的代码将被测试。在主要代码文件中有许多 assert()语句。另外,源文件 test1.c 通过 test5.c 和 md5.c 执行只为测试用的扩展名。os_test.c 向后的接口程序通过模拟断电来验证页面调度程序中的系统性事故恢复机制。

2. SQLite 数据库的特点

- 更加适用于嵌入式系统,嵌入到使用它的应用程序中。
- 占用非常少,运行高效可靠,可移植性好。
- 提供了零配置(Zero-Configuration)运行模式。
- SQLite 数据库不仅提高了运行效率,而且屏蔽了数据库使用和管理的复杂性,程序仅需要进行最基本的数据操作,其他操作可以交给进程内部的数据库引擎完成。
- SQLite 数据库具有很强的移植性,可以运行在 Windows、Linux、BSD、Mac OS X 和一些商用 UNIX 系统,比如 Sun 的 Solaris,IBM 的 AIX。SQLite 数据库也可以工作在许多嵌入式操作系统下,如 QNX、VxWorks、Palm OS、Symbin 和 Windows CE。

3. SQLite 数据库和其他数据库的区别

SQLite 和其他数据库最大的不同就是对数据类型的支持,SQLite3 支持 NULL、INTEGER、REAL(浮点数字)、TEXT(字符串文本)和 BLOB(二进制对象)数据类型。虽然它支持的类型只有 5 种,但实际上 SQLite3 也接受 varchar(n)、char(n)和 decimal(p,s)等数据类型,只不过在运算或保存时会转成对应的 5 种数据类型。创建一个表时,可以在 CREATE TABLE 语句中指定某列的数据类型,但是可以把任何数据类型放入任何列中。当某个值插入数据库时,SQLite 将检查它的类型。如果该类型与关联的列不匹配,SQLite 会尝试将该值转换成该列的类型。如果不能转换,该值将作为其本身具有的类型存储。比如可以把一个字符串(String)放入 INTEGER 列。SQLite 称这为“弱类型(Manifest Typing)”。此外,SQLite 不支持一些标准的 SQL 功能,特别是外键约束(FOREIGN KEY Constrains),嵌套 transcaction 和 RIGHT OUTER JOIN 和 FULL OUTER JOIN 以及一些 ALTER TABLE 的功能。除了上述功能外,SQLite 是一个完整的 SQL 系统,拥有完整的触发器、交易等。

注意:定义为 INTEGER PRIMARY KEY 的字段只能存储 64 位整数,当向这种字段中保存除整数以外的数据时将会产生错误。

另外,SQLite 在解析 CREATE TABLE 语句时会忽略 CREATE TABLE 语句中跟在字段名后面的数据类型信息。

9.3.2 创建 SQLite 数据库方式

创建 SQLite 数据库的方式有两种,分别是使用 sqlite3 工具命令行方式和使用程序编码方式,下面就分别介绍它们创建数据库的过程。

1. sqlite3 工具命令行方式(适合调试用)

sqlite3 是 SQLite 数据库自带的一个基于命令行的 SQL 命令执行工具,并可以显示命令执行结果。sqlite3 工具被集成在 Android 系统中,用户在 Linux 的命令行界面中输入 sqlite3 可启动 sqlite3 工具,并得到工具的版本信息。在 CMD 中输入 adb shell 命令可以启动 Linux 的命令行界面,过程如下所示:

(1) 首先用命令或在 Eclipse 中启动模拟器,然后在 cmd 下输入命令“adb shell”进入设备 Linux 控制台,出现提示符“#”后,输入命令 sqlite3,如图 9-20 所示。

```
H:\android-sdk-3\platform-tools>adb shell
# sqlite3
sqlite3
SQLite version 3.6.22
Enter ".help" for instructions
Enter SQL statements terminated with a ";"
sqlite>
```

图 9-20 进入 sqlite

在启动 sqlite3 工具后,提示符从“#”变为“sqlite>”,表示命令行界面进入与 SQLite 数据库的交互模式,此时可以输入命令建立、删除或修改数据库的内容。正确退出 sqlite3 工具的方法是使用命令.exit,如图 9-21 所示。

(2) 命令行方式手动创建 sqlite 数据库,步骤如下:

① cmd 下输入命令 adb shell 进入设备 Linux 控制台。

② # cd /data/data,进入应用 data 目录,如图 9-22 所示。

③ # ls,列表目录,查看文件,如图 9-23 所示。

```
sqlite> .exit
```

图 9-21 sqlite3 退出命令

```
# cd /data/data
```

图 9-22 进入 data 目录

```
# ls
```

图 9-23 查看目录

找到自己的项目包目录并进入,如图 9-24 所示。

④ 使用 ls 命令查看有无 databases 目录,如果没有,则创建一个,命令如下,如图 9-25 所示。

```
mkdir databases
cd databases                                              //进入并创建数据库
sqlite3 mydb.db
sqlite3 friends.db
SQLite version 3.5.9
Enter ".help" for instructions
sqlite>
//ctrl+d 或 .exit 退出 sqlite 提示符
```

⑤ 使用 ls 命令查看列表目录会看到有一个文件为 mydb. db,即 sqlite 数据库,如图 9-26 所示。

```
# cd com.hisoft.activity
cd com.hisoft.activity
#
```

图 9-24 进入项目包目录

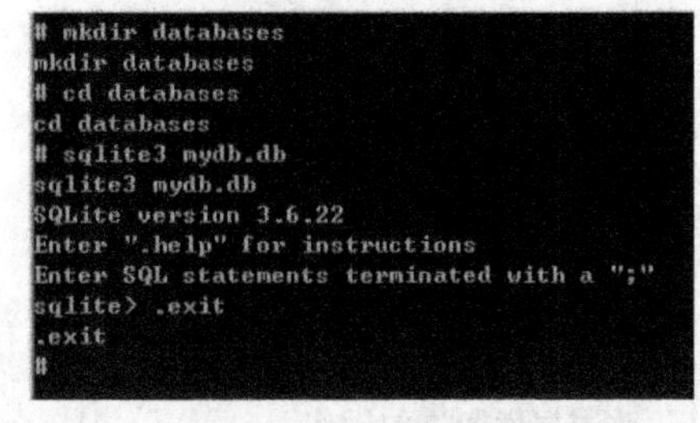

图 9-25 创建数据库

图 9-26 查看 sqlite 数据库文件

2. 使用程序编码方式(常用方式)

在程序代码中动态建立 sqlite 数据库是比较常用的方法。在程序运行过程中,当需要进行数据库操作时,应用程序会首先尝试打开数据库,此时如果数据库并不存在,程序会自动建立数据库,然后再打开数据库。

在 Android 应用程序中创建使用 SQLite 数据库有两种方式:一种是自定义类继承 SQLiteOpenHelper;另外一种是调用 openOrCreateDatabases()方法创建数据库。下面分别进行介绍。

1) 自定义类继承 SQLiteOpenHelper,创建数据库

在 Android 应用程序中使用 SQLite,必须自己创建数据库,然后创建表、索引,填充数据。Android 提供了 SQLiteOpenHelper 帮助创建一个数据库,只要继承 SQLiteOpenHelper 类就可以轻松的创建数据库。SQLiteOpenHelper 类根据开发应用程序的需要,封装了创建和更新数据库使用的逻辑。

创建 SQLiteOpenHelper 的子类至少需要实现三个方法:

(1) 构造函数,调用父类 SQLiteOpenHelper 的构造函数。这个方法需要 4 个参数:上下文环境(例如一个 Activity),数据库名字,一个可选的游标工厂(通常是 Null),一个代表正在使用的数据库模型版本的整数。

(2) onCreate()方法。它需要一个 SQLiteDatabase 对象作为参数,根据需要对这个对象填充表和初始化数据。

(3) onUpgrage()方法。它需要三个参数:一个 SQLiteDatabase 对象,一个旧的版本号和一个新的版本号,这样就可以知道如何把一个数据库从旧的模型转变到新的模型。

应用程序编码创建 sqlite 数据库的通常步骤如下:

① 创建自己的类 DatabaseHelper 继承 SQLiteOpenHelper,并实现上述三个方法,代码如下:

```
1.    public class DatabaseHelper extends SQLiteOpenHelper {
2.      DatabaseHelper(Context context, String name, CursorFactory cursorFactory,
        int version)
3.      {
4.        super(context, name, cursorFactory, version);
5.        }
6.
7.        @Override
8.        public void onCreate(SQLiteDatabase db) {
9.            //TODO 创建数据库后对数据库的操作
10.       }
11.
12.       @Override
13.    public void onUpgrade(SQLiteDatabase db, int oldVersion, int newVersion) {
14.           //TODO 更改数据库版本的操作
15.       }
16.
17.    @Override
```

```
18.   public void onOpen(SQLiteDatabase db) {
19.           super.onOpen(db);
20.           //TODO 每次成功打开数据库后首先被执行
21.       }
22.   }
```

② 获取 SQLiteDatabase 类对象实例。

根据需要改变数据库的内容，决定是调用 getReadableDatabase()或 getWriteableDatabase()方法，获取 SQLiteDatabase 实例。例如：

```
db=(new DatabaseHelper(getContext())).getWritableDatabase();
```

上面这段代码会返回一个 SQLiteDatabase 类的实例，使用这个对象就可以查询或者修改数据库。当完成了对数据库的操作(如 Activity 已经关闭)，需要调用 SQLiteDatabase 的 Close()方法来释放掉数据库连接。

2) 调用 openOrCreateDatabase ()方法创建数据库

android.content.Context 中提供了方法 openOrCreateDatabase ()来创建数据库。

```
db=context .openOrCreateDatabase(String DATABASE _NAME, int Context. MODE_PRIVATE,
null);
```

- DATABASE _NAME：数据库的名字。
- MODE：操作模式，如 Context.MODE_PRIVATE 等。
- CursorFactory：指针工厂，本例中传入 null，暂不用。

9.3.3 SQLite 数据库操作

在编程实现时，一般将所有对数据库的操作都封装在一个类中，因此只要调用这个类，就可以完成对数据库的添加、更新、删除和查询等操作。上节已经讲述了如何创建数据库，下面就在数据库中对创建表、索引，给表添加数据等操作进行介绍。

1. 创建表和索引

为了创建表和索引，需要调用 SQLiteDatabase 的 execSQL()方法来执行 DDL 语句。如果没有异常，这个方法没有返回值。

```
db.execSQL("CREATE TABLE mytable (_id INTEGER PRIMARY KEY AUTOINCREMENT, title TEXT,
value REAL);");
```

上述语句创建表名为 mytable，表有一个列名为_id，并且是主键，列值是会自动增长的整数。另外还有两列：title(字符)和 value(浮点数)。SQLite 会自动为主键列创建索引。通常情况下，第一次创建数据库时创建了表和索引。

另外，SQLiteDatabase 类提供了一个重载后的 execSQL(String sql, Object[] bindArgs)方法。

使用这个方法支持使用占位符参数(?)。使用例子如下：

```
SQLiteDatabase db=…;
db.execSQL("insert into person(name, age) values(?,?)",new Object[]{"Tom", 4});
```

```
db.close();
```

第一个参数为 SQL 语句,第二个参数为 SQL 语句中占位符参数的值,参数值在数组中的顺序要和占位符的位置对应。

如果不需要改变表的 schema,不需要删除表和索引。删除表和索引需要使用 execSQL()方法调用 DROP INDEX 和 DROP TABLE 语句。

2. 给表添加数据

给数据库中表添加数据有两种方法。

(1) 使用 execSQL()方法执行 INSERT、UPDATE 和 DELETE 等语句来更新表的数据。execSQL()方法适用于所有不返回结果的 SQL 语句。例如:

```
db.execSQL("INSERT INTO widgets (name, inventory)"+"VALUES ('Sprocket', 5)");
```

(2) 使用 SQLiteDatabase 对象的 insert()、update()和 delete()方法。这些方法把 SQL 语句的一部分作为参数。

① insert()方法

insert()方法用于添加数据,各个字段的数据使用 ContentValues 进行存放。

ContentValues 类似于 MAP,相对于 MAP,它提供了存取数据对应的 put(String key,Xxx value)和 getAsXxx(String key)方法,key 为字段名称,value 为字段值。

例如:

```
SQLiteDatabase db=databaseHelper.getWritableDatabase();
ContentValues values=new ContentValues();
values.put("name", "Tom");
values.put("age", 4);
long rowid=db.insert("person", null, values);      //返回新添记录的行号,与主键 id 无关
```

不管第三个参数是否包含数据,执行 insert()方法必然会添加一条记录。如果第三个参数为空,会添加一条除主键之外其他字段值为 Null 的记录。

② update()方法

update()方法有 4 个参数,分别是表名,表示列名和值的 ContentValues 对象,可选的 WHERE 条件和可选的填充 WHERE 语句的字符串,这些字符串会替换 WHERE 条件中的"?"标记。

update()根据条件更新指定列的值,所以用 execSQL()方法可以达到同样的目的。WHERE 条件及其参数和用过的其他 SQL APIs 类似。

例如:

```
String[] parms=new String[] {"this is a string"};
db.update("widgets", replacements, "name=?", parms);
```

③ delete()方法

delete()方法的使用和 update()类似,使用表名,可选的 WHERE 条件和相应的填充 WHERE 条件的字符串。例如:

```
db.delete("person", "personid<?", new String[]{"2"});
```

```
db.close();
```

3. 查询数据库

在 Android 系统中，数据库查询结果的返回值并不是数据集合的完整备份，而是返回数据集的指针，这个指针就是 Cursor 类。

Cursor 类支持在查询的数据集合中多种方式移动，并能够获取数据集合的属性名称和序号。

查询数据库使用 SELECT 从 SQLite 数据库检索数据有两种方法：一种是使用 rawQuery() 直接调用 SELECT 语句；另外一种是使用 query()方法构建一个查询。

(1) 使用 rawQuery()直接调用 SELECT 语句。

调用 SQLiteDatabase 类的 rawQuery()方法用于执行 select 语句。

例如：

```
Cursor c=db.rawQuery("SELECT name FROM sqlite_master WHERE type='table' AND name=
'mytable'", null);
```

rawQuery()方法的第一个参数为 select 语句；第二个参数为 select 语句中占位符参数的值，如果 select 语句没有使用占位符，该参数可以设置为 null。

带占位符参数的 select 语句使用例子如下：

```
Cursor cursor=db.rawQuery("select * from person where name like ?and age=?", new
String[]{"%Tom%", "4"});
```

在上面例子中，查询 SQLite 系统表(sqlite_master)检查 table 表是否存在。返回值是一个 cursor 对象，这个对象的方法可以迭代查询结果。如果查询是动态的，使用这个方法就会非常复杂。

例如，当需要查询的列在程序编译的时候不能确定，这时使用 query()方法会方便很多。

(2) 使用 query()方法构建一个查询。

调用 SQLiteDatabase 类的 query()，query()的语法如下：

```
Cursor android.database.sqlite.SQLiteDatabase.query(String table, String[]
columns, String selection, String[] selectionArgs, String groupBy, String having,
String orderBy,String limit)
```

query()的参数说明如表 9-2 所示。

表 9-2 query()的参数说明

位置	类型+名称	说　明
1	String table	表名称
2	String[] columns	返回的属性列名称
3	String selection	查询条件子句
4	String[] selectionArgs	如果在查询条件中使用的是问号，则需要定义替换符的具体内容

续表

位置	类型＋名称	说　明
5	String groupBy	分组方式
6	String having	定义组的过滤器
7	String limit	指定偏移量和获取的记录数

例如：

```
1.   SQLiteDatabase db=databaseHelper.getWritableDatabase();
2.   Cursor cursor=db.query("person", new String[]{"personid,name,age"}, "name
     like ?", new String[]{"%Tom%"}, null, null, "personid desc", "1,2");
3.   while (cursor.moveToNext()) {
4.         int personid=cursor.getInt(0);      //获取第一列的值,第一列的索引从 0 开始
5.         String name=cursor.getString(1);  //获取第二列的值
6.         int age=cursor.getInt(2);           //获取第三列的值
7.   }
8.   cursor.close();
9.   db.close();
```

在 Android 的 SQLite 数据库使用游标，不论如何执行查询，都会返回一个 Cursor 对象。

Cursor 类的常用方法和说明如表 9-3 所示。

表 9-3　Cursor 类的常用方法和说明

方　法	说　明
moveToFirst	将指针移动到第一条数据上
moveToNext	将指针移动到下一条数据上
moveToPrevious	将指针移动到上一条数据上
getCount	获取集合的数据数量
getColumnIndexOrThrow	返回指定属性名称的序号。如果属性不存在，则产生异常
getColumnName	返回指定序号的属性名称
getColumnNames	返回属性名称的字符串数组
getColumnIndex	根据属性名称返回序号
moveToPosition	将指针移动到指定的数据上
getPosition	返回当前指针的位置
getString，getInt 等	获取给定字段当前记录的值
requery	重新执行查询得到游标
close	释放游标资源

9.3.4 SQLite 数据库管理

在 Android 系统中，针对 sqlite 数据库的查看和管理有两种方式：一种是使用 Eclipse 插件 DDMS 查看和管理；另外一种是使用 Android 工具包中的 adb 工具来查看和管理。如前所述，Android 项目中 sqlite 数据库位置：/data/data/<package-name>/databases/。

1. 使用 Eclipse 插件 DDMS 查看和管理 SQLite 数据库

(1) 在 Eclipse 中打开 DDMS 视图，如图 9-27 所示。

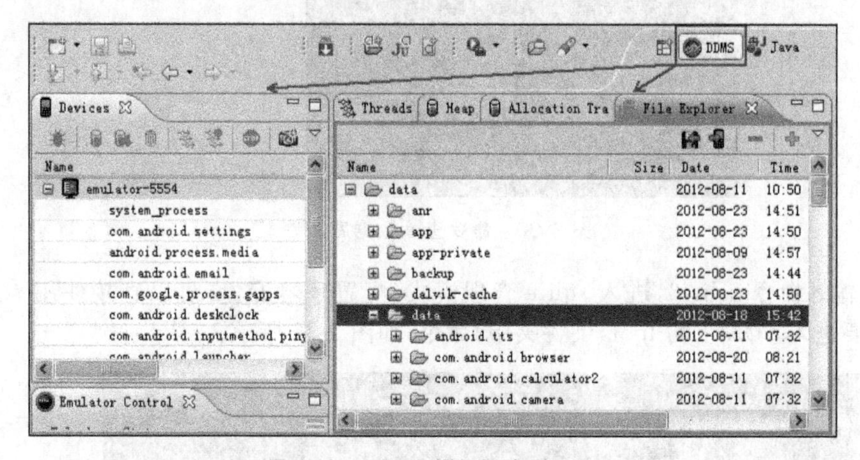

图 9-27 DDMS 视图下查看 data

注意：如果是模拟器进行项目调试，必须先启动模拟器，打开 DDMS 视图才能有内容。

(2) 选择 File Explorer 窗口，然后在/data/data/<package-name>/目录下打开 databases 文件即可看见 sqlite 数据库文件，如图 9-28 所示。选择 Pull a file from the device 按钮，可以导出 sqlite 数据库文件，然后可以选择 sqlite 界面管理工具，如 sqlite man、sqlite administrator 等打开操作，如图 9-29 所示。

名称	大小	日期	时间	权限
com.hisoft.activity		2012-08-23	14:50	drwxr-x--x
databases		2012-08-22	14:05	drwxrwxrwx
mydb.db —— sqlite数据库文件	0	2012-08-22	14:05	-rw-r--r--
lib		2012-08-23	14:50	drwxr-xr-x

图 9-28 sqlite 数据库文件路径

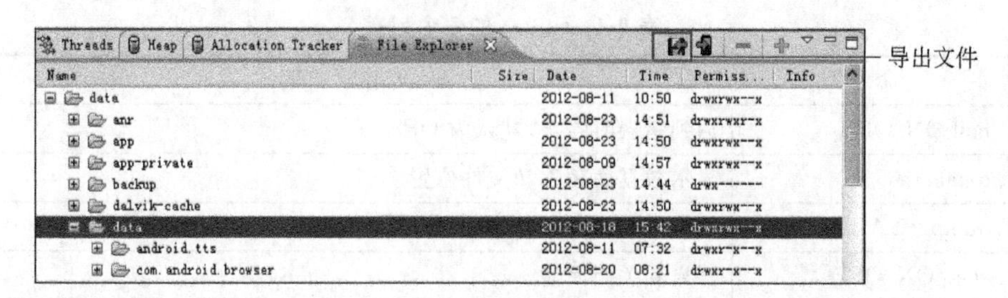

图 9-29 导出数据库文件

2. 使用 adb 工具管理 sqlite 数据库

(1) 在控制台窗口中(在"运行"中输入 cmd)输入命令 adb shell,进入设备 Linux 控制台,出现提示符"#"后,输入命令"cd/data/data/<package-name>/databases"进入目录。

(2) 使用 ls 命令查看数据库文件是否存在,如图 9-30 所示。

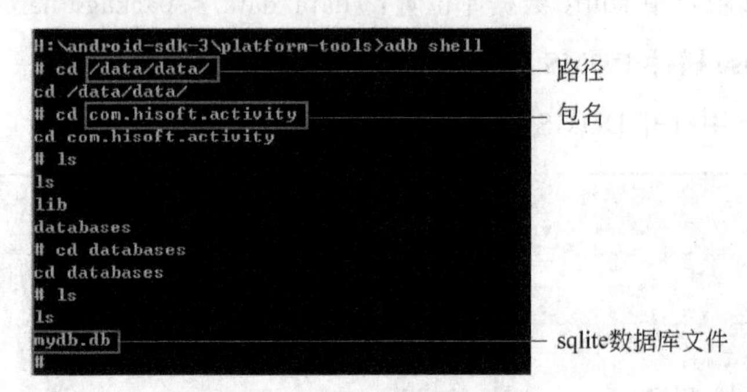

图 9-30 命令查看数据库文件

(3) 输入命令 sqlite3,进入 sqlite 管理模式(配置环境变量,可以直接使用 sqlite3 工具,否则要进入 SDK 下的 tools 文件夹中)命令,如图 9-31 所示。

```
# sqlite3
sqlite3
SQLite version 3.6.22
Enter ".help" for instructions
Enter SQL statements terminated with a ";"
sqlite> .tables
.tables
sqlite> .databases
.databases
seq  name             file
---  ---------------  ----------------------------------------------------------
0    main
1    temp
sqlite> _
```

命令

命令

图 9-31 sqlite 工具命令

sqlite 命令行工具默认是以";"结束语句的。所以如果只是一行语句,要在末尾加";",或者在下一行中输入,这样 sqlite 命令才会被执行。具体 sqlite3 的命令如表 9-4 所示。

表 9-4 sqlite3 的命令列表

编号	命令	说　明
1	. bail ON|OFF	遇到错误时停止,默认为 OFF
2	. databases	显示数据库名称和文件位置
3	. dump ?TABLE?...	将数据库以 SQL 文本形式导出
4	. echo ON|OFF	开启和关闭回显
5	. exit	退出
6	. explain ON|OFF	开启或关闭适当输出模式。如果开启模式将更改为 column,并自动设置宽度
7	. tables	查看数据库的表列表

其他命令可随时使用命令用法. help 来查看帮助。SQL 命令直接在此命令行上执行即可。

9.3.5 SQLite 数据库应用案例

上述内容介绍了 SQLite 数据库创建方式、操作、管理及访问方法，下面通过一个案例详细介绍访问 SQLite 数据库的应用。

(1) 创建一个新的 Android 工程，工程名为 SQLiteDemo，目标 API 选择 10(即 Android 2.3.3 版本)，应用程序名为 SQLiteDemo，包名为 com.hisoft.activity，创建的 Activity 的名字为 MainActivity，最小 SDK 版本根据选择的目标 API 会自动添加为 10，创建项目工程如图 9-32 所示。

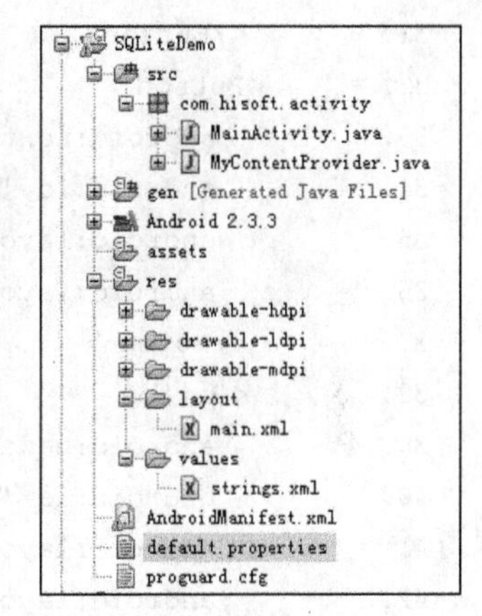

图 9-32 SQLiteDemo 工程目录结构

(2) 修改 res 目录下 layout 文件夹中的 main.xml 文件，设置线性布局，添加一个 EditText 控件、6 个 Button 控件和一个 ScrollView 控件描述，并设置相关属性，代码如下：

```
1.  <?xml version="1.0" encoding="utf-8"?>
2.  <LinearLayout xmlns:android="http://schemas.android.com/apk/res/android"
3.      android:orientation="vertical"
4.      android:layout_width="fill_parent"
5.      android:layout_height="fill_parent"
6.      >
7.          <Button
8.        android:text="创建/打开数据库"
9.        android:id="@+id/create_open"
10.       android:layout_width="120dip"
11.       android:layout_height="wrap_content">
12.     </Button>
13.     <Button
14.       android:text="关闭数据库"
15.       android:id="@+id/close"
16.       android:layout_width="120dip"
17.       android:layout_height="wrap_content">
18.     </Button>
19.
20.     <Button
21.       android:text="添加记录"
22.       android:id="@+id/insert"
23.       android:layout_width="wrap_content"
24.       android:layout_height="wrap_content">
25.     </Button>
```

```
26.         <Button
27.           android:text="删除记录"
28.           android:id="@+id/delete"
29.           android:layout_width="wrap_content"
30.           android:layout_height="wrap_content">
31.         </Button>
32.        <Button
33.           android:text="修改记录"
34.           android:id="@+id/update"
35.           android:layout_width="wrap_content"
36.           android:layout_height="wrap_content">
37.         </Button>
38.         <Button
39.           android:text="查询记录"
40.           android:id="@+id/query"
41.           android:layout_width="wrap_content"
42.           android:layout_height="wrap_content">
43.         </Button>
44.
45.       <ScrollView
46.         android:id="@+id/ScrollView01"
47.         android:layout_width="fill_parent"
48.         android:layout_height="wrap_content">
49.         <EditText
50.           android:id="@+id/EditText01"
51.           android:layout_width="fill_parent"
52.           android:layout_height="wrap_content">
53.       </EditText>
54.       </ScrollView>
55. </LinearLayout>
```

(3) 修改 src 目录中 com.hisoft.activity 包下的 MainActivity.java 文件，代码如下：

```
1.  package com.hisoft.activity;
2.
3.  import android.app.Activity;
4.  import android.database.Cursor;
5.  import android.database.sqlite.SQLiteDatabase;
6.  import android.os.Bundle;
7.  import android.view.View;
8.  import android.view.View.OnClickListener;
9.  import android.widget.Button;
10. import android.widget.EditText;
11. import android.widget.Toast;
12.
13. public class MainActivity extends Activity {
```

```
14.
15.  private SQLiteDatabase sld;
16.  private Button create_open, close, insert, delete, update, query;
17.
18.    @Override
19.    public void onCreate(Bundle savedInstanceState) {
20.        super.onCreate(savedInstanceState);
21.        setContentView(R.layout.main);
22.
23.        //初始化创建数据库按钮
24.        this.create_open=(Button)this.findViewById(R.id.create_open);
25.        this.create_open.setOnClickListener(
26.          new OnClickListener()
27.          {
28.            @Override
29.            public void onClick(View v) {
30.                createOrOpenDatabase();
31.          }
32.         }
33.        );
34.
35.        //初始化关闭数据库按钮
36.        this.close=(Button)this.findViewById(R.id.close);
37.        this.close.setOnClickListener(
38.          new OnClickListener()
39.          {
40.            @Override
41.            public void onClick(View v) {
42.              closeDatabase();
43.            }
44.          }
45.        );
46.
47.        //初始化添加记录按钮
48.        this.insert=(Button)this.findViewById(R.id.insert);
49.        this.insert.setOnClickListener(
50.          new OnClickListener()
51.          {
52.            @Override
53.            public void onClick(View v) {
54.              insert();
55.            }
56.          }
57.        );
58.
```

```
59.         //初始化删除记录按钮
60.         this.delete=(Button)this.findViewById(R.id.delete);
61.         this.delete.setOnClickListener(
62.           new OnClickListener()
63.           {
64.             @Override
65.             public void onClick(View v) {
66.               delete();
67.             }
68.           }
69.         );
70.
71.         //初始化修改记录按钮
72.         this.update=(Button)this.findViewById(R.id.update);
73.         this.update.setOnClickListener(
74.           new OnClickListener()
75.           {
76.             @Override
77.             public void onClick(View v) {
78.               update();
79.             }
80.           }
81.         );
82.
83.         //初始化查询记录按钮
84.         this.query=(Button)this.findViewById(R.id.query);
85.         this.query.setOnClickListener(
86.           new OnClickListener()
87.           {
88.             @Override
89.             public void onClick(View v) {
90.               query();
91.             }
92.           }
93.         );
94.     }
95.
96.     //创建或打开数据库的方法
97.     public void createOrOpenDatabase()
98.     {
99.       try
100.      {
101.        sld=SQLiteDatabase.openDatabase
102.        (
103.            "/data/data/com.hisoft.activity/mydb",      //数据库所在路径
```

```
104.            null,                                   //CursorFactory
105.            SQLiteDatabase.OPEN_READWRITE|SQLiteDatabase.CREATE_IF_NECESSARY
                                                        //读写,若不存在则创建
106.          );
107.          appendMessage("数据库已经成功打开!");
108.          String sql="create table if not exists student(stuno char(5),stuname
              varchar(20),stuage integer,stuclass char(5))";
109.          sld.execSQL(sql);
110.          appendMessage("student 已经成功创建!");
111.        }
112.        catch(Exception e)
113.        {
114.          Toast.makeText(this, "数据库错误:"+e.toString(), Toast.LENGTH_
              SHORT).show();
115.        }
116.      }
117.
118.      //关闭数据库的方法
119.      public void closeDatabase()
120.      {
121.        try
122.        {
123.          sld.close();
124.          appendMessage("数据库已经成功关闭!");
125.        }
126.      catch(Exception e)
127.      {
128.        Toast.makeText(this, "数据库错误:"+e.toString(), Toast.LENGTH_
            SHORT).show();;
129.        }
130.      }
131.
132.      //插入记录的方法
133.      public void insert()
134.      {
135.        try
136.        {
137.            String sql="insert into student values('10001','张三',10,'11010')";
138.            sld.execSQL(sql);
139.            appendMessage("成功插入一条记录!");
140.        }
141.        catch(Exception e)
142.        {
143.          Toast.makeText(this, "数据库错误:"+e.toString(), Toast.LENGTH_
              SHORT).show();;
```

```
144.        }
145.    }
146.
147.    //删除记录的方法
148.    public void delete()
149.    {
150.      try
151.      {
152.          String sql="delete from student;";
153.          sld.execSQL(sql);
154.          appendMessage("成功删除所有记录！");
155.      }
156.      catch(Exception e)
157.      {
158.         Toast.makeText(this, "数据库错误:"+e.toString(), Toast.LENGTH_
         SHORT).show();;
159.      }
160.    }
161.
162.    //修改记录的方法
163.    public void update()
164.    {
165.      try
166.      {
167.        String sql="update student set stuname='李四'";
168.          sld.execSQL(sql);
169.          appendMessage("成功更新记录！");
170.      }
171.      catch(Exception e)
172.      {
173.         Toast.makeText(this, "数据库错误:"+e.toString(), Toast.LENGTH_
         SHORT).show();;
174.      }
175.    }
176.
177.    //查询的方法
178.    public void query()
179.    {
180.      try
181.      {
182.          String sql="select * from student where stuage>?";
183.          Cursor cur=sld.rawQuery(sql, new String[]{"5"});
184.          appendMessage("学号\t\t姓名\t\t年龄\t班级");
185.          while(cur.moveToNext())
186.          {
```

```
187.              String sno=cur.getString(0);
188.              String sname=cur.getString(1);
189.              int sage=cur.getInt(2);
190.              String sclass=cur.getString(3);
191.              appendMessage(sno+"\t"+sname+"\t\t"+sage+"\t"+sclass);
192.            }
193.            cur.close();
194.        }
195.        catch(Exception e)
196.        {
197.            Toast.makeText(this, "数据库错误:"+e.toString(), Toast.LENGTH_
                SHORT).show();;
198.        }
199.      }
200.
201.      //向文本区中添加文本
202.      public void appendMessage(String msg)
203.      {
204.        EditText et=(EditText)this.findViewById(R.id.EditText01);
205.        et.append(msg+"\n");
206.      }
207.  }
```

(4) 在 src 目录中 com. hisoft. activity 包下创建 MyContentProvider. java 文件,代码如下:

```
1.   package com.hisoft.activity;
2.
3.   import android.content.ContentProvider;
4.   import android.content.ContentValues;
5.   import android.content.UriMatcher;
6.   import android.database.Cursor;
7.   import android.database.sqlite.SQLiteDatabase;
8.   import android.net.Uri;
9.
10.
11.  public class MyContentProvider extends ContentProvider {
12.
13.    private static final UriMatcher um;
14.    static
15.    {
16.      um=new UriMatcher(UriMatcher.NO_MATCH);
17.      um.addURI("com.hisoft.provider.student", "stu", 1);
18.    }
19.
20.    SQLiteDatabase sld;
```

```
21.
22.    @Override
23.    public String getType(Uri uri) {
24.      return null;
25.    }
26.
27.    @Override
28.    public Cursor query(Uri uri, String[] projection, String selection,
29.        String[] selectionArgs, String sortOrder) {
30.
31.      switch(um.match(uri))
32.      {
33.        case 1:
34.
35.          Cursor cur=sld.query
36.          (
37.              "student",
38.              projection,
39.              selection,
40.              selectionArgs,
41.              null,
42.              null,
43.              sortOrder
44.          );
45.          return cur;
46.      }
47.      return null;
48.    }
49.
50.    @Override
51.    public int delete(Uri arg0, String arg1, String[] arg2) {
52.      //TODO Auto-generated method stub
53.      return 0;
54.    }
55.
56.    @Override
57.    public Uri insert(Uri uri, ContentValues values) {
58.      //TODO Auto-generated method stub
59.      return null;
60.    }
61.
62.    @Override
63.    public boolean onCreate() {
64.
```

```
65.        sld=SQLiteDatabase.openDatabase
66.          (
67.              "/data/data/com.hisoft.activity/mydb",           //数据库所在路径
68.              null,                                            //CursorFactory
69.              SQLiteDatabase.OPEN_READWRITE|SQLiteDatabase.CREATE_IF_NECESSARY
                                                                  //读写,若不存在则创建
70.          );
71.
72.        return false;
73.     }
74.
75.     @Override
76.     public int update(Uri uri, ContentValues values, String selection,
77.          String[] selectionArgs) {
78.        //TODO Auto-generated method stub
79.        return 0;
80.     }
81.  }
```

(5) 在 AndroidManifest. xml 文件中的<application>根结点下添加<provider>节点标签,添加权限为后续的 9.4 节案例程序应用提供数据接口,暴露数据的内容,代码如下:

```
1.  <provider
2.              android:name=".MyContentProvider"
3.              android:authorities="com.hisoft.provider.student"
4.          />
```

(6) 部署 SQLiteDemo 工程,程序运行结果如图 9-33 所示。

单击"创建/打开数据库"按钮,如果数据库存在,则打开数据库;如果不存在,则创建数据库,并同时在数据库中创建 student 表。运行结果如图 9-34 所示。

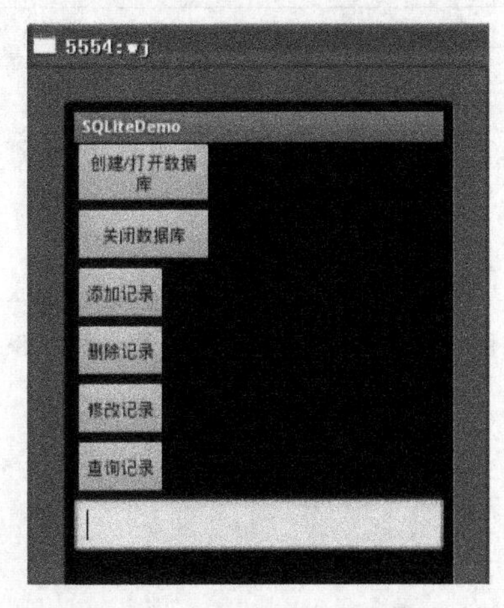

图 9-33 SQLiteDemo 运行效果

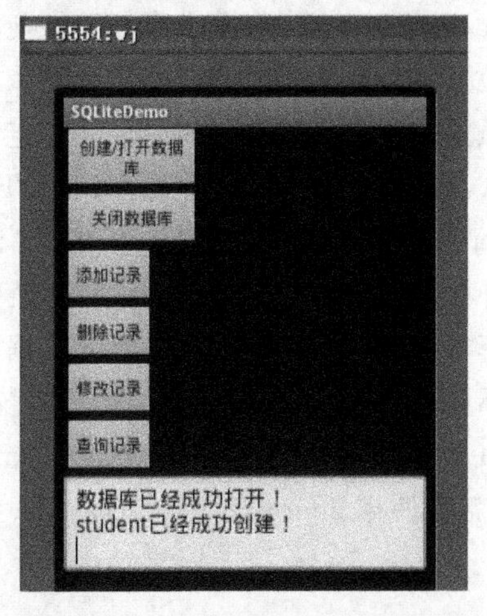

图 9-34 打开数据库并创建数据库表

单击“添加记录”按钮，程序中代码默认设置的 SQL 语句添加一条学生编号为 10001 的记录，如添加成功，显示“成功插入一条记录！”信息，如图 9-35 所示。

单击“查询记录”按钮，继续在信息后面添加数据库存在的记录，如图 9-36 所示。

数据库已经成功打开！
student已经成功创建！
成功插入一条记录！

图 9-35　插入记录

数据库已经成功打开！
student已经成功创建！
成功插入一条记录！
学号　姓名　年龄　班级
10001　张三　10 11010

图 9-36　查询记录

单击“修改记录”按钮，如修改成功，显示“成功更新记录！”信息，数据库中学生姓名更新为“李四”，单击“查询”按钮，如图 9-37 所示。

单击“删除记录”按钮，删除数据库中所有的记录数据，如删除成功，显示“成功删除所有记录！”，如图 9-38 所示。

单击“关闭数据库”按钮，显示如图 9-39 所示，表示数据库关闭成功。

数据库已经成功打开！
student已经成功创建！
成功插入一条记录！
学号　姓名　年龄　班级
10001　张三　10 11010
成功更新记录！
学号　姓名　年龄　班级
10001　李四　10 11010

图 9-37　修改记录

数据库已经成功打开！
student已经成功创建！
成功插入一条记录！
学号　姓名　年龄　班级
10001　张三　10 11010
成功更新记录！
学号　姓名　年龄　班级
10001　李四　10 11010
成功删除所有记录！

图 9-38　删除记录

student已经成功创建！
成功插入一条记录！
学号　姓名　年龄　班级
10001　张三　10 11010
成功更新记录！
学号　姓名　年龄　班级
10001　李四　10 11010
成功删除所有记录！
学号　姓名　年龄　班级
数据库已经成功关闭！

图 9-39　关闭数据库

9.4　数据共享

9.4.1　ContentProvider 简介

ContentProvider 类位于 android. content 包下，其类的继承结构如图 9-40 所示，ContentProvider(数据提供者)是在应用程序间共享数据的一种接口机制。

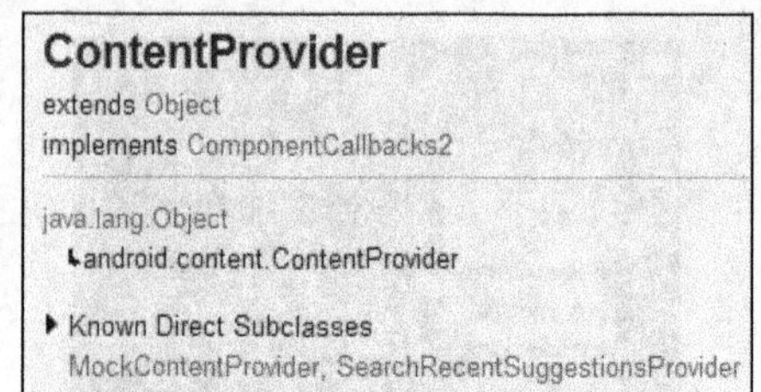

图 9-40　ContentProvider 类继承关系

虽然在前面章节讲述中，通过指定文件的操作模式为 Context. MODE_WORLD_READABLE 或 Context. MODE_WORLD_WRITEABLE 也可以对外共享数据。但如果采用文件操作模式对外共享数据，数据的访问方式会因数据存储的方式而不同，导致数据的访问方式无法统一，如采用 xml 文件对外共享数据，需要进行 xml 解析才能读取数据；采用 sharedpreferences 共享数据，需要使用 sharedpreferences API 读取数据。

1. ContentProvider 的作用

在 Android 系统中，ContentProvider 的作用是对外共享数据，也就是说 ContentProvider

提供了在多个应用程序之间统一的数据共享方法，将需要共享的数据封装起来，提供了一组供其他应用程序调用的接口，通过 ContentResolver 来操作数据。应用程序可以指定需要共享的数据，而其他应用程序则可以在不知数据来源、路径的情况下对共享数据进行查询、添加、删除和更新等操作。使用 ContentProvider 对外共享数据的好处是统一了数据的访问方式。如果用户不需要在多个应用程序之间共享数据，可以通过上节讲述 SQLiteDatabase 创建数据库的方式实现数据内部共享。

2. ContentProvider 调用原理

ContentProvider 创建和使用前，需要先通过数据库、文件系统或网络实现底层数据存储功能，然后自定义类继承 ContentProvider 类，并在其中实现基本数据操作的接口函数，包括添加、删除、查找和更新等功能。

ContentProvider 的接口函数不能直接使用，需要使用 ContentResolver 对象，通过 URI 间接调用 ContentProvider。

用户使用 ContentResolver 对象与 ContentProvider 进行交互，而 ContentResolver 则通过 URI 确定需要访问的 ContentProvider 的数据集。ContentResolver 对象与 ContentProvider 的调用关系如图 9-41 所示。

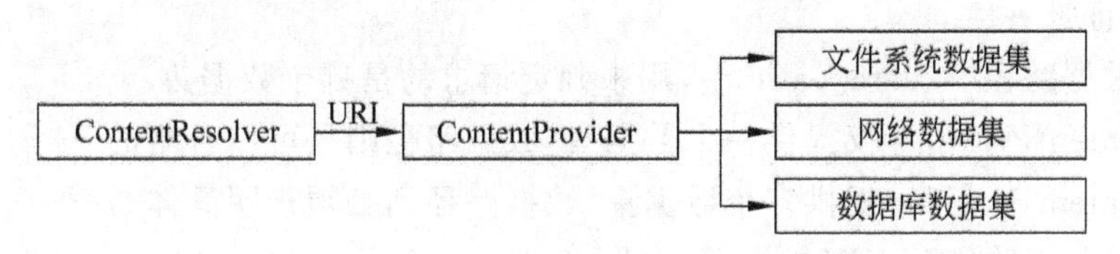

图 9-41 ContentResolver 与 ContentProvider 调用关系

其中 ContentProvider 负责组织应用程序的数据；向其他应用程序提供数据。

ContentResolver 则负责获取 ContentProvider 提供的数据；修改/添加/删除更新数据等。

9.4.2 Uri、UriMatcher 和 ContentUris 简介

1. Uri 简介

上述 Uri 代表了要操作的数据 Uri 的信息，主要有两部分：

(1) 需要操作的 ContentProvider。

(2) 对 ContentProvider 中的什么数据进行操作通过 Uri 来确定。

下面分别就上述 Uri 包含的两部分进行介绍。

(1) ContentProvider 数据模式。

ContentProvider 的数据模式类似于数据库的数据表，每行是一条记录，每列具有相同的数据类型，每条记录都包含一个长型的字段_ID 用来唯一标识每条记录。

ContentProvider 可以提供多个数据集，调用者使用 URI 对不同的数据集的数据进行操作。

ContentProvider 数据模式如表 9-5 所示。

表 9-5　ContentProvider 数据模式

_ID	NAME	_ID	NAME
1	John	2	Sam

(2) Uri(Uniform Resource Identifier,通用资源标志符)。

Uri 用来定位任何远程或本地的可用资源。在 ContentProvider 中使用的 Uri 通常由以下几部分组成,如图 9-42 所示。

图 9-42　Uri 的组成结构

ContentProvider(数据提供者)的 scheme 已经由 Android 所规定,scheme 为 content://。content://是通用前缀,表示该 URI 用于 ContentProvider 定位资源,无须修改。

主机名或＜authority＞是授权者名称,用来确定具体由哪一个 ContentProvider 提供资源,外部调用者可以根据这个标识来找到它。因此,一般＜authority＞都由类的小写全称组成,以保证唯一性。

路径是数据路径(＜data_path＞),用来确定请求的是哪个数据集。

如果 ContentProvider 仅提供一个数据集,数据路径则是可以省略的。

如果 ContentProvider 提供多个数据集,数据路径则必须指明具体是哪一个数据集。

数据集的数据路径可以写成多段格式,例如/wjj/house 和/wjj/tea。＜id＞是数据编号,用来唯一确定数据集中的一条记录,用来匹配数据集中_ID 字段的值。

如果请求的数据并不只限于一条数据,则＜id＞可以省略。

android SDK 推荐的方法是:在提供数据表字段中包含一个 ID,在创建表时 INTEGER PRIMARY KEY AUTOINCREMENT 标识此 ID 字段。

例如:

wjj/1 表示要操作 wjj 表中 id 为 1 的记录。

wjj/1/name 表示要操作 wjj 表中 id 为 1 的记录的 name 字段。

/wjj 表示要操作 wjj 表中的所有记录。

注意:如上述调用关系中所述,要操作的数据不一定来自数据库,也可以是文件系统、xml 或网络等其他存储方式。例如要操作 xml 文件中 wjj 节点下的 name 节点,构建的路径为/wjj/name。

如果要把一个字符串转换成 Uri,可以使用 Uri 类中的 parse()方法,如下:

```
Uri uri=Uri.parse("content://com.hisoft. provider.helloprovider/wjj ")
```

2. UriMatcher 类简介

上述 Uri 代表了要操作的数据,需要解析 Uri 并从 Uri 中获取数据。

UriMatcher 类是 Android 系统提供的用于操作 Uri 的工具类。它用于匹配 Uri,用法如下:

(1) 注册需要匹配 Uri 路径,如下:

```
UriMatcher  sMatcher=new UriMatcher(UriMatcher.NO_MATCH);
//常量 UriMatcher.NO_MATCH 表示不匹配任何路径的返回码
//如果 match()方法匹配 content://com.hisoft. provider.helloprovider/wjj 路径,返回
//匹配码为 1
sMatcher.addURI("com.hisoft. provider.helloprovider ", "wjj", 1);
                                        //添加需要匹配 uri,如果匹配就会返回匹配码
//如果 match()方法匹配 content://com.hisoft. provider.helloprovider/wjj/1 路径,返
//回匹配码为 2
```

上述代码中 addURI ()方法的声明语法:

```
public void addURI (String authority, String path, int code)
```

authority 表示匹配的授权者名称。

path 表示数据路径。

可以代表任何数字。

code 表示返回代码。

(2) 使用 sMatcher. match(uri)方法对输入的 Uri 进行匹配。

如果匹配就返回匹配码,匹配码是调用 addURI()方法传入的第三个参数。假设匹配 content://com. hisoft. provider. helloprovider/wjj 路径,返回的匹配码为 1。代码如下:

```
sMatcher.addURI("com.hisoft. provider.helloprovider ", "wjj/#", 2);  //#号为通配符
switch (sMatcher.match(Uri.parse("content://com.hisoft. provider.helloprovider/
wjj/1"))) {
  case 1
    break;
  case 2
    break;
  default:                //不匹配
    break;
}
```

3. ContentUris 类简介

ContentUris 类也是 Android 系统提供的用于操作 Uri 的工具类,用于操作 Uri 路径后面的 ID 部分。它有两个比较常用的方法: withAppendedId(uri,id)和 parseId(uri)方法。

withAppendedId(uri,id)用于为路径加上 ID 部分,代码如下:

```
Uri uri=Uri.parse("content://com.hisoft. provider.helloprovider/wjj")
Uri resultUri=ContentUris.withAppendedId(uri, 1);
//生成后的 Uri 为 content://com.hisoft. provider.helloprovider/wjj/1
```

parseId(uri)方法用于从路径中获取 ID 部分:

```
Uri uri=Uri.parse("content://com.hisoft. provider.helloprovider/wjj/1")
```

```
long personid=ContentUris.parseId(uri);              //获取的结果为 1
```

9.4.3 创建 ContentProvider

ContentProvider 的创建分为三步：

(1) 自定义类继承 ContentProvider，并重载 ContentProvider 的 6 个方法。

新创建的自定义类继承 ContentProvider 后，需要重载 6 个方法，代码如下：

```
1.   public class ContentProviderDemo extends ContentProvider{
2.      public boolean onCreate();            //初始化底层数据集和建立数据连接等工作
3.      public Uri insert(Uri uri, ContentValues values);        //添加数据集
4.      public int delete(Uri uri, String selection, String[] selectionArgs);
5.                                                                //删除数据集
6.       public int update(Uri uri, ContentValues values, String selection, String[]
         selectionArgs);                                          //更新数据集
7.      public Cursor query(Uri uri, String[] projection,String selection, String[]
        selectionArgs, String sortOrder);                         //查询数据集
8.      public String getType(Uri uri)              //返回指定 URI 的 MIME 数据类型
9.   }
```

注意：如果 URI 是单条数据，则返回的 MIME 数据类型应以 vnd.android.cursor.item 开头。

如果 URI 是多条数据，则返回的 MIME 数据类型应以 vnd.android.cursor.dir/开头。

(2) 实现 UriMatcher。

在新创建的 ContentProvider 类中，通过创建一个 UriMatcher 用于判断 URI 是单条数据还是多条数据。通常为了便于判断和使用 URI，一般将 URI 的授权者名称和数据路径等内容声明为静态常量，并声明 CONTENT_URI。

```
1.   public static final String AUTHORITY=" com.hisoft.helloprovider ";
2.   public static final String PATH_SINGLE="wjj/#";
3.   public static final String PATH_MULTIPLE="wjj";
4.   public static final String CONTENT_URI_STRING="content://"+AUTHORITY+"/"+
     PATH_MULTIPLE;
5.   public static final Uri CONTENT_URI=Uri.parse(CONTENT_URI_STRING);
6.   private static final int MULTIPLE_WJJ=1;
7.   private static final int SINGLE_WJJ=2;
8.
9.   private static final UriMatcher uriMatcher;
10.  static {
11.    uriMatcher=new UriMatcher(UriMatcher.NO_MATCH);
12.    uriMatcher.addURI(AUTHORITY, PATH_SINGLE, MULTIPLE_WJJ);
13.    uriMatcher.addURI(AUTHORITY, PATH_MULTIPLE, SINGLE_WJJ);
14.  }
```

然后，在使用 UriMatcher 时可以直接调用 match()，对指定的 URI 进行判断，代码如下：

```
switch(uriMatcher.match(uri)){
  case MULTIPLE_WJJ:
    //多条数据的处理过程
    break;
  case SINGLE_WJJ:
    //单条数据的处理过程
    break;
  default:
    throw new IllegalArgumentException("非法的 URI:"+uri);
}
```

(3) 在 AndroidManifest.xml 文件中注册 ContentProvider。

实现完成上述 ContentProvider 类的代码后，需要在 AndroidManifest.xml 文件中进行注册，在<application>根节点下添加<provider>标签，并设置属性，代码如下：

```
<provider android:name=".HelloProvider"
    android:authorities="com.hisoft.helloprovider"/>
<!--注册了一个授权者名称为"com.hisoft.helloprovider 的 ContentProvider,其实现类是
HelloProvider-->
```

9.4.4 ContentResolver 操作数据

使用 ContentResolver 类可以完成外部应用对 ContentProvider 中的数据进行添加、删除、修改和查询操作。ContentResolver 对象的创建可以使用 Activity 提供的 getContentResolver()方法。ContentResolver 类的方法有：

- public Uri insert(Uri uri,ContentValues values)：用于向 ContentProvider 添加数据。
- public int delete(Uri uri, String selection, String[] selectionArgs)：用于从 ContentProvider 删除数据。
- public int update(Uri uri, ContentValues values, String selection, String[] selectionArgs)：用于更新 ContentProvider 中的数据。
- public Cursor query(Uri uri, String[] projection, String selection, String[] selectionArgs,String sortOrder)：用于从 ContentProvider 中获取数据。

这些方法的第一个参数为 Uri，代表要操作的 ContentProvider 和对其中的什么数据进行操作。示例代码如下：

```
1.   ContentResolver resolver=getContentResolver();
2.   Uri uri=Uri.parse("content://com.hisoft.helloprovider/wjj ");
3.   //添加一条记录
4.   ContentValues values=new ContentValues();
5.   values.put("name", "John");
6.   values.put("age", 20);
7.   resolver.insert(uri, values);
8.   //获取 wjj 表中所有记录
```

```
9.   Cursor cursor=resolver.query(uri, null, null, null, "usrid desc");
10.  while(cursor.moveToNext()){
11.      Log.i("ContentTest", "usrid="+cursor.getInt(0)+", name="+cursor.
         getString(1));
12.  }
13.  //把 id 为 1 的记录的 name 字段值更新为 lisi
14.  ContentValues updateValues=new ContentValues();
15.  updateValues.put("name", "lisi");
16.  Uri updateIdUri=ContentUris.withAppendedId(uri, 2);
17.  resolver.update(updateIdUri, updateValues, null, null);
18.  //删除 id 为 2 的记录
19.  Uri deleteIdUri=ContentUris.withAppendedId(uri, 2);
20.  resolver.delete(deleteIdUri, null, null);
```

9.4.5 ContentProvider 应用案例

前面介绍了 ContentProvider 的调用关系、创建 ContentProvider 的步骤，以及 ContentResolver 操作数据的方法，下面通过 9.3.5 节的案例详细介绍 ContentProvider 的应用。

(1) 创建一个新的 Android 工程，工程名为 ContentProviderDemo，目标 API 选择 10(即 Android 2.3.3 版本)，应用程序名为 ContentProviderDemo，包名为 com.hisoft.activity，创建的 Activity 的名字为 MainActivity，最小 SDK 版本根据选择的目标 API 会自动添加为 10，创建项目工程如图 9-43 所示。

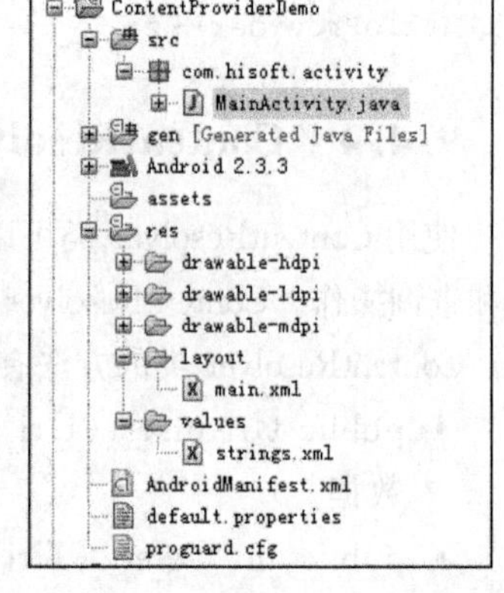

图 9-43 ContentProviderDemo 工程目录结构

(2) 修改 res 目录下 layout 文件夹中的 main.xml 文件，设置线性布局中嵌套线性布局，添加两个 EditText 控件、一个 Button 控件、一个 TextView 和一个 ScrollView 控件描述，并设置相关属性，代码如下：

```
1.   <?xml version="1.0" encoding="utf-8"?>
2.   <LinearLayout xmlns:android="http://schemas.android.com/apk/res/android"
3.       android:orientation="vertical"
4.       android:layout_width="fill_parent"
5.       android:layout_height="fill_parent"
6.       >
7.       <LinearLayout
8.       android:orientation="horizontal"
9.       android:layout_width="fill_parent"
10.      android:layout_height="wrap_content"
11.      >
12.        <TextView
13.          android:layout_width="wrap_content"
14.          android:layout_height="wrap_content"
```

```
15.          android:text="请输入姓名:"
16.          android:textColor="@android:color/white"
17.          android:textSize="18dip"
18.          android:paddingRight="3dip"
19.        />
20.        <EditText
21.          android:text=""
22.          android:id="@+id/EditText01"
23.          android:layout_width="150dip"
24.          android:layout_height="wrap_content">
25.        </EditText>
26.        <Button
27.          android:text="查询"
28.          android:id="@+id/Button01"
29.          android:layout_width="wrap_content"
30.          android:layout_height="wrap_content">
31.        </Button>
32.     </LinearLayout>
33.     <ScrollView
34.       android:id="@+id/ScrollView01"
35.       android:layout_width="fill_parent"
36.       android:layout_height="wrap_content">
37.      <EditText
38.        android:id="@+id/EditText02"
39.        android:layout_width="fill_parent"
40.        android:layout_height="wrap_content">
41.      </EditText>
42.     </ScrollView>
43.  </LinearLayout>
```

(3) 修改 src 目录中 com. hisoft. activity 包下的 MainActivity. java 文件,读取上一个案例创建的数据库数据,代码如下:

```
1.   package com.hisoft.activity;
2.   import android.app.Activity;
3.   import android.content.ContentResolver;
4.   import android.database.Cursor;
5.   import android.net.Uri;
6.   import android.os.Bundle;
7.   import android.view.View;
8.   import android.view.View.OnClickListener;
9.   import android.widget.Button;
10.  import android.widget.EditText;
11.
12.  public class MainActivity extends Activity {
13.
```

```
14.        private ContentResolver cr;
15.
16.        @Override
17.        public void onCreate(Bundle savedInstanceState) {
18.          super.onCreate(savedInstanceState);
19.          setContentView(R.layout.main);
20.
21.          cr=this.getContentResolver();
22.
23.          //初始化查询按钮
24.          Button b=(Button)this.findViewById(R.id.Button01);
25.          b.setOnClickListener(
26.            new OnClickListener()
27.            {
28.              @Override
29.              public void onClick(View v) {
30.                EditText et=(EditText)findViewById(R.id.EditText01);
31.                String stuname=et.getText().toString().trim();
32.
33.                Cursor cur=cr.query
34.                (
35.        Uri.parse("content://com.hisoft.activity.mycontentprovider/stu "),
36.               new String[]{"stuno","stuname","stuage","stuclass"},
37.               "stuname=?",
38.               new String[]{stuname},
39.               "stuage ASC"
40.            );
41.
42.                appendMessage("学号\t\t姓名\t\t年龄\t班级");
43.                while(cur.moveToNext())
44.                {
45.                  String stuno=cur.getString(0);
46.                  String sname=cur.getString(1);
47.                  int stuage=cur.getInt(2);
48.                  String stuclass=cur.getString(3);
49.                     appendMessage(stuno+"\t"+sname+"\t\t"+stuage+"\t"+
                     stuclass);
50.                }
51.                cur.close();
52.              }
53.            }
54.          );
55.        }
56.
57.        //向文本区中添加文本
```

```
58.        public void appendMessage(String msg)
59.        {
60.          EditText et=(EditText)this.findViewById(R.id.EditText02);
61.          et.append(msg+"\n");
62.        }
63.    }
```

(4) 部署 ContentProviderDemo 工程，程序运行后如图 9-44 所示。在编辑框中输入查询的姓名，然后单击“查询”按钮，程序从 9.3.5 节案例中读取数据，结果如图 9-45 所示。

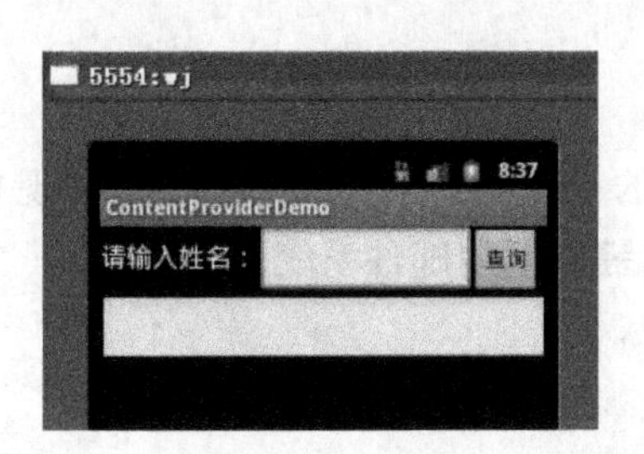

图 9-44　ContentProviderDemo 运行效果

图 9-45　查询结果

9.5　网络存储

9.5.1　网络存储简介

前面介绍的 4 种存储都是将数据存储在本地设备上，本节介绍的是另外一种存储(获取)数据的方式，通过网络来实现数据的存储和获取。通过网络来获取和保存数据资源需要设备保持网络连接状态，所以相对存在一些限制。经常用于相关操作的两个类，分别是 java.net.* 和 android.net.*。

9.5.2　网络存储应用案例

下面通过一个使用模拟器给 gmail 邮箱发邮件案例，详细介绍网络存储的应用。

(1) 创建一个新的 Android 工程，工程名为 NetWorkDemo，目标 API 选择 10(即 Android 2.3.3 版本)，应用程序名为 NetWorkDemo，包名为 com.hisoft.activity，创建的 Activity 的名字为 MainActivity，最小 SDK 版本根据选择的目标 API 会自动添加为 10，创建项目工程。

(2) 修改 res 目录下 layout 文件夹中的 main.xml 文件，设置线性布局中嵌套线性布局，添加一个 EditText 控件、一个 TextView 描述，并设置相关属性，代码如下：

```
1.   <?xml version="1.0" encoding="utf-8"?>
2.   <LinearLayout xmlns:android="http://schemas.android.com/apk/res/android"
3.       android:orientation="vertical"
4.       android:layout_width="fill_parent"
5.       android:layout_height="fill_parent"
```

```
6.     >
7.   <TextView
8.     android:layout_width="fill_parent"
9.     android:layout_height="wrap_content"
10.    android:text="@string/hello"
11.    />
12.  <EditText
13.  android:id="@+id/EditText01"
14.  android:layout_width="fill_parent"
15.  android:layout_height="wrap_content">
16. </EditText>
17. </LinearLayout>
```

(3) 修改 src 目录中 com. hisoft. activity 包下的 MainActivity. java 文件,实现在文本框中输入邮件内容,邮件主题为"网络存储",按返回键则调用邮件系统,代码如下:

```
1.  package com.hisoft.activity;
2.
3.  import android.app.Activity;
4.  import android.content.Intent;
5.  import android.net.Uri;
6.  import android.os.Bundle;
7.  import android.view.KeyEvent;
8.  import android.widget.EditText;
9.
10.
11.
12. public class NetWorkDemoActivity extends Activity {
13.   private EditText mEditText;
14.
15.   /**Called when the activity is first created. */
16.   public void onCreate(Bundle savedInstanceState) {
17.   super.onCreate(savedInstanceState);
18.   setContentView(R.layout.main);
19.   mEditText=(EditText) findViewById(R.id.EditText01);
20.   }
21.
22.   @Override
23.   public boolean onKeyDown(int keyCode, KeyEvent event) {
24.   //TODO Auto-generated method stub
25.   if (keyCode==KeyEvent.KEYCODE_BACK) {
26.     final Intent intent=new Intent(android.content.Intent.ACTION_SEND);
27.     intent.putExtra(android.content.Intent.EXTRA_EMAIL, new String[]{
        "wjjyu@gmail.com"});
28.   intent.setType("plain/text");
```

```
29.    intent.putExtra(android.content.Intent.EXTRA_SUBJECT, "网络存储");
30.    intent.putExtra(android.content.Intent.EXTRA_TEXT,
31.    String.valueOf(mEditText.getText()));
32.    startActivity(Intent.createChooser(intent, "Send mail..."));
33.    this.finish();
34.    return true;
35.    }
36.    return super.onKeyDown(keyCode, event);
37.    }
38. }
```

(4) 设置模拟器邮件系统配置,选择 E-mail 图标,设置 gmail 账户的用户名和密码,连接 gmail 服务器,测试联通。

(5) 部署运行 NetWorkDemo 工程,运行后在文本框中输入邮件内容“测试邮件”,如图 9-46 所示。

然后按返回键,自动调用邮件系统,如图 9-47 所示。然后单击 Send 按钮,即可发送邮件到设定的 gmail 邮箱。

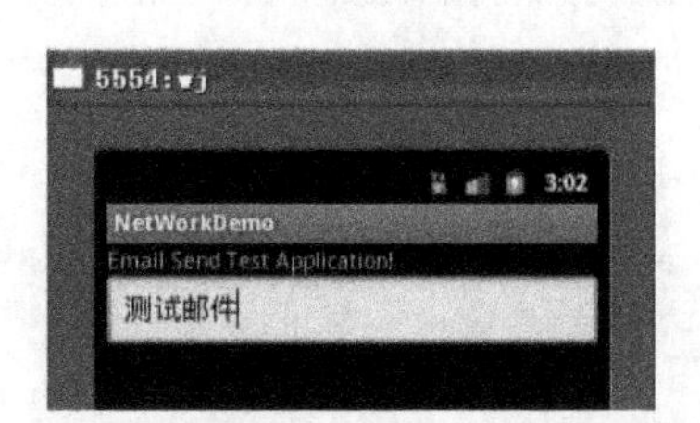

图 9-46 测试邮件

图 9-47 发送邮件

拓展提示：在项目开发中,不同的数据存储方式有不同的应用条件,它们之间各自的优劣需要进行对比和分析。此外,尤其是面对比较大型的数据时,如何进行数据的有效存储和提取,以及是否进行缓存都是当前比较重要的应用。

9.6 数据存储项目案例

学习目标：学习、掌握控件及界面布局,SQLite 数据库创建、操作、管理及应用。

案例描述：使用 LinearLayout 线性布局等不同的布局文件,添加 TableRow、TextView 和 ListView 等控件,并通过 Intent、SimpleCursorAdapter、onListItemClick、onCreateContextMenu、onOptionsItemSelected、onPrepareOptionsMenu 和 onCreateOptionsMenu 等方法,然后再调用 IntentFilter 的匹配规则,实现个人通讯录。

案例要点：IntentFilter 规则设置、SQLite 数据库操作方法。

案例实施：

(1) 创建工程 Android_Contacts，选择 Android 2.3.3 作为目标平台，如图 9-48 所示。

(2) 在 res 目录中的 layout 文件下创建 contact_list.xml 文件，代码如下：

```
1.   <?xml version="1.0" encoding="utf-8"?>
2.    < LinearLayout  xmlns: android =" http://
      schemas.android.com/apk/res/android"
3.        android:orientation="vertical"
4.        android:layout_width="fill_parent"
5.        android:layout_height="fill_parent"
6.        android:background="@drawable/bg">
7.    < ListView android: id =" @ + id/ListView01"
     android: layout _ width =" fill _ parent "
     android:layout_height="wrap_content">
     </ListView>
8.   </LinearLayout>
```

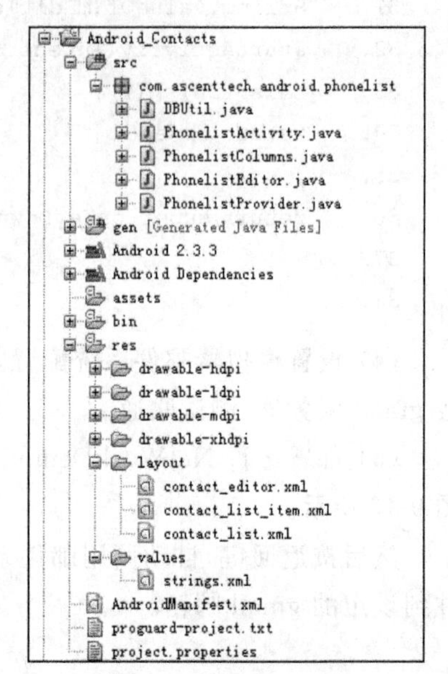

图 9-48 Android_Contacts 工程目录结构

(3) 在 res 目录中的 layout 文件下创建 contact_list_item.xml 文件，代码如下：

```
1.   <?xml version="1.0" encoding="utf-8"?>
2.   <LinearLayout xmlns:android="http://schemas.android.com/apk/res/android"
3.       android:orientation="vertical"
4.       android:layout_width="fill_parent"
5.       android:layout_height="fill_parent">
6.   <TextView xmlns:android="http://schemas.android.com/apk/res/android"
7.       android:id="@android:id/text1"
8.       android:layout_width="fill_parent"
9.       android:layout_height="fill_parent"
10.      android:textStyle="bold"
11.      android:textSize="18px"
12.      android:gravity="center_vertical"
13.      android:paddingLeft="10px"
14.      android:singleLine="true"
15.  />
16.  <TextView xmlns:android="http://schemas.android.com/apk/res/android"
17.      android:id="@android:id/text2"
18.      android:layout_width="fill_parent"
19.      android:layout_height="fill_parent"
20.      android:textStyle="normal"
21.      android:textSize="14px"
22.      android:gravity="center_vertical"
23.      android:paddingLeft="10px"
24.      android:singleLine="true"
25.  />
```

```
26.  </LinearLayout>
```

(4) 在 res 目录中的 layout 文件下创建 contact_edit.xml 文件,代码如下:

```
1.   <?xml version="1.0" encoding="utf-8"?>
2.   <LinearLayout xmlns:android="http://schemas.android.com/apk/res/android"
3.       android:layout_width="fill_parent"
4.       android:layout_height="fill_parent"
5.       android:background="@drawable/bg"
6.       android:orientation="vertical">
7.
8.       <TableRow
9.           android:id="@+id/TableRow01"
10.          android:layout_width="fill_parent"
11.          android:layout_height="wrap_content">
12.
13.          <TextView
14.              android:id="@+id/TextView01"
15.              android:layout_width="wrap_content"
16.              android:layout_height="wrap_content"
17.              android:text="@string/name"
18.              android:textSize="16px">
19.          </TextView>
20.
21.          <EditText
22.              android:id="@+id/EditText01"
23.              android:layout_width="fill_parent"
24.              android:layout_height="wrap_content"/>
25.      </TableRow>
26.
27.      <TableRow
28.          android:id="@+id/TableRow02"
29.          android:layout_width="fill_parent"
30.          android:layout_height="wrap_content">
31.
32.          <TextView
33.              android:id="@+id/TextView02"
34.              android:layout_width="wrap_content"
35.              android:layout_height="wrap_content"
36.              android:text="@string/mobile"
37.              android:textSize="16px">
38.          </TextView>
39.
40.          <EditText
41.              android:id="@+id/EditText02"
42.              android:layout_width="fill_parent"
```

```
43.                android:layout_height="wrap_content">
44.            </EditText>
45.        </TableRow>
46.
47.        <TableRow
48.            android:id="@+id/TableRow03"
49.            android:layout_width="fill_parent"
50.            android:layout_height="wrap_content">
51.
52.            <TextView
53.                android:id="@+id/TextView03"
54.                android:layout_width="wrap_content"
55.                android:layout_height="wrap_content"
56.                android:text="@string/email"
57.                android:textSize="16px">
58.            </TextView>
59.
60.            <EditText
61.                android:id="@+id/EditText03"
62.                android:layout_width="fill_parent"
63.                android:layout_height="wrap_content">
64.            </EditText>
65.        </TableRow>
66.
67.        <TableRow
68.            android:id="@+id/TableRow04"
69.            android:layout_width="fill_parent"
70.            android:layout_height="wrap_content">
71.
72.            <TextView
73.                android:id="@+id/TextView04"
74.                android:layout_width="wrap_content"
75.                android:layout_height="wrap_content"
76.                android:text="@string/addr"
77.                android:textSize="16px">
78.            </TextView>
79.
80.            <EditText
81.                android:id="@+id/EditText04"
82.                android:layout_width="fill_parent"
83.                android:layout_height="wrap_content">
84.            </EditText>
85.        </TableRow>
86.
87.        <TableRow
```

```
88.         android:id="@+id/TableRow05"
89.         android:layout_width="fill_parent"
90.         android:layout_height="wrap_content">
91.
92.         <Button
93.             android:id="@+id/Button01"
94.             android:layout_width="wrap_content"
95.             android:layout_height="wrap_content"
96.             android:text="@string/save"
97.             android:textSize="16px">
98.         </Button>
99.
100.         <Button
101.             android:id="@+id/Button02"
102.             android:layout_width="wrap_content"
103.             android:layout_height="wrap_content"
104.             android:text="@string/cancel"
105.             android:textSize="16px">
106.         </Button>
107.     </TableRow>
108. </LinearLayout>
```

(5) 修改在 res 目录中 values 文件中的 strings. xml 文件，代码如下：

```
1.  <?xml version="1.0" encoding="utf-8"?>
2.  <resources>
3.  <string name="app_name">手机通讯录</string>
4.  <string name="name">姓名:</string>
5.  <string name="mobile">电话:</string>
6.  <string name="email">邮箱:</string>
7.  <string name="addr">地址:</string>
8.  <string name="group">分组:</string>
9.  <string name="save">保存</string>
10. <string name="cancel">取消</string>
11. <string name="contact_edit">编辑联系人</string>
12. <string name="contact_create">添加新的联系人</string>
13. <string name="error_msg">发生错误</string>
14. <string name="menu_revert">返回</string>
15. <string name="menu_delete">删除</string>
16. <string name="menu_discard">丢弃</string>
17. <string name="menu_add">添加联系人</string>
18. <string name="menu_edit">编辑联系人</string>
19. </resources>
```

(6) 修改 AndroidManifest. xml 文件，添加<provider>标签和<intent-filter>标签，代码如下：

```
1.   <?xml version="1.0" encoding="utf-8"?>
2.   <manifest xmlns:android="http://schemas.android.com/apk/res/android"
3.        package="com.ascenttech.android.phonelist"
4.        android:versionCode="1"
5.        android:versionName="1.0">
6.         <uses-sdk android:minSdkVersion="10"/>
7.      <application android:icon="@drawable/icon" android:label="@string/app_
     name">
8.      <provider android:name="PhonelistProvider"
9.                android:authorities="com.ascenttech.android.phonelist.
                  provider.contact"/>
10.
11.         <activity android:name=".PhonelistActivity"
12.                 android:label="@string/app_name">
13.             <intent-filter>
14.                 <action android:name="android.intent.action.MAIN"/>
15.                 <category android:name="android.intent.category.LAUNCHER"/>
16.             </intent-filter>
17.             <intent-filter>
18.                 <action android:name="android.intent.action.VIEW"/>
19.                 <action android:name="android.intent.action.EDIT"/>
20.                 <action android:name="android.intent.action.PICK"/>
21.                 <category android:name="android.intent.category.DEFAULT"/>
22.                     <data android:mimeType="vnd.android.cursor.dir/com.
                        ascenttech.android.phonelist.gcontacts"/>
23.             </intent-filter>
24.             <intent-filter>
25.                 <action android:name="android.intent.action.GET_CONTENT"/>
26.                 <category android:name="android.intent.category.DEFAULT"/>
27.                 <data android:mimeType="vnd.android.cursor.item/com.
                    ascenttech.android.phonelist.gcontacts"/>
28.             </intent-filter>
29.         </activity>
30.
31.
32.         <activity android:name=".PhonelistEditor"
33.             android:theme="@android:style/Theme.Light"
34.             android:label="PhonelistEditor">
35.             <intent-filter android:label="@string/menu_edit">
36.                 <action android:name="android.intent.action.VIEW"/>
37.                 <action android:name="android.intent.action.EDIT"/>
38.                 <action android:name="com.android.notepad.action.EDIT_NOTE"/>
39.                 <category android:name="android.intent.category.DEFAULT"/>
40.                    <data android:mimeType="vnd.android.cursor.item/com.
                       ascenttech.android.phonelist.gcontacts"/>
```

```
41.            </intent-filter>
42.
43.            <intent-filter>
44.                <action android:name="android.intent.action.INSERT"/>
45.                <category android:name="android.intent.category.DEFAULT"/>
46.                <data android:mimeType="vnd.android.cursor.dir/com.
                   ascenttech.android.phonelist.gcontacts"/>
47.            </intent-filter>
48.
49.        </activity>
50.
51.    </application>
52.
53. </manifest>
```

(7) 在 src 目录下的 com. ascenttech. android. phonelist 包中创建 DBUtil. java,代码如下:

```
1.  public class DBUtil extends SQLiteOpenHelper {
2.
3.    public static final String DATABASE_NAME="gcontacts.db";
4.    public static final int DATABASE_VERSION=2;
5.    public static final String CONTACTS_TABLE="contacts";
6.    //创建数据库
7.     private static final String DATABASE_CREATE="CREATE TABLE "+CONTACTS_
       TABLE+" ("+PhonelistColumns._ID+" integer primary key autoincrement,"
8.  +PhonelistColumns.NAME+" text,"
9.  +PhonelistColumns.MOBILE+" text,"+PhonelistColumns.EMAIL+" text,"+
PhonelistColumns.ADDR+" text,"+PhonelistColumns.CREATED+" long,"+
PhonelistColumns.MODIFIED+" long);";
10.   public DBUtil(Context context) {
11.     super(context, DATABASE_NAME, null, DATABASE_VERSION);
12.
13.   }
14.
15.   @Override
16.   public void onCreate(SQLiteDatabase db) {
17.
18.     db.execSQL(DATABASE_CREATE);
19.   }
20.
21.   @Override
22.   public void onUpgrade(SQLiteDatabase db, int oldVersion, int newVersion) {
23.
24.     db.execSQL("DROP TABLE IF EXISTS "+CONTACTS_TABLE);
25.     onCreate(db);
26.   }
```

```
27.
28. }
```

（8）在 src 目录下的 com. ascenttech. android. phonelist 包中创建 PhonelistActivity. java,代码如下：

```
1.   public class PhonelistActivity extends ListActivity {
2.     private static final String TAG="Contacts";
3.     private static final int AddContact_ID=Menu.FIRST;
4.     private static final int EditContact_ID=Menu.FIRST+1;
5.
6.       @Override
7.       public void onCreate(Bundle savedInstanceState) {
8.           super.onCreate(savedInstanceState);
9.           setDefaultKeyMode(DEFAULT_KEYS_SHORTCUT);
10.
11.          Intent intent=getIntent();
12.          if (intent.getData()==null) {
13.              intent.setData(PhonelistProvider.CONTENT_URI);
14.          }
15.
16.          getListView().setOnCreateContextMenuListener(this);
17.          Cursor cursor=managedQuery(getIntent().getData(),
             PhonelistColumns.PROJECTION, null, null,null);
18.          //注册每个列表表示形式:姓名+手机号码
19.          SimpleCursorAdapter adapter=new SimpleCursorAdapter(this, R.layout.
             contact_list_item, cursor,
20.                  new String[] { PhonelistColumns.NAME,PhonelistColumns.MOBILE
                     }, new int[] { android.R.id.text1,android.R.id.text2 });
21.          setListAdapter(adapter);
22.          Log.e(TAG+"onCreate"," is ok");
23.      }
24.
25.      @Override
26.      public boolean onCreateOptionsMenu(Menu menu) {
27.          super.onCreateOptionsMenu(menu);
28.
29.          menu.add(0, AddContact_ID, 0, R.string.menu_add)
30.             .setShortcut('3', 'a')
31.             .setIcon(android.R.drawable.ic_menu_add);
32.
33.          Intent intent=new Intent(null, getIntent().getData());
34.          intent.addCategory(Intent.CATEGORY_ALTERNATIVE);
35.          menu.addIntentOptions(Menu.CATEGORY_ALTERNATIVE, 0, 0,
36.        new ComponentName(this, PhonelistActivity.class), null, intent, 0, null);
37.          return true;
```

```
38.
39.     }
40.
41.     @Override
42.     public boolean onPrepareOptionsMenu(Menu menu) {
43.         super.onPrepareOptionsMenu(menu);
44.         final boolean haveItems=getListAdapter().getCount()>0;
45.
46.         if (haveItems) {
47.               Uri uri = ContentUris.withAppendedId(getIntent().getData(),
                  getSelectedItemId());
48.
49.             Intent[] specifics=new Intent[1];
50.             specifics[0]=new Intent(Intent.ACTION_EDIT, uri);
51.             MenuItem[] items=new MenuItem[1];
52.
53.             Intent intent=new Intent(null, uri);
54.             intent.addCategory(Intent.CATEGORY_ALTERNATIVE);
55.              menu.addIntentOptions(Menu.CATEGORY_ALTERNATIVE, 0, 0, null,
                 specifics, intent, 0,items);
56.
57.             if (items[0] !=null) {
58.                 items[0].setShortcut('1', 'e');
59.             }
60.         } else {
61.             menu.removeGroup(Menu.CATEGORY_ALTERNATIVE);
62.         }
63.
64.         return true;
65.     }
66.
67.     @Override
68.     public boolean onOptionsItemSelected(MenuItem item) {
69.         switch (item.getItemId()) {
70.         case AddContact_ID:
71.             //添加联系人
72.               startActivity(new Intent(Intent.ACTION_INSERT, getIntent().
                  getData()));
73.             return true;
74.         }
75.         return super.onOptionsItemSelected(item);
76.     }
77.
78.     @Override
79.     public void onCreateContextMenu(ContextMenu menu, View view, ContextMenuInfo
```

```
        menuInfo) {
80.         AdapterView.AdapterContextMenuInfo info;
81.         try {
82.             info= (AdapterView.AdapterContextMenuInfo) menuInfo;
83.         } catch (ClassCastException e) {
84.            return;
85.         }
86.
87.         Cursor cursor= (Cursor) getListAdapter().getItem(info.position);
88.         if (cursor==null) {
89.            return;
90.         }
91.
92.         menu.setHeaderTitle(cursor.getString(1));
93.
94.         menu.add(0, EditContact_ID, 0, R.string.menu_delete);
95.     }
96.
97.     @Override
98.     public boolean onContextItemSelected(MenuItem item) {
99.         AdapterView.AdapterContextMenuInfo info;
100.         try {
101.              info= (AdapterView.AdapterContextMenuInfo) item.getMenuInfo();
102.         } catch (ClassCastException e) {
103.             return false;
104.         }
105.
106.         switch (item.getItemId()) {
107.             case EditContact_ID: {
108.
109.                 Uri noteUri=ContentUris.withAppendedId(getIntent().
                     getData(), info.id);
110.                 getContentResolver().delete(noteUri, null, null);
111.                 return true;
112.             }
113.         }
114.         return false;
115.      }
116.
117.      @Override
118.      protected void onListItemClick(ListView l, View v, int position, long id) {
119.          Uri uri=ContentUris.withAppendedId(getIntent().getData(), id);
120.
121.          String action=getIntent().getAction();
122.          if (Intent.ACTION_PICK.equals(action)||Intent.ACTION_GET_CONTENT.
```

```
        equals(action)) {
123.
124.            setResult(RESULT_OK, new Intent().setData(uri));
125.        } else {
126.            //编辑联系人
127.            startActivity(new Intent(Intent.ACTION_EDIT, uri));
128.        }
129.    }
130. }
```

(9) 在 src 目录下的 com. ascenttech. android. phonelist 包中创建 PhonelistColumns. java,代码如下:

```
1.  public class PhonelistColumns implements BaseColumns {
2.    public PhonelistColumns(){
3.    }
4.    //列名
5.    public static final String NAME="name";
6.    public static final String MOBILE="mobileNumber";
7.    public static final String EMAIL="email";
8.    public static final String ADDR="addr";
9.    public static final String CREATED="createdDate";
10.   public static final String MODIFIED="modifiedDate";
11.   //列索引值
12.   public static final int _ID_COLUMN=0;
13.   public static final int NAME_COLUMN=1;
14.   public static final int MOBILE_COLUMN=2;
15.   public static final int EMAIL_COLUMN=3;
16.   public static final int ADDR_COLUMN=4;
17.   public static final int CREATED_COLUMN=5;
18.   public static final int MODIFIED_COLUMN=6;
19.
20.   //查询结果
21.   public static final String[] PROJECTION={
22.     _ID,                                        //0
23.     NAME,                                       //1
24.     MOBILE,                                     //2
25.     EMAIL,                                      //3
26.     ADDR                                        //4
27.   };
28.
29. }
```

(10) 在 src 目录下的 com. ascenttech. android. phonelist 包中创建 PhonelistEditor. java,代码如下:

```
1.  public class PhonelistEditor extends Activity {
```

```
2.        private static final String TAG="ContactEditor";
3.        private static final int STATE_EDIT=0;
4.        private static final int STATE_INSERT=1;
5.
6.        private static final int REVERT_ID=Menu.FIRST;
7.        private static final int DISCARD_ID=Menu.FIRST+1;
8.        private static final int DELETE_ID=Menu.FIRST+2;
9.
10.       private int mState;
11.       private Uri mUri;
12.       private Cursor mCursor;
13.
14.       private EditText nameText;
15.       private EditText mPhoneText;
16.       private EditText emailText;
17.       private EditText addrText;
18.       private Button saveButton;
19.       private Button cancelButton;
20.
21.       private String originalNameText="";
22.       private String originalMPhoneText="";
23.       private String originalEmailText="";
24.       private String originalAddrText="";
25.
26.       public void onCreate(Bundle savedInstanceState) {
27.           super.onCreate(savedInstanceState);
28.
29.           final Intent intent=getIntent();
30.           final String action=intent.getAction();
31.           Log.e(TAG+":onCreate", action);
32.           if (Intent.ACTION_EDIT.equals(action)) {
33.               mState=STATE_EDIT;
34.               mUri=intent.getData();
35.           } else if (Intent.ACTION_INSERT.equals(action)) {
36.               mState=STATE_INSERT;
37.               mUri=getContentResolver().insert(intent.getData(), null);
38.
39.               if (mUri==null) {
40.                   Log.e(TAG+":onCreate", "Failed to insert new Contact into "+
                      getIntent().getData());
41.                   finish();
42.                   return;
43.               }
44.               setResult(RESULT_OK, (new Intent()).setAction(mUri.toString()));
45.
```

```
46.          } else {
47.             Log.e(TAG+":onCreate", " unknown action");
48.             finish();
49.             return;
50.          }
51.
52.          setContentView(R.layout.contact_editor);
53.          nameText=(EditText) findViewById(R.id.EditText01);
54.          mPhoneText=(EditText) findViewById(R.id.EditText02);
55.          emailText=(EditText) findViewById(R.id.EditText03);
56.          addrText=(EditText) findViewById(R.id.EditText04);
57.
58.          saveButton=(Button)findViewById(R.id.Button01);
59.          cancelButton=(Button)findViewById(R.id.Button02);
60.
61.          saveButton.setOnClickListener(new OnClickListener(){
62.
63.            public void onClick(View v) {
64.              String text=nameText.getText().toString();
65.              if(text.length()==0){
66.                setResult(RESULT_CANCELED);
67.                deleteContact();
68.                finish();
69.              }else{
70.                updateContact();
71.              }
72.            }
73.
74.          });
75.          cancelButton.setOnClickListener(new OnClickListener(){
76.
77.            public void onClick(View v) {
78.              if(mState==STATE_INSERT){
79.                setResult(RESULT_CANCELED);
80.                deleteContact();
81.                finish();
82.              }else{
83.                backupContact();
84.              }
85.
86.            }
87.
88.          });
89.
90.          Log.e(TAG+":onCreate", mUri.toString());
```

```
91.            //获得并保存原始联系人信息
92.             mCursor= managedQuery(mUri, PhonelistColumns.PROJECTION, null, null,
                null);
93.            mCursor.moveToFirst();
94.            originalNameText=mCursor.getString(PhonelistColumns.NAME_COLUMN);
95.            originalMPhoneText=mCursor.getString(PhonelistColumns.MOBILE_
               COLUMN);
96.            originalEmailText=mCursor.getString(PhonelistColumns.EMAIL_
               COLUMN);
97.            originalAddrText=mCursor.getString(PhonelistColumns.ADDR_COLUMN);
98.
99.            Log.e(TAG, "end of onCreate()");
100.       }
101.
102.       @Override
103.       protected void onResume() {
104.           super.onResume();
105.
106.           if (mCursor !=null) {
107.             Log.e(TAG+":onResume","count:"+mCursor.getColumnCount());
108.               //读取并显示联系人信息
109.               mCursor.moveToFirst();
110.               if (mState==STATE_EDIT) {
111.                   setTitle(getText(R.string.contact_edit));
112.               } else if (mState==STATE_INSERT) {
113.                   setTitle(getText(R.string.contact_create));
114.               }
115.               String name=mCursor.getString(PhonelistColumns.NAME_COLUMN);
116.               String mPhone=mCursor.getString(PhonelistColumns.MOBILE_
                   COLUMN);
117.               String email=mCursor.getString(PhonelistColumns.EMAIL_
                   COLUMN);
118.               String addr=mCursor.getString(PhonelistColumns.ADDR_COLUMN);
119.
120.               Log.e(TAG+":onResume","name:"+name+"mPhone:"+mPhone+"email:"+
                   email+"addr:"+addr);
121.
122.               nameText.setText(name);
123.               mPhoneText.setText(mPhone);
124.               emailText.setText(email);
125.               addrText.setText(addr);
126.
127.           }else{
128.             setTitle(getText(R.string.error_msg));
129.           }
```

```
130.
131.  }
132.
133.      @Override
134.      protected void onPause() {
135.          super.onPause();
136.
137.          if (mCursor !=null) {
138.              String text=nameText.getText().toString();
139.
140.              if (text.length()==0) {
141.                  Log.e(TAG+":onPause","nameText is null ");
142.                  setResult(RESULT_CANCELED);
143.                  deleteContact();
144.
145.              //更新信息
146.              } else {
147.                ContentValues values=new ContentValues();
148.                  values.put(PhonelistColumns.NAME, nameText.getText().
                      toString());
149.                  values.put(PhonelistColumns.MOBILE, mPhoneText.getText().
                      toString());
150.                  values.put(PhonelistColumns.EMAIL, emailText.getText().
                      toString());
151.                  values.put(PhonelistColumns.ADDR, addrText.getText().
                      toString());
152.      Log.e(TAG+":onPause",mUri.toString());
153.      Log.e(TAG+":onPause",values.toString());
154.                  getContentResolver().update(mUri, values, null, null);
155.              }
156.          }
157.      }
158.
159.  @Override
160.      public boolean onCreateOptionsMenu(Menu menu) {
161.          super.onCreateOptionsMenu(menu);
162.
163.          if (mState==STATE_EDIT) {
164.              menu.add(0, REVERT_ID, 0, R.string.menu_revert)
165.                      .setShortcut('0', 'r')
166.                      .setIcon(android.R.drawable.ic_menu_revert);
167.              menu.add(0, DELETE_ID, 0, R.string.menu_delete)
168.              .setShortcut('0', 'd')
169.              .setIcon(android.R.drawable.ic_menu_delete);
170.
```

```
171.         } else {
172.             menu.add(0, DISCARD_ID, 0, R.string.menu_discard)
173.                     .setShortcut('0', 'd')
174.                     .setIcon(android.R.drawable.ic_menu_delete);
175.         }
176.         return true;
177.     }
178.   @Override
179.     public boolean onOptionsItemSelected(MenuItem item) {
180.         switch (item.getItemId()) {
181.         case DELETE_ID:
182.             deleteContact();
183.             finish();
184.             break;
185.         case DISCARD_ID:
186.             cancelContact();
187.             break;
188.         case REVERT_ID:
189.             backupContact();
190.             break;
191.         }
192.         return super.onOptionsItemSelected(item);
193.     }
194.   //删除联系人信息
195.   private void deleteContact() {
196.       if (mCursor !=null) {
197.             mCursor.close();
198.             mCursor=null;
199.             getContentResolver().delete(mUri, null, null);
200.             nameText.setText("");
201.         }
202. 
203.   }
204.   //丢弃信息
205.   private void cancelContact() {
206.         if (mCursor !=null) {
207.           deleteContact();
208.         }
209.         setResult(RESULT_CANCELED);
210.         finish();
211. 
212.   }
213.   //更新变更的信息
214.   private void updateContact() {
215.     if (mCursor !=null) {
```

```
216.        mCursor.close();
217.              mCursor=null;
218.              ContentValues values=new ContentValues();
219.              values.put(PhonelistColumns.NAME, nameText.getText().toString());
220.              values.put(PhonelistColumns.MOBILE, mPhoneText.getText().
              toString());
221.              values.put(PhonelistColumns.EMAIL, emailText.getText().
              toString());
222.              values.put(PhonelistColumns.ADDR, addrText.getText().
              toString());
223.  Log.e(TAG+":onPause",mUri.toString());
224.  Log.e(TAG+":onPause",values.toString());
225.              getContentResolver().update(mUri, values, null, null);
226.          }
227.          setResult(RESULT_CANCELED);
228.          finish();
229.
230.    }
231.    //取消时用,回退到最初的信息
232.    private void backupContact() {
233.      if (mCursor !=null) {
234.              mCursor.close();
235.              mCursor=null;
236.              ContentValues values=new ContentValues();
237.              values.put(PhonelistColumns.NAME, this.originalNameText);
238.              values.put(PhonelistColumns.MOBILE,this.originalMPhoneText);
239.              values.put(PhonelistColumns.EMAIL, this.originalEmailText);
240.              values.put(PhonelistColumns.ADDR, this.originalAddrText);
241.  Log.e(TAG+":onPause",mUri.toString());
242.  Log.e(TAG+":onPause",values.toString());
243.              getContentResolver().update(mUri, values, null, null);
244.          }
245.          setResult(RESULT_CANCELED);
246.          finish();
247.    }
248.  }
```

(11) 在 src 目录下的 com. ascenttech. android. phonelist 包中创建 PhonelistProvider. java,代码如下:

```
1.    public class PhonelistProvider extends ContentProvider {
2.      private static final String TAG="PhonelistProvider";
3.      private DBUtil dbUtil;
4.      private SQLiteDatabase phonelistDB;
5.
6.      public static final String AUTHORITY="com.ascenttech.android.phonelist.
```

```
        provider.contact";
7.      public static final String CONTACTS_TABLE="contacts";
8.      public static final Uri CONTENT_URI=Uri.parse("content://"+AUTHORITY+
        "/contacts");
9.
10.     public static final int CONTACTS=1;
11.     public static final int CONTACT_ID=2;
12.     private static final UriMatcher uriMatcher;
13.
14.     static{
15.       uriMatcher=new UriMatcher(UriMatcher.NO_MATCH);
16.       uriMatcher.addURI(AUTHORITY,"contacts",CONTACTS);
17.       //单独列
18.       uriMatcher.addURI(AUTHORITY,"contacts/#",CONTACT_ID);
19.     }
20.
21.     @Override
22.     public boolean onCreate() {
23.       dbUtil=new DBUtil(getContext());
24.       phonelistDB=dbUtil.getWritableDatabase();
25.       return (phonelistDB==null)?false:true;
26.     }
27.     //删除指定数据列
28.     @Override
29.     public int delete(Uri uri, String where, String[] selectionArgs) {
30.       //TODO Auto-generated method stub
31.       int count;
32.       switch (uriMatcher.match(uri)) {
33.         case CONTACTS:
34.           count=phonelistDB.delete(CONTACTS_TABLE, where, selectionArgs);
35.           break;
36.         case CONTACT_ID:
37.           String contactID=uri.getPathSegments().get(1);
38.           count=phonelistDB.delete(CONTACTS_TABLE,
39.               PhonelistColumns._ID+"="+contactID
40.               +(!TextUtils.isEmpty(where)?" AND ("+where+")" : ""),
41.               selectionArgs);
42.           break;
43.         default: throw new IllegalArgumentException("Unsupported URI: "+uri);
44.       }
45.       getContext().getContentResolver().notifyChange(uri, null);
46.       return count;
47.     }
48.     //URI 类型转换
49.     @Override
```

```
50.    public String getType(Uri uri) {
51.      //TODO Auto-generated method stub
52.      switch (uriMatcher.match(uri)) {
53.        case CONTACTS:
54.        return "vnd.android.cursor.dir/com.ascenttech.android.phonelist.
           gcontacts";
55.        case CONTACT_ID:
56.        return "vnd.android.cursor.item/com.ascenttech.android.phonelist.
           gcontacts";
57.        default:
58.        throw new IllegalArgumentException("Unsupported URI: "+uri);
59.      }
60.    }
61.    //插入数据
62.    @Override
63.    public Uri insert(Uri uri, ContentValues initialValues) {
64.
65.      if (uriMatcher.match(uri) !=CONTACTS) {
66.        throw new IllegalArgumentException("Unknown URI "+uri);
67.      }
68.
69.      ContentValues values;
70.      if (initialValues !=null) {
71.        values=new ContentValues(initialValues);
72.        Log.e(TAG+"insert","initialValues is not null");
73.      } else {
74.        values=new ContentValues();
75.      }
76.      Long now=Long.valueOf(System.currentTimeMillis());
77.      //设置默认值
78.      if (values.containsKey(PhonelistColumns.CREATED)==false) {
79.        values.put(PhonelistColumns.CREATED, now);
80.      }
81.      if (values.containsKey(PhonelistColumns.MODIFIED)==false) {
82.        values.put(PhonelistColumns.MODIFIED, now);
83.      }
84.      if (values.containsKey(PhonelistColumns.NAME)==false) {
85.        values.put(PhonelistColumns.NAME, "");
86.        Log.e(TAG+"insert","NAME is null");
87.      }
88.      if (values.containsKey(PhonelistColumns.MOBILE)==false) {
89.        values.put(PhonelistColumns.MOBILE, "");
90.      }
91.      if (values.containsKey(PhonelistColumns.EMAIL)==false) {
92.        values.put(PhonelistColumns.EMAIL, "");
93.      }
94.      if (values.containsKey(PhonelistColumns.ADDR)==false) {
```

```
95.        values.put(PhonelistColumns.ADDR, "");
96.      }
97.      Log.e(TAG+"insert",values.toString());
98.      long rowId=phonelistDB.insert(CONTACTS_TABLE, null, values);
99.      if (rowId>0) {
100.        Uri noteUri=ContentUris.withAppendedId(CONTENT_URI,rowId);
101.        getContext().getContentResolver().notifyChange(noteUri, null);
102.        Log.e(TAG+"insert",noteUri.toString());
103.        return noteUri;
104.      }
105.
106.      throw new SQLException("Failed to insert row into "+uri);
107.
108.    }
109.    //查询数据
110.    @Override
111.    public Cursor query(Uri uri, String[] projection, String selection,
112.        String[] selectionArgs, String sortOrder) {
113.      Log.e(TAG+":query"," in Query");
114.      SQLiteQueryBuilder qb=new SQLiteQueryBuilder();
115.      qb.setTables(CONTACTS_TABLE);
116.
117.      switch (uriMatcher.match(uri)) {
118.        case CONTACT_ID:
119.          qb.appendWhere(PhonelistColumns._ID+"="+uri.getPathSegments().
          get(1));
120.          break;
121.        default: break;
122.      }
123.      String orderBy;
124.      if (TextUtils.isEmpty(sortOrder)) {
125.        orderBy=PhonelistColumns._ID;
126.      } else {
127.        orderBy=sortOrder;
128.      }
129.      Cursor c=qb.query(phonelistDB,projection,
130.              selection, selectionArgs,
131.              null, null,orderBy);
132.
133.      c.setNotificationUri(getContext().getContentResolver(), uri);
134.
135.      return c;
136.
137.    }
138.    //更新数据库
139.    @Override
140.    public int update(Uri uri, ContentValues values, String where, String[]
```

```
        selectionArgs) {
141.
142.      int count;
143.      Log.e(TAG+"update",values.toString());
144.      Log.e(TAG+"update",uri.toString());
145.      Log.e(TAG+"update :match",""+uriMatcher.match(uri));
146.      switch (uriMatcher.match(uri)) {
147.        case CONTACTS:
148.          Log.e(TAG+"update",CONTACTS+"");
149.          count=phonelistDB.update(CONTACTS_TABLE, values, where, selectionArgs);
150.          break;
151.        case CONTACT_ID:
152.          String contactID=uri.getPathSegments().get(1);
153.          Log.e(TAG+"update",contactID+"");
154.          count=phonelistDB.update(CONTACTS_TABLE,values,
155.              PhonelistColumns._ID+"="+contactID
156.              +(!TextUtils.isEmpty(where) ?" AND ("+where+")" : ""),
157.              selectionArgs);
158.          break;
159.        default: throw new IllegalArgumentException("Unsupported URI: "+uri);
160.      }
161.      getContext().getContentResolver().notifyChange(uri, null);
162.      return count;
163.    }
164.
165.  }
```

(12) 部署 Android_Contacts 项目工程，程序运行后，单击手机上的 menu 按钮添加联系人，结果如图 9-49 所示。

单击“添加联系人”按钮后，添加联系人的基本信息如图 9-50 所示。

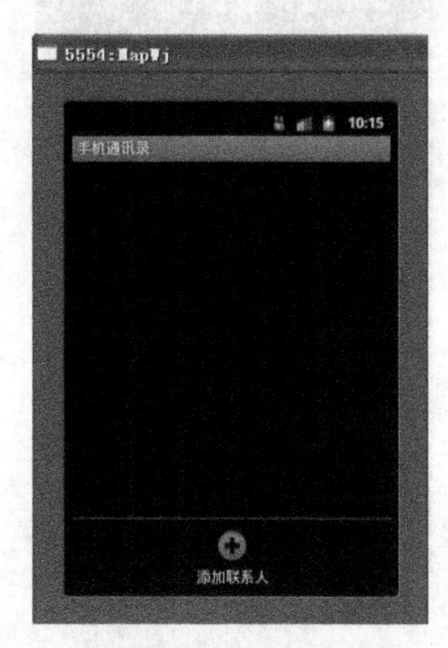

图 9-49 添加联系人

图 9-50 添加联系人基本信息

单击“保存”按钮，程序把联系人信息保存到数据库中，如图 9-51 所示。

然后选中保存后的联系人信息，程序自动跳转到编辑界面，编辑修改联系人信息后，单击 menu 菜单，在界面下面弹出“返回”和“删除”菜单，可以选择“返回”，也可以选择“删除”，如图 9-52 所示。

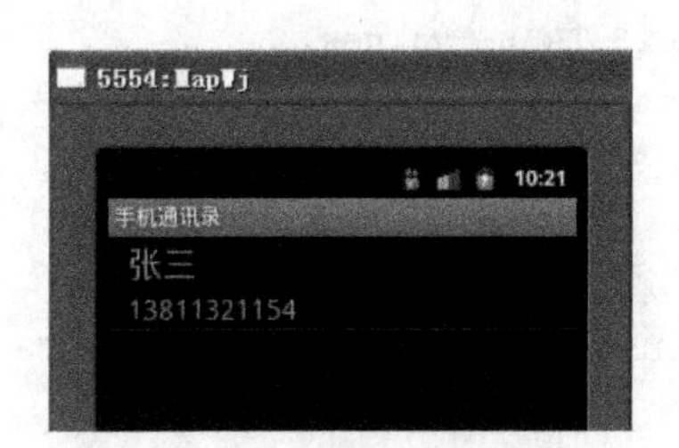

图 9-51　保存后的信息显示

图 9-52　编辑联系人信息

单击“返回”菜单后，用上下键选中任一联系人后单击 munu 菜单，在界面下部弹出“添加联系人”和“删除联系人”菜单，如图 9-53 所示。

在添加联系人信息的中间，如果不想进行添加，想进行丢弃，可以直接单击 menu 按钮，在界面下部出现“丢弃”菜单，选择“丢弃”即可，如图 9-54 所示。

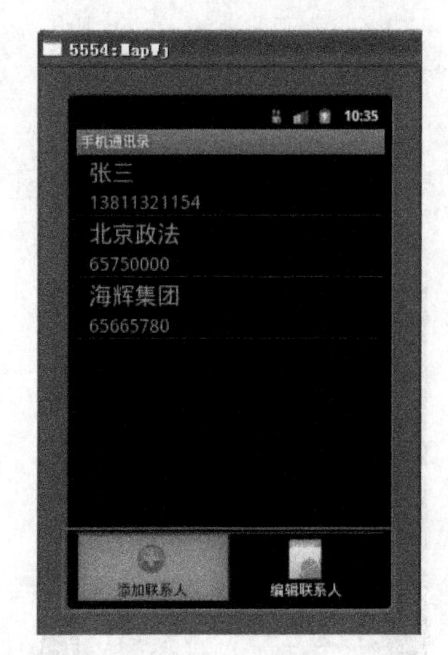

图 9-53　添加、编辑联系人信息

图 9-54　丢弃联系人信息

1. 简答题

(1) SharedPreferences 的访问模式有几种？它们分别是什么？

(2) SharedPreferences 读取应用程序和其他应用程序的区别是什么？

(3) Android 系统支持的文件操作模式有哪些？

(4) SQLite 数据库体系结构由哪些部分组成？

(5) 在 AndroidManifest.xml 文件中注册 ContentProvider 的目的是什么？如何进行注册？

2. 完成下面的实训项目

要求：在本章项目案例的基础上完成手机通讯录的姓名查询及自动提示、设置、根据姓名字母快速定位用户个人信息功能。

第10章 手机通信服务

学习目标

本章主要介绍短信服务、SmsManager 管理短信息操作、电话服务、TeleListener、邮件协议(SMTP)、邮件客户端设计开发方式、访问网络资源的方式、URL 读取网络资源、HTTP 访问网络资源(HttpURLConnection)等。使读者通过本章的学习,能够深入熟悉 Android 系统通信服务,掌握以下知识要点:

(1) SmsManager 管理短信息操作及应用。

(2) TelephonyManager、onCallStateChanged、onServiceStateChanged、onSignalStrengthChanged 等类及常用方法。

(3) 邮件客户端设计开发方式(调用 Android 系统自带的邮件服务、javamail 功能包)。

(4) 访问因特网的权限设置及添加。

(5) Android 系统中访问网络资源的方式及方法。

(6) URL、HttpURLConnection 访问网络资源及方法。

10.1 短信服务

10.1.1 短信服务简介

短信服务是当前任何一款手机都不可缺少的程序应用之一,而且是用户手机使用频率最高的应用之一。在 Android 系统中,以前与短信应用相关的类主要位于 android.telephony.gsm 包中,android.telephony.gsm 中包含的类有 GsmCellLocation、SmsManager、SmsMessage 和 SmsMessage.SubmitPdu,具体如表 10-1 所示。

表 10-1 android. telephony. gsm 包中的类

类名	描述
GsmCellLocation	表示 GSM 手机的位置
SmsManager	管理各种短信操作。这个类已经不再推荐使用，被 android. telephony. SmsManager 替代，以支持 GSM 和 CDMA
SmsMessage	表示具体的短信息。这个类已经不再推荐使用，被 android. telephony. SmsMessage 替代，以支持 GSM 和 CDMA
SmsMessage. SubmitPdu	这个类已经不再推荐使用，用 Use android. telephony. SmsMessage

现在与短信应用相关的类主要位于 android. telephony 包中。从表 10-1 中也可以看出，原来的一些位于 android. telephony. gsm 包中的类已经不再推荐使用，被 android. telephony 包的类替代。

如上所述，在 Android 系统中类 SmsManager 管理短信息操作，用户利用它可以完成手机的短信发送与接收工作。其中 sendTextMessage() 方法需要传入 5 个值，依次是收件人地址(String)、发送人地址(String)、正文内容(String)、发送服务(PendingIntent)、送达服务(PendingIntent)，其中收件人地址与正文内容是不能为 NULL 的参数。跟打电话一样，涉及重要的必须在配置文件分配权限，权限代码如下：<uses-permission android: name="android. permission. SEND_SMS"/>。

10.1.2 短信发送与提示案例

下面通过一个使用 SmsManager 的应用案例详细介绍短信发送与提示的应用。

(1) 创建一个新的 Android 工程，工程名为 SendSMSDemo，目标 API 选择 10(即 Android 2.3.3 版本)，应用程序名为 SendSMSDemo，包名为 com. hisoft. activity，创建的 Activity 的名字为 MainActivity，最小 SDK 版本根据选择的目标 API 会自动添加为 10，创建项目工程如图 10-1 所示。

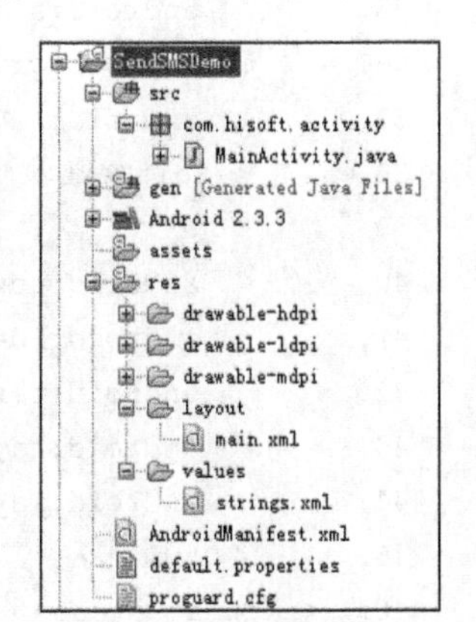

图 10-1 SendSMSDemo 工程目录结构

(2) 修改 res 目录下 layout 文件夹中的 main. xml 文件，设置线性布局，添加两个 EditText 控件、两个 TextView 和一个 Button 控件描述，并设置相关属性，代码如下：

```
1.  <?xml version="1.0" encoding="utf-8"?>
2.  <LinearLayout xmlns:android="http://schemas.
    android.com/apk/res/android"
3.      android:orientation="vertical"
4.      android:layout_width="fill_parent"
5.      android:layout_height="fill_parent"
6.      >
7.      <TextView
8.          android:text="接收号码:"
9.          android:id="@+id/TextView02"
10.         android:textSize="20dip"
11.         android:textStyle="bold"
```

```
12.             android:layout_width="wrap_content"
13.             android:layout_height="wrap_content"
14.             android:paddingLeft="5dip">
15.         </TextView>
16.         <EditText
17.             android:text=""
18.             android:id="@+id/dial_num"
19.             android:layout_width="fill_parent"
20.             android:layout_height="wrap_content">
21.         </EditText>
22.         <TextView
23.             android:text="短信内容:"
24.             android:id="@+id/TextView01"
25.             android:layout_width="wrap_content"
26.             android:textSize="20dip"
27.             android:textStyle="bold"
28.             android:paddingLeft="5dip"
29.             android:layout_height="wrap_content">
30.         </TextView>
31.         <EditText
32.             android:text=""
33.             android:id="@+id/sms_content"
34.             android:layout_width="fill_parent"
35.             android:singleLine="false"
36.             android:gravity="top|left"
37.             android:layout_height="100dip">
38.         </EditText>
39.         <Button
40.             android:text="发送短信"
41.             android:id="@+id/send"
42.             android:textSize="20dip"
43.             android:layout_width="fill_parent"
44.             android:layout_height="wrap_content">
45.         </Button>
46.     </LinearLayout>
```

(3) 修改 src 目录中 com. hisoft. activity 包下的 MainActivity. java 文件,代码如下:

```
1.  package com.hisoft.activity;
2.
3.  import android.app.Activity;
4.  import android.app.PendingIntent;
5.  import android.content.Intent;
6.  import android.os.Bundle;
7.  import android.telephony.PhoneNumberUtils;
8.  import android.telephony.SmsManager;
```

```
9.   import android.view.View;
10.  import android.widget.Button;
11.  import android.widget.EditText;
12.  import android.widget.Toast;
13.
14.  public class MainActivity extends Activity {
15.    /**Called when the activity is first created. * /
16.    @Override
17.    public void onCreate(Bundle savedInstanceState) {
18.      super.onCreate(savedInstanceState);
19.      setContentView(R.layout.main);
20.
21.      Button bdial= (Button) this.findViewById(R.id.send);
22.      bdial.setOnClickListener(                      //为拨号按钮添加监听器
23.      //OnClickListener 为 View 的内部接口,其实现者负责监听鼠标单击事件
24.      new View.OnClickListener() {
25.        public void onClick(View v) {
26.          //获取输入的电话号码
27.          EditText etTel= (EditText) findViewById(R.id.dial_num);
28.          String telStr=etTel.getText().toString();
29.
30.          //获取输入的短信内容
31.          EditText etSms= (EditText) findViewById(R.id.sms_content);
32.          String smsStr=etSms.getText().toString();
33.          //判断号码字符串是否合法
34.          if (PhoneNumberUtils.isGlobalPhoneNumber(telStr)) {    //合法则发送短信
35.            v.setEnabled(false);            //短信发送完成前将发送按钮设置为不可用
36.            sendSMS(telStr, smsStr, v);
37.          } else {                                  //不合法则提示
38.            Toast.makeText(MainActivity.this,       //上下文
39.                "电话号码不符合格式!!!",             //提示内容
40.                Toast.LENGTH_SHORT                  //信息显示时间
41.            ).show();
42.          }
43.        }
44.      });
45.    }
46.
47.    //自己开发的直接发送短信的方法
48.    private void sendSMS(String telNo, String smsStr, View v) {
49.      PendingIntent pi=                             //创建 PendingIntent 对象
50.      PendingIntent.getActivity(this, 0,
51.          new Intent(this, MainActivity.class), 0);
52.      SmsManager sms=SmsManager.getDefault();
53.      sms.sendTextMessage(telNo, null, smsStr, pi, null);
             //收件人,发送人,正文,发送服务,送达服务,其中收件人和正文不可为空
54.      //短信发送成功给予提示
```

```
55.        Toast.makeText(MainActivity.this,    //上下文
56.            "恭喜你,短信发送成功!",              //提示内容
57.            Toast.LENGTH_SHORT               //信息显示时间
58.        ).show();
59.        v.setEnabled(true);                  //短信发送完成后恢复发送按钮的可用状态
60.    }
61. }
```

(4) 部署 SendSMSDemo 项目工程,程序运行后,在“接收号码”文本框中输入另外一个模拟器号码 5556,在“短信内容”文本框中输入图 10-2 所示内容,单击“发送信息”按钮,信息发送成功,提示“恭喜你,短信发送成功!”,如图 10-2 所示。然后在模拟器 5556 中查看接收的信息,如图 10-3 和图 10-4 所示。如果在“接收号码”文本框中输入不符合要求的手机号码,单击“发送短信”按钮会提示“电话号码不符合格式”信息,如图 10-5 所示。

图 10-2　发送信息

图 10-3　查看信息

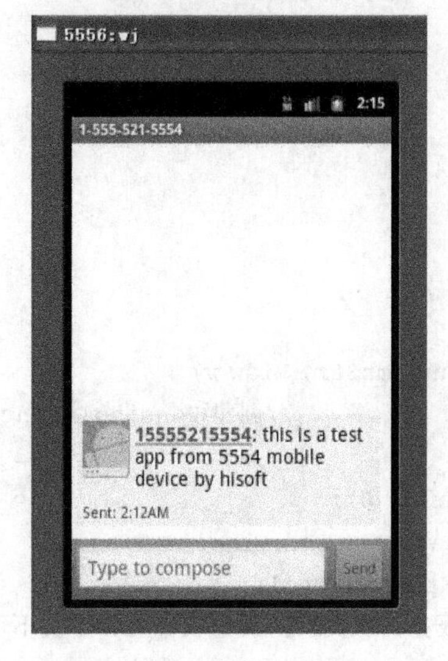

图 10-4　接收信息内容

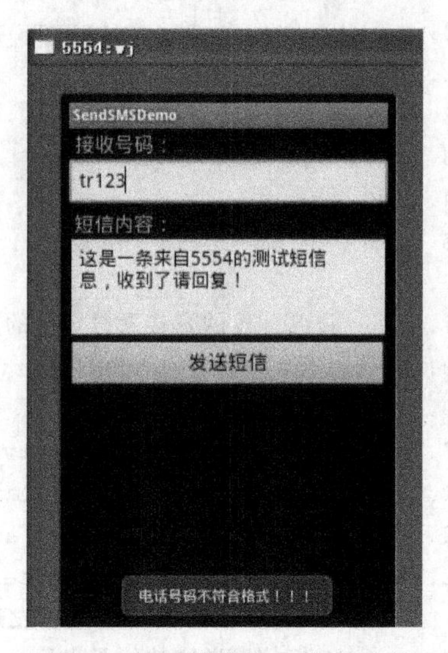

图 10-5　验证接收信息的电话号码

然后如前面章节讲述步骤，在 Eclipse 中创建或者在命令行中使用命令创建 5556 模拟器。打开模拟器 5556 后，单击 Messaging 图标即可查见刚才模拟器 5554 发送的短信息，如图 10-3 和图 10-4 所示。

注意：如果短信内容过长，可以使用 SmsManager. divideMessage(String text)方法把短信内容自动拆分成一个 ArrayList 数组，再根据数组长度循环发送，或者直接用 sendMultipartTextMessage 方法发送，参数与 sendTextMessage 类似，也就是短信内容转变为用 divideMessage 拆成的 ArrayList 数组。

10.1.3 短信发送状态查询案例

上面介绍了使用 SmsManager 发送短信息的应用，下面在上面案例的基础上来介绍短信息发送后其状态状况，帮助用户了解短信息的发送状态，以供用户决定下一步的短信息操作。

(1) 创建一个新的 Android 工程，工程名为 GetSendSMSStateDemo，目标 API 选择 10(即 Android 2.3.3 版本)，应用程序名为 GetSendSMSStateDemo，包名为 com. hisoft. activity，创建的 Activity 的名字为 MainActivity，最小 SDK 版本根据选择的目标 API 会自动添加为 10，创建项目工程如图 10-6 所示。

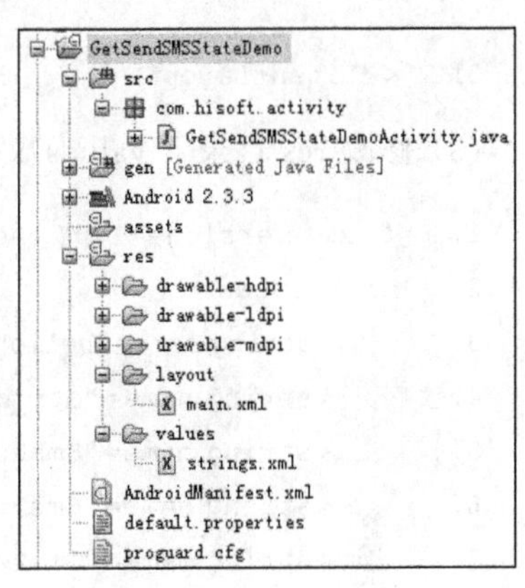

图 10-6 GetSendSMSStateDemo 工程目录结构

(2) 修改 res 目录下 layout 文件夹中的 main. xml 文件，设置线性布局，添加两个 EditText 控件、一个 TextView 和一个 Button 控件描述，并设置相关属性，代码如下：

```
1.  <?xml version="1.0" encoding="utf-8"?>
2.  <LinearLayout xmlns:android="http://schemas.android.com/apk/res/android"
3.      android:orientation="vertical"
4.      android:layout_width="fill_parent"
5.      android:layout_height="fill_parent"
6.      >
7.    <TextView
8.        android:layout_width="fill_parent"
9.        android:layout_height="wrap_content"
10.       android:text="@string/SmsNumber"/>
11.     <EditText
12.       android:id="@+id/number"
13.       android:layout_width="fill_parent"
14.       android:layout_height="wrap_content"
15.       android:text="@string/SmstempNumber"/>
16.   <TextView
17.       android:layout_width="fill_parent"
18.       android:layout_height="wrap_content"
```

```
19.        android:text="@string/SmsBody"/>
20.      <EditText
21.        android:id="@+id/body"
22.        android:layout_width="fill_parent"
23.        android:layout_height="wrap_content"
24.        android:text="@string/SmstempBody"/>
25.      <Button
26.        android:id="@+id/send"
27.        android:layout_width="fill_parent"
28.        android:layout_height="wrap_content"
29.        android:text="@string/Smssend"/>
30.
31.  </LinearLayout>
```

(3) 修改 res 目录下 values 文件夹中的 strings. xml 文件,代码如下:

```
1.  <?xml version="1.0" encoding="utf-8"?>
2.  <resources>
3.      <string name="hello">Hello World, GetSendSMSStateDemoActivity!</string>
4.      <string name="app_name">GetSendSMSStateDemo</string>
5.      <string name="SmsNumber">收件人</string>
6.      <string name="SmsBody">信息内容</string>
7.      <string name="SmstempNumber">5556</string>
8.      <string name="SmstempBody">测试短信发送状态!</string>
9.      <string name="Smssend">发送信息</string>
10. </resources>
```

(4) 在 AndroidManifest. xml 文件中的<manifest>根节点标签下添加权限,代码如下:

```
1.  <uses-permission android:name="android.permission.SEND_SMS"/>
```

(5) 修改 src 目录中 com. hisoft. activity 包下的 MainActivity. java 文件,代码如下:

```
1.  package com.hisoft.activity;
2.  import android.app.Activity;
3.  import android.app.PendingIntent;
4.  import android.content.BroadcastReceiver;
5.  import android.content.Context;
6.  import android.content.Intent;
7.  import android.content.IntentFilter;
8.  import android.os.Bundle;
9.  import android.telephony.SmsManager;
10. import android.view.View;
11. import android.view.View.OnClickListener;
12. import android.widget.Button;
13. import android.widget.EditText;
```

```
14.  import android.widget.Toast;
15.  public class GetSendSMSStateDemoActivity extends Activity implements
     OnClickListener{
16.    EditText number;                                              //电话号码
17.    EditText body;                                                //短信内容
18.    Button send;                                                  //发送按钮
19.      @Override
20.     public void onCreate(Bundle savedInstanceState) {         //重写的onCreate方法
21.         super.onCreate(savedInstanceState);
22.         setContentView(R.layout.main);
23.         send=(Button) this.findViewById(R.id.send);
24.         number=(EditText) this.findViewById(R.id.number);
25.         body=(EditText) this.findViewById(R.id.body);
26.         send.setOnClickListener(this);                           //给按钮添加监听
27.     IntentFilter myIntentFilter=new IntentFilter("SMS_SEND_ACTION");
                                                                     //创建过滤器
28.     MySmsReceiver mySmsReceiver=new MySmsReceiver();         //创建广播接收
29.     registerReceiver(mySmsReceiver, myIntentFilter);         //注册广播接收
30.      }
31.    @Override
32.    public void onClick(View v) {                                 //监听方法
33.      if(v==send){                                                //如果是按下发送按钮
34.       send.setEnabled(false);                                    //设置按钮为不可用
35.       String strNumber=number.getText().toString();            //得到发送电话号码
36.       String strBody=body.getText().toString();             //得到需要发送的信息内容
37.       SmsManager smsManager=SmsManager.getDefault();       //获取SmsManager对象
38.       Intent intentSend=new Intent("SMS_SEND_ACTION");     //创建Intent
39.
40.       PendingIntent sendPI=PendingIntent.getBroadcast(getApplicationContext(),
          0, intentSend, 0);
41.       smsManager.sendTextMessage(strNumber, null, strBody, sendPI, null);
                                                                //发送短信
42.       send.setEnabled(true);                                //设置按钮为可用
43.      }
44.    }
45.    public class MySmsReceiver extends BroadcastReceiver{  //自定义的广播接收类
46.      @Override
47.      public void onReceive(Context context, Intent intent) {
                                                                //重写的onReceive方法
48.       switch(getResultCode()){
49.       case Activity.RESULT_OK://如发送成功
50.         Toast.makeText(context, "信息发送成功", Toast.LENGTH_LONG).show();
                                                                //信息提示
```

```
51.         break;
52.       case SmsManager.RESULT_ERROR_GENERIC_FAILURE:  //发送失败
53.         Toast.makeText(context, "信息发送失败", Toast.LENGTH_LONG).show();
54.         break;
55.       default:           //其他情况
56.         Toast.makeText(context, "状态未知",
            Toast.LENGTH_LONG).show();
57.         break;
58.       }
59.     }
60.   }
61. }
```

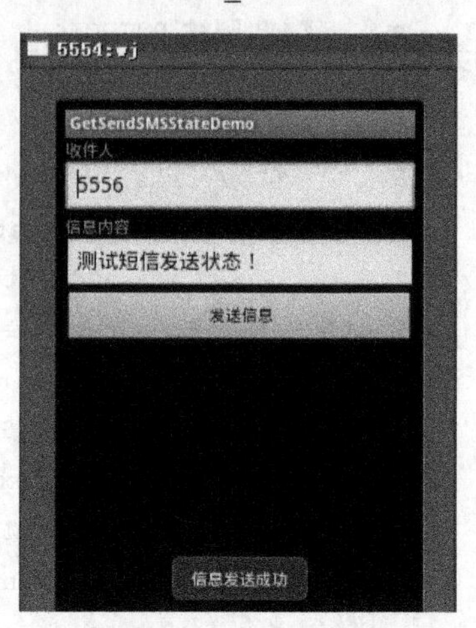

图 10-7　发送信息成功

(6) 部署 GetSendSMSStateDemo 项目工程，程序运行后，在“收件人”文本框中填入收件人的电话号码，在“信息内容”文本框中写入需要发送的信息内容，单击“发送信息”按钮，如发送成功，结果如图 10-7 所示。

10.2　电话服务

10.2.1　电话服务简介

Android 系统中，电话功能主要通过 Radio Interface Layer (RIL)来提供电话服务以及各个相关硬件之间的抽象层等。Radio Interface Layer RIL(Radio Interface Layer)负责数据的可靠传输、AT 命令的发送以及 response 的解析。应用处理器通过 AT 命令集与带 GPRS 功能的无线通信模块通信。

在 Android 系统中，与电话应用相关的类主要位于 android. telephony 包下。在 android. telephony 包中与电话应用相关的类具体如表 10-2 所示。

表 10-2　android. telephony 包中与电话应用相关的类

类　名	描　述
CellLocation	抽象类，表示设备的位置
PhoneNumberFormattingTextWatcher	监听一个 TextView 控件，如果有电话号码进入，则调用 formatNumber()方法处理电话号码
PhoneNumberUtils	处理电话号码字符串的各种工具
PhoneStateListener	监听手机设备中电话状态变化的监听类，监听的包含有服务的状态、信号的强弱、短信息的等待提示等
TelephonyManager	提供对手机设备中电话服务信息的访问

使用 Android 系统调用电话拨打，代码如下：

```
Intent intent=new Intent("android.intent.action.DIAL", Uri.parse("tel:10086"));
```

```
startActivity(intent);
```

在执行上述代码后，即可进入拨打电话呼叫界面。

在 Android 系统中，监听电话状态的功能如下：

监听电话状态首先建立一个继承于 PhoneStateListener 电话监听类（如 TeleListener），并调用 TelephonyManager 进行监听，关键代码如下：

```
TelephonyManager mTelephonyMgr = (TelephonyManager) getSystemService (Context.
TELEPHONY_SERVICE);
mTelephonyMgr.listen(new TeleListener(), PhoneStateListener.LISTEN_CALL_STATE|
PhoneStateListener.LISTEN_SERVICE_STATE | PhoneStateListener.LISTEN_SIGNAL_
STRENGTH);
```

此外，TeleListener 需要实现父类 onCallStateChanged、onServiceStateChanged 和 onSignalStrengthChanged 几个方法，根据方法名便可知其功能，这里就不再详述。

10.2.2 接打电话案例

上面介绍了电话服务常用的包及类，下面通过一个使用电话服务拨打电话的应用案例详细介绍电话服务的应用。

（1）创建一个新的 Android 工程，工程名为 PhoneCallDemo，目标 API 选择 10（即 Android 2.3.3 版本），应用程序名为 PhoneCallDemo，包名为 com.hisoft.activity，创建的 Activity 的名字为 MainActivity，最小 SDK 版本根据选择的目标 API 会自动添加为 10，创建项目工程如图 10-8 所示。

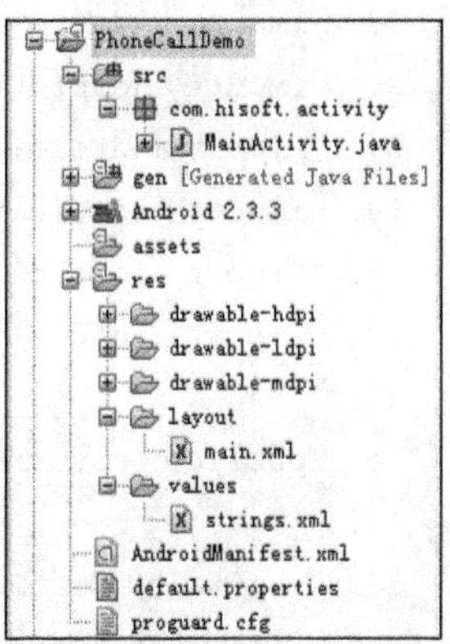

图 10-8 PhoneCallDemo 工程目录结构

（2）修改 res 目录下 layout 文件夹中的 main.xml 文件，设置线性布局，添加一个 EditText 控件、一个 TextView 和一个 Button 控件描述，并设置相关属性，代码如下：

```
1.  <?xml version="1.0" encoding="utf-8"?>
2.  <LinearLayout xmlns:android="http://schemas.
    android.com/apk/res/android"
3.      android:orientation="vertical"
4.      android:layout_width="fill_parent"
5.      android:layout_height="fill_parent"
6.      >
7.    <TextView
8.       android:layout_width="fill_parent"
9.       android:layout_height="wrap_content"
10.      android:text="请您输入电话号码:"
11.     />
12.     <EditText
13.      android:id="@+id/phone_num"
14.      android:layout_width="fill_parent"
15.      android:layout_height="wrap_content">
```

```
16.      </EditText>
17.
18.    <Button
19.     android:text="拨打"
20.     android:id="@+id/call"
21.     android:layout_width="wrap_content"
22.     android:layout_height="wrap_content">
23.    </Button>
24.  </LinearLayout>
```

(3) 修改 src 目录中 com. hisoft. activity 包下的 MainActivity. java 文件,代码如下:

```
1.   package com.hisoft.activity;
2.   import java.util.regex.Matcher;
3.   import java.util.regex.Pattern;
4.   import android.app.Activity;
5.   import android.content.Intent;
6.   import android.net.Uri;
7.   import android.os.Bundle;
8.   import android.view.View;
9.   import android.view.View.OnClickListener;
10.  import android.widget.Button;
11.  import android.widget.EditText;
12.  import android.widget.Toast;
13.
14.  public class MainActivity extends Activity {
15.
16.    private EditText phone_num;
17.    private Button call;
18.
19.    @Override
20.    public void onCreate(Bundle savedInstanceState) {
21.        super.onCreate(savedInstanceState);
22.        setContentView(R.layout.main);
23.
24.        this.phone_num=(EditText)this.findViewById(R.id.phone_num);
25.        this.call=(Button)this.findViewById(R.id.call);
26.
27.        this.call.setOnClickListener
28.        (
29.            new OnClickListener()
30.        {
31.        @Override
32.        public void onClick(View v) {
```

```
33.          String number=phone_num.getText().toString().trim();
                                                    //获取输入的手机号码
34.          boolean flag=phoneNumber(number);
35.          if(flag)
36.          {
37.            Intent intent=new Intent("android.intent.action.CALL",Uri.
               parse("tel:"+number));
38.            startActivity(intent);
39.            phone_num.setText("");
40.          }else
41.          {
42.            Toast.makeText(MainActivity.this,"您输入的电话号码格式不正确",
               Toast.LENGTH_SHORT).show();
43.            phone_num.setText("");              //将 EditText 字符设为空
44.          }
45.        }
46.
47.      }
48.      );
49.    }
50.    public boolean phoneNumber(String number)
51.    {
52.      boolean flag=false;
53.      String pare="\\d{11}";                   //11 个整数的手机号码正则式
54.      String pare2="\\d{12}";                  //12 个整数的座机号码正则式
55.      CharSequence num=number;                 //获取电话号码
56.      Pattern pattern=Pattern.compile(pare);   //判断是否为手机号码
57.      Matcher matcher=pattern.matcher(num);
58.      Pattern pattern2=Pattern.compile(pare2); //判断是否为座机号码
59.      Matcher matcher2=pattern2.matcher(num);
60.      if(matcher.matches()||matcher2.matches()) //如果符合格式
61.      {
62.        flag=true;                             //标志位设为 true
63.      }
64.      return flag;
65.    }
66.  }
```

(4) 部署 PhoneCallDemo 项目工程,程序运行后如图 10-9 所示。在文本框中输入电话号码,如果输入的电话号码格式不符合要求,单击“拨打”按钮,会显示“您输入的电话号码格式不正确”提示信息,然后文本框中的内容会置空,等待重新输入;如果输入座机号,长度符合 12 位条件要求,如 010627770312,单击“拨打”按钮,则调用系统电话程序,拨打电话,如图 10-10 所示。

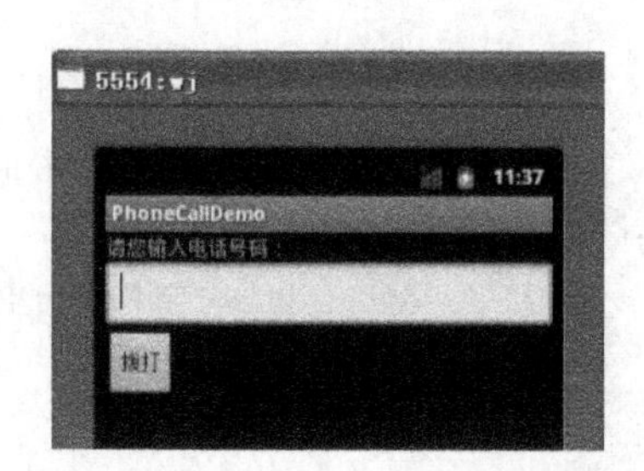

图 10-9　PhoneCallDemo 运行效果

图 10-10　拨打电话界面

10.3　E-mail 服务

10.3.1　SMTP 简介

SMTP(Simple Mail Transfer Protocol,简单邮件传输协议)是一组用于由源地址到目的地址传送邮件的规则,或者说是由它来控制信件传输的一种中转方式。SMTP 协议属于 TCP/IP 协议族,它帮助每台计算机或移动终端设备在发送或中转信件时找到下一个目的地。通过 SMTP 协议所指定的服务器,可以把 E-mail 寄到收信人的服务器上。SMTP 服务器则是遵循 SMTP 协议的发送邮件服务器,用来发送或中转电子邮件。

在 Android 系统中,邮件的发送都通过内置的 Gmail 程序或者设置邮件服务器的方式完成,而系统平台的底层通信则是采用 SMTP 协议传输信息。

在 Android 系统里进行邮件客户端设计开发可以有两种方式:一种是调用 Android 系统自带的邮件服务;另外一种是采用 javamail 功能包的方式,下面分别进行介绍。

1. 调用 Android 系统自带的邮件服务

```
//建立 Intent 对象
    Intent intent=new Intent();
    //设置对象动作
    intent.setAction(Intent.ACTION_SEND);
    //设置对方邮件地址
    intent.putExtra(Intent.EXTRA_EMAIL, new String[]
    { "abc@com.cn","edf@com.cn" });
    //设置标题内容
```

```
intent.putExtra(Intent.EXTRA_SUBJECT, "test");
//设置邮件文本内容
intent.putExtra(Intent.EXTRA_TEXT, "test mail");
启动一个新的 ACTIVITY,"Sending mail..."是在启动这个
  ACTIVITY 的等待时间时所显示的文字
startActivity(Intent.createChooser(intent, "Sending mail..."));
```

其优点是比较简单易用，而缺点是发送邮件的账号必须是 gmail 账号。

只有上面的代码有可能还会出现异常，运行的时候会提示一个错误：no application can perform this action，这是由于没有在模拟器上配置 gmail 邮箱，输入自己的 gmail 账号和密码，默认使用的是你的 gmail 账户发信。

2. 采用 javamail 功能包

其优点可以设置邮件服务器地址，不必局限于 gmail 邮箱。而缺点是用法比较复杂。

在 Android 里使用 javamail 需要依赖三个包：activation. jar（JDK1. 6 之前需要）、additionnal. jar 和 mail. jar。

同时还要注意在 AndroidManifest. xml 中添加访问因特网的权限，代码如下：

```
<uses-permission android:name="android.permission.INTERNET"></uses-permission>
```

对于 JavaMail，最基础的功能就是邮件的发送和接收。发送邮件主要包括三个部分：创建连接、创建邮件体和发送邮件。

JavaMail 中是使用会话（Session）来管理连接的，创建一个连接就需要创建一个会话。在会话中有两个重要的因素：一是会话的属性，二是会话的认证。在使用 Hotmail 等邮件工具的时候就要设置"SMTP 服务器身份验证"，也就是这里的会话认证。

首先创建一个连接属性。

```
Properties props=new Properties();
props.put("mail.smtp.host","smtp.126.com");    //设置 smtp 的服务器地址是 smtp.126.com
props.put("mail.smtp.auth","true");                    //设置 smtp 服务器要身份验证
```

再创建一个身份验证。身份验证稍微复杂一点，要创建一个 Authenticator 的子类，并重载 getPasswordAuthentication()方法，代码如下：

```
class PopupAuthenticator extends Authenticator {
      public PasswordAuthentication getPasswordAuthentication() {
    String username="cqhcp";                    //126 邮箱登录账号
    String pwd="12345";                         //登录密码
    return new PasswordAuthentication(username, pwd);
  }
}
```

创建身份验证的实例：

```
PopupAuthenticator auth=new PopupAuthenticator();
```

创建会话：

关于会话的创建有两种方法，具体请参看后续的文章，这里只简单使用一种。

```
Session session=Session.getInstance(props, auth);
```

定义邮件地址：

```
//发送人地址
Address addressFrom=new InternetAddress("cqhcp@126.com", "George Bush");
//收件人地址
Address addressTo=new InternetAddress("webmaster@javazy.com", "George Bush");
//抄送地址
Address addressCopy=new InternetAddress("haocongping@gmail.com", "George Bush");
```

创建邮件体：

```
message.setContent("Hello", "text/plain");      //或者使用 message.setText("Hello");
message.setSubject("Title");
message.setFrom(addressFrom);
message.addRecipient(Message.RecipientType.TO,addressTo);
message.addRecipient(Message.RecipientType.CC,addressCopy);
message.saveChanges();
```

发送邮件的过程：

```
Transport transport=session.getTransport("smtp");                   //创建连接
transport.connect("smtp.126.com", "cqhcp", "12345");                //连接服务器
transport.send(message);                                            //发送信息
transport.close();                                                  //关闭连接
```

整体程序的代码如下：

```
class PopupAuthenticator extends Authenticator {
    public PasswordAuthentication getPasswordAuthentication() {
        String username="cqhcp";                                    //163 邮箱登录账号
        String pwd="12345";                                         //登录密码
        return new PasswordAuthentication(username, pwd);
    }

}
Properties props=new Properties();
props.put("mail.smtp.host","smtp.126.com");
props.put("mail.smtp.auth","true");
PopupAuthenticator auth=new PopupAuthenticator();
Session session=Session.getInstance(props, auth);
MimeMessage message=new MimeMessage(session);
Address addressFrom=new InternetAddress("cqhcp@126.com", "George Bush");
Address addressTo=new InternetAddress("webmaster@javazy.com", "George Bush");
Address addressCopy=new InternetAddress("haocongping@gmail.com", "George Bush");
message.setContent("Hello", "text/plain");      //或者使用 message.setText("Hello");
```

```
message.setSubject("Title");
message.setFrom(addressFrom);
message.addRecipient(Message.RecipientType.TO,addressTo);
message.addRecipient(Message.RecipientType.CC,addressCopy);
message.saveChanges();
Transport transport=session.getTransport("smtp");
transport.connect("smtp.126.com","cqhcp", "12345");
transport.send(message);
transport.close();
```

若想在登录时判断输入的用户名和密码是否正确，正确时登录，不正确时提示出错而不登录，只需像下面这样实现：

```
try {
        session.setDebug(true);
        Transport trans=session.getTransport("smtp");
        trans.connect("smtp.126.com",account, password);

  } catch (AuthenticationFailedException ae) {
        ae.printStackTrace();
        DisplayToast("用户名或者密码错误!");
                                        //其中 DisplayToast 是笔者自己写的一个 Toast

   } catch (MessagingException mex) {
         mex.printStackTrace();
         Exception ex=null;
         if ((ex=mex.getNextException()) !=null) {
           ex.printStackTrace();
         }
   }
```

10.3.2 发送邮件案例

上面介绍了在 Android 中发送邮件常用的包及类和方法，下面通过一个使用发送邮件的应用案例详细介绍邮件的应用。

(1) 创建一个新的 Android 工程，工程名为 EMailDemo，目标 API 选择 10(即 Android 2.3.3 版本)，应用程序名为 EMailDemo，包名为 com.hisoft.activity，创建的 Activity 的名字为 MainActivity，最小 SDK 版本根据选择的目标 API 会自动添加为 10，创建项目工程如图 10-11 所示。

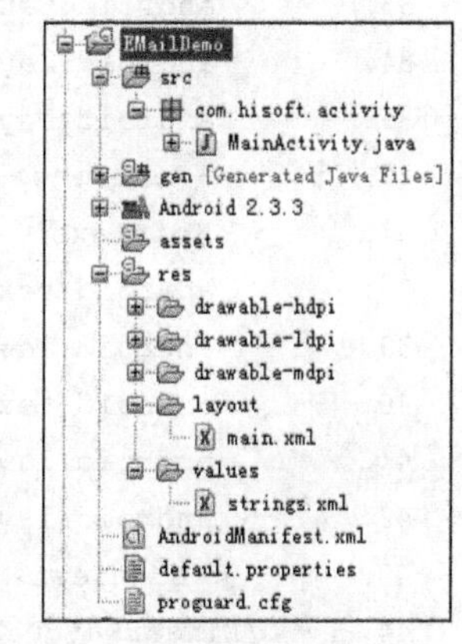

图 10-11 EMailDemo 工程目录结构

(2) 修改 res 目录下 layout 文件夹中的 main.xml 文件，设置线性布局并嵌套线性布局，添加 4 个 EditText 控件、4 个 TextView 和一个 Button 控件描述，并设置相关属性，代码如下：

```
1.   <?xml version="1.0" encoding="utf-8"?>
2.   <LinearLayout xmlns:android="http://schemas.
```

```
    android.com/apk/res/android"
3.      android:orientation="vertical"
4.      android:layout_width="fill_parent"
5.      android:layout_height="fill_parent"
6.      >
7.    <LinearLayout
8.     android:layout_width="fill_parent"
9.     android:layout_height="wrap_content"
10.    android:orientation="horizontal">
11.      <TextView
12.       android:text="收件人地址:"
13.       android:id="@+id/TextView01"
14.       android:textColor="@android:color/white"
15.       android:layout_width="wrap_content"
16.       android:layout_height="wrap_content">
17.      </TextView>
18.      <EditText
19.       android:text=""
20.       android:id="@+id/EditText01"
21.       android:textColor="#222222"
22.       android:layout_width="fill_parent"
23.       android:layout_height="wrap_content">
24.      </EditText>
25.    </LinearLayout>
26.    <LinearLayout
27.     android:layout_width="fill_parent"
28.     android:layout_height="wrap_content"
29.     android:orientation="horizontal">
30.      <TextView
31.       android:text="发件人地址:"
32.       android:id="@+id/TextView04"
33.       android:textColor="@android:color/white"
34.       android:layout_width="wrap_content"
35.       android:layout_height="wrap_content">
36.      </TextView>
37.      <EditText
38.       android:text=""
39.       android:id="@+id/EditText04"
40.       android:textColor="#222222"
41.       android:layout_width="fill_parent"
42.       android:layout_height="wrap_content">
43.      </EditText>
44.    </LinearLayout>
45.    <LinearLayout
46.     android:layout_width="fill_parent"
```

```
47.     android:layout_height="wrap_content"
48.     android:orientation="horizontal">
49.       <TextView
50.         android:text="邮件主题:"
51.         android:id="@+id/TextView02"
52.         android:textColor="@android:color/white"
53.         android:layout_width="wrap_content"
54.         android:layout_height="wrap_content">
55.       </TextView>
56.       <EditText
57.         android:id="@+id/EditText02"
58.         android:textColor="#222222"
59.         android:layout_width="fill_parent"
60.         android:layout_height="wrap_content">
61.       </EditText>
62.    </LinearLayout>
63.    <TextView
64.         android:text="邮件内容:"
65.         android:textColor="@android:color/white"
66.         android:id="@+id/TextView03"
67.         android:layout_width="wrap_content"
68.         android:layout_height="wrap_content">
69.     </TextView>
70.     <EditText
71.         android:id="@+id/EditText03"
72.         android:textColor="#222222"
73.         android:layout_width="fill_parent"
74.         android:layout_height="100dip"
75.         android:gravity="top|left">
76.     </EditText>
77.     <Button
78.         android:text="发送"
79.         android:textColor="#222222"
80.         android:id="@+id/Button01"
81.         android:layout_width="wrap_content"
82.         android:layout_height="wrap_content">
83.     </Button>
84.  </LinearLayout>
```

(3) 修改 res 目录下 values 文件夹中的 strings. xml 文件，代码如下：

```
1.  <?xml version="1.0" encoding="utf-8"?>
2.  <resources>
3.      <string name="hello">Hello World, MainActivity!</string>
4.      <string name="app_name">EMailDemo</string>
5.      <string name="start">邮件发送中......</string>
```

```
6.   </resources>
```

(4) 修改 src 目录中 com. hisoft. activity 包下的 MainActivity. java 文件，代码如下：

```
1.   package com.hisoft.activity;
2.
3.   import android.app.Activity;
4.   import android.content.Intent;
5.   import android.os.Bundle;
6.   import android.view.View;
7.   import android.view.View.OnClickListener;
8.   import android.widget.Button;
9.   import android.widget.EditText;
10.  import android.widget.Toast;
11.
12.  public class MainActivity extends Activity {
13.
14.    EditText etReceiver;                    //收件人
15.    EditText etSender;                      //发件人
16.    EditText etTheme;                       //主题
17.    EditText etMessage;                     //内容
18.    Button bSend;                           //发送按钮
19.    String strReceiver;                     //收件人信息
20.    String strSender;                       //发件人信息
21.    String strTheme;                        //主题信息
22.    String strMessage;                      //内容信息
23.
24.      @Override
25.      public void onCreate(Bundle savedInstanceState) {
26.          super.onCreate(savedInstanceState);
27.          setContentView(R.layout.main);
28.          etReceiver=(EditText)this.findViewById(R.id.EditText01);   //获取对象
29.          etSender=(EditText)this.findViewById(R.id.EditText04);  //获取对象
30.          etTheme=(EditText)this.findViewById(R.id.EditText02);   //获取对象
31.          etMessage=(EditText)this.findViewById(R.id.EditText03);   //获取对象
32.          bSend=(Button)this.findViewById(R.id.Button01);           //发送按钮
33.
34.
35.          bSend.setOnClickListener
36.          (
37.              new OnClickListener()
38.              {
39.                @Override
40.                public void onClick(View v) {
41.        strReceiver=etReceiver.getText().toString().trim();      //获取收件人
42.        strSender=etSender.getText().toString().trim();          //获取发件人
43.        strTheme=etTheme.getText().toString().trim();            //获取主题
44.        strMessage=etMessage.getText().toString().trim();        //获取内容
```

```
45.         String parent="^[a-zA-Z][\\w\\.-]*[a-zA-Z0-9]@[a-zA-Z0-9][\\w\\.-]
            *[a-zA-Z0-9]\\.[a-zA-Z][a-zA-Z\\.]*[a-zA-Z]$";
46.       if(!strReceiver.matches(parent))            //查看收件人地址是否符合格式
47.               {
48.                 Toast.makeText(MainActivity.this, "收件人地址格式错误", Toast.
                    LENGTH_SHORT).show();
49.           }else if(!strSender.matches(parent))    //查看发件人地址是否符合格式
50.               {
51.                 Toast.makeText(MainActivity.this, "发件人地址格式错误", Toast.
                    LENGTH_SHORT).show();
52.               }else                               //若都符合格式,则发送邮件
53.               {
54.                 Intent intent=new Intent(android.content.Intent.ACTION_SEND);
                                                      //发送邮件功能
55.                 intent.setType("plain/text");
56.         intent.putExtra(android.content.Intent.EXTRA_EMAIL, strReceiver);
57.         intent.putExtra(android.content.Intent.EXTRA_CC, strSender);
58.         intent.putExtra(android.content.Intent.EXTRA_SUBJECT, strTheme);
59.         intent.putExtra(android.content.Intent.EXTRA_TEXT, strMessage);
60.                   startActivity(Intent.createChooser(intent, getResources().
                      getString(R.string.start)));
61.               }
62.           }


63.         }
64.       );
65.   }
66. }
```

(5) 部署 EMailDemo 项目工程,程序运行后如图 10-12 所示。填写发件人地址、收件人地址、邮件主题、邮件内容,然后单击“发送”按钮。如前面所讲,如果使用的是系统自带的 gmail 程序,需要先配置好 gmail 账户和密码,然后才能调用发送成功,否则会提示“No application can perform this action”信息,如图 10-13 所示。如果使用的是 SMTP 服务器,需要设置好邮件服务器才能正确发送成功。

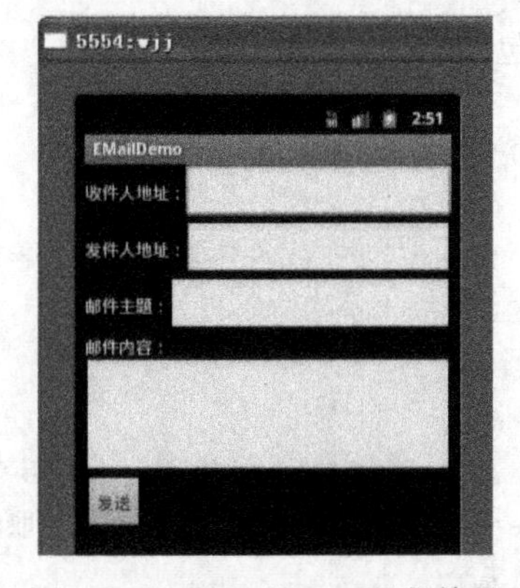

图 10-12 EMailDemo 运行效果

图 10-13 发送邮件

10.4 网络资源访问与处理

本节介绍 Android 系统进行网络访问与数据处理的知识和方法。Android 系统中与访问网络相关的包主要如表 10-3 所示。

表 10-3 Android 系统中与访问网络相关的包

序号	包 名	包 描 述
1	java.net	包含访问网络有关的类，包括流和数据包（datagram）sockets、Internet 协议和常见 HTTP 处理。该包是一个多功能网络资源
2	org.apache.*	包含许多为 HTTP 通信提供精确控制和功能的包。可以将 Apache 视为流行的开源 Web 服务器
3	android.net	除核心 java.net.* 类以外，包含额外的网络访问 socket。该包包括 URI 类
4	android.net.http	包含处理 SSL 证书的类
5	android.net.wifi	包含在 Android 平台上管理有关 WiFi（802.11 无线 Ethernet）所有方面的类
6	android.telephony	包含用于管理和发送 SMS（文本）消息的类。支持 CDMA 或 android.telephony.cdma 等网络
7	java.io	该包中的类由其他 Java 包中提供的 socket 和连接使用。它们还用于与本地文件（在与网络进行交互时会经常出现）的交互
8	java.nio	包含表示特定数据类型的缓冲区的类。适合用于两个基于 Java 语言的端点之间的通信

Android 系统中访问网络资源的方式有 4 种，分别是：

（1）使用 URL/HttpURLConnection 访问网络（URL/HttpURLConnection 位于 java.net 包中）。

```
//创建 URL 对象
URL url=new URL("http://www.google.com/");
//打开连接
HttpURLConnection http=(HttpURLConnection) url.openConnection();
```

（2）使用 Socket 访问网络（Socket 位于 java.net 包中）。

服务器端：

```
ServerSocket ser=new ServerSocket(8888);                    //设置监听端口号
Socket socket=ser.accept();                                 //获取连接的 socket 对象
DataOutputStream sout=new DataOutputStream(socket.getOutputStream());
```

客户端：

```
Socket socket=new Socket("192.168.1.14",8888);              //创建 Socket 对象
DataInputStream sout=new DataInputStream(socket.getInputStream()); //获取输入流
```

（3）使用 HttpClient 应用 Post 和 Get 方式访问网络（HttpClient 位于 org.apache.

http 包中）。

```
//DefaultHttpClient 表示默认属性
HttpClient httpClient=new DefaultHttpClient();
//使用 HttpGet 创建对象实例
HttpGet get=new HttpGet("http://www.google.com/");
HttpResponse rp=httpClient.execute(get);
```

(4) 使用 InetAddress 访问网络(InetAddress 位于 android. net 包中)。

```
InetAddress inetAddress=InetAddress.getByName("192.168.1.1");
//端口
Socket client=new Socket(inetAddress,61203,true);
//取得数据
InputStream in=client.getInputStream();
OutputStream out=client.getOutputStream();
```

10.4.1 使用 URL 读取网络资源

URL(Uniform Resource Locator,统一资源定位符)在 Android 系统中,URL 类位于 java. net 包下,使用资源可以是简单的文件或目录,也可以是对更复杂的对象的引用。URL 可以由协议名、主机、端口和资源路径组成。

URL 的 openConnection()方法将返回一个 URLConnection 对象,该对象表示应用程序和 URL 之间的通信连接,程序可以通过 URLConnection 实例向该 URL 发送请求,读取 URL 引用的资源。

通常创建一个和 URL 的连接,并发送请求。读取此 URL 引用的资源需要如下步骤:

(1) 创建 URL 对象,如下:

```
URL myURL=new URL(HTTP://www.baidu.com/hello.txt);
```

(2) 通过调用 Url 对象 openConnection()方法来创建 URLConnection 对象,用类 URLConnection 表示一个打开的网络连接。

```
URLConnection ur=myURL.openConnection();
```

(3) 创建输入流,从网络上读到的数据用字节流的形式表示,如下:

```
InputStream is=ucon.getInputStream();
```

为了避免频繁读取字节流,提高读取效率,用 BufferedInputStream 缓存读到的字节流。

```
InputStream is=ur.getInputStream();
BufferedInputStream bis=new BufferedInputStream(is);
```

(4) 创建 BufferdInputStream 后,就可以用 read 方法读入网络数据。

```
ByteArrayBuffer baf=new ByteArrayBuffer(50);
      int current=0;
```

```
while((current=bis.read())!=-1)
{
    baf.append((byte)current);
}
```

(5) 将字节流转换为可读取的字符串,并设置编码 UTF-8。

```
myString=EncodingUtils.getString(baf.toByteArray(),"UTF-8");
```

如果读取的是.txt 等文件,是 UTF-8 格式的,就需要对数据进行专门的转换。

(6) 在 AndroidManifest.xml 中加入访问因特网服务的权限。

```
<uses-permission android:name="android.permission.INTERNET"/>
```

如果不加入,程序运行就会出现 permission denied 的异常。

10.4.2 使用 URL 访问网络应用案例

上面介绍了 URL 读取网络资源常用的包及类和方法,下面通过一个访问网络资源的应用案例详细介绍 URL 的应用。

(1) 创建一个新的 Android 工程,工程名为 NetWorkDemo,目标 API 选择 10(即 Android 2.3.3 版本),应用程序名为 NetWorkDemo,包名为 com.hisoft.activity,创建的 Activity 的名字为 MainActivity,最小 SDK 版本根据选择的目标 API 会自动添加为 10,创建项目工程如图 10-14 所示。

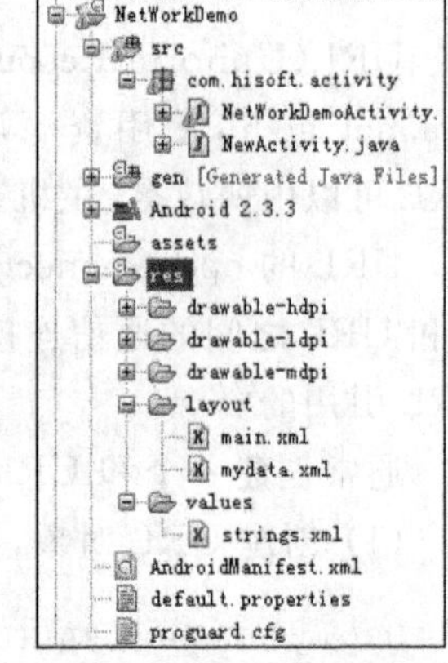

图 10-14 NetWorkDemo 工程目录结构

(2) 在 res 目录下的 layout 文件夹中新建 mydata.xml 文件,设置线性布局,添加一个 ScrollView 控件、一个 TextView 描述,并设置相关属性,代码如下:

```
1.  <LinearLayout
2.  android:id="@+id/LinearLayout01"
3.  android:layout_width="fill_parent"
4.  android:layout_height="fill_parent"
5.  xmlns:android="http://schemas.android.com/apk/res/android">
6.  <ScrollView
7.  android:id="@+id/ScrollView01"
8.  android:layout_width="fill_parent"
9.  android:layout_height="wrap_content">
10. <TextView
11. android:id="@+id/TextView01"
12. android:layout_width="wrap_content"
13. android:layout_height="wrap_content"></TextView>
14. </ScrollView>
15. </LinearLayout>
```

(3) 修改 src 目录中 com.hisoft.activity 包下的 NewActivity.java 文件,代码如下:

```
1.  package com.hisoft.activity;
```

```
2.  import java.io.BufferedInputStream;
3.  import java.io.InputStream;
4.  import java.net.URL;
5.  import java.net.URLConnection;
6.  import org.apache.http.util.ByteArrayBuffer;
7.  import android.app.Activity;
8.  import android.os.Bundle;
9.  import android.widget.TextView;
10. public class NewActivity extends Activity {
11. @Override
12. protected void onCreate(Bundle savedInstanceState) {
13. //TODO Auto-generated method stub
14. super.onCreate(savedInstanceState);
15. setContentView(R.layout.mydata);
16.
17. TextView tv= (TextView)findViewById(R.id.TextView01);
18.
19. String msg="";
20. try {
21. URL url=new URL("http://linux.chinaitlab.com/c/896411.html");
22. URLConnection con=url.openConnection();
23. InputStream is=con.getInputStream();
24. BufferedInputStream bis=new BufferedInputStream(is);
25.
26. ByteArrayBuffer baf=new ByteArrayBuffer(100);
27. int current=0;
28. while ((current=bis.read())!=-1) {
29. baf.append((byte)current);
30. }
31. msg=new String(baf.toByteArray(),"GBK");
32.
33. } catch (Exception e) {
34. msg=e.getMessage();
35. }
36. tv.setText(msg);
37. }
38. }
```

(4) 在 AndroidManifest.xml 中的<manifest>根节点标签下添加访问网络的权限，代码如下：

```
1.  <uses-permission android:name="android.
    permission.INTERNET"/>
```

(5) 部署 NetWorkDemo 项目工程，程序运行后，获取 URL 指定地址的网页资源，并显示出来，如图 10-15

图 10-15 NetWorkDemo 提取网页内容

所示。

10.4.3 使用 HTTP 访问网络资源(HttpURLConnection)

Android 中使用 Http 访问网络资源主要是通过 POST 方式和 GET 方式进行网络请求。Http 访问通信中的 POST 和 GET 请求方式不同。GET 可以获得静态页面,也可以把参数放在 URL 字符串后面传递给服务器。而 POST 方法的参数是放在 Http 请求中。因此,在编程之前,应当首先明确使用的请求方法,然后再根据所使用的方式选择相应的编程方式。

使用 HttpURLConnection 位于 java.net 包下,是继承于 URLConnection 类,二者都是抽象类。其对象主要通过 URL 的 openConnection 方法获得。与上述的 URLConnection 方法相比较,HttpURLConnection 用于预先不知数据长度的数据的接收与发送。

(1) HttpURLConnection 使用 POST 方式请求访问资源的通常用法如下面代码所示:

```
1.  URL url=new URL("http://www.zfjsjx.cn/ ");
2.  HttpURLConnection urlConn=(HttpURLConnection)url.openConnection();
3.  通过以下方法可以对请求的属性进行一些设置
4.  //设置输入和输出流
5.  urlConn.setDoOutput(true);
6.  urlConn.setDoInput(true);
7.  //设置请求方式为 POST
8.  urlConn.setRequestMethod("POST");
9.  //POST 请求不能使用缓存
10. urlConn.setUseCaches(false);
11. //关闭连接
12. urlConn.disConnection();
```

(2) HttpURLConnection 默认使用 GET 方式请求访问资源,代码如下:

```
1.  //使用 HttpURLConnection 打开连接
2.         HttpURLConnection urlConn=(HttpURLConnection) url.openConnection();
3.         //得到读取的内容(流)
4.         InputStreamReader in=new InputStreamReader(urlConn.getInputStream());
5.         //为输出创建 BufferedReader
6.         BufferedReader buffer=new BufferedReader(in);
7.         String inputLine=null;
8.         //使用循环来读取获得的数据
9.         while (((inputLine=buffer.readLine()) !=null))
10.        {
11.            //在每一行后面加上一个"\n"来换行
12.            resultData+=inputLine+"\n";
13.        }
14.        //关闭 InputStreamReader
15.        in.close();
16.        //关闭 http 连接
```

```
17.        urlConn.disconnect();
```

如果需要使用 POST 方式，则需要 setRequestMethod 设置。代码如下：

```
1.    String httpUrl="http://www.zfjsjx.cn:8080/a.jsp";
2.            //获得的数据
3.            String resultData="";
4.            URL url=null;
5.            try
6.            {
7.                //构造一个 URL 对象
8.                url=new URL(httpUrl);
9.            }
10.           catch (MalformedURLException e)
11.           {
12.               Log.e(DEBUG_TAG, "MalformedURLException");
13.           }
14.           if (url !=null)
15.           {
16.               try
17.               {
18.                   //使用 HttpURLConnection 打开连接
19.                   HttpURLConnection urlConn=(HttpURLConnection) url.
                      openConnection();
20.                   //因为这个是 post 请求,设立需要设置为 true
21.                   urlConn.setDoOutput(true);
22.                   urlConn.setDoInput(true);
23.                   //设置以 POST 方式
24.                   urlConn.setRequestMethod("POST");
25.                   //Post 请求不能使用缓存
26.                   urlConn.setUseCaches(false);
27.                   urlConn.setInstanceFollowRedirects(true);
28.                   /*配置本次连接的 Content-type,配置为 application/x-www-form
                        -urlencoded 的*/
29.                   urlConn.setRequestProperty("Content-Type","application/x-
                     www-form-urlencoded");
30.                   //连接,从 postUrl.openConnection()至此的配置必须要在 connect 之前
                      //完成
31.                   //要注意的是,connection.getOutputStream 会隐含地进行 connect
32.                   urlConn.connect();
33.                   //DataOutputStream 流
34.                   DataOutputStream out=new DataOutputStream(urlConn.
                      getOutputStream());
35.                   //要上传的参数
36.                   String content="par="+URLEncoder.encode("ABCDEFG", "gb2312");
37.                   //将要上传的内容写入流中
```

```
38.                 out.writeBytes(content);
39.                 //刷新、关闭
40.                 out.flush();
41.                 out.close();
```

10.4.4 使用 HTTP 访问网络应用案例

上面介绍了 URL 读取网络资源常用的包及类和方法,下面通过一个访问网络资源的应用案例详细介绍 URL 的应用。

(1) 创建一个新的 Android 工程,工程名为 HttpURLConnectionDemo,目标 API 选择 10(即 Android 2.3.3 版本),应用程序名为 HttpURLConnectionDemo,包名为 com. hisoft. activity,创建的 Activity 的名字为 HttpURLConnectionDemoActivity,最小 SDK 版本根据选择的目标 API 会自动添加为 10,创建项目工程如图 10-16 所示。

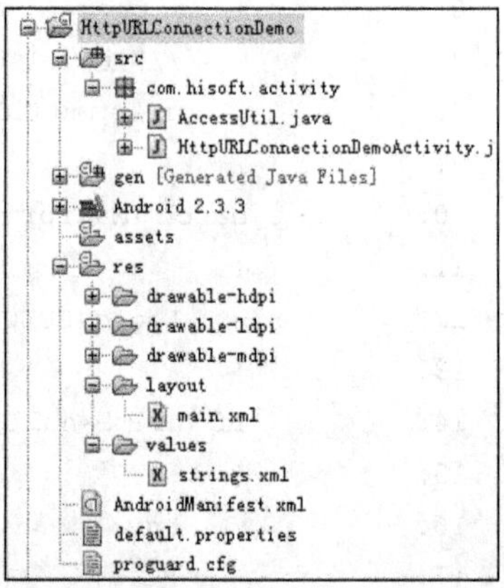

图 10-16 HttpURLConnectionDemo 工程目录结构

(2) 修改 res 目录下 layout 文件夹中的 main. xml 文件,设置线性布局,添加一个 EditText 控件、一个 TextView 和一个 Button 控件描述,并设置相关属性,代码如下:

```
1.  <?xml version="1.0" encoding="utf-8"?>
2.  <LinearLayout xmlns:android="http://schemas.android.com/apk/res/android"
3.      android:orientation="vertical"
4.      android:layout_width="fill_parent"
5.      android:layout_height="fill_parent"
6.      >
7.  <TextView
8.      android:layout_width="fill_parent"
9.      android:layout_height="wrap_content"
10.     android:text="@string/resource"
11.     />
12. <EditText
13.     android:id="@+id/url"
14.     android:layout_width="fill_parent"
15.     android:layout_height="wrap_content"
16.     android:text="@string/surl"
17.     />
18. <TextView
19.     android:layout_width="fill_parent"
20.     android:layout_height="wrap_content"
21.     android:text="@string/goalfile"
22.     />
23. <EditText
```

```
24.     android:id="@+id/target"
25.     android:layout_width="fill_parent"
26.     android:layout_height="wrap_content"
27.     android:text="/mnt/sdcard/bjzf.rar"
28.     />
29. <Button
30.     android:id="@+id/down"
31.     android:layout_width="fill_parent"
32.     android:layout_height="wrap_content"
33.     android:text="@string/down"
34.     />
35. <!--定义一个水平进度条,用于显示下载进度-->
36. <ProgressBar
37.     android:id="@+id/bar"
38.     android:layout_width="fill_parent"
39.     android:layout_height="wrap_content"
40.     android:max="100"
41.     style="@android:style/Widget.ProgressBar.Horizontal"
42.     />
43. </LinearLayout>
```

(3) 修改 res 目录下 values 文件中的 strings.xml 文件,代码如下:

```
1.  <?xml version="1.0" encoding="utf-8"?>
2. <resources>
3.     <string name="hello">Hello World, HttpURLConnectionDemoActivity!
       </string>
4.     <string name="app_name">HttpURLConnectionDemo</string>
5.     <string name="down">点击下载网络资源</string>
6.      < string name =" surl" > http://rsdownload. rising. com. cn/for _ down/
       rsfree2011/ravf/set1137225.exe</string>
7.     <string name="resource">网络资源 URL 路径:</string>
8.     <string name="goalfile">目标保存文件:</string>
9. </resources>
```

(4) 修改 AndroidManifest.xml 文件,在<manifest>根节点下添加访问网络权限、SD 卡写入数据权限、SD 卡创建删除文件权限,代码如下:

```
1. <!--在 SD 卡中创建与删除文件权限-->
2. <uses-permission android:name="android.permission.MOUNT_UNMOUNT_
   FILESYSTEMS"/>
3. <!--向 SD 卡写入数据权限-->
4. <uses-permission android:name="android.permission.WRITE_EXTERNAL
   _STORAGE"/>
5. <!--授权访问网络-->
6. <uses-permission android:name="android.permission.INTERNET"/>
```

(5) 在 src 目录中的 com.hisoft.activity 包下创建 AccessUtil.java 文件,代码如下:

```
1.  package com.hisoft.activity;
2.  import java.io.InputStream;
3.  import java.io.RandomAccessFile;
4.  import java.net.HttpURLConnection;
5.  import java.net.URL;
6.
7.  public class AccessUtil
8.  {
9.      //定义下载资源的路径
10.     private String path;
11.     //指定所下载文件的保存位置
12.     private String targetFile;
13.     //定义需要使用多少线程下载资源
14.     private int threadNum;
15.     //定义下载的线程对象
16.     private DownloadThread[] threads;
17.     //定义下载文件的总大小
18.     private int fileSize;
19.
20.     public AccessUtil(String path, String targetFile, int threadNum)
21.     {
22.         this.path=path;
23.         this.threadNum=threadNum;
24.         //初始化 threads 数组
25.         threads=new DownloadThread[threadNum];
26.         this.targetFile=targetFile;
27.     }
28.
29.     public void download() throws Exception
30.     {
31.         URL url=new URL(path);
32.         HttpURLConnection conn=(HttpURLConnection) url.openConnection();
33.         conn.setConnectTimeout(5 * 1000);
34.         conn.setRequestMethod("GET");
35.         conn.setRequestProperty(
36.             "Accept",
37.             "image/gif, image/jpeg, image/pjpeg, image/pjpeg, application/x-
                shockwave-flash, application/xaml+xml, application/vnd.ms-
                xpsdocument, application/x-ms-xbap, application/x-ms-
                application, application/vnd.ms-excel, application/vnd.ms-
                powerpoint, application/msword, * / * ");
38.         conn.setRequestProperty("Accept-Language", "zh-CN");
39.         conn.setRequestProperty("Charset", "UTF-8");
40.         conn.setRequestProperty(
```

```
41.             "User-Agent",
42.             "Mozilla/4.0 (compatible; MSIE 7.0; Windows NT 5.2; Trident/4.0;
                .NET CLR 1.1.4322; .NET CLR 2.0.50727; .NET CLR 3.0.04506.30; .NET
                CLR 3.0.4506.2152; .NET CLR 3.5.30729)");
43.         conn.setRequestProperty("Connection", "Keep-Alive");
44.         //得到文件大小
45.         fileSize=conn.getContentLength();
46.         conn.disconnect();
47.         int currentPartSize=fileSize/threadNum+1;
48.         RandomAccessFile file=new RandomAccessFile(targetFile, "rw");
49.         //设置本地文件的大小
50.         file.setLength(fileSize);
51.         file.close();
52.         for (int i=0; i<threadNum; i++)
53.         {
54.             //计算每条线程下载的开始位置
55.             int startPos=i * currentPartSize;
56.             //每个线程使用一个 RandomAccessFile 进行下载
57.             RandomAccessFile currentPart=new RandomAccessFile(targetFile,
58.                 "rw");
59.             //定位该线程的下载位置
60.             currentPart.seek(startPos);
61.             //创建下载线程
62.             threads[i]=new DownloadThread(startPos, currentPartSize,
63.                 currentPart);
64.             //启动下载线程
65.             threads[i].start();
66.         }
67.     }
68.
69.     //获取下载的完成百分比
70.     public double getCompleteRate()
71.     {
72.         //统计多条线程已经下载的总大小
73.         int sumSize=0;
74.         for (int i=0; i<threadNum; i++)
75.         {
76.             sumSize+=threads[i].length;
77.         }
78.         //返回已经完成的百分比
79.         return sumSize * 1.0/fileSize;
80.     }
81.
82.     private class DownloadThread extends Thread
```

```
83.      {
84.          //当前线程的下载位置
85.          private int startPos;
86.          //定义当前线程负责下载的文件大小
87.          private int currentPartSize;
88.          //当前线程需要下载的文件块
89.          private RandomAccessFile currentPart;
90.          //定义该线程已下载的字节数
91.          public int length;
92.
93.          public DownloadThread(int startPos, int currentPartSize,
94.              RandomAccessFile currentPart)
95.          {
96.              this.startPos=startPos;
97.              this.currentPartSize=currentPartSize;
98.              this.currentPart=currentPart;
99.          }
100.
101.          @Override
102.          public void run()
103.          {
104.              try
105.              {
106.                  URL url=new URL(path);
107.                  HttpURLConnection conn=(HttpURLConnection) url
108.                      .openConnection();
109.                  conn.setConnectTimeout(5*1000);
110.                  conn.setRequestMethod("GET");
111.                  conn.setRequestProperty(
112.                          "Accept","image/gif, image/jpeg, image/pjpeg, image/
                            pjpeg,"+
113.                          "application/x-shockwave-flash, application/xaml+
                            xml, application/vnd.ms-xpsdocument, application/x-
                            ms-xbap, application/x-ms-application, application/
                            vnd.ms-excel, application/vnd.ms-powerpoint,
                            application/msword, */*");
114.                  conn.setRequestProperty("Accept-Language", "zh-CN");
115.                  conn.setRequestProperty("Charset", "UTF-8");
116.                  InputStream inStream=conn.getInputStream();
117.                  //跳过 startPos 个字节,表明该线程只下载自己负责的那部分文件
118.                  inStream.skip(this.startPos);
119.                  byte[] buffer=new byte[1024];
120.                  int hasRead=0;
121.                  //读取网络数据,并写入本地文件
```

```
122.                 while (length<currentPartSize
123.                     && (hasRead=inStream.read(buffer)) !=-1)
124.                 {
125.                     currentPart.write(buffer, 0, hasRead);
126.                     //累计该线程下载的总大小
127.                     length+=hasRead;
128.                 }
129.                 currentPart.close();
130.                 inStream.close();
131.             }
132.             catch (Exception e)
133.             {
134.                 e.printStackTrace();
135.             }
136.         }
137.     }
138. }
```

(6) 修改 src 目录中 com. hisoft. activity 包下的 HttpURLConnectionDemoActivity. java 文件,代码如下:

```
1.  package com.hisoft.activity;
2.  import java.util.Timer;
3.  import java.util.TimerTask;
4.
5.  import android.app.Activity;
6.  import android.os.Bundle;
7.  import android.os.Handler;
8.  import android.os.Message;
9.  import android.view.View;
10. import android.view.View.OnClickListener;
11. import android.widget.Button;
12. import android.widget.EditText;
13. import android.widget.ProgressBar;
14.
15. public class HttpURLConnectionDemoActivity extends Activity
16. {
17.     EditText url;
18.     EditText target;
19.     Button downBn;
20.     ProgressBar bar;
21.     AccessUtil downUtil;
22.     private int mDownStatus;
23.
24.     @Override
```

```
25.        public void onCreate(Bundle savedInstanceState)
26.        {
27.            super.onCreate(savedInstanceState);
28.            setContentView(R.layout.main);
29.            //获取程序界面中的三个界面控件
30.            url=(EditText) findViewById(R.id.url);
31.            target=(EditText) findViewById(R.id.target);
32.            downBn=(Button) findViewById(R.id.down);
33.            bar=(ProgressBar) findViewById(R.id.bar);
34.            //创建一个 Handler 对象
35.            final Handler handler=new Handler()
36.            {
37.                @Override
38.                public void handleMessage(Message msg)
39.                {
40.                    if (msg.what==0x123)
41.                    {
42.                        bar.setProgress(mDownStatus);
43.                    }
44.                }
45.            };
46.            downBn.setOnClickListener(new OnClickListener()
47.            {
48.                @Override
49.                public void onClick(View v)
50.                {
51.                    //初始化 DownUtil 对象
52.                    downUtil=new AccessUtil(url.getText().toString(),
53.                        target.getText().toString(), 4);
54.                    try
55.                    {
56.                        //开始下载
57.                        downUtil.download();
58.                    }
59.                    catch (Exception e)
60.                    {
61.                        e.printStackTrace();
62.                    }
63.                    //定义每秒调度获取一次系统的完成进度
64.                    final Timer timer=new Timer();
65.                    timer.schedule(new TimerTask()
66.                    {
67.                        @Override
68.                        public void run()
```

```
69.                     {
70.                         //获取下载任务的完成比率
71.                         double completeRate=downUtil.getCompleteRate();
72.                         mDownStatus=(int) (completeRate*100);
73.                         //发送消息通知界面更新进度条
74.                         handler.sendEmptyMessage(0x123);
75.                         //下载完全后取消任务调度
76.                         if (mDownStatus>=100)
77.                         {
78.                             timer.cancel();
79.                         }
80.                     }
81.                 }, 0, 100);
82.             }
83.         });
84.     }
85. }
```

(7) 部署 HttpURLConnectionDemo 项目工程,程序运行如图 10-17 所示。单击"点击下载网络资源"按钮开始下载,下载完成后的存储路径如图 10-18 和图 10-19 所示。

图 10-17 HttpURLConnectionDemo 运行效果

图 10-18 下载网络资源

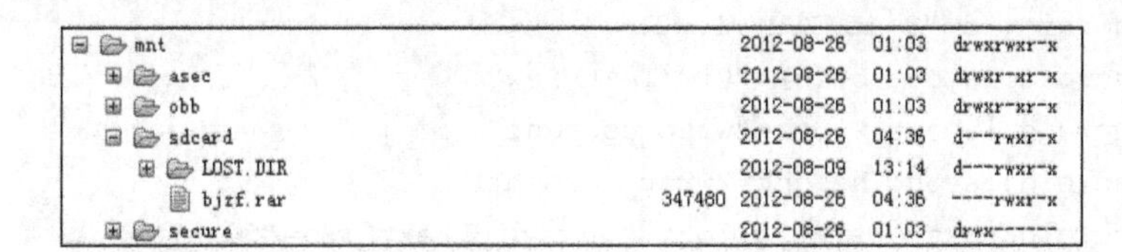

图 10-19 下载文件存储路径

拓展提示:由于篇幅和项目侧重点不同的原因,本书只是简单介绍了 Socket 访问网络资源的方式。但实际项目开发中,Android 平台下的 Socket 通信程序的开发是非常重要的知识点,通常采用 Socket 服务器端和客户端开发的方式,利用手机客户端通过网络方式从服务器端进行数据信息的读取和应用。

10.5 项目案例

学习目标：学习远程访问网络资源及读取数据库信息、提交数据到服务器的应用。

案例描述：使用 RelativeLayout 相对布局、TextView 控件、EditText 控件、Button 按钮，通过 HttpURLConnection、URL、OnClickListener、Intent、Bundle、onCreateDialog、Menu、onOptionsItemSelected 等类及方法的应用，实现医药商品信息显示、选择、购买以及添加到购物车。

案例要点：HttpURLConnection、URL、Bundle、Menu。

案例实施：

(1) 创建工程 Project_Chapter_10，选择 Android 2.3.3 作为目标平台，如图 10-20 所示。

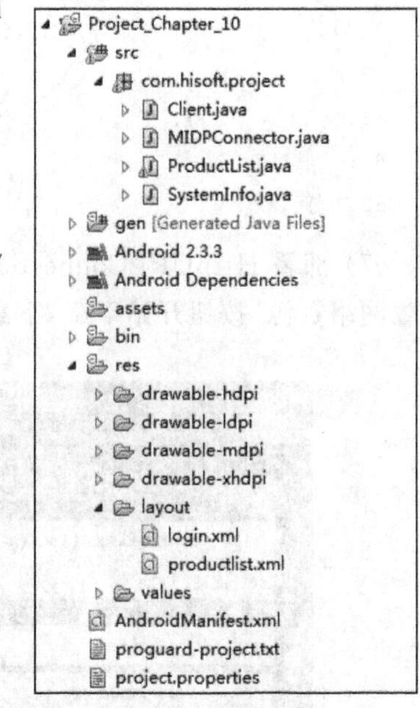

图 10-20 Project_Chapter_10 工程目录结构

(2) 创建 login. xml 文件并存放在 res/layout 目录下，代码如下：

```
1.  <?xml version="1.0" encoding="utf-8"?>
2.   < RelativeLayout  xmlns: android =" http://
    schemas.android.com/apk/res/android"
3.     android:orientation="vertical"
4.     android:layout_width="fill_parent"
5.     android:layout_height="fill_parent"
6.     >
7.  <TextView
8.    android:id="@+id/TextView"
9.     android:layout_width="fill_parent"
10.    android:layout_height="wrap_content"
11.    android:text=""
12.
13.    />
14.
15. <TextView
16.   android:id="@+id/TextView01"
17.   android:layout_below="@id/TextView"
18.   android:layout_width="wrap_content"
19.   android:layout_height="wrap_content"
20.   android:text="@string/login" android:textSize="30px">
21. </TextView>
22.
23. <TextView
24.   android:id="@+id/TextView02"
25.   android:layout_marginTop="12px"
26.   android:layout_marginLeft="5dip"
27.   android:layout_below="@id/TextView01"
28.   android:layout_width="wrap_content"
```

```
29.     android:layout_height="wrap_content"
30.     android:text="@string/input_username" android:textSize="20px">
31.     </TextView>
32.
33.  <EditText
34.     android:id="@+id/username"
35.     android:layout_alignTop="@id/TextView02"
36.     android:layout_toRightOf="@id/TextView02"
37.     android:layout_width="fill_parent"
38.     android:layout_height="wrap_content"
39.     android:singleLine="true">
40.     </EditText>
41.
42.  <TextView
43.     android:id="@+id/TextView03"
44.     android:layout_below="@id/username"
45.     android:layout_alignLeft="@id/TextView02"
46.     android:layout_width="wrap_content"
47.     android:layout_height="wrap_content"
48.     android:text="@string/input_userpwd" android:textSize="20px">
49.  </TextView>
50.
51.  <EditText
52.     android:id="@+id/password"
53.     android:layout_alignTop="@id/TextView03"
54.     android:layout_toRightOf="@id/TextView03"
55.     android:layout_alignLeft="@id/username"
56.     android:layout_width="fill_parent"
57.     android:layout_height="wrap_content" android:password="true" android:
    singleLine="true">
58.  </EditText>
59.
60.  <Button
61.     android:id="@+id/login"
62.     android:layout_marginTop="12px"
63.     android:layout_width="100px"
64.     android:layout_height="wrap_content"
65.     android:layout_below="@id/password"
66.     android:layout_alignLeft="@id/password"
67.     android:text="@string/bt_login"
68.     android:textSize="16px"
69.     >
70.  </Button>
71.
72.  <Button
```

```
73.    android:id="@+id/exit"
74.    android:layout_marginLeft="15px"
75.    android:layout_width="100px"
76.    android:layout_height="wrap_content"
77.    android:text="@string/bt_exit"
78.    android:layout_toRightOf="@id/login"
79.    android:layout_alignTop="@id/login"
80.    android:textSize="16px"
81.    >
82.  </Button>
83.  </RelativeLayout>
```

(3) 创建 productlist.xml 文件并存放在 res/layout 目录下，代码如下：

```
1.   <?xml version="1.0" encoding="utf-8"?>
2.   <RelativeLayout
3.     xmlns:android="http://schemas.android.com/apk/res/android"
4.     android:orientation="vertical"
5.     android:layout_width="wrap_content"
6.     android:layout_height="wrap_content">
7.     <TextView
8.       android:text="TextView01"
9.       android:id="@+id/temp"
10.      android:layout_width="wrap_content"
11.      android:layout_height="wrap_content"
12.    ></TextView>
13.
14.     <ListView
15.      android:id="@+id/productlist"
16.      android:layout_below="@id/TextView01"
17.      android:layout_width="fill_parent"
18.      android:layout_height="wrap_content"
19.      android:focusable="true">
20.    </ListView>
21.    <TextView
22.      android:id="@+id/pageinfo"
23.      android:layout_alignParentRight="true"
24.      android:layout_width="wrap_content"
25.      android:layout_height="wrap_content"
26.      android:text="TextView02"
27.    ></TextView>
28.  </RelativeLayout>
```

(4) 在 src 目录下的 com.hisoft.project 包下创建 SystemInfo.java 文件，代码如下：

```
1.   ***********
2.     * 程序名称:SystemInfo.java                                    *
```

```
3.    * 功能:定义一些常量,包括显示的字符信息和连接字符串的信息                    *
4.    * 作者:                                                                    *
5.    * 日期:                                                                    *
6.  /*************************************************************************/
7.  public interface SystemInfo {
8.    String serverURL="http://10.0.2.2:8080/ESysAndroidServer/";  //服务器地址
9.    String formtitle="Ascent 移动版医药商务系统";
10.   String errorPng="/ErrorClaim.png";
11.   String logo="/ascent.png";
12.   String login="登录";
13.   String exit="离开";
14.   String user="用户";
15.   String password="密码";
16.   String loginValidate="用户名密码不能为空";
17.   String listtitle="请选择医药商品";
18.   String purchase="添加到购物车";
19.   String order="提交订单";
20.   String tel="电话";
21.   String address="地址";
22.   String familyname="姓名";
23.   String email="邮箱";
24.   String previous="上一层";
25.   String failinfo="登录失败";
26.   String orderurl="order";
27.   String loginurl="login";
28.   String producturl="product";
29.
30. }
```

(5) 在 src 目录下的 com. hisoft. project 包下创建 MIDPConnector. java 文件,代码如下:

```
1.  package com.hisoft.project;
2.  /************************************************************/**
3.  /**
4.    * A class that is responsible for establishing the HTTP connection between
5.    * client and server.
6.    */
7.  package com.hisoft.project;
8.  /***********************************************************************
9.    * 程序名称:MIDPConnector.java                                             *
10.   * 功能:连接后台,获取与后台的连接                                           *
11.   * 作者:                                                                   *
12.   * 日期:                                                                   *
13. /***********************************************************************/
14.
```

```
15.  import java.io.*;
16.  import java.net.*;
17.
18.
19.  public class MIDPConnector {
20.
21.    /**
22.     * 构造方法
23.     */
24.    public MIDPConnector() {
25.    }
26.
27.    /**
28.     * 创建到后台的连接
29.     * @param url 请求的地址
30.     * @param queryString 请求字符串
31.     * @return 返回连接
32.     * @throws IOException
33.     */
34.    public static HttpURLConnection open(String url,String queryString) throws
       IOException {
35.      /**
36.       *  Below is the Server URL when server is running on local host.
37.       *  If you run the server on a different machine, then change the URL to the
38.       *  corresponding server URL.
39.       */
40.
41.      String medicalServerURL=SystemInfo.serverURL+url;
42.      HttpURLConnection conn=null;
43.
44.      String urlString=medicalServerURL+"?"+   queryString;
45.      /**Connection is opened with a READ_WRITE flag.
46.       *  It indicates that this connection is used for sending message to and
47.       *  receiving message from the server.
48.       *  The default request method is GET.
49.       */
50.      URL url2=new URL(urlString);
51.      conn=(HttpURLConnection)url2.openConnection();
52.      return conn;
53.    }
54.    /**
55.     * 创建到后天的链接
56.     * @param url 链接地址
57.     * @param queryString 请求字符串
58.     * @param searchString 请求字符串
```

```
59.     * @return 返回到后台的连接
60.     * @throws IOException
61.     */
62.   public static HttpURLConnection open(String url,String queryString, String
searchString) throws IOException {
63.      /**
64.       *  Below is the bookstoreServer URL when server is running on local host.
65.       *  If you run the server on a different machine, then change the URL to the
66.       *  corresponding server URL.
67.       */
68.
69.      String medicalServerURL=SystemInfo.serverURL+url;
70.      HttpURLConnection conn=null;
71.      //拼接请求字符串
72.      String urlString=medicalServerURL+"?"+  queryString+searchString;
73.      /**Connection is opened with a READ_WRITE flag.
74.       *  It indicates that this connection is used for sending message to and
75.       *  receiving message from the server.
76.       *  The default request method is GET.
77.       */
78.      URL url2=new URL(urlString);
79.      conn=(HttpURLConnection)url2.openConnection();
80.      return conn;
81.   }
82.
83. }
```

(6) 在 src 目录下的 com. hisoft. project 包下创建 Client. java 文件,代码如下:

```
1.    package com.hisoft.project;
2.   /*********************************************************************
3.    * 程序名称:Client.java                                                 *
4.    * 功能:显示用户登录窗口,可登录和退出系统                                 *
5.    * 作者:                                                              *
6.    * 日期:                                                              *
7.   *********************************************************************/
8.
9.   import java.io.IOException;
10.  import java.net.HttpURLConnection;
11.  import java.net.UnknownHostException;
12.  import android.app.Activity;
13.  import android.app.Dialog;
14.  import android.content.Context;
15.  import android.content.Intent;
16.  import android.os.Bundle;
17.  import android.view.View;
```

```
18.  import android.widget.Button;
19.  import android.widget.EditText;
20.
21.
22.  public class Client extends Activity {
23.    /**Called when the activity is first created. * /
24.
25.    //登录事件监听
26.    public Button.OnClickListener bt_login_ocl;
27.    //退出事件监听
28.    public Button.OnClickListener bt_exit_ocl;
29.    //登录用户名
30.    public EditText edit_name;
31.    //登录密码
32.    public EditText edit_pwd;
33.    public String username;
34.    public String password;
35.    private Button bt_login;
36.    private Button bt_exit;
37.
38.    int status=-1;
39.    private boolean isValid=false;
40.
41.    public String getUsername() {
42.      return username;
43.    }
44.
45.    public void setUsername(String username) {
46.      this.username=username;
47.    }
48.
49.    public String getPassword() {
50.      return password;
51.    }
52.
53.    public void setPassword(String password) {
54.      this.password=password;
55.    }
56.    /**
57.     * 显示登录框页面
58.     * /
59.    @Override
60.      public void onCreate(Bundle savedInstanceState) {
61.          super.onCreate(savedInstanceState);
62.          setContentView(R.layout.login);
```

```
63.         //获取输入框对象
64.         edit_name=(EditText)findViewById(R.id.username);
65.         edit_pwd=(EditText)findViewById(R.id.password);
66.         //获得按钮对象
67.         bt_login=(Button)findViewById(R.id.login);
68.         bt_login_ocl=new Button.OnClickListener()
69.         {
70.           public void onClick(View v) {
71.             //登录
72.             login();
73.
74.             }
75.         };
76.         //设置登录按钮监听
77.         bt_login.setOnClickListener(bt_login_ocl);
78.         //获得退出按钮对象
79.         bt_exit=(Button)findViewById(R.id.exit);
80.         bt_exit_ocl=new Button.OnClickListener()
81.         {
82.           //@Override
83.           public void onClick(View v){
84.             //TODO Auto-generated method stub
85.             //退出操作
86.             finish();
87.             }
88.         };
89.         //设置退出按钮监听
90.         bt_exit.setOnClickListener(bt_exit_ocl);
91.   }
92.   /**
93.    * 用户登录操作
94.    */
95.   public void login(){
96.
97.         //检测用户名密码有效性
98.         isValid=loginCheck();
99.         if (isValid) {
100.
101.            Intent intent=new Intent();
102.            intent.setClass(Client.this, ProductList.class);
103.            Bundle bundle=new Bundle();
104.
105.            bundle.putString("loginuser",Client.this.username);
106.            bundle.putString("tempPid","");
107.            /*将登录用户传给 resultscreen*/
```

```
108.
109.            intent.putExtras(bundle);
110.            //登录成功,启动新的画面
111.            startActivity(intent);
112.
113.            this.finish();
114.
115.          } else {
116.
117.            //登录失败,显示出错信息
118.            showDialog(1);
119.          }
120.        }
121.
122.
123.      /**
124.       * 检测用户有效性
125.       */
126.     private boolean loginCheck() {
127.       //实例化登录用户实例
128.       setUsername(edit_name.getText().toString());
129.       setPassword(edit_pwd.getText().toString());
130.
131.       HttpURLConnection con=null;
132.       try {
133.         con=MIDPConnector.open(SystemInfo.loginurl, "username="
134.             +Client.this.username+"&password="
135.             +Client.this.password+"&a=login");
136.
137.         //获得请求状态码
138.         int code=con.getResponseCode();
139.
140.         if (code==200) {
141.           return true;
142.         }
143.
144.       } catch (IOException e) {
145.
146.         e.printStackTrace();
147.         return false;
148.       }
149.       return false;
150.
151.     }
152.     /**
```

```
153.     * 创建提示对话框
154.     */
155.    protected Dialog onCreateDialog(int id) {
156.      switch (id) {
157.      case 1:
158.        try {
159.            return buildDialog(Client.this);
160.        } catch (UnknownHostException e) {
161.            //TODO Auto-generated catch block
162.            e.printStackTrace();
163.        } catch (IOException e) {
164.            //TODO Auto-generated catch block
165.            e.printStackTrace();
166.        }
167.      }
168.      return null;
169.    }
170.      /**
171.       * 创建提示框
172.       * @param context
173.       * @return 返回创建的对话框对象
174.       * @throws UnknownHostException
175.       * @throws IOException
176.       */
177.    public Dialog buildDialog(Context context) throws UnknownHostException,
178.        IOException {
179.      Dialog dialog=new Dialog(context);
180.      dialog.setTitle("用户名/密码错误,请重新登录");
181.      dialog.setCanceledOnTouchOutside(true);
182.      //返回创建的对话框对象
183.      return dialog;
184.    }
185.  }
```

(7) 在 src 目录下的 com. hisoft. project 包下创建 ProductList. java 文件,代码如下:

```
1.   package com.hisoft.project;
2.   /*********************************************************************
3.    * 程序名称:ProductList.java                                          *
4.    * 功能:显示所有的商品信息,可选择需要的商品添加到购物车              *
5.    * 作者:                                                              *
6.    * 日期:                                                              *
7.   *********************************************************************/
8.   import java.io.DataInputStream;
9.   import java.io.IOException;
10.  import java.io.InputStream;
```

```
11.  import java.net.HttpURLConnection;
12.  import android.app.Activity;
13.  import android.app.Dialog;
14.  import android.content.Context;
15.  import android.content.Intent;
16.  import android.os.Bundle;
17.  import android.view.Menu;
18.  import android.view.MenuItem;
19.  import android.widget.ArrayAdapter;
20.  import android.widget.ListView;
21.  import android.widget.TextView;
22.  import android.widget.Toast;
23.
24.  public class ProductList extends Activity implements Runnable {
25.
26.    public String[] optionStrings=null;
27.    //存放购买的商品信息
28.    public static StringBuffer cartBuffer=new StringBuffer();
29.    public Thread thread;
30.    int currentPage=1;                                    //当前页面
31.    int total;                                            //总页数
32.    int pageCount;
33.    int topage=currentPage;
34.
35.    int amount=0;
36.    String loginuser="";
37.    String tempPid="";
38.
39.      //定义静态变量存放列表信息
40.    public static ProductList pList=null;
41.    private TextView temp;
42.    private TextView pageInfo;                            //页面显示信息
43.    private ListView productList;                         //商品列表展示控件
44.    String status="";
45.    ArrayAdapter<String>adapter;
46.
47.    String queryString="";
48.
49.    //menu 菜单
50.    public Menu menu;
51.    public static final int HANDLE=0;                     //添加到购物车
52.    public static final int PRE=1;                        //上一页
53.    public static final int NEXT=2;                       //下一页
54.    public static final int CART=3;                       //查看购物车
55.    public static final int ORDER=5;
```

```
56.
57.       /**
58.         * 创建产品列表页面,初始化信息
59.         * /
60.     @Override
61.     public void onCreate(Bundle savedInstanceState) {
62.       super.onCreate(savedInstanceState);
63.       setContentView(R.layout.productlist);
64.       //取得前一个页面传过来的参数
65.       Bundle bunde=this.getIntent().getExtras();
66.       loginuser=bunde.getString("loginuser");
67.       tempPid=bunde.getString("tempPid");
68.       //静态变量赋值
69.       pList=this;
70.       //拿到页面控件对象
71.       temp=(TextView) findViewById(R.id.temp);
72.       temp.setText(" 请选择商品");
73.       temp.setText(queryString+"tempid:"+tempPid);
74.       pageInfo=(TextView) findViewById(R.id.pageinfo);
75.       //获取 ListView 对象
76.       productList=(ListView) findViewById(R.id.productlist);
77.       pList.getInformation();
78.
79.       productList.setItemsCanFocus(true);
80.       //设置商品可多选
81.       productList.setChoiceMode(ListView.CHOICE_MODE_MULTIPLE);
82.       productList.setTextFilterEnabled(true);
83.       //设置分页信息
84.       pageInfo.setText(status);
85.       //配置适配器
86.       productList.setAdapter(adapter);
87.     }
88.
89.
90.       /**
91.         * 获取商品信息
92.         * /
93.     public void getInformation() {
94.       run();
95.     }
96.       /**
97.         * 获取选中项信息
98.         * @return
99.         * /
100.      private String getQueryString() {
```

```
101.
102.        for (int i=0; i<productList.getCount(); i++) {
103.          //逐个判断是否被选中
104.          if (productList.isItemChecked(i)) {
105.              //被选中则放入缓存
106.              String s=optionStrings[i];
107.              cartBuffer.append(s.substring(0, s.indexOf(" "))+",");
108.          }
109.        }
110.        if (cartBuffer.length()>0) {
111.          return cartBuffer.toString().substring(0, cartBuffer.length()-1);
112.        } else {
113.          return "";
114.        }
115.
116.      }
117.
118.      /**
119.       * 恢复各个选项到初始状态
120.       */
121.      private void clearOptions() {
122.
123.        for (int i=0; i<productList.getCount(); i++) {
124.
125.          if (productList.isItemChecked(i)) {
126.              //有被选择的选项则置为 false
127.              productList.setItemChecked(i, false);
128.          }
129.        }
130.      }
131.
132.      /**
133.       * 判断是否被选择
134.       * @return
135.       */
136.      private boolean isSelected() {
137.
138.        boolean isSelected=false;
139.
140.        for (int i=0; i<productList.getCount(); i++) {
141.
142.          if (productList.isItemChecked(i)) {
143.            //被选中
144.            isSelected=true;
145.            break;
```

```
146.         }
147.       }
148.       return isSelected;
149.     }
150.       /**
151.        * 创建 MENU 菜单
152.        * /
153.     public boolean onCreateOptionsMenu(Menu menu) {
154.
155.
156.       menu.add(0, HANDLE, 0, SystemInfo.purchase);                //添加到购物车 0
157.
158.       menu.add(0, PRE, 0, "上一页");                              //1 翻到上一页
159.
160.       menu.add(0, NEXT, 0, "下一页");                             //2 翻到下一页
161.
162.       menu.add(0, CART, 0, "我的购物车");                         //3 查看购物车
163.
164.       return true;
165.
166.     }
167.
168.     /**
169.      * MENU 菜单的选择事件
170.      * /
171.     @Override
172.     public boolean onOptionsItemSelected(MenuItem item) {
173.       switch (item.getItemId()) {
174.       case HANDLE:
175.         //添加到购物车
176.         if (this.isSelected()) {
177.           queryString=this.getQueryString();
178.           this.clearOptions();
179.           temp.setText(this.getQueryString());
180.
181.         } else {
182.         }
183.         return true;
184.       case PRE:
185.         //上一页
186.         if(currentPage>1){
187.           currentPage=currentPage-1;
188.           topage=currentPage;
189.           //获得该页信息
190.           this.getInformation();
```

```
191.          //设置适配信息
192.          productList.setAdapter(adapter);
193.          //设置提示信息
194.          pageInfo.setText(status);
195.        }
196.        else{
197.
198.        }
199.        return true;
200.      case NEXT:
201.        //下一页
202.        if(currentPage<pageCount){
203.          currentPage=currentPage+1;
204.          topage=currentPage;
205.          //获取下一页面信息
206.          this.getInformation();
207.
208.          productList.setAdapter(adapter);
209.          pageInfo.setText(status);
210.        }
211.        else{
212.
213.        }
214.        return true;
215.
216.      case CART:
217.        //查看购物车
218.         Toast.makeText(ProductList.this, "所选择的是:"+item.getTitle(),
             Toast.LENGTH_SHORT).show();
219.        return true;
220.      }
221.      return false;                                      //should never happen
222.
223.    }
224.
225.    /**
226.     * 查询获得产品数据
227.     */
228.    public void run() {
229.
230.      HttpURLConnection con=null;
231.      try {
232.        //建立与后台的链接
233.        con=MIDPConnector.open(SystemInfo.producturl, "a=all"+"&cpage="
```

```
234.              +topage+"&style=");
235.
236.          InputStream is=con.getInputStream();
237.          DataInputStream dis=new DataInputStream(is);
238.                 //读取分页信息
239.          total=dis.readInt();
240.          pageCount=dis.readInt();
241.          amount=dis.readInt();
242.
243.          optionStrings=new String[amount];
244.          for (int i=0; i<amount; i++) {
245.
246.            optionStrings[i]=dis.readUTF()+"元";
247.
248.          }
249.                 //建立适配信息
250.          adapter=new ArrayAdapter<String>(pList,
251.              android.R.layout.simple_list_item_multiple_choice,
252.              optionStrings);
253.          //设置提示信息
254.          status="共 "+total+" 件商品"+"      第"+currentPage+"页    "
255.              +"  共 "+pageCount+" 页 ";
256.
257.
258.          is.reset();
259.          dis.close();
260.          con.disconnect();
261.
262.        } catch (IOException e) {
263.          e.printStackTrace();
264.
265.        } finally {
266.
267.          if (con !=null)
268.            try {
269.              con.disconnect();
270.            } catch (Exception e) {
271.              e.printStackTrace();
272.            }
273.        }
274.
275.      }
276.
277.      /**
278.       * 创建进度提示框
```

```
279.      */
280.    protected Dialog onCreateDialog(int id) {
281.      switch (id) {
282.      case 1:
283.        return buildDialog(ProductList.this);
284.      }
285.      return null;
286.    }
287.      /**
288.        * 创建提示框
289.        * @param context
290.        * @return
291.        */
292.    public Dialog buildDialog(Context context) {
293.      Dialog dialog=new Dialog(context);
294.      dialog.setTitle("正在连接到系统");
295.      return dialog;
296.    }
297.
298.  }
```

(8) 修改 AndroidManifest.xml 文件,代码如下:

```
1.   <?xml version="1.0" encoding="utf-8"?>
2.   <manifest xmlns:android="http://schemas.android.com/apk/res/android"
3.       package="com.hisoft.project"
4.       android:versionCode="1"
5.       android:versionName="1.0">
6.
7.       <uses-sdk android:minSdkVersion="10"/>
8.       <uses-permission android:name="android.permission.INTERNET"/>
9.
10.      <application
11.          android:icon="@drawable/ic_launcher"
12.          android:label="@string/app_name">
13.          <activity android:name=".Client"
14.                    android:label="@string/app_name">
15.              <intent-filter>
16.                <action android:name="android.intent.action.MAIN"/>
17.              <category android:name="android.intent.category.LAUNCHER"/>
18.              </intent-filter>
19.          </activity>
20.
21.            <activity android:name=".ProductList" android:label="@string/SysM_
               Label">
22.          </activity>
```

```
23.     </application>
24. </manifest>
```

(9) 部署工程 Project_Chapter_10，程序运行效果如图 10-21 所示。

图 10-21　Project_Chapter_10 运行效果

1. 简答题

(1) 在 Android 系统中,短信息服务主要使用哪些包及类？如何添加权限？

(2) 尝试用代码实现直接调用 Android 系统自带的电话程序。

(3) 在 Android 系统里进行邮件客户端设计开发的方式有几种？它们之间有什么区别？

(4) Android 系统中访问网络资源的方式有哪些？如何注册并添加它们的权限？

(5) 分析使用 URL 和 HttpURLConnection 访问网络资源的优劣,并描述它们各自的应用条件。

2. 完成下面的实训项目

要求：

(1) 使用 Socket 开发服务器和客户端,服务器端 IP 为 192.168.14.2,端口为 8888,启动服务器后会在控制台(Console)显示“Waiting Connection…”。

(2) 启动客户端,程序自动连接服务器,服务器收到客户端的请求后会在控制台显示提示信息“Client Connection Success”。

(3) 客户端在成功连接到服务器端后会自动读取服务器发送的消息“this is a message from server”,并显示在客户端的 TextView 中。

第11章 Google API服务

学习目标

本章主要深入介绍 Google Map API 服务的应用，主要介绍 LocationManager、LocationProvider、Criteria、Location、LocationListener 等常用的方法、属性设置，以及 Google Maps、Google Map API Key 的申请方法及过程、创建 Google Map API 目标平台的方法。使读者通过本章的学习，能够深入熟悉 Google Map API 服务，掌握以下知识要点：

(1) LocationManager 类继承关系、常用属性及设置描述。

(2) LocationProvider 类继承关系、属性设置及描述、常用的方法。

(3) Criteria 类和 Location 类属性及设置、常用方法。

(4) LocationListener 类继承关系、属性描述及通常引用方法。

(5) MapView、Map API 类的关系及继承。

(6) Google Map API Key 申请过程、设置及通常用法。

(7) 创建 Google Map API AVD 方法。

11.1 地理位置定位服务

地理位置定位服务(Location-Based Service，LBS)，又称为移动定位服务，是指通过移动运营商提供的无线电通信网络(如 GSM 网、CDMA 网)或外部定位方式(如 GPS)获取移动终端用户的位置信息(地理坐标)。在 GIS (Geographic Information System，地理信息系统)平台的支持下为用户提供相应服务的一种增值业务。它通常包含两方面服务：一是确定移动设备或用户所在的地理位置；二是提供与位置相关的各类信息服务。

地理位置定位服务在移动服务方面的基本原理：当移动用户需要信息服务或监控管理中心需要对某移动终端进行移动计算时，首先移动终端通过内嵌的定位设备如 GNSS (GPS/GLONASS)获得终端本身当前的空间位置数据，并实时地通过无线通信把数据上传送到服务中心，然后服务中心 GIS 服务器根据终端的地理位置、服务要求进行空间分析、决策，进而再下传移动终端或中心的计算机。

地理位置定位服务已经被广泛地应用到健康、工作、个人生活等许多行业领域，其中包括一些高端或低端的用户，也包括公共事业或私人用途上。目前主要有以下一些应用：

(1) 紧急救援服务。

(2) 与地图坐标有关的信息服务及交互式地图信息服务。

(3) 导航服务。提供交通路况及最佳行车路线；帮助用户寻找最近的目的地及相关信息；帮助查找某公司的位置、电话号码和邮政编码等。

(4) 定位服务。跟踪车队、船队及贵重物品的运输，了解用户所在位置及移动情况。

(5) 移动广告，无线广告，移动黄页，旅游信息。

(6) 银行财务的虚拟支付。

(7) 个人安全服务。

(8) 位置计费。室内、室外不同位置打电话的不同费用和计费。

(9) 费率的不同组合。

目前，LBS 开发应用的版本有 Web 版、移动客户端版、平板电脑版，如人人网的人人报到、Firefox 地理位置定位、Google 的 Google Maps Coordinate 等，提供地图定位服务的有 Google 地图、百度地图和搜狗地图等。本书主要介绍 Android 平台下对 Google 位置定位服务和 Map 的应用开发。

11.1.1 Android Location API 简介

Google 对于 Android 平台提供了 Location 包和 Maps 扩展库，以开发基于移动客户端的 Google Map 地图应用。在 Android 中，android. location 包提供了访问设备位置信息的服务。在 location 包中，其主要是通过核心组件 LocationManager 类来实现设备的定位、跟踪和趋近提示。提供上述系统服务，LocationManager 类不能被直接用来创建对象实例化，而是通过调用 Context 对象的 getSystemService(Context. LOCATION_SERVICE)方法来获取对象句柄，来创建 LocationManager 对象实例。下面就上述的 LocationManager 类以及相关类的常用属性和方法进行介绍。

1. LocationManager 类

通过 LocationManager 可以实现设备的定位、跟踪和趋近提示。上述内容已讲解了它可以通过 Context. getSystemService(Context. LOCATION_SERVICE)方法来获得 LocationManager 类对象实例。其常用的属性、方法如表 11-1 所示。

2. LocationProvider 类

LocationProvider 位于 android. location 包下，是位置提供者的抽象超类，用来描述位置提供者，设置位置提供者的一些属性，周期性报告设备的地理位置信息。可以通过 Criteria 类来为 LocationProvider 对象设置条件，获得用户定义的 LocationProvider 对象。其常用的方法和属性如表 11-2 所示。

表 11-1　LocationManager 类对象常用的属性、方法

属性和方法	描　述
GPS_PROVIDER	静态字符串常量，表明 LocationProvider 是 GPS
NETWORK_PROVIDER	静态字符串常量，表明 LocationProvider 是网络
addGpsStatusListener(GpsStatus. Listener listener)	添加一个 GPS 状态监听器
addProximityAlert(double latitude, double longitude, float radius, long expiration, PendingIntent intent)	添加一个趋近警告
getAllProviders()	获得所有的 LocationProvider 列表
getBestProvider(Criteria criteria, boolean enabledOnly)	根据 Criteria 返回最适合的 LocationProvider
getLastKnownLocation(String provider)	根据 Provider 获得位置信息
getProvider(String name)	获得指定名称的 LocationProvider
getProvider(boolean enableOnly)	获得可利用的 LocationProvider 列表
removeProximityAlert(PendingIntent intent)	删除趋近警告
requestLocationUpdates(String provider, long minTime, float minDistance, PendingIntent intent)	通过给定的 Provider 名称，周期性地通知当前 Activity
requestLocationUpdates(String provider, long minTime, float minDistance, LocationListener listener)	通过给定的 Provider 名称，并将其绑定指定的 LocationListener 监听器

表 11-2　LocationProvider 对象常用的方法和属性

属性或方法名称	描　述
AVAILABLE	静态整型常量，标示是否可利用
OUT_OF_SERVICE	静态整型常量，不在服务区
TEMPORAILY_UNAVAILABLE	静态整型常量，临时不可利用
getAccuarcy()	获得精度
getName()	获得位置提供者的名称
getPowerRequirement()	获得电源需求
hasMonetaryCost()	如果 provider 收费返回 true，免费返回 false
requiresCell()	是否需要访问基站网络
requiresNetWork()	是否需要网络数据
requiresSatelite()	是否需要访问卫星
supportsAltitude()	是否能够提供高度信息
supportsBearing()	是否能够提供方向信息
supportsSpeed()	是否能够提供速度信息

3. Criteria 类

Criteria 类也位于 android. location 包下,它封装了用于获得 LocationProvider 的条件,可以根据指定的 Criteria 条件来过滤获得 LocationProvider。其常用属性和方法如表 11-3 所示。

表 11-3 Criteria 类常用属性和方法

属性或访求名称	描　述
ACCERACY_COARSE	粗略精确度
ACCURACY_FINE	较高精确度
POWER_HING	用电消耗高
POWER_LOW	用电消耗低
isAlititudeRequried()	返回 Provider 是否需要高度信息
isBearingRequired()	返回 Provider 是否需要方位信息
isSpeedRequried()	返回 Provider 是否需要速度信息
isCostAllowed()	是否允许产生费用
setAccuracy(int accuracy)	设置 Provider 的精确度
setAltitudeRequired (boolean altitudeRequired)	设置 Provider 是否需要高度信息
setBearingRequired (boolean bearingRequired)	设置 Provider 是否需要方位信息
setCostAllowed (boolean costAllowed)	设置 Provider 是否产生费用
setSpeedAccuracy (int accuracy)	设置 Provider 是否需要速度信息
getAccuracy()	获得精度

4. Location 类

Location 类也位于 android. location 包下,它用于描述当前设备的地理位置信息,包括经纬度、方向、高度和速度等。开发者可以通过 LocationManager. getLastKnownLocation (String provider)方法获得 Location 实例。其常用的方法、属性如表 11-4 所示。

表 11-4 Location 类常用的方法、属性

方　法	描　述
public float getAccuracy ()	获得精确度
public double getAltitude ()	获得高度
public float getBearing ()	获得方向
public double getLatitude ()	获取经度
public double getLongitude ()	获得纬度
public float getSpeed ()	获得速度

从上述内容得知,开发者借助 LocationManager 类的对象生成的系统服务来调用或使

用 Android 平台下的 GPS 服务，通过 Context. getSystemService(Context. LOCATION_SERVIER)方法来获取 LocationManager 对象实例，然后可以通过 LocationProvider 来描述位置提供者。此外，可以用上述的 Criteria 类来设置自定义满足用户的最佳要求，最后通过 LocationManager. getLastKnownLocation(String provider)方法获得 Location 实例，然后用 Location 类获取自己所在的位置信息，如经纬度等信息。

5. LocationListener 类

LocationListener 类位于 android. location 包下，用于接收从 LocationManager 的位置发生改变时的通知。如果 LocationListener 被注册添加到 LocationManager 对象，并且此 LocationManager 对象调用了 requestLocationUpdates(String, long, float, LocationListener)方法，那么接口中的相关方法将会被调用。

LocationListener 类常用的方法如表 11-5 所示。

表 11-5 LocationListener 类常用的方法

方法名	描述
onLocationChanged (Location location)	此方法在当位置发生改变后被调用。这里可以没有限制地使用 Location 对象。参数为位置发生变化后的新位置
onProviderDisabled(String provider)	此方法在 provider 被用户关闭后被调用。如果基于一个已经关闭了的 provider 来调用 requestLocationUpdates 方法，那么这个方法被调用。 参数为与之关联的 location provider 名称
onPorviderEnabled (Location location)	此方法在 provider 被用户开启后调用。 参数为 provider 与之关联的 location provider 名称
onStatusChanged (String provider, int Status, Bundle extras)	此方法在 Provider 的状态在可用、暂时不可用和无服务三个状态直接切换时被调用。 参数： provider：与变化相关的 location provider 名称。 Status：如果服务已停止，并且在短时间内不会改变，状态码为 OUT_OF_SERVICE；如果服务暂时停止，并且在短时间内会恢复，状态码为 TEMPORARILY_UNAVAILABLE；如果服务正常有效，状态码为 AVAILABLE。 Extras：一组可选参数，其包含 provider 的特定状态

11.1.2 获取位置定位案例

上面介绍了在 Android 中应用地理位置定位服务常用的包及类和方法，下面通过一个获取地理位置定位服务的应用案例详细介绍 Location API 的应用。由于 Android 2.3.3 系统的 bug 问题，使用 Location 类获取地理位置在模拟器 AVD 或真机测试中会出现不稳定现象，通常会出现问题或者无法调用 GPS 服务，问题在 Logcat 中的显示如图 11-1 所示。

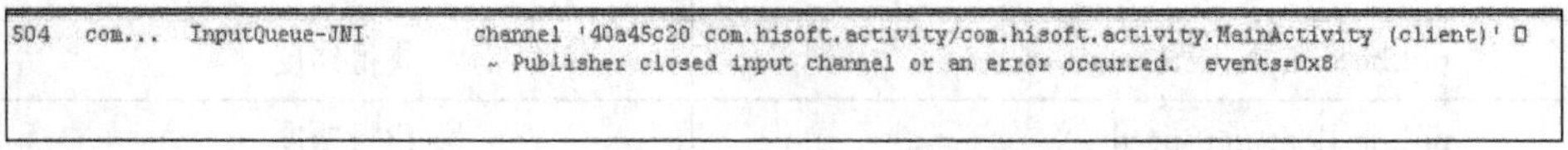

```
504   com...   InputQueue-JNI        channel '40a45c20 com.hisoft.activity/com.hisoft.activity.MainActivity (client)' 0
                                       ~ Publisher closed input channel or an error occurred.  events=0x8
```

图 11-1 Android 2.3.3 中 GPS 调用错误

基于以上使用 Location 类中存在的问题，本节案例单独采用了 Android 2.2 版本，以保证案例稳定正常运行。

(1) 创建一个新的 Android 工程，工程名为 LocationDemo，目标 API 选择 8（即 Android 2.2 版本），应用程序名为 LocationDemo1，包名为 com.hisoft，创建的 Activity 的名字为 MainActivity，最小 SDK 版本根据选择的目标 API 会自动添加为 8，创建项目工程如图 11-2 所示。

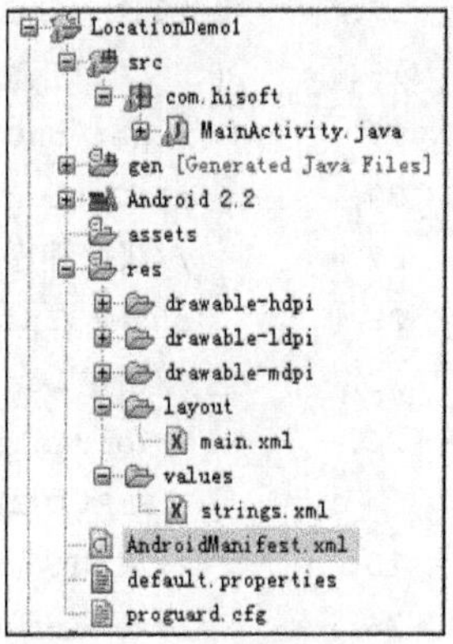

图 11-2 LocationDemo 工程目录结构

(2) 修改 res 目录下 layout 文件夹中的 main.xml 文件，设置线性布局，添加一个 EditText 控件及描述，并设置相关属性，代码如下：

```
1.  <?xml version="1.0" encoding="utf-8"?>
2.   <LinearLayout xmlns:android="http://schemas.
    android.com/apk/res/android"
3.      android:orientation="vertical"
4.      android:layout_width="fill_parent"
5.      android:layout_height="fill_parent"
6.      >
7.  <TextView
8.      android:layout_width="fill_parent"
9.      android:layout_height="wrap_content"
10.     android:text="@string/hello"
11.      android:id="@+id/textView1"
12.
13.     />
14. </LinearLayout>
```

(3) 修改 src 目录中 com.hisoft.activity 包下的 MainActivity.java 文件，代码如下：

```
1.  package com.hisoft.activity;
2.  import android.app.Activity;
3.  import android.content.Context;
4.  import android.location.Location;
5.  import android.location.LocationListener;
6.  import android.location.LocationManager;
7.  import android.os.Bundle;
8.  import android.widget.EditText;
9.
10. public class MainActivity extends Activity {
11.   //定义 LocationManager 对象
12.     private LocationManager locManager;
13.     //定义程序界面中的 EditText 组件
14.     private EditText show;
15.
```

```
16.     @Override
17.     public void onCreate(Bundle savedInstanceState)
18.     {
19.       super.onCreate(savedInstanceState);
20.       setContentView(R.layout.main);
21.       //获取程序界面上的 EditText 组件
22.       show=(EditText) findViewById(R.id.show);
23.       //创建 LocationManager 对象
24.         locManager = (LocationManager) getSystemService(Context.LOCATION_
          SERVICE);
25.       //从 GPS 获取最近的定位信息
26.       Location location=locManager.getLastKnownLocation(
27.         LocationManager.GPS_PROVIDER);
28.       //使用 location 更新 EditText 的显示
29.       updateView(location);
30.       //设置每 3s 获取一次 GPS 的定位信息
31.       locManager.requestLocationUpdates(LocationManager.GPS_PROVIDER
32.         , 3000, 8, new LocationListener()
33.       {
34.         @Override
35.         public void onLocationChanged(Location location)
36.         {
37.           //当 GPS 定位信息发生改变时更新位置
38.           updateView(location);
39.         }
40.
41.         @Override
42.         public void onProviderDisabled(String provider)
43.         {
44.           updateView(null);
45.         }
46.
47.         @Override
48.         public void onProviderEnabled(String provider)
49.         {
50.           //当 GPS LocationProvider 可用时更新位置
51.           updateView(locManager
52.             .getLastKnownLocation(provider));
53.         }
54.
55.         @Override
56.         public void onStatusChanged(String provider, int status,
57.           Bundle extras)
58.         {
```

```
59.         }
60.       });
61.     }
62.
63.     //更新 EditText 中显示的内容
64.     public void updateView(Location newLocation)
65.     {
66.       if (newLocation !=null)
67.       {
68.         StringBuilder sb=new StringBuilder();
69.         sb.append("实时的位置信息:\n");
70.         sb.append("经度:");
71.         sb.append(newLocation.getLongitude());
72.         sb.append("\n 纬度:");
73.         sb.append(newLocation.getLatitude());
74.         sb.append("\n 高度:");
75.         sb.append(newLocation.getAltitude());
76.         sb.append("\n 速度:");
77.         sb.append(newLocation.getSpeed());
78.         sb.append("\n 方向:");
79.         sb.append(newLocation.getBearing());
80.         show.setText(sb.toString());
81.       }
82.       else
83.       {
84.         //如果传入的 Location 对象为空,则清空 EditText
85.         show.setText("");
86.       }
87.     }
88. }
```

(4) 修改 AndroidManifest. xml 文件,在<manifest>根目录下添加获取定位信息权限,代码如下:

```
1.   <!--授权获取定位信息-->
2.   <uses-permission android:name="android.permission.ACCESS_FINE_LOCATION"/>
3.   <uses-permission android:name="android.permission.ACCESS_COARSE_LOCATION"/>
```

(5) 部署 LocationDemo 项目工程,程序运行如图 11-3 所示。

调试经纬度,打开 DDMS,在 Emulator Control 面板中选择 Location CONTROLS→Manual→Decimal,然后在 Longitude、Latitude 中输入经纬度,单击 Send 按钮,如图 11-4 所示,运行后界面显示如图 11-3 所示。由于使用模拟器的缘故,高度、速度、方向为 0。如果在真机测试,则会显示真值。

图 11-3　读取经纬度、高度、速度、方向数据

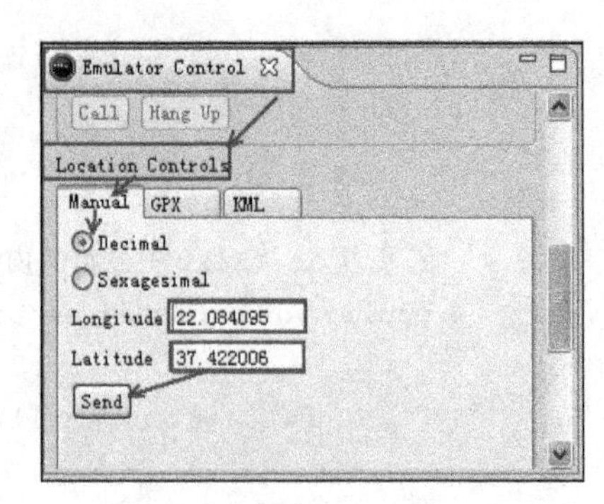

图 11-4　输入经纬度并发送

11.2　Google Map 服务

11.2.1　Google Map API 简介

Google 提供的 Maps 扩展库位于 com. google. android. maps 包中，它不是 Android SDK 标准库，扩展包 com. google. android. maps 中包含了一系列用于在 Google Map 上显示、控制和层叠信息的功能类，在 Maps 库中关键的类是 MapView，在 Android 标准库中它是 ViewGroup 的子类，MapView 显示的地图和数据来自于 Google Maps 服务器，当 MapView 成为焦点时，它能捕获按键、触摸来自动拉伸缩放地图，它能够提供基本的 UI 元素供用户控制、操作地图。

在应用程序中使用 Maps 扩展库，需要安装关于 Google APIs 的 add-on 文件夹。如果使用 Android SDK，则不需要安装 add-on，因为它已经被预装在 Android SDK 包中。

通常 MapView 类提供封装了 Google Maps API 的适配器，供开发者在应用程序中通过类的方法操纵 Google Maps 的数据。MapView 对象以并列块的形式显示从 Google 地图服务器上下载的地图，但在应用 Google 地图服务器数据之前，必须先进行注册并申请 Maps API Key。下面就申请 Maps API Key，对 Google Map 应用进行讲解。

11.2.2　申请 Map API KEY

申请 Google Map API Key 的过程如下：

(1) 在使用 JDK 内置的 keytool 工具之前，必须先确定默认的 keystore 存储位置，在 Eclipse 中选择 Windows→Preferences 命令，然后在打开的对话框中选择 Android 节点下的 Build，然后在右侧的 Default debug keystore 文本框中查找存放路径，如图 11-5 所示。

(2) 在"运行"中输入 cmd，打开控制台，输入命令使用 keytool 工具生成 MD5 指纹，命令如下：

```
C:\>keytool-list-alias androiddebugkey-keystore "D:\android\eclipse_android_
win7\.android\debug.keystore"-storepass android-keypass android
```

执行上述命令后，生成的 MD5 指纹如图 11-6 所示。

(3) 生成 MD5 指纹后，在浏览器地址栏中输入：https://developers. google. com/

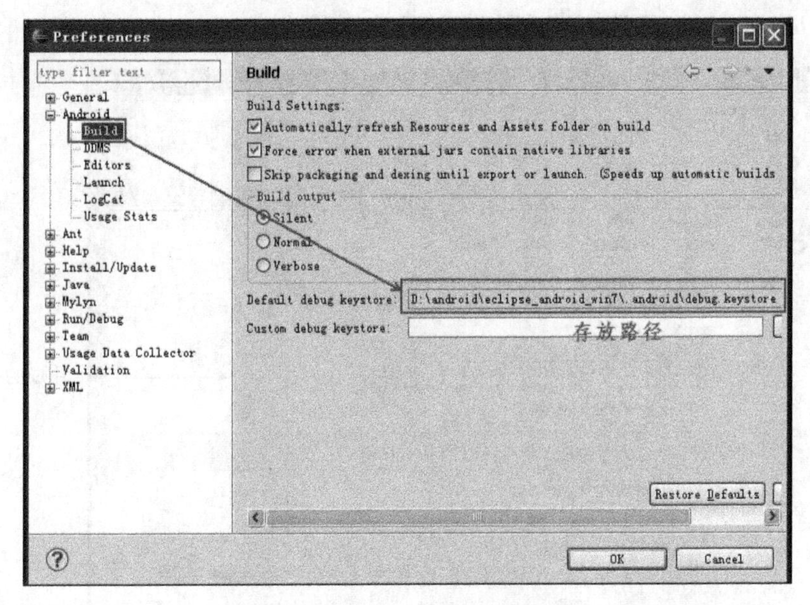

图 11-5 编辑 keystore 存放路径

```
C:\>keytool -list -alias androiddebugkey -keystore "D:\android\eclipse_android_win7\.android\debug.keystore" -storepass android -keypass android
androiddebugkey, 2012-8-9, PrivateKeyEntry,
认证指纹 (MD5): 63:53:AD:3F:47:5E:A9:08:7C:AF:FC:62:9D:22:24:CC
```

命令

生成指纹

图 11-6 命令生成 MD5 指纹

android/maps-api-signup，打开 Map API Key 申请页面，输入上述生成的 MD5 指纹。在申请 Map API Key 之前，必须先申请 Google 账户，然后单击 Generate API Key 按钮，如图 11-7 所示。

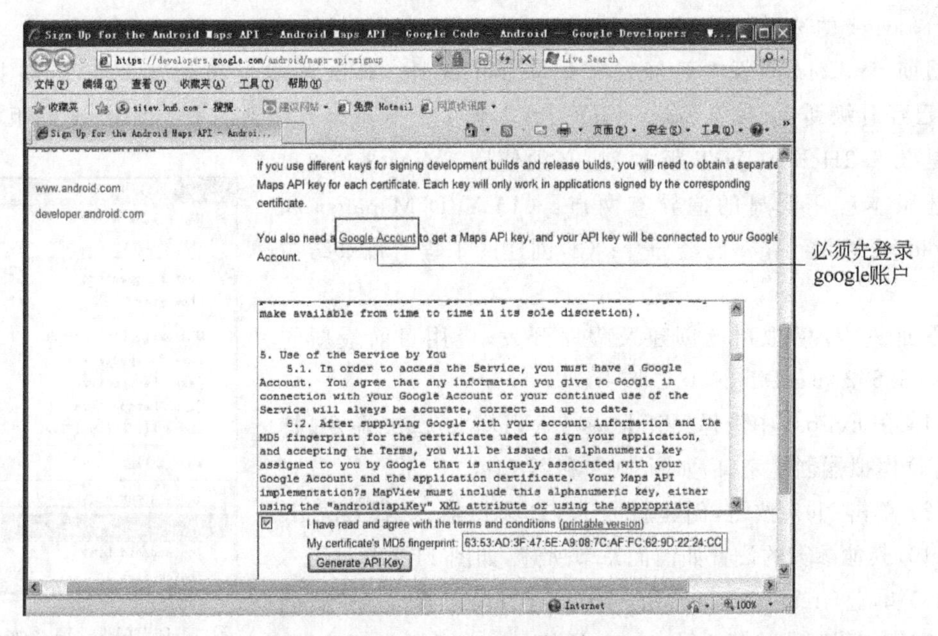

图 11-7 申请 Map API Key

(4) 弹出 Google API Key 生成页面,如图 11-8 所示。

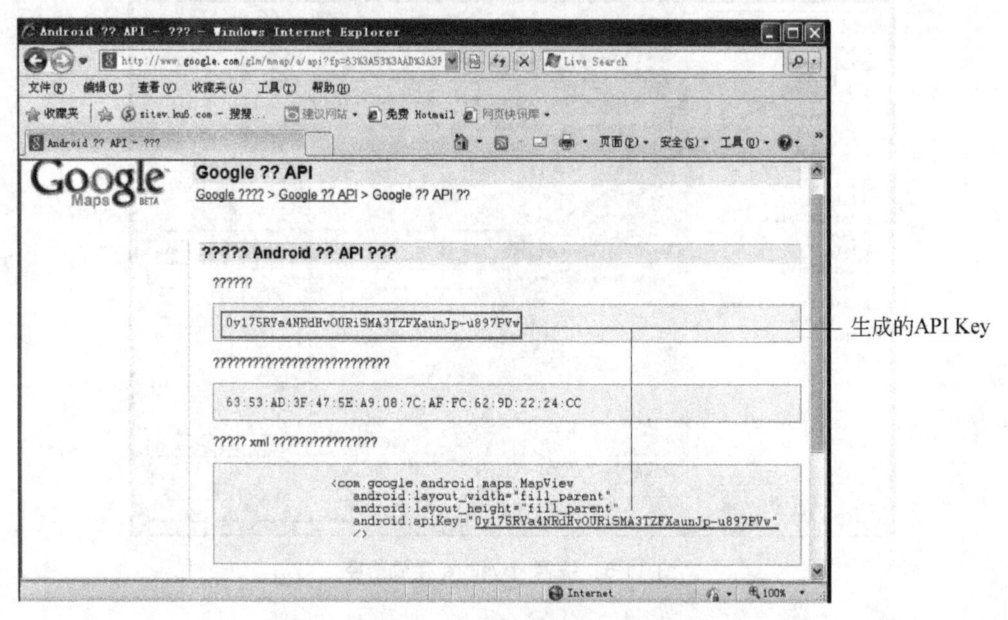

图 11-8　生成 Map API Key 页面

注意:目前国内用户登录申请 Google Map Key,由于网络的原因,经常难以打开网页,需要使用网页代理。此外,目前也有采用 HTML5＋PhoneGap 开发 Android 移动应用的趋势。

11.2.3　使用 Map API 创建 AVD 应用

创建 Map API 的 AVD 模拟器与创建通常的 AVD 模拟器的步骤主要的不同在于目标平台 target 的选择不同,应该选择 Google APIs 版本,具体步骤如下:

目前 Android 的版本已经发布到 4.1,Google APIs 版本发布到 3.0,Eclipse 插件 ADT 已经升级到 20.0.2,Eclipse IDE for Java Developers 升级到 1.3.2,Eclipse 的版本采用的是 3.6.2Helios(ADT 版本 20,必须使用 Eclipse3.6.2),与前述版本应用明显的差异有两点:(1) AVD Manager 和 Android SDK Manager 已经分开;(2)创建的工程的目录结构不同。

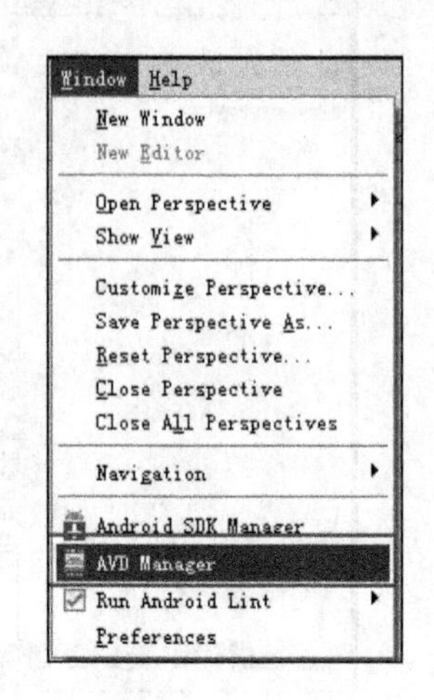

图 11-9　AVD Manager

下面 AVD 模拟器的创建及程序开发,应用目前最新的 Eclipse 3.6.2 和 ADT 20.0.2 如下:

(1) 在 Eclipse 中选择 Windows→AVD Manager 命令,打开 AVD 模拟器创建窗口,如图 11-9 和图 11-10 所示。

(2) 单击 New 按钮,创建新的 AVD 模拟器,选择 Google APIs 10,其他参数的设置如前面章节所述,如图 11-11 所示。

(3) 单击 Create AVD 按钮,生成 AVD 模拟器。选中新创建的 AVD 模拟器,然后单击 Start 按钮,启动新创建的 AVD

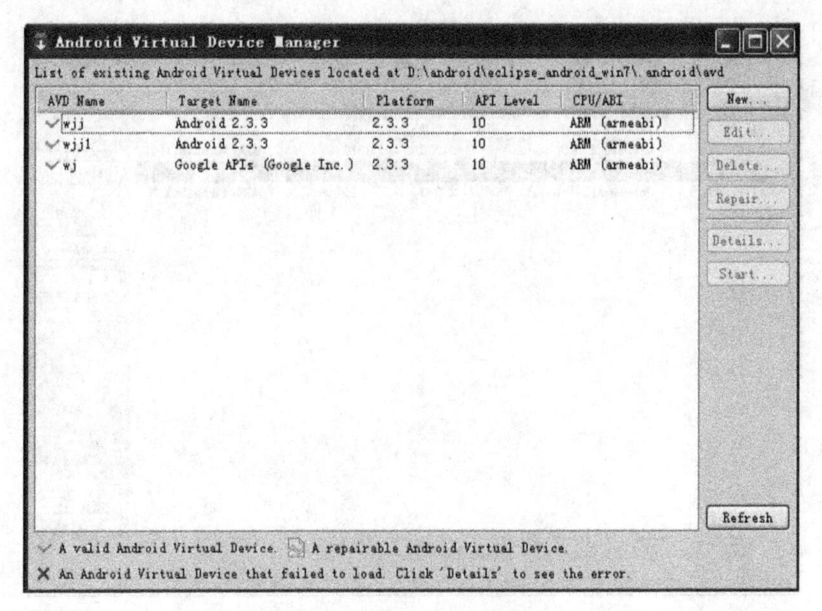

图 11-10 创建 AVD

图 11-11 选择 Google APIs 作为目标平台

模拟器，如图 11-12 所示。

拓展提示：围绕 Google Map API 有许多方面的应用，其核心都是调用控件进行应用。由于 Google 退出了中国市场，导致 Google Map 更新及配套应用开发与其他 Map 相比较为滞后，考虑在 Android 系统中调用其他的 Map 应用进行基于 Map 的应用开发。

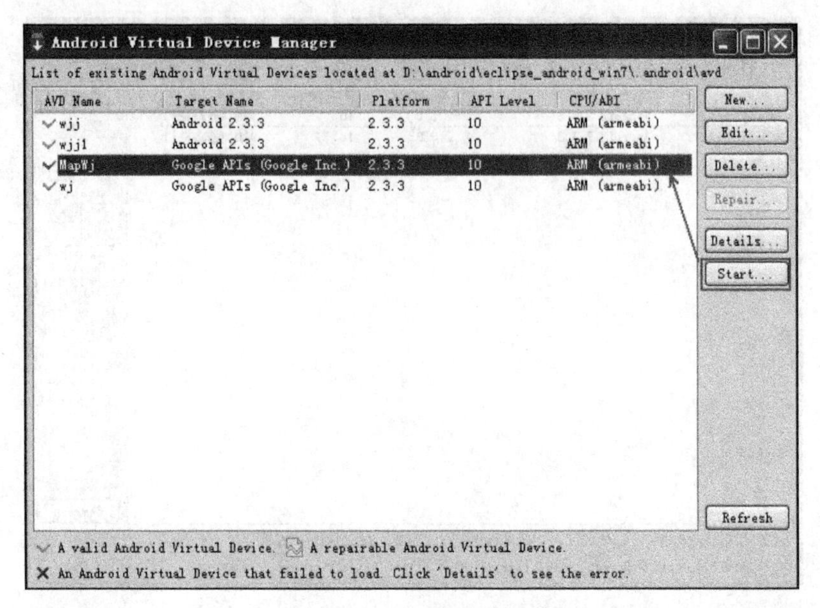

图 11-12　AVD创建与启动

11.3　项目案例

学习目标：学习 Google Map 的基本方法、属性的设置等应用。

案例描述：申请 Google Map KEY，使用 AbsoluteLayout 布局和 MapView 控件、Button 按钮，调用 Map 应用实现公司位置的准确定位和地图显示。

案例要点：MapView 控件描述、Google Map KEY 申请及添加、Map 相关方法。

案例实施：

(1) 创建工程 Project_Chapter_11，选择目标平台为 Google APIs(API Level 10)，工程目录结构如图 11-13 所示。

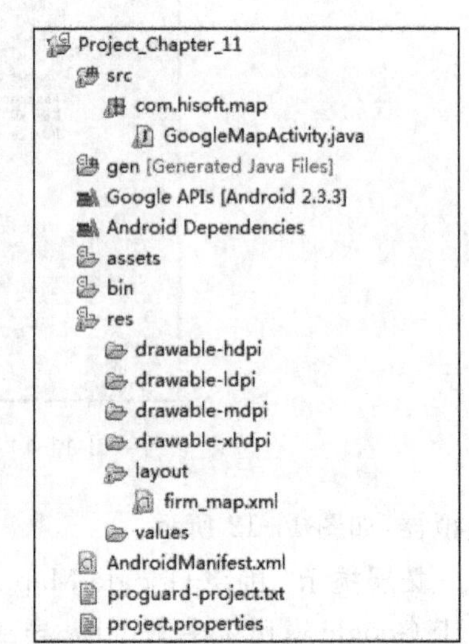

图 11-13　Project_Chapter_11 工程目录结构

(2) 创建 firm_map.xml 文件，将文件存放在 res/layout 目录下，注意把自己申请的 Map API KEY 添加到代码第 17 行，代码如下：

```
1.   <?xml version="1.0" encoding="utf-8"?>
2.   <AbsoluteLayout
3.     android:id="@+id/layout1"
4.     android:layout_width="fill_parent"
5.     android:layout_height="fill_parent"
6.     android:background="#ffffff"
7.     xmlns:android="http://schemas.android.
     com/apk/res/android"
8.   >
9.     <!--Google MapView Widget-->
```

```
10.
11.    <com.google.android.maps.MapView
12.        android:id="@+id/mapView1"
13.        android:layout_width="fill_parent"
14.        android:layout_height="463dp"
15.        android:layout_x="0px"
16.        android:layout_y="68dp"
17.        android:apiKey="0j0qyQ9cSwEKhQM_I28rZSFn_ZZn5XiojmoK5Gg"
18.        android:clickable="true"
19.        android:enabled="true"/>
20.
21.    <Button
22.        android:id="@+id/zoom_out"
23.        android:layout_width="60px"
24.        android:layout_height="40px"
25.        android:layout_x="73dp"
26.        android:layout_y="22dp"
27.        android:text="@string/btn_zoom_out"
28.        android:textSize="10sp"/>
29.
30.    <Button
31.        android:id="@+id/zoom_in"
32.        android:layout_width="60px"
33.        android:layout_height="40px"
34.        android:layout_x="170dp"
35.        android:layout_y="22dp"
36.        android:text="@string/btn_zoom_in"
37.        android:textSize="10sp"/>
38. </AbsoluteLayout>
```

(3) 在 src 目录下的 com. hisoft. map 包下创建 GoogleMapActivity. java 文件，代码如下：

```
1.    package com.hisoft.map;
2.
3.  import android.app.AlertDialog;
4.  import android.content.DialogInterface;
5.  import android.os.Bundle;
6.  import android.view.View;
7.  import android.widget.Button;
8.
9.  import com.google.android.maps.GeoPoint;
10. import com.google.android.maps.MapActivity;
11. import com.google.android.maps.MapController;
12. import com.google.android.maps.MapView;
13.
```

```
14.  public class GoogleMapActivity extends MapActivity {
15.
16.    private MapController mapController01;
17.    private MapView mapView01;
18.    private Button zoom_out, zoom_in;
19.    private int intZoomLevel=0;
20.    /*Map 启动时的预设坐标*/
21.    private double dLat=40.053728;
22.    private double dLng=116.302648;
23.
24.    @Override
25.    protected void onCreate(Bundle icicle) {
26.      super.onCreate(icicle);
27.      setContentView(R.layout.firm_map);
28.
29.      /*建立 MapView 对象*/
30.      mapView01=(MapView) findViewById(R.id.mapView1);
31.      mapController01=mapView01.getController();
32.      /*设定 MapView 的显示选项(卫星、街道)*/
33.      mapView01.setSatellite(false);
34.      mapView01.setStreetView(true);
35.      /*预设放缩的层级*/
36.      intZoomLevel=17;
37.      mapController01.setZoom(intZoomLevel);
38.      /*设定 Map 的中点为预设经纬度*/
39.      refreshMapView();
40.
41.      /*放大 Map 的 Button*/
42.      this.zoom_in=(Button) findViewById(R.id.zoom_in);
43.      this.zoom_in.setOnClickListener(new Button.OnClickListener() {
44.        @Override
45.        public void onClick(View v) {
46.          intZoomLevel++;
47.          if (intZoomLevel>mapView01.getMaxZoomLevel()) {
48.            intZoomLevel=mapView01.getMaxZoomLevel();
49.          }
50.          mapController01.setZoom(intZoomLevel);
51.        }
52.      });
53.
54.      /*缩小 Map 的 Button*/
55.      this.zoom_out=(Button) findViewById(R.id.zoom_out);
56.      this.zoom_out.setOnClickListener(new Button.OnClickListener() {
57.        @Override
58.        public void onClick(View v) {
```

```
59.            intZoomLevel--;
60.            if (intZoomLevel<1) {
61.              intZoomLevel=1;
62.            }
63.            mapController01.setZoom(intZoomLevel);
64.          }
65.        });
66.      }
67.
68.      /*重整 Map 的 method*/
69.      public void refreshMapView() {
70.        GeoPoint p=new GeoPoint((int) (dLat*1E6), (int) (dLng*1E6));
71.        mapView01.displayZoomControls(true);
72.        /*将 Map 的中点移动 GeoPoint*/
73.        mapController01.animateTo(p);
74.        mapController01.setZoom(intZoomLevel);
75.      }
76.
77.      @Override
78.      protected boolean isRouteDisplayed() {
79.        return false;
80.      }
81.
82.      /*显示 Dialog 的 method*/
83.      private void showDialog(String mess) {
84.        new AlertDialog.Builder(GoogleMapActivity.this).setTitle("Message")
85.            .setMessage(mess)
86.            .setNegativeButton("确定", new DialogInterface.OnClickListener() {
87.              public void onClick(DialogInterface dialog, int which) {
88.              }
89.            }).show();
90.      }
91.
92.    }
```

(4) 修改 AndroidManifest.xml 文件,在代码第 24 行添加 Google Map 的使用权限,代码如下:

```
1.     <?xml version="1.0" encoding="utf-8"?>
2.    <manifest xmlns:android="http://schemas.android.com/apk/res/android"
3.        package="com.hisoft.map"
4.        android:versionCode="1"
5.        android:versionName="1.0">
6.
7.        <uses-permission android:name="android.permission.INTERNET"></uses-
      permission>
```

```
8.
9.        <uses-sdk android:minSdkVersion="10"/>
10.
11.       <application
12.           android:icon="@drawable/ic_launcher"
13.           android:label="@string/app_name">
14.           <activity
15.               android:name=".GoogleMapActivity"
16.               android:label="@string/app_name">
17.               <intent-filter>
18.                 <action android:name="android.intent.action.MAIN"/>
19.
20.                 <category android:name="android.intent.category.LAUNCHER"/>
21.               </intent-filter>
22.           </activity>
23.
24.           <uses-library android:name="com.google.android.maps"/>
25.       </application>
26.   </manifest>
```

(5) 部署工程 Project_Chapter_11,程序运行效果如图 11-14 所示。

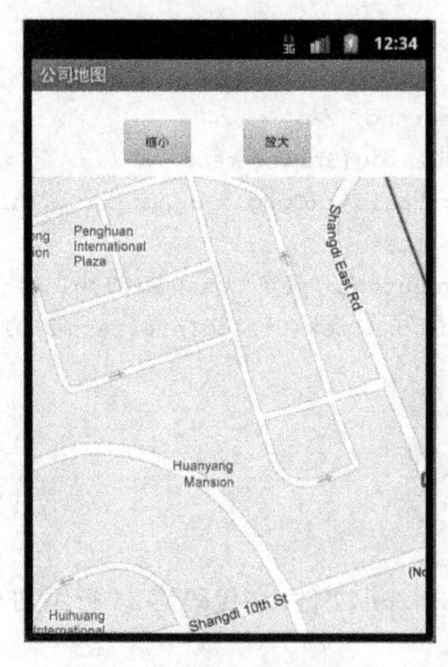

图 11-14 Map 运行效果

1. 简答题

(1) 简述 LBS 在移动服务方面的基本原理及应用领域。

(2) LocationManager 类的主要作用是什么？如何进行实例化？

(3) LocationManager、LocationProvider、Criteria 和 Location 这 4 个类之间的应用关系是什么？

(4) MapView 类的作用是什么？如何使用它？

(5) 在 Android 系统中如何生成 MD5 指纹？使用 MD5 指纹主要用于哪些方面？

2. 完成下面的实训项目

要求：在本章项目案例的基础上,要求完成基于 Google Map 的导航应用,用户输入起始位置和终点位置后,单击“查询”按钮,给用户在地图中显示合理的行程路线图。

致　谢

本套系列教材的编写得到了北京亚思晟科技有限公司(以下简称亚思晟)的大力协助。亚思晟的资深专家梁立新老师协同系统设计师、高级培训经理和软件工程师针对软件行业各个职业岗位所需关键技术能力、职业素质、可持续发展能力进行深入的调查研究,并在此基础上构建了软件工程专业学生的知识、技能和素质体系结构,为本套系列教材设计了“项目驱动式”教学法和提供了真实项目案例。

亚思晟作为教育部首家“软件工程专业大学生实习实训基地”和跨国企业高端 IT 人才实训基地,力主为大学生建立起从学校到跨国企业的桥梁,解决高校学子就业难的社会化问题。通过亚思晟人才培养方略以及 IT 高端人才订单培养方案,以“项目驱动”作为基本实训方式,为 IT 行业培养优秀 IT 人才。

亚思晟自主研发并持续更新着国际化软件教育产品和实训体系,公开出版的系列教材得到学校、企业和社会的一致认可。公司团队以雄厚的海外人才为核心,平均拥有 10 年以上管理和技术经验,多数具有海外著名大学管理或技术背景。亚思晟还研发和打造了远程教育平台,将高端精品课程体系网络化,采用线上学习和线下实训相结合的模式,保证可扩展性和持续性发展。另外,亚思晟一直开展软件项目的研发和外包服务,具有针对金融、企业信息化、政府信息化、电信等多个行业的国内外项目经验和真实项目案例。